Advances in Intelligent Systems and Computing

Volume 889

Series editor

Janusz Kacprzyk, Systems Research Institute, Polish Academy of Sciences,
Warsaw, Poland
e-mail: kacprzyk@ibspan.waw.pl

The series "Advances in Intelligent Systems and Computing" contains publications on theory, applications, and design methods of Intelligent Systems and Intelligent Computing. Virtually all disciplines such as engineering, natural sciences, computer and information science, ICT, economics, business, e-commerce, environment, healthcare, life science are covered. The list of topics spans all the areas of modern intelligent systems and computing such as: computational intelligence, soft computing including neural networks, fuzzy systems, evolutionary computing and the fusion of these paradigms, social intelligence, ambient intelligence, computational neuroscience, artificial life, virtual worlds and society, cognitive science and systems, Perception and Vision, DNA and immune based systems, self-organizing and adaptive systems, e-Learning and teaching, human-centered and human-centric computing, recommender systems, intelligent control, robotics and mechatronics including human-machine teaming, knowledge-based paradigms, learning paradigms, machine ethics, intelligent data analysis, knowledge management, intelligent agents, intelligent decision making and support, intelligent network security, trust management, interactive entertainment, Web intelligence and multimedia.

The publications within "Advances in Intelligent Systems and Computing" are primarily proceedings of important conferences, symposia and congresses. They cover significant recent developments in the field, both of a foundational and applicable character. An important characteristic feature of the series is the short publication time and world-wide distribution. This permits a rapid and broad dissemination of research results.

More information about this series at http://www.springer.com/series/11156

Jerzy Pejaś · Imed El Fray
Tomasz Hyla · Janusz Kacprzyk
Editors

Advances in Soft and Hard Computing

 Springer

Editors
Jerzy Pejaś
West Pomeranian University of Technology
in Szczecin
Szczecin, Poland

Tomasz Hyla
West Pomeranian University of Technology
in Szczecin
Szczecin, Poland

Imed El Fray
West Pomeranian University of Technology
in Szczecin
Szczecin, Poland

Janusz Kacprzyk
Polish Academy of Sciences
Systems Research Institute
Warsaw, Poland

ISSN 2194-5357 ISSN 2194-5365 (electronic)
Advances in Intelligent Systems and Computing
ISBN 978-3-030-03313-2 ISBN 978-3-030-03314-9 (eBook)
https://doi.org/10.1007/978-3-030-03314-9

Library of Congress Control Number: 2018960424

This Springer imprint is published by the registered company Springer Nature Switzerland AG
The registered company address is: Gewerbestrasse 11, 6330 Cham, Switzerland

Preface

Advanced Computer System 2018 (ACS 2018) conference was the 21st in the series of conferences organized by the Faculty of Computer Science and Information Technology of the West Pomeranian University of Technology in Szczecin, Poland. That event could not be possible without scientific cooperation with Warsaw University of Technology, Faculty of Mathematics and Information Science, Poland; Warsaw University of Life Sciences (SGGW), Poland; AGH University of Science and Technology, Faculty of Physics and Applied Computer Science, Poland; Polish Academy of Sciences (IPIPAN), Institute of Computer Science, Poland; Kuban State University of Technology, Institute of Information Technology and Safety, Russia; Bialystok University of Technology, Poland; and —last but not least—Ehime University in Matsuyama, Japan. As usual, the conference was held in Międzyzdroje, Poland, on 24–26 September 2018.

This volume contains a collection of carefully selected, peer-reviewed papers presented during the conference sessions. The main topics covered by the chapters in this book are artificial intelligence, software technologies, information technology security and multimedia systems.

It has been a tradition since the first conference that the organizers have always invited top specialists in the fields. Many top scientists and scholars, who have presented keynote talks over the years, have always provided an inspiration for future research and for young and experienced participants.

The book places a great emphasis both on theory and practice. The contributions not only reflect the invaluable experience of eminent researchers in relevant areas but also point new methods, approaches and interesting direction for the future researches.

In keeping with ACS mission over the last twenty years, this 21st conference, ACS 2018, was also an event providing a comprehensive state-of-the-art summary from keynote speakers as well as a look forward towards future research priorities. We believe that the keynote talks provided an inspiration for all attendees. This year authors of the keynote talks were professors: Nabendu Chaki from University of Calcutta (India), Akira Imada from Brest State Technical University (Belarus), Keiichi Endo and Shinya Kobayashi from Ehime University (Japan), Ryszard

Kozera from Warsaw University of Life Sciences SGGW (Poland), Jacek Pomykała from the University of Warsaw (Poland) and Marian Srebrny from Polish Academy of Sciences (Poland).

We would like to give a proof of appreciation to all members of the International Programme Committee for their time and effort in reviewing the papers, helping us to shape the scope and topics of the conference and providing us with much advice and support. Moreover, we want to express a gratitude to all of the organizers from the Faculty of Computer Science and Information Technology, West Pomeranian University of Technology in Szczecin for their enthusiasm and hard work, notably Ms. Hardej, Secretary of the Conference, and all other members of Organizing Committee including Luiza Fabisiak, Tomasz Hyla and Witold Maćków.

We expect this book to shed new light on unresolved issues and inspire the reader to greater challenges. We also hope that the book will provide tools or ideas for their creation that will be more effective in solving increasingly complex research problems and reaching common scientific goals.

September 2018   Imed El Fray
                                                                 Tomasz Hyla
                                                              Janusz Kacprzyk
                                                                  Jerzy Pejaś

Organization

Advanced Computer System 2018 (ACS 2018) was organized by the West Pomeranian University of Technology in Szczecin, Faculty of Computer Science and Information Technology (Poland), in cooperation with Warsaw University of Technology, Faculty of Mathematics and Information Science (Poland); AGH University of Science and Technology, Faculty of Physics and Applied Computer Science (Poland); Ehime University (Japan); Polish Academy of Sciences IPIPAN (Poland); Kuban State University of Technology, Institute of Information Technology and Safety (Russia); and Bialystok University of Technology (Poland).

Organizing Committee

Tomasz Hyla (Chair)	West Pomeranian University of Technology, Szczecin, Poland
Sylwia Hardej (Secretary)	West Pomeranian University of Technology, Szczecin, Poland
Witold Maćków	West Pomeranian University of Technology, Szczecin, Poland
Luiza Fabisiak	West Pomeranian University of Technology, Szczecin, Poland

Programme Committee Chairs

Jerzy Pejaś	West Pomeranian University of Technology, Szczecin, Poland
Imed El Fray	West Pomeranian University of Technology, Szczecin, Poland

Tomasz Hyla   West Pomeranian University of Technology,
                                     Szczecin, Poland

International Programming Committee

Costin Badica   University of Craiova, Romania
Zbigniew Banaszak   Warsaw University of Technology, Poland
Anna Bartkowiak   Wroclaw University, Poland
Włodzimierz Bielecki   West Pomeranian University of Technology,
                                     Szczecin, Poland
Leon Bobrowski   Bialystok Technical University, Poland
Grzegorz Bocewicz   Koszalin University of Technology, Poland
Robert Burduk   Wroclaw University of Technology, Poland
Andrzej Cader   Academy of Humanities and Economics in Lodz,
                                     Poland
Aleksandr Cariow   West Pomeranian University of Technology,
                                     Szczecin, Poland
Nabendu Chaki   Calcutta University, India
Krzysztof Chmiel   Poznan University of Technology, Poland
Ryszard S. Choraś   University of Technology and Life Sciences,
                                     Poland
Krzysztof Ciesielski   Polish Academy of Sciences, Poland
Nicolas Tadeusz Courtois   University College London, UK
Albert Dipanda   Le Centre National de la Recherche Scientifique,
                                     France
Bernard Dumont   European Commission, Information Society
                                     and Media Directorate General, France
Jos Dumortier   KU Leuven University, Belgium
Keiichi Endo   Ehime University, Japan
Özgür Ertuğ   Gazi University, Turkey
Oleg Fińko   Kuban State University of Technology, Russia
Paweł Forczmański   West Pomeranian University of Technology,
                                     Szczecin, Poland
Dariusz Frejlichowski   West Pomeranian University of Technology,
                                     Szczecin, Poland
Jerzy August Gawinecki   Military University of Technology, Poland
Larisa Globa   National Technical University of Ukraine,
                                     Ukraine
Janusz Górski   Technical University of Gdansk, Poland
Władysław Homenda   Warsaw University of Technology, Poland
Akira Imada   Brest State Technical University, Belarus
Michelle Joab   LIRMM, Universite Montpellier 2, France
Jason T. J. Jung   Yeungnam University, Korea

Janusz Kacprzyk	Systems Research Institute, Polish Academy of Sciences, Poland
Andrzej Kasiński	Poznan University of Technology, Poland
Shinya Kobayashi	Ehime University, Japan
Marcin Korzeń	West Pomeranian University of Technology, Szczecin, Poland
Zbigniew Adam Kotulski	Polish Academy of Sciences, Poland
Piotr Andrzej Kowalski	AGH University of Science and Technology and SRI Polish Academy of Sciences, Poland
Ryszard Kozera	Warsaw University of Life Sciences—SGGW, Poland
Mariusz Kubanek	Częstochowa University of Technology, Poland
Mieczysław Kula	University of Silesia, Poland
Eugeniusz Kuriata	University of Zielona Gora, Poland
Mirosław Kurkowski	Cardinal Stefan Wyszyński University in Warsaw, Poland
Jonathan Lawry	University of Bristol, UK
Javier Lopez	University of Malaga, Spain
Andriy Luntovskyy	BA Dresden University of Coop. Education, Germany
Kurosh Madani	Paris XII University, France
Przemysław Mazurek	West Pomeranian University of Technology, Szczecin, Poland
Andrzej Niesler	Wroclaw University of Economics, Poland
Arkadiusz Orłowski	Warsaw University of Life Sciences—SGGW, Poland
Marcin Paprzycki	Systems Research Institute, Polish Academy of Sciences, Poland
Paweł Pawlewski	Poznań University of Technology, Poland
Witold Pedrycz	University of Alberta, Canada
Andrzej Piegat	West Pomeranian University of Technology, Szczecin, Poland
Josef Pieprzyk	Macquarie University, Australia
Jacek Pomykała	Warsaw University, Poland
Alexander Prokopenya	Warsaw University of Life Sciences—SGGW, Poland
Elisabeth Rakus-Andersson	Blekinge Institute of Technology, School of Engineering, Sweden
Izabela Rejer	West Pomeranian University of Technology, Szczecin, Poland
Vincent Rijmen	Graz University of Technology, Austria
Valery Rogoza	West Pomeranian University of Technology, Szczecin, Poland
Leszek Rutkowski	Częstochowa University of Technology, Poland
Khalid Saeed	Warsaw University of Technology, Poland

Kurt Sandkuhl	University of Rostock, Germany
Albert Sangrá	Universitat Oberta de Catalunya, Spain
Władysław Skarbek	Warsaw University of Technology, Poland
Vaclav Snašel	Technical University of Ostrava, Czech Republic
Jerzy Sołdek	West Pomeranian University of Technology, Szczecin, Poland
Zenon Sosnowski	Białystok University of Technology, Poland
Marian Srebrny	Institute of Computer Science, Polish Academy of Sciences, Poland
Peter Stavroulakis	Technical University of Crete, Greece
Janusz Stokłosa	Poznan University of Technology, Poland
Marcin Szpyrka	AGH University of Science and Technology, Poland
Ryszard Tadeusiewicz	AGH University of Science and Technology, Poland
Oleg Tikhonenko	University of K. Wyszynski, Warsaw, Poland
Natalia Wawrzyniak	Maritime University of Szczecin, Poland
Jan Węglarz	Poznan University of Technology, Poland
Sławomir Wierzchoń	Institute of Computer Science, Polish Academy of Sciences, Poland
Antoni Wiliński	West Pomeranian University of Technology, Szczecin, Poland
Toru Yamaguchi	Tokyo Metropolitan University, Japan

Additional Reviewers

Bilski, Adrian
Bobulski, Janusz
Chmielewski, Leszek
Fabisiak, Luiza
Goszczyńska, Hanna
Grocholewska-Czuryło, Anna
Hoser, Paweł
Jaroszewicz, Szymon
Jodłowski, Andrzej
Karwański, Marek
Klęsk, Przemysław
Kurek, Jarosław

Landowski, Marek
Maleika, Wojciech
Mantiuk, Radosław
Maćków, Witold
Okarma, Krzysztof
Olejnik, Remigiusz
Radliński, Lukasz
Rozenberg, Leonard
Różewski, Przemysław
Siedlecka-Lamch, Olga
Steingartner, William
Świderski, Bartosz

Contents

Software Technology

Invited Paper

Fitting Dense and Sparse Reduced Data

Ryszard Kozera[1,2(✉)] and Artur Wiliński[1]

[1] Faculty of Applied Informatics and Mathematics, Warsaw University of Life
Sciences - SGGW, ul. Nowoursynowska 159, 02-776 Warsaw, Poland
ryszard.kozera@gmail.com
[2] Department of Computer Science and Software Engineering, The University of
Western Australia, 35 Stirling Highway, Crawley, Perth, WA 6009, Australia

Abstract. This paper addresses the topic of fitting reduced data represented by the sequence of interpolation points $\mathcal{M} = \{q_i\}_{i=0}^n$ in arbitrary Euclidean space $\mathbb{E}^m$. The parametric curve γ together with its knots $\mathcal{T} = \{t_i\}_{i=0}^n$ (for which $\gamma(t_i) = q_i$) are both assumed to be unknown. We look at some recipes to estimate $\mathcal{T}$ in the context of dense versus sparse $\mathcal{M}$ for various choices of interpolation schemes $\hat{\gamma}$. For $\mathcal{M}$ dense, the convergence rate to approximate γ with $\hat{\gamma}$ is considered as a possible criterion to force a proper choice of new knots $\hat{\mathcal{T}} = \{\hat{t}_i\}_{i=0}^n \approx \mathcal{T}$. The latter incorporates the so-called exponential parameterization "retrieving" the missing knots $\mathcal{T}$ from the geometrical spread of $\mathcal{M}$. We examine the convergence rate in approximating γ by commonly used interpolants $\hat{\gamma}$ based here on $\mathcal{M}$ and exponential parameterization. In contrast, for $\mathcal{M}$ sparse, a possible optional strategy is to select $\hat{\mathcal{T}}$ which optimizes a certain cost function depending on the family of admissible knots $\hat{\mathcal{T}}$. This paper focuses on minimizing "an average acceleration" within the family of natural splines $\hat{\gamma} = \hat{\gamma}^{NS}$ fitting $\mathcal{M}$ with $\hat{\mathcal{T}}$ admitted freely in the ascending order. Illustrative examples and some applications listed supplement theoretical component of this work.

Keywords: Interpolation · Reduced data
Computer vision and graphics

1 Introduction

Let $\gamma : [0, T] \to \mathbb{E}^m$ be a smooth regular curve (i.e. $\dot{\gamma}(t) \neq \mathbf{0}$) defined over $t \in [0, T]$, for $0 < T < \infty$ - see e.g. [1]. The term *reduced data* (denoted by $\mathcal{M}$) represents the sequence of $n + 1$ interpolation points $\{q_i\}_{i=0}^n$ in arbitrary Euclidean space $\mathbb{E}^m$. Here, each point from $\mathcal{M}$ satisfies the condition $q_i = \gamma(t_i)$ with extra constraint $q_{i+1} \neq q_i$ ($i = 0, 1, \ldots, n - 1$). The respective knots $\mathcal{T} = \{t_i\}_{i=0}^n$ are assumed to be unavailable. The latter stands in contrast with the classical problem of fitting *non-reduced data* where both $\mathcal{M}$ and $\mathcal{T}$ are given. Naturally, any interpolation scheme $\hat{\gamma}$ fitting $\mathcal{M}$ relies on the provision of some $\hat{\mathcal{T}} = \{\hat{t}_i\}_{i=0}^n$ at best "well approximating" the unknown knots $\mathcal{T}$. This paper discusses two different approaches in selecting the substitutes $\hat{\mathcal{T}}$ of $\mathcal{T}$ (subject

© Springer Nature Switzerland AG 2019
J. Pejaś et al. (Eds.): ACS 2018, AISC 889, pp. 3–17, 2019.
https://doi.org/10.1007/978-3-030-03314-9_1

to $\hat{\gamma}(\hat{t}_i) = q_i$ and $\hat{t}_i < \hat{t}_{i+1}$) for either dense or sparse reduced data $\mathcal{M}$. The theoretical component of this work is also complemented by several indicative examples. The relevant discussion on the topic in question can be found e.g. in [2–5,7,9–15,17,19,22,23,26]. The problem of interpolating reduced or non-reduced data arises in computer graphics and vision (e.g. for trajectory modelling and image compression or segmentation), in engineering (like robotics: path planning or motion modelling) in physics (e.g. for trajectory modelling) and in medical image processing (e.g. in image segmentation and area estimation) - see [27–30]. More literature on the above topic can be found among all in [2,27,31].

2   Interpolating Dense Reduced Data

For $\mathcal{M}$ forming *dense reduced data* the intrinsic assumption admits n as sufficiently large. Thus upon selecting specific interpolation scheme $\hat{\gamma} : [0, \hat{T}] \to \mathbb{E}^m$ together with a particular choice of $\hat{\mathcal{T}} \approx \mathcal{T}$ the question of convergence rate α in approximating γ with $\hat{\gamma}$ (for $n \to \infty$) arises naturally. Furthermore, an equally intriguing matter refers to the existence of such $\hat{\mathcal{T}}$ so that the respective convergence rates α in $\hat{\gamma} \approx \gamma$ coincide once $\hat{\gamma}$ is taken either with $\hat{\mathcal{T}}$ or with $\mathcal{T}$.

This section addresses both issues raised above. In doing so, recall first some preliminaries (see e.g. [2,3]):

Definition 1. *The sampling* $\mathcal{T} = \{t_i\}_{i=0}^n$ *is called* admissible *provided:*

$$\lim_{n \to \infty} \delta_n = 0, \text{ where } \delta_n = \max_{1 \leq i \leq n} \{t_i - t_{i-1} : \quad i = 1, 2, \dots, n\}. \tag{1}$$

In addition, $\mathcal{T}$ *represents* more-or-less uniform sampling *if there exist some constants* $0 < K_l \leq K_u$ *such that for sufficiently large* n:

$$\frac{K_l}{n} \leq t_i - t_{i-1} \leq \frac{K_u}{n} \tag{2}$$

holds, for all $i = 1, 2, \dots, n$. *Alternatively, more-or-less uniformity requires the existence of a constant* $0 < \beta \leq 1$ *fulfilling asymptotically* $\beta \delta_n \leq t_i - t_{i-1} \leq \delta_n$, *for all* $i = 1, 2, \dots, n$. *Noticeably, the case of* $K_l = K_u = \beta = 1$ *yields* $\mathcal{T}$ *as a* uniform sampling. *Lastly we call* $\mathcal{T}$ *as* ε-uniformly sampled *(with* $\varepsilon > 0$*) if:*

$$t_i = \phi(\frac{iT}{n}) + O(\frac{1}{n^{1+\varepsilon}}), \tag{3}$$

holds for sufficiently large n *and* $i = 1, 2, \dots, n$. *Here the function* $\phi : [0, T] \to [0, T]$ *is an order preserving re-parameterization (i.e. with* $\dot{\phi} > 0$*).*

Note that both (2) and (3) are genuine subfamilies of (1). We formulate now the notion of convergence order (see again e.g. [3]):

Definition 2. *Consider a family* $\{f_{\delta_n}, \delta_n > 0\}$ *of functions* $f_{\delta_n} : [0, T] \to \mathbb{E}$. *We say that* f_{δ_n} *is of order* $O(\delta_n^\alpha)$ *(denoted as* $f_{\delta_n} = O(\delta_n^\alpha)$*), if there is a constant* $K > 0$ *such that, for some* $\bar{\delta} > 0$ *the inequality* $|f_{\delta_n}(t)| < K\delta_n^\alpha$ *holds for all* $\delta_n \in (0, \bar{\delta})$, *uniformly over* $[0, T]$. *In case of vector-valued functions* $F_{\delta_n} : [0, T] \to \mathbb{E}^n$ *by* $F_{\delta_n} = O(\delta_n^\alpha)$ *we understand* $\|F_{\delta_n}\| = O(\delta_n^\alpha)$.

In case of non-reduced data represented by $\mathcal{M}$ and $\mathcal{T}$, in Definition 2, one sets $F_{\delta_n} = \gamma - \hat{\gamma}$ as both domains of γ and $\hat{\gamma}$ coincide with $[0, T]$. If only $\mathcal{M}$ is available (with somehow guessed $\hat{\mathcal{T}}$), the domain of the interpolant $\hat{\gamma} : [0, \hat{T}] \to \mathbb{E}^m$ should be re-mapped (at best reparameterized with $\dot{\psi} > 0$) with $\psi : [0, T] \to [0, \hat{T}]$ so that the convergence analysis of $\gamma - \hat{\gamma} \circ \psi$ can be performed. In fact here, the function F_{δ_n} from Definition 2 reads as $F_{\delta_n} = \gamma - \hat{\gamma} \circ \psi$.

Finally, the notion of *sharpness* of convergence rates α is recalled:

Definition 3. *For a given interpolation scheme $\hat{\gamma}$ based on $\mathcal{M}$ and some $\hat{\mathcal{T}} \approx \mathcal{T}$ (subject to some mapping $\phi : [0, T] \to [0, \hat{T}]$) the asymptotics $\gamma - \hat{\gamma} \circ \phi = O(\delta_n^\alpha)$ over $[0, T]$ is sharp within the predefined family of curves $\gamma \in \mathcal{J}$ and family of samplings $\mathcal{T} \in \mathcal{K}$, if for some $\gamma \in \mathcal{J}$ and some sampling from $\mathcal{K}$, there exists $t^* \in [0, T]$ and some positive constant K such that $\|\gamma(t^*) - (\hat{\gamma} \circ \phi)(t^*)\| = K\delta_n^\alpha + O(\delta_n^\rho)$, where $\rho > \alpha$. A similar definition applies to non-reduced data $\mathcal{M}$ and $\mathcal{T}$ with ψ omitted.*

Suppose the unknown knots $\mathcal{T}$ are estimated by $\hat{\mathcal{T}}_\lambda$ with the so-called *exponential parameterization* (see e.g. [27]):

$$\hat{t}_0 = 0 \quad \text{and} \quad \hat{t}_i = \hat{t}_{i-1} + \|q_i - q_{i-1}\|^\lambda, \tag{4}$$

for $i = 1, 2, \ldots, n$, where $\lambda \in [0, 1]$ is a free parameter. The technical condition $q_i \neq q_{i+1}$ assumed in Sect. 1 guarantees $\hat{t}_i < \hat{t}_{i+1}$. The case $\lambda = 0$ renders for $\hat{\mathcal{T}}_0$ uniform knots $\hat{t}_i = i$ which represents a "blind guess" of $\mathcal{T}$. In contrast $\lambda = 1$ yields the so-called *cumulative chord parameterization* $\hat{\mathcal{T}}_1$ (see e.g. [12, 27]):

$$\hat{t}_i = \hat{t}_{i-1} + \|q_i - q_{i-1}\|. \tag{5}$$

Visibly, the latter accounts for the geometrical layout of reduced data $\mathcal{M}$. For $\lambda = 1$ the last node $\hat{T}$ from now on is denoted by $\hat{T}_c = \hat{t}_n$.

We pass now to different classes of splines $\hat{\gamma}$ (see e.g. [2]) which at junction points in $\mathcal{M}$ (where consecutive local interpolants are glued together) are of class C^l (for $l = 0, 1, 2$) and are C^∞ over sub-interval (t_i, t_{i+1}), with $i = 0, 1, \ldots, n-1$.

2.1 Continuous Splines at Junction Points

To fit $\mathcal{M}$ with $\mathcal{T}$ given (i.e. for *non-reduced data*) one can apply piecewise-r-degree Lagrange polynomials $\gamma_{L(r)}$ (see [2]) for which if $\gamma \in C^{r+1}$ then:

$$\gamma_{L(r)} = \gamma + O(\delta_n^{r+1}), \tag{6}$$

uniformly over $[0, T]$. By (6) and Definition 2 for any samplings (1) the convergence order $\alpha = r + 1$ prevails in $\gamma \approx \gamma_{L(r)}$. Noticeably (6) is sharp (see Definition 3).

Surprisingly, for reduced data $\mathcal{M}$, if $\gamma_{L(r)}$ is used with (5) (i.e. for $\hat{\gamma} = \hat{\gamma}_{L(r)}$) the resulting asymptotics in $\gamma \approx \hat{\gamma}_{L(r)}$ matches (6) for $r = 2, 3$. At this point recall that Newton Interpolation formula [2] (based on divided differences) yields over each consecutive sub-interval $I_i = [\hat{t}_i, \hat{t}_{i+2}]$ the *quadratic* $\hat{\gamma}^i_{L(2)} = \hat{\gamma}_{L(2)}|_{I_i}$ defined as:

$$\hat{\gamma}^i_{L(2)}(\hat{t}) = \gamma[\hat{t}_i] + \gamma[\hat{t}_i, \hat{t}_{i+1}](\hat{t} - \hat{t}_i) + \gamma[\hat{t}_i, \hat{t}_{i+1}, \hat{t}_{i+2}](\hat{t} - \hat{t}_i)(\hat{t} - \hat{t}_{i+1}) \qquad (7)$$

and also over each consecutive sub-interval $\bar{I}_i = [\hat{t}_i, \hat{t}_{i+3}]$ the *cubic* $\hat{\gamma}^i_{L(3)} = \hat{\gamma}_{L(3)}|_{\bar{I}_i}$ defined as:

$$\hat{\gamma}^i_{L(3)}(\hat{t}) = \hat{\gamma}^i_{L(2)}(\hat{t}) + \gamma[\hat{t}_i, \hat{t}_{i+1}, \hat{t}_{i+2}, \hat{t}_{i+3}](\hat{t} - \hat{t}_i)(\hat{t} - \hat{t}_{i+1})(\hat{t} - \hat{t}_{i+2}). \qquad (8)$$

For (7) and (8) the following result is established in [3,4]:

Theorem 1. *Suppose γ is a regular C^r curve in $\mathbb{E}^m$, where $r \geq k+1$ and k is either 2 or 3. Let $\hat{\gamma}_{L(k)} : [0, \hat{T}] \to \mathbb{E}^m$ be the cumulative chord based piecewise-degree-k interpolant defined by $\mathcal{M}$ (sampled admissibly (1)) with $\hat{T}_1 \approx \mathcal{T}$ defined by (5). Then there is a piecewise reparameterization $\psi : [0, T] \to [0, \hat{T}]$ such that:*

$$\hat{\gamma}_{L(k)} \circ \psi = \gamma + O(\delta_n^{k+1}), \qquad (9)$$

holds uniformly over $[0, T]$ (i.e. here $\alpha = 3, 4$). The asymptotics in (9) is sharp.

Thus for either *piecewise-quadratic* or *piecewise-cubic Lagrange interpolants* based on reduced data $\mathcal{M}$ and cumulative chords (5) the missing knots $\mathcal{T}$ can be well compensated by $\hat{T}_1$. Indeed, to approximate γ with $\hat{\gamma}_{L(2,3)}$, Theorem 1 guarantees identical convergence orders as compared to those from (6). Note also that for $r = 1$ the trajectories of both piecewise-linear interpolants $\gamma_{L(1)}$ (based on $\mathcal{T}$) and $\hat{\gamma}_{L(1)}$ (based on any $\hat{T}$) coincide as they are uniquely determined by $\mathcal{M}$. Therefore by (6), for both $\gamma \approx \gamma_{L(1)}$ and $\gamma \approx \hat{\gamma}_{L(1)}$ the convergence rate $\alpha = 2$.

Interestingly, raising the polynomial degree $r \geq 4$ in $\hat{\gamma}_{L(r)}$ (used with (5)) does not further accelerate α in (9) - see [3,6]. The latter stands in contrast with (6) for which any r in $\gamma_{L(r)}$ renders extra speed-up in $\alpha(r) = r + 1$.

The remaining cases of exponential parameterization (4) lead to another unexpected result (see [7–9]) which extends Theorem 1 to all $\lambda \in [0, 1)$:

Theorem 2. *Suppose γ is a regular C^{k+1} curve in $\mathbb{E}^m$ sampled more-or-less uniformly (2) (here $k = 2, 3$). Let $\mathcal{M}$ form reduced data and the unknown knots $\mathcal{T}$ are estimated by $\hat{T}_\lambda$ according to (4) for $\lambda \in [0, 1)$. Then there exists a mapping $\psi : [0, T] \to [0, \hat{T}]$ such that (see also (7) and (8)):*

$$\hat{\gamma}_{L(k)} = \gamma + O(\delta_n), \qquad (10)$$

which holds uniformly over $[0, T]$. The convergence rate $\alpha(\lambda) = 1$ in (10) is sharp. Additionally, a sharp accelerated $\alpha(\lambda)$ follows for $\mathcal{M}$ sampled ε-uniformly (3), with $\varepsilon > 0$ and $\lambda \in [0, 1)$:

$$\hat{\gamma}_{L(2)} = \gamma + O(\delta_n^{\max\{3, 1+2\varepsilon\}}). \qquad (11)$$

The more-or-less uniformity (2) cannot be dropped in Theorem 2. Noticeably the mapping ψ forms a genuine reparameterization only for special $\lambda \in [0,1)$ - see [11]. Both Theorems 1 and 2 underline the substantial discontinuous deceleration effect in $\alpha(\lambda)$ dropping abruptly from $\alpha(1) = 3$ for $k = 2$ (or from $\alpha(1) = 4$ for $k = 3$) to the linear one $\alpha(\lambda) = 1$, for all $\lambda \in [0,1)$. A possible advantage to deal with $\lambda \in [0,1)$ in (4) is to retain a certain degree of freedom (controlled by a single parameter $\lambda \in [0,1)$) at the cost of keeping much slower linear convergence order in $\gamma \approx \hat{\gamma}_{L(2,3)}$. Such relaxation of $\lambda \in [0,1)$ can be exploited if on top of securing even a slow convergence in $\gamma \approx \hat{\gamma}$, some other extra shape-preserving properties of $\hat{\gamma}_{L(2,3)}$ are stipulated - see e.g. [28].

2.2 C^1 Splines at Junction Points

In order to fit reduced data with C^1 interpolant at all junction points (coinciding here with $\mathcal{M} \setminus \{q_0, q_m\}$) a *modified Hermite interpolation* $\hat{\gamma}_H$ can be applied (see [2,3,13] or the next Sect. 3). The latter defines a piecewise-cubic $\hat{\gamma}_H$ which over each sub-interval $[\hat{t}_i, \hat{t}_{i+1}]$ satisfies (19). It also relies on the provision of the estimates of the missing velocities $\mathcal{V} = \{\hat{\gamma}(t_i)\}_{i=0}^n$ over $\mathcal{M}$ (for $i = 0, 1 \ldots, n$). Such estimates $\{v_i\}_{i=0}^n$ of $\mathcal{V}$ can be possibly obtained upon exploiting Lagrange piecewise-cubic $\hat{\gamma}_{L(3)}$ from (8) over each sub-interval $\bar{I}_i = [\hat{t}_i, \hat{t}_{i+3}]$ with $v_i = \hat{\gamma}_{L(3)}^{i'}(\hat{t}_i)$. Here to compute the next v_{i+1} we consider $\hat{\gamma}_{L(3)}^{i+1}$ defined over $\bar{I}_{i+1}$. The last four velocities $\{v_j\}_{j=n-3}^n$ are the derivatives of $\hat{\gamma}_{L(3)}$ (defined over $[\hat{t}_{n-3}, \hat{t}_n]$) calculated at $\{\hat{t}_j\}_{j=n-3}^n$. The following result holds (see [3,13,14]):

Theorem 3. *Let γ be a regular $C^4([0,T])$ curve in $\mathbb{E}^m$ sampled according to (1). Given reduced data $\mathcal{M}$ and knots' estimates (5) (i.e. for $\lambda = 1$ in (4)) there exists a piecewise-cubic C^1 reparameterization $\phi_H : [0,T] \to [0,\hat{T}]$ such that:*

$$\hat{\gamma}_H \circ \phi_H = \gamma + O(\delta_n^4), \tag{12}$$

uniformly over $[0,T]$. If additionally (1) is also more-or-less uniform (2) then for $\mathcal{M}$ and (4) (with $\lambda \in [0,1)$) there exists a mapping $\phi_H : [0,T] \to [0,\hat{T}]$ such that (uniformly over $[0,T]$) we have:

$$\hat{\gamma}_H \circ \phi_H = \gamma + O(\delta_n). \tag{13}$$

Both (12) and (13) are sharp.

Similarly to Subsect. 2.1, both (12) and (13) imply an abrupt left-hand side discontinuity of $\alpha(\lambda)$ at $\lambda = 1$ once $\hat{\gamma}_H$ is used. In addition, by (12) cumulative chords (5) combined with $\mathcal{M}$ and $\hat{\gamma}_H$ yield the same quartic convergence order $\alpha(1) = 4$ as established for classical case of non-reduced data $\mathcal{M}$ combined with $\mathcal{T}$ and with exact velocities $\mathcal{V} = \{\gamma(t_i)\}_{i=0}^n$, for which we also have $\gamma_H = \gamma + O(\delta_n^4)$ (see e.g. [2]). Here γ_H is a standard Hermite interpolant based on $\mathcal{M}$, $\mathcal{T}$ and $\mathcal{V}$ - see Sect. 3. Consequently fitting $\mathcal{M}$ with modified Hermite interpolant $\hat{\gamma}_H$ based on (5) compensates the unavailable $\mathcal{T}$ and $\mathcal{V}$ without decelerating the asymptotic rate in trajectory estimation. For the remaining $\lambda \in [0,1)$ in (4), by (13) a slow linear convergence order prevails in exchange of retaining some flexibility (controlled by $\lambda \in [0,1)$ in modelling the trajectory of $\hat{\gamma}_H$.

2.3 C^2 Splines at Junction Points

In order to fit $\mathcal{M}$ with some C^2 interpolant $\hat{\gamma}$ at all junction points $\mathcal{M} \setminus \{q_0, q_n\}$ (and elsewhere C^∞) one can apply e.g. *a complete spline* $\hat{\gamma} = \hat{\gamma}_{CS}$ or *a natural spline* $\hat{\gamma} = \hat{\gamma}_{NS}$ - see [2] or the next Sect. 3. The first one relies on the additional provision of exact initial and terminal velocities $v_0 = \dot{\gamma}(0)$ and $v_n = \dot{\gamma}(t_n)$. The following result holds (see [10]):

Theorem 4. *Let γ be a regular $C^4([0,T])$ curve in $\mathbb{E}^m$ sampled according to (1). Given reduced data $\mathcal{M}$, v_0, v_m and cumulative chord based knots' estimates (5) there exists a piecewise-cubic C^2 reparameterization $\phi_{CS} : [0,T] \to [0,\hat{T}]$ such that (uniformly over $[0,T]$):*

$$\hat{\gamma}_{CS} \circ \phi_{CS} = \gamma + O(\delta_n^4). \tag{14}$$

The asymptotics in (14) is sharp.

The case of natural spline $\hat{\gamma}_{NS}$ combined with $\mathcal{M}$ and (5) yields decelerated $\alpha(1)$ which upon repeating the argument in [10] leads to a sharp asymptotic estimate:

$$\hat{\gamma}_{NS} \circ \phi_{NS} = \gamma + O(\delta_n^2). \tag{15}$$

Indeed, for the natural spline $\hat{\gamma}_{NS}$ the unknown $\ddot{\gamma}(t_0)$ and $\ddot{\gamma}(t_n)$ are substituted by *ad hock* taken null vectors which ultimately results in slower asymptotics (15) over both sub-intervals $[t_0, t_1]$ and $[t_{n-1}, t_n]$. The latter pulls down a fast quartic order $\alpha(1) = 4$ from (14) (holding for $\hat{\gamma}_{CS}$) to $\alpha(1) = 2$ claimed in (15) for $\hat{\gamma}_{NS}$. As previously, by (14) and (15) and [2] both C^2 interpolants $\hat{\gamma}_{CS}$ and $\hat{\gamma}_{NS}$ coupled with (5) yield exactly the same asymptotics in γ approximation as compared to γ_{CS} and γ_{NS} used with $\mathcal{T}$ given.

The numerical tests for $\hat{\gamma}_{CS}$ and $\hat{\gamma}_{NS}$ combined for $\lambda \in [0,1)$ in (4) indicate the same asymptotic effects as claimed in (10) and (13). In practice, the terminal velocities v_0 and v_n do not accompany reduced data $\mathcal{M}$. However, they can still be well estimated with $w_1 = \hat{\gamma}'_{L(3)}(0)$ and $w_n = \hat{\gamma}'_{L(3)}(\hat{t}_n)$. The interpolant based on $\mathcal{M}$, w_0, w_n and (4) is called *modified complete spline* and is denoted by $\hat{\gamma}_{CS_m}$. It is numerically verified in [15,19] that for $\mathcal{M}$ sampled more-or-less uniformly (2), $\lambda \in [0,1)$ and $\gamma \in C^4$ the following holds:

$$\hat{\gamma}_{NS} \circ \phi_{NS} = \gamma + O(\delta_n) \qquad \hat{\gamma}_{CS_m} \circ \phi_{CM_m} = \gamma + O(\delta_n), \tag{16}$$

for some C^2 mappings $\phi_{NS}, \phi_{CM_m} : [0,T] \to [0,\hat{T}]$.

The discussion for the alternative schemes retrieving the estimates of $\mathcal{T}$ can be found e.g. in [2,16,18,20,21].

3 Fitting Sparse Reduced Data

In this section a possible alternative to fit *sparse reduced data* $\mathcal{M}$ is discussed. Since here $n << \infty$, an arbitrary interpolation scheme $\hat{\gamma}$ coupled with any knots' estimates $\hat{\mathcal{T}} \approx \mathcal{T}$ cannot result in $\hat{\gamma}$ converging to γ. Thus for $\mathcal{M}$ sparse the need for new knots' selection criterion arises. This section addresses this issue.

Consider now a class $\mathcal{I}$ of *admissible curves* $\hat{\gamma}$ forming piecewise C^2 curves $\hat{\gamma} : [0, \hat{T}] \to \mathbb{E}^m$ (where $0 < \hat{T} < \infty$ is fixed) interpolating $\mathcal{M}$ together with the *free unknown knots* $\hat{\mathcal{T}}$ in ascending order $\hat{t}_i < \hat{t}_{i+1}$ and satisfying $\hat{\gamma}(\hat{t}_i) = q_i$. It is assumed here that curves $\hat{\gamma} \in \mathcal{I}$ can possibly be only C^1 at $\mathcal{M}$.

The task is now to find $\hat{\gamma}_{opt} \in \mathcal{I}$ which minimizes the following cost function (with $\{\hat{t}_i\}_{i=1}^{n-1}$ freed as $\hat{t}_0 = 0$ and $\hat{T}$ is fixed):

$$\mathcal{J}(\hat{\gamma}) = \sum_{i=0}^{n-1} \int_{\hat{t}_i}^{\hat{t}_{i+1}} \|\ddot{\hat{\gamma}}(t)\|^2 dt. \tag{17}$$

Note that $\mathcal{J}$ from (17) measures the integrated squared norm of the $\hat{\gamma}$ acceleration. The requirement for $\hat{\mathcal{T}}$ to preserve the ascending order stipulates (17) to be optimized over *an admissible zone of knots*:

$$\Omega = \{(\hat{t}_0, \hat{t}_1, \ldots, \hat{t}_{n-1}, \hat{t}_n) \in \mathbb{R}^{n+1} : \hat{t}_0 = 0 < \hat{t}_1 < \ldots < \hat{t}_{n-1} < \hat{t}_n = \hat{T}\}. \tag{18}$$

Consequently, the latter reformulates (17) into *constrained optimization problem*.

For further consideration we recall now the construction of *a piecewise-cubic spline interpolant* $\hat{\gamma}_C$ based on data points $\mathcal{M}$ and knots $\hat{\mathcal{T}}$ (see e.g. [2]). Here over each sub-interval $\hat{I}_i = [\hat{t}_i, \hat{t}_{i+1}]$ one introduces a cubic $\hat{\gamma}_{C_i} = \hat{\gamma}_C|_{[\hat{t}_i, \hat{t}_{i+1}]}$:

$$\hat{\gamma}_{C_i}(\hat{t}) = c_{1,i} + c_{2,i}(\hat{t} - \hat{t}_i) + c_{3,i}(\hat{t} - \hat{t}_i)^2 + c_{4,i}(\hat{t} - \hat{t}_i)^3, \tag{19}$$

which (for $i = 0, 1, 2, \ldots, n-1$; $c_{j,i} \in \mathbb{R}^m$, where $j = 1, 2, 3, 4$) fulfills:

$$\hat{\gamma}_{C_i}(\hat{t}_{i+k}) = q_{i+k}, \quad \dot{\hat{\gamma}}_{C_i}(\hat{t}_{i+k}) = v_{i+k}, \quad k = 0, 1. \tag{20}$$

The respective velocities $v_0, v_1, v_2, \ldots, v_{n-1}, v_n \in \mathbb{R}^m$ in (20) are treated temporarily as free parameters (*if unknown*). The coefficients $c_{j,i}$ from (19) (with $\Delta \hat{t}_i = \hat{t}_{i+1} - \hat{t}_i$) by [2] read as:

$$c_{1,i} = x_i, \quad c_{2,i} = v_i,$$

$$c_{4,i} = \frac{v_i + v_{i+1} - 2\frac{q_{i+1} - q_i}{\Delta \hat{t}_i}}{(\Delta \hat{t}_i)^2} \quad \text{and} \quad c_{3,i} = \frac{\frac{(q_{i+1} - q_i)}{\Delta \hat{t}_i} - v_i}{\Delta \hat{t}_i} - c_{4,i}\Delta \hat{t}_i. \tag{21}$$

Assuming that both velocities v_0 and v_n are *a priori* given, a cubic spline $\hat{\gamma}_C$ called *a complete spline* (denoted as $\hat{\gamma}_{CS}$) can e.g. be used to fit $\mathcal{M}$ with $\hat{\mathcal{T}}$.

If in turn, the velocities v_0 and v_n are unavailable a possible alternative to interpolate $\mathcal{M}$ with $\hat{\mathcal{T}}$ is to apply the so-called *natural splines* (denoted as $\hat{\gamma}_C = \hat{\gamma}_{NS}$) - see e.g. [2]. In fact, both missing v_0 and v_n can be easily expressed in terms of the initial and terminal accelerations a_0 and a_n. The latter as usually not supplied are commonly substituted by *ad hock* admitted values $a_0 = a_n = 0 \in \mathbb{R}^m$. In this case the combination of nullified a_0 and a_n with (19) and (20) leads to the following two linear (m dimensional vector) equations:

$$2v_0 + v_1 = 3\frac{q_1 - q_0}{\Delta \hat{t}_0} \quad \text{and} \quad v_{n-1} + 2v_n = 3\frac{q_n - q_{n-1}}{\Delta \hat{t}_{n-1}}. \tag{22}$$

The remaining velocities $\{v_i\}_{i=1}^{n-1}$ can be computed from (19) and (20) which leads to the tridiagonal m linear systems (strictly diagonally dominant - see [2]):

$$v_{i-1}\Delta\hat{t}_i + 2v_i(\Delta\hat{t}_{i-1} + \Delta\hat{t}_i) + v_{i+1}\Delta\hat{t}_{i-1} = b_i, \tag{23}$$

where

$$b_i = 3(\Delta\hat{t}_i \frac{q_i - q_{i-1}}{\Delta\hat{t}_{i-1}} + \Delta\hat{t}_{i-1}\frac{q_{i+1} - q_i}{\Delta\hat{t}_i}), \tag{24}$$

and $1 \le i \le n - 1$. Equations (22), (23) and (24) render m linear systems, each of size $(n + 1) \times (n + 1)$, strictly diagonally dominant which can be solved e.g. by Gauss elimination without pivoting [2]. The latter renders a unique sequence of missing velocities $\{v_i\}_{i=0}^n$ all expressed in terms of $\mathcal{M}$ and $\hat{\mathcal{T}}$. Ultimately, the substitution of computed $\{v_i\}_{i=0}^n$ into (21) yields the explicit formula for *the natural cubic spline* $\hat{\gamma}_{NS}$ determined unambiguously by $\mathcal{M}$ and $\hat{\mathcal{T}}$.

In case of natural splines forming the subfamily of $\mathcal{I}$ the cost function (17) for $\hat{\gamma}_{NS}$ and (19) reformulates into (see [22,23]):

$$\mathcal{J}(\hat{\gamma}_{NS}) = \sum_{i=0}^{n-1} \int_{\hat{t}_i}^{\hat{t}_{i+1}} \|\ddot{\hat{\gamma}}_{NS_i}(t)\|^2 dt$$

$$= 4\sum_{i=0}^{n-1} (\|c_{3,i}\|^2 \Delta\hat{t}_i + 3\|c_{4,i}\|^2(\Delta\hat{t}_i)^3 + 3\langle c_{3,i}|c_{4,i}\rangle(\Delta\hat{t}_i)^2), \tag{25}$$

where $\langle \cdot | \cdot \rangle$ is a standard dot product in $\mathbb{R}^m$. Coupling (21) with (25) yields:

$$\mathcal{J}(\hat{\gamma}_{NS}) = 4\sum_{i=0}^{n-1} \Big(\frac{-1}{(\Delta\hat{t}_i)^3}(-3\|q_{i+1} - q_i\|^2 + 3\langle v_i + v_{i+1}|q_{i+1} - q_i\rangle\Delta\hat{t}_i$$

$$-(\|v_i\|^2 + \|v_{i+1}\|^2 + \langle v_i|v_{i+1}\rangle)(\Delta\hat{t}_i)^2 \Big).$$

For arbitrary fixed knots $\hat{\mathcal{T}}$ satisfying $\hat{t}_i < \hat{t}_{i+1}$ we have (see e.g. [2]):

Lemma 1. *For a given reduced data $\mathcal{M}$ and fixed $\hat{\mathcal{T}}$ in arbitrary Euclidean space $\mathbb{E}^m$ the subclass of natural splines $\mathcal{I}_{\hat{\mathcal{T}}}^{NS} \subset \mathcal{I}_{\hat{\mathcal{T}}}$ satisfies:*

$$\min_{\hat{\gamma}\in\mathcal{I}_{\hat{\mathcal{T}}}} \mathcal{J}(\hat{\gamma}) = \min_{\gamma^{NS}\in\mathcal{I}_{\hat{\mathcal{T}}}^{NS}} \mathcal{J}(\hat{\gamma}_{NS}). \tag{26}$$

The last lemma reformulates the problem of minimizing (17) over $\mathcal{I}_{\hat{\mathcal{T}}}$ (with $\hat{\mathcal{T}}$ temporarily fixed) to the class of natural splines $\mathcal{I}_{\hat{\mathcal{T}}}^{NS}$. Hence by (26), once the knots $\hat{\mathcal{T}}$ are permitted to vary subject to (18) the problem (17) reduces into:

$$\min_{\hat{\gamma}\in\mathcal{I}} \mathcal{J}(\hat{\gamma}) = \min_{\gamma^{NS}\in\mathcal{I}^{NS}} \mathcal{J}(\hat{\gamma}_{NS}). \tag{27}$$

As demonstrated above since each $\hat{\gamma}_{NS}$ is uniquely determined by $\hat{\mathcal{T}}$ and $\mathcal{M}$ the optimization of (17) reformulates into *a constrained finite dimensional optimization task* (28) over $(\hat{t}_1, \hat{t}_2, \ldots, \hat{t}_{n-1}) \in \Omega$ which satisfies $\mathcal{J}(\hat{\gamma}_{NS}^{opt}) = \mathcal{J}^F(\hat{t}_0, \hat{t}_1^{opt}, \hat{t}_2^{opt}, \ldots, \hat{t}_{n-1}^{opt}, \hat{t}_n)$ (with $\hat{t}_0$ and $\hat{t}_n$ fixed). Here for the optimal knots $\hat{\mathcal{T}}^{opt} = (\hat{t}_0, \hat{t}_1^{opt}, \hat{t}_2^{opt}, \ldots, \hat{t}_{n-1}^{opt}, \hat{t}_n) \in \Omega$ (see (18)) we arrive at (see also [22, 23]):

$$\mathcal{J}^F(\hat{\mathcal{T}}^{opt}) = 4 \min_{\hat{\mathcal{T}} \in \Omega} \sum_{i=0}^{n-1} \Big(\frac{-1}{(\Delta \hat{t}_i)^3} (-3\|q_{i+1} - q_i\|^2 + 3\langle v_i + v_{i+1} | q_{i+1} - q_i \rangle \Delta \hat{t}_i$$

$$-(\|v_i\|^2 + \|v_{i+1}\|^2 + \langle v_i | v_{i+1} \rangle)(\Delta \hat{t}_i)^2 \Big), \tag{28}$$

where the velocities $\{v_i\}_{i=0}^n$ are expressed in terms of $\hat{\mathcal{T}}$ as outlined above.

The function $\mathcal{J}^F$ is shown to be highly non-linear and non-convex [24]. The latter makes it hard to optimize upon applying standard numerical schemes. In addition the explicit formula (28) depends on velocities $\{v_i\}_{i=0}^n$ which are implicitly given by (22) and (23) in terms of a given $\mathcal{M}$ and $\hat{\mathcal{T}}$ relaxed to be free. In fact even for n small the expression for (28) treated as a function of $\hat{\mathcal{T}}$ (with $\mathcal{M}$ fixed) gets quite complicated [24] (see also Example 3 below). It can be shown however that $\mathcal{J}^F$ is a continuous function over $\hat{\mathcal{T}} \in \Omega$ (see [24]). Furthermore, though Ω is not compact, the existence of a global minimizer $\hat{\mathcal{T}}^{opt}$ to $\mathcal{J}^F$ can still be justified as claimed below (see [25]):

Theorem 5. *There exists at least one global minimum to (28) (and thus by (27) to (17)) with the some optimal knots* $(\hat{t}_0 = 0, \hat{t}_1^{opt}, \hat{t}_2^{opt}, \ldots, \hat{t}_{n-1}^{opt}, \hat{t}_n = \hat{T}) \in \Omega$.

More literature on fitting sparse reduced data can be found e.g. in [2, 3, 26]. In the next step we report on implementing selected schemes from Sects. 2 and 3.

4 Examples for Fitting Reduced Data

This section includes two examples to fit dense and sparse 2D and 3D reduced data $\mathcal{M}$ based on computation performed in *Mathematica* - see [32]. For more examples covering different n and m see [3, 14, 22, 23]. Noticeably, all $\hat{\gamma}$ from Sects. 2 and 3 are applicable to multi-dimensional $\mathcal{M}$. For the detailed discussion of the numerical optimization schemes used in Subsect. 4.2 see e.g. [23, 33–35].

4.1 Examples for $\mathcal{M}$ dense

We use two types of more-or-less uniform samplings (2) to simulate $\mathcal{M}$:

$$t_i = \begin{cases} \frac{i}{n}, & \text{if } i = 2k; \\ \frac{i}{n} + \frac{1}{2n}, & \text{if } i = 4k+1; \\ \frac{i}{n} - \frac{1}{2n}, & \text{if } i = 4k+3, \end{cases} \qquad (ii) \ \ t_i = \frac{i}{n} + \frac{(-1)^{i+1}}{3n}. \tag{29}$$

For a selected scheme $\hat{\gamma}$ the numerical estimate $\bar{\alpha}(\lambda) \approx \alpha(\lambda)$ for γ approximation is computed from the regression line $y(x) = \bar{\alpha}(\lambda)x + b$ applied to the collection of pairs of points $\{(\log(n), -\log(\mathcal{E}_n))\}_{n_{min}}^{n_{max}}$ (see [3]), where the error $\mathcal{E}_n = \sup_{t \in [0,T]} \|(\hat{\gamma} \circ \psi)(t) - \gamma(t)\| = \max_{t \in [0,T]} \|(\hat{\gamma} \circ \psi)(t) - \gamma(t)\|$.

Example 1. Take the following 3D curve $\gamma_{st} : [0, 1] \to \mathbb{E}^3$:

$$\gamma_{st}(t) = (\cos(2\pi t), \sin(2\pi t), \sqrt{(1.2)^2 - (\sin(2\pi t))^2}, \qquad (30)$$

sampled along (29). Assume *Lagrange piecewise-cubic* $\hat{\gamma}_{L(3)}$ is used to fit $\mathcal{M}$ - see Subsect. 2.1. The Table 1 confirm *the sharpness* of (10) claimed in Theorem 2. More experiments can be found (including piecewise-quadratics $\hat{\gamma}_{L(2)}$) in [7–9].

Table 1. Computed $\bar{a}(\lambda) \approx \alpha(\lambda)$ for γ_{st} (30) sampled as in (29), for $n \in \{72, \dots, 162\}$.

λ	0.0	0.1	0.3	0.5	0.7	0.9	1.0
$\bar{a}(\lambda)$ for (29)(i)	1.000	1.000	1.003	1.001	1.008	1.200	4.981
$\bar{a}(\lambda)$ for (29)(ii)	0.993	0.994	0.996	0.998	1.009	1.186	3.961
$\alpha(\lambda)$ in Theorem 2	1.0	1.0	1.0	1.0	1.0	1.0	4.0

Example 2. Consider *a planar spiral* $\gamma_{sp} : [0, 1] \to \mathbb{E}^2$ defined by:

$$\gamma_{sp}(t) = ((0.2 + t)\cos(\pi(1 - t)), (0.2 + t)\sin(\pi(1 - t))), \qquad (31)$$

sampled according to (29). Assume *a modified Hermite interpolant* $\hat{\gamma}_H$ is used to fit $\mathcal{M}$ - see Subsect. 2.2. The experimental results from Table 2 confirm numerically the *sharpness* of Theorem 3.

Table 2. Computed $\bar{a}(\lambda) \approx \alpha(\lambda)$ for γ_{sp} (31) sampled as in (29), for $n \in \{96, \dots, 144\}$.

λ	0.0	0.1	0.3	0.5	0.7	0.9	1.0
$\bar{a}(\lambda)$ for (29)(i)	0.968	0.969	0.970	0.973	0.979	1.021	3.918
$\bar{a}(\lambda)$ for (29)(ii)	0.969	0.970	0.974	0.978	0.986	1.038	3.987
$\alpha(\lambda)$ in Theorem 3	1.0	1.0	1.0	1.0	1.0	1.0	4.0

Due to the page limitation, the examples verifying the sharpness of the asymptotics from Subsect. 2.3 are omitted. For more see e.g. [10, 15, 19].

4.2   Examples for $\mathcal{M}$ sparse

Example 3. Consider eight 2D data points (i.e. here $n = 7$) in $\mathbb{E}^2$ with

$$\mathcal{M}_7 = \{(0,0), (-0.5, -4), (0.5, -4), (-0.5, 4), (0.5, 0.4), (-1, 3.8), (0.3, 0.3), (0.5, 0.5)\}$$

The *FindMinimum (Newton Method)* [32] causes computer to hang and no result was reported within 60 min. The respective initial guesses for the knots $\hat{\mathcal{T}}$ coincide with cumulative chords (5) calculated in terms of $\mathcal{M}_7$. On the other hand *FindMinimum (Secant Method)* yields the corresponding optimal knots:

$$\mathcal{T}^{opt} = \{0, 2.67713, 4.69731, 10.3221, 12.3943, 14.8132, 19.0316, 19.6231\}.$$

The respective execution time amounts to $T_{\mathcal{M}_7}^{SM} = 35.708519$ s. Recall that *Secant Method* requires two sequences of initial knots to initialize the iteration process. For the latter cumulative chords (5) are perturbed here by taking their left-shift $\hat{t}_i^c - \delta_i$ and right-shift $\hat{t}_i^c + \delta_i$, accordingly, to serve as two initial guesses.

Example 3 illustrates that for $n \geq 7$ *FindMinimum (Secant Method)* offers a feasible computational scheme to optimize (28). The next example compares the latter with an alternative scheme called *Leap-Frog* (see e.g. [23]).

We pass now to the example with $\mathcal{M}$ representing reduced data in $\mathbb{E}^3$.

Example 4. Consider for $n = 6$ the 3D data points (see dotted points in Fig. 1):

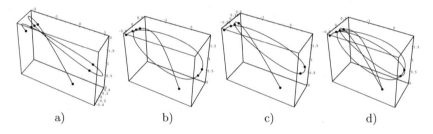

a)   b)   c)   d)

Fig. 1. Natural splines $\hat{\gamma}_{NS}$ interpolating $\mathcal{M}_6$ from (32) with (a) uniform knots $\hat{\mathcal{T}}_{uni}$, (b) cumulative chords $\hat{\mathcal{T}}_c$, (c) optimal knots $\hat{\mathcal{T}}_{LF}^{opt} = \hat{\mathcal{T}}_{SM}^{opt2}$ (d) plotted together for (a)–(c).

$$\mathcal{M}_6 = \{\gamma(0), \gamma(3), \gamma(3.1), \gamma(0.4), \gamma(0.6), \gamma(3.2), \gamma(3.4)\}, \tag{32}$$

where $\gamma(t) = (\sqrt{t}\cos t, \sqrt{t}\sin t, \sqrt{t})$. The *uniform interpolation knots* (re-scaled to $\hat{\mathcal{T}}_c$) coincide with:

$$\hat{\mathcal{T}}_{uni} = \{0, 1.42223, 2.84446, 4.26669, 5.68892, 7.11115, 8.53338\}. \tag{33}$$

The initial guess derived from *cumulative chords* $\hat{\mathcal{T}}_c$ (see (5)) is represented by:

$$\hat{\mathcal{T}}_c = \{0, 2.44949, 2.62868, 5.23375, 5.47857, 8.16249, 8.53338\}. \tag{34}$$

The natural splines $\hat{\gamma}^{NS}$ (based either on $\hat{\mathcal{T}}_{uni}$ or on $\hat{\mathcal{T}}_c$) render the respective energies (28) as $\mathcal{J}^F(\hat{\mathcal{T}}_{uni}) = 18.7237 > \mathcal{J}^F(\hat{\mathcal{T}}_c) = 13.8136$. The trajectories of both interpolants are presented in Fig. 1(a) and (b).

The *Secant Method* yields (for (28)) the optimal knots (augmented by terminal knots $\hat{t}_0 = 0$ and $\hat{t}_6 = \hat{T}_c$ - see (5)) as:

$$\hat{\mathcal{T}}_{SM}^{opt_1} = \{0, 1.75693, 2.33172, 4.89617, 5.49792, 8.12181, 8.53338\}$$

with the corresponding optimal energy $\mathcal{J}^F(\hat{\mathcal{T}}_{SM}^{opt_1}) = 9.21932$. The *execution time* equals to $T_1^{SM} = 33.79$ s. For each free variable the *Secant Method* uses here two initial numbers $\hat{t}_i^c \pm 0.1$ (i.e. perturbed cumulative chord numbers). For other initial guesses $\hat{t}_i^c \pm 0.2$ marginally more precise knots (compatible with *Leap-Frog* - see below) are generated:

$$\hat{\mathcal{T}}_{SM}^{opt_2} = \{0, 1.76066, 2.35289, 4.90326, 5.50495, 8.12262, 8.53338\} \qquad (35)$$

with more accurate optimal energy $\mathcal{J}^F(\hat{\mathcal{T}}_{SM}^{opt_2}) = 9.21787$. Here *the execution time* reads as $T_2^{SM} = 51.88436$ s and gets longer if accuracy is improved. The resulting curve $\hat{\gamma}^{NS}$ is plotted in Fig. 1(c).

The *Leap-Frog Algorithm* decreases the energy to $\mathcal{J}^F(\hat{\mathcal{T}}_{LF}^{opt}) = \mathcal{J}^F(\hat{\mathcal{T}}_{SM}^{opt_2})$ (as for the *Secant Method*) with the iteration stopping conditions $\hat{\mathcal{T}}_{LF}^{opt} = \hat{\mathcal{T}}_{SM}^{opt_2}$ (up to 6th decimal point) upon 79 iterations. The respective *execution time* is equal to $T^{LF} = 8.595979$ s. $< T_2^{SM} < T_1^{SM}$. The 0th (i.e. $\mathcal{J}^F(\hat{\mathcal{T}}_c)$), 1st, 2nd, 3rd, 10th, 20th, 40nd and 79th iterations of *Leap-Frog* decrease the energy to:

$$\{13.8136, 11.3619, 10.3619, 9.88584, 9.25689, 9.21987, 9.21787, 9.21787\} \qquad (36)$$

with only the first three iterations substantially correcting the initial guess knots $\hat{\mathcal{T}}_c$. Since $\hat{\mathcal{T}}_{LF}^{opt} = \hat{\mathcal{T}}_{SM}^{opt_2}$ both natural splines $\hat{\gamma}^{NS}$ are identical - see in Fig. 1(c). The graphical comparison between $\hat{\gamma}_{NS}$ based on either (33) or (34) or (35) is shown in Fig. 1(d).

Note that if the *Leap-Frog* iteration bound condition is adjusted e.g. to ensure the current *Leap-Frog* energy to coincide with $\mathcal{J}^F(\hat{\mathcal{T}}_c^{SM})$ (say up to 5th decimal place) then only 40 iterations are needed which speeds-up the execution time to $T_E^{LF} = 4.789121$ s. $< T^{SM}$ with adjusted optimal knots

$$\hat{\mathcal{T}}_{LF_E}^{opt} = \{0, 1.76153, 2.35384, 4.90451, 5.50603, 8.12278, 8.53338\}.$$

Evidently, at the cost of losing marginal accuracy in optimal knots' estimation the acceleration in *Leap-Frog* execution time is achieved with almost identical interpolating curve as the optimal one - here $\hat{\mathcal{T}}_{LF_E}^{opt} \approx \hat{\mathcal{T}}_{SM}^{opt_2}$. Similar acceleration follows for other *a posteriori* selected stopping conditions like e.g. a bound on a relative decrease of the $\mathcal{J}^F$.

5   Conclusions

In this paper we discuss several methods of fitting reduced data $\mathcal{M}$ forming ordered collection of $n + 1$ points in arbitrary Euclidean space $\mathbb{E}^m$. The points in $\mathcal{M}$ are generated from the interpolation condition $\gamma(t_i) = q_i$ with the corresponding knots $\mathcal{T} = \{t_i\}_{i=0}^n$ assumed to be unknown. Different criteria of estimating the missing knots $\mathcal{T}$ are discussed here in the context of sparse or dense $\mathcal{M}$ fitted with various interpolation schemes.

The first part of this work addresses the problem of interpolating $\mathcal{M}$ when n is large. Different interpolants $\hat{\gamma}$ combined with exponential parameterization (4) are discussed to determine the underlying speed of convergence in $\hat{\gamma} \approx \gamma$. It is also demonstrated that cumulative chords (5) yield identical convergence orders to approximate γ as if the genuine knots $\mathcal{T}$ were given. The annotated experiments conducted in *Mathematica* confirm the asymptotics obtained by theoretical analysis.

The second part of this work deals with the case of fitting $\mathcal{M}$, when $n << \infty$. Under such scenario the convergence of $\hat{\gamma}$ to γ with any $\hat{\mathcal{T}}$ (or $\mathcal{T}$) is obviously excluded. Different knots' selection strategy is therefore required. In doing so, an infinite dimensional optimization task (17) is reformulated into a corresponding finite dimensional minimization problem (28) - see Lemma 1. The existence of global optimal knots $\mathcal{T}^{opt}$ minimizing the "average acceleration" of $\hat{\gamma}$ in (28) (and given in the ascending order) justifies Theorem 5. Two numerical schemes for optimizing (28) (i.e. *Leap-Frog* and *Secant Method*) are also implemented and compared. The computations conducted in *Mathematica* indicate the superiority of *Leap-Frog* over *Secant Method* upon accounting the execution times.

References

1. do Carmo, M.P.: Differential Geometry of Curves and Surfaces. Prentice-Hall, Engelwood Cliffs (1976)
2. de Boor, C.: A Practical Guide to Spline. Springer, Heidelberg (1985)
3. Kozera, R.: Curve modeling via interpolation based on multidimensional reduced data. Studia Informatica **25**(4B(61)), 1–140 (2004)
4. Noakes, L., Kozera, R.: Cumulative chords piecewise-quadratics and piecewise-cubics. In: Klette, R., Kozera, R., Noakes, L., Wieckert, J. (eds.) Geometric Properties from Incomplete Data, Computational Imaging and Vision, vol. 31, pp. 59–75. Kluver Academic Publisher (2006)
5. Epstein, M.P.: On the influence of parameterization in parametric interpolation. SIAM J. Numer. Anal. **13**, 261–268 (1976)
6. Kozera, R., Noakes, L.: Asymptotics for length and trajectory from cumulative chord piecewise-quartics. Fund. Inform. **61**(3–4), 267–283 (2004)
7. Kozera, R., Noakes, L.: Piecewise-quadratics and exponential parameterization for reduced data. Appl. Math. Comput. **221**, 620–638 (2013)
8. Kozera, R., Wilkołazka, M.: Convergence order in trajectory estimation by piecewise cubics and exponential parameterization. Math. Model. Anal. (in press)
9. Kozera, R., Noakes, L.: Piecewise-quadratics and ε-uniformly sampled reduced data. Appl. Math. Inf. Sci. **10**(1), 33–48 (2016)
10. Floater, M.S.: Chordal cubic spline interpolation is fourth order accurate. IMA J. Numer. Anal. **26**(1), 25–33 (2005)
11. Kozera, R., Noakes, L.: Piecewise-quadratics and reparameterizations for interpolating reduced data. In: 17th International Workshop on Computer Algebra in Scientific Computing. LNCS, vol. 9301, pp. 260–274. Springer, Cham (2015). https://doi.org/10.1007/978-3-319-24021-3_20

12. Lee, E.T.Y.: Choosing nodes in parametric curve interpolation. Comput. Aided Geom. Des. **21**(6), 363–370 (1989)
13. Kozera, R., Noakes, L.: C^1 interpolation with cumulative chord cubics. Fund. Inform. **61**(3–4), 285–301 (2004)
14. Kozera, R., Wilkołazka, M.: A modfied Hermite interpolation and exponential parameterization. Math. Comput. Sci. (2018). https://doi.org/10.1007/s11786-018-0362-4
15. Kozera, R., Noakes, L., Wilkołazka, M.: A modified complete spline interpolation and exponential parameterization. In: 14th International Conference on Computer Information Systems and Industrial Management. LNCS, vol. 9339, pp. 98–110. Springer, Cham (2015). https://doi.org/10.1007/978-3-319-24369-6_8
16. Mørken, K., Scherer, K.: A general framework for high-accuracy parametric interpolation. Math. Comput. **66**(217), 237–260 (1997)
17. Rababah, A.: High order approximation methods for curves. Comput. Aided Geom. Des. **12**(1), 89–102 (1995)
18. Kocić, L.J.M., Simoncelli, A.C., Della, V.B.: Blending parameterization of polynomial and spline interpolants. Facta Univ. Ser. Math. Inform. **5**, 95–107 (1990)
19. Kozera, R., Noakes, L., Wilkołazka, M.: A natural spline interpolation and exponential parameterization. In: Proceedings of American Institute of Physics, vol. 1738, pp. 180003-1–183003-4. AIP Publishing (2016). https://doi.org/10.1063/1.4992579
20. Noakes, L., Kozera, R.: More-or-less uniform samplings and lengths of curves. Q. Appl. Math. Des. **61**(3), 475–483 (2003)
21. Kozera, R., Noakes, L.: Interpolating sporadic data. In: 7th European Conference on Computer Vision. LNCS, vol. 2351/II, pp. 613–625. Springer, Heidelberg (2002). https://doi.org/10.1007/3-540-47967-8_41
22. Kozera, R., Noakes, L.: Optimal knots selection for sparse reduced data. In Workshops of 7th Pacific-Rim Symposium on Image and Video Technology. LNCS, vol. 9555, pp. 3–14. Springer, Cham (2016). https://doi.org/10.1007/978-3-319-30285-0_1
23. Kozera, R., Noakes, L.: Modelling reduced sparse data. In: Photonics Applications in Astronomy, Communications, Industry and High Energy Physics Experiments, vol. 10031, pp. 100314V-1–100314V-8. SPIE Proceedings (2016). https://doi.org/10.1117/12.2249260
24. Kozera, R., Noakes, L.: Non-linearity and non-convexity in optimal knots selection for sparse reduced data. In: 19th International Workshop on Computer Algebra in Scientific Computing. LNCS, vol. 10490, pp. 257–271. Springer, Cham (2017). https://doi.org/10.1007/978-3-319-66320-3_19
25. Kozera, R., Noakes, L.: Natural cubic splines and and optimal knots selection for sparse reduced data. Submitted
26. Kuznetsov, E.B., Yakimovich, A.Y.: The best parameterization for parametric interpolation. J. Appl. Math. **191**, 239–245 (2006)
27. Kvasov, B.I.: Methods of Shape-Preserving Spline Approximation. World Scientific Publishing Company, Singapore (2000)
28. Piegl, L., Tiller, W.: The NURBS Book. Springer, Heidelberg (1997)
29. Farin, G.: Curves and Surfaces for Computer Aided Geometric Design. Academic Press, San Diego (1993)
30. Janik, M., Kozera, R., Kozioł, P.: Reduced data for curve modeling - applications in graphics, computer vision and physics. Adv. Sci. Technol. **7**(18), 28–35 (2013)
31. Rogers, D.F., Adams, J.A.: Mathematical Elements for Computer Graphics. McGraw-Hill, New York (1976)

32. Wolfram, S.: The Mathematica Book, 5th. edn. Wolfram Media (2003)
33. Boyd, S., Vandenberghe, L.: Convex Optimization. Cambridge University Press, Cambridge (2004)
34. Leal, M.M., Raydan, M.: Newton's method and secant methods: a longstanding relationship from vector to matrices. Port. Math. **68**(4), 431–475 (2011)
35. Ralston, A.: A First Course in Numerical Analysis. McGraw-Hill, New York (1965)

Artificial Intelligence

Survey of AI Methods for the Purpose of Geotechnical Profile Creation

Adrian Bilski[✉]

Warsaw University of Life Sciences, Nowoursynowska 159,
02-776 Warsaw, Poland
adrian_bilski1@sggw.pl

Abstract. The goal of this paper is to present methodology of unsupervised learning application in geotechnical data categorization. Geotechnical layers identification is conducted based on measurement from the Dilatometer of Marchetti Test taken at the campus of Warsaw University of Life Sciences. To cluster data, the Ant Clustering Algorithm, k-means, fuzzy sets and Self Organizing Map algorithms were introduced. All methods are adjusted to the presented problem and their efficiency compared. The paper is concluded with comments about the applications of computer intelligence methods for the geotechnical data analysis.

Keywords: Dilatometer of Marchetti Test · Geotechnical data clustering
K-means · Fuzzy c-means · Ant Colony Algorithm · Self Organizing Map

1 Introduction

Geotechnical data provides vital information from site investigations carried out for civil engineering purposes. Usually conducted with the use of bore-holes, it is now supplemented with geotechnical probes (DMT and CPT), which are less expensive and deliver measurement information faster. Their usage is now standard in site conditions and structure foundations assessment. Determination of soil categories is usually conducted using diagrams, which require additional computation of indexes and are of limited accuracy. As they are prepared for specific geographical location (like North America or Western Europe), their utilization in other sites is restricted.

Clustering is the process of grouping data into classes of similar objects based on the proximity measure. The clustering process is unsupervised, because no predefined categories for data samples exist [1]. The quality of obtained clustering solution depends on the type of the used algorithm. A good clustering produces clusters, in which the intra-class similarity is high, while the inter-class similarity is low.

The purpose of this paper is the examination of various clustering algorithms, such as k-means, FCM (fuzzy c-means), Ant Clustering Algorithm (ACA) and Self Organizing Map (SOM) to the geotechnical data analysis. Two of the latter were selected because of their ability to automatically adjust the number of groups based on the examples' attributes. As implemented methods are heuristic and relatively new, more studies on their characteristics are justified.

© Springer Nature Switzerland AG 2019
J. Pejaś et al. (Eds.): ACS 2018, AISC 889, pp. 21–33, 2019.
https://doi.org/10.1007/978-3-030-03314-9_2

Each clustering algorithm may be used for making decisions about the category of the soil based on the measurements taken by the probes at the particular depth. This way the soil profile can be constructed. In order to improve clustering results by noise reduction, PCA (Principal Component Analysis) is typically used before the initial grouping takes place. Additionally it allows for data visualization.

Figure 1 presents the scheme of the clustering approach [2], which was used for the geotechnical profile generation based on the data from the probes acquires in the Warsaw University of Life Sciences campus. This work is a continuation of studies presented in [3–5].

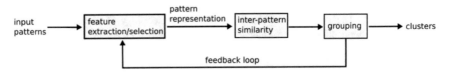

Fig. 1. Different stages of pattern clustering [2].

The paper is organized as follows. Section 2 presents the probe-based methods and test site description. In Sect. 3 the clustering algorithms utilized for the purpose of these studies are introduced. Section 4 focuses on the clustering principles for the given geotechnical data. Simulation results and conclusions are in Sects. 5 and 6, respectively.

2 Probe-Based Methods and Test Site Description

This section provides the description of the test site along with the data acquisition methodology. Various quantitative measures (like material index I_D), introduced for the purpose of soil classification quality assessment, are defined here.

The Dilatometer of Marchetti Test (DMT) is a probe-based invasive method of soil categorization generating continuous data profiles of high reproducibility. It measures working gas pressure values required to move the head of the probe (membrane) against the soil. The acquired pressure readings A, B must be corrected by the calibration values ΔA, ΔB to consider the membrane stiffness and converted into p_0, p_1, further used to calculate indexes E_D and K_D. These are primary soil characteristics of stiffness and stress history, respectively. Classification of soils is made through the quantitative measure of material index I_D [6]. It is defined as the difference between two basic pressures which is then normalized in terms of the effective lift-off pressure, $p_0 - u_0$:

$$I_D = \frac{p_1 - p_0}{p_0 - u_0} \tag{1}$$

where u_0 is the pre-insertion in-situ pore pressure.

The traditional chart for estimating the soil type based on the DMT is presented in Fig. 2, where letters A–D denote boundaries between different soil categories, while

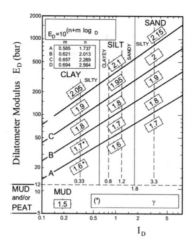

Fig. 2. Chart for estimating soil types based on DMT measurements (1 bar = 100 kPa) [7].

vertical lines separate them [6]. Because I_D describes mechanical behavior and is not designed for the purpose of sieve analysis, the material index is prone to erroneous classification of the examined objects. This might lead to I_D confusing silt with clay (and vice versa) or categorize the mixture of clay and sand as silt. This justifies formulation of a more precise, automatic method of soil categorization based on artificial intelligence [7].

The speed of inserting these probes into the ground is constant (two centimeters per second). The readings are done every twenty cm. Two parameters are acquired during DMT implementation: the gas pressure A, obtained in the first phase of the membrane movement and the gas pressure B, obtained by the additional inclination of the centre of the membrane toward the ground [8].

The processed data was obtained from the test site, i.e. Warsaw University of Life Sciences campus (the area where currently the building No. 34 is established). The dimensions of the investigated area are 57 by 120 m (Fig. 3). Multiple boreholes were drilled, among them CPT and DMT performed.

3 Clustering Methods

This section provides information concerning clustering validity techniques and a brief overview of four clustering algorithms utilized in studies presented here: ACA, k-means, FCM and SOM, and the related work on clustering methods and results visualization techniques.

The centroid-based method (like k-means or it's fuzzy variant) is deterministic, i.e. its repeated utilization on the same data sets produces a fixed number of groupings and the same geotechnical profile is always obtained. This is due to the fact that the method does not depend on any random factors. The same can't however be said about the Neural Network (NN) and Ant Clustering Algorithms.

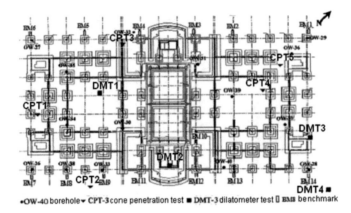

Fig. 3. Location of boreholes in the foundation of building no. 34.

This is because initial values of neuron weights are generated randomly each time the clustering process is conducted by the NN algorithm (in this case SOM). It is however probable that the weight of only one (or several) neurons will be close to the input data, so we will receive one (or several) categories after the process of network learning is completed. The introduction of the neuron conscience mechanism can prevent this effect.

3.1   Cluster Validity

The aim of the cluster evaluation is estimation of data randomness and its categorization. Subsequently, it provides the information about quality of the selected clustering algorithm. The most effective measures of cluster validity assessment for unlabeled data are the internal/relative indices, which determine if the acquired data structure is intrinsically appropriate for the given data [9].

Davies-Bouldin Index (DBI) is one of the most popular such method, often used when no means of external evaluation are available. It evaluates intra-cluster similarity (examples within a cluster are similar) and inter-cluster differences (examples from different clusters are dissimilar) [10]. By calculating the quality of clustering scheme, it measures the ration of the distance between the centers of clusters (Euclidean) to the measure of scatter within the cluster. The best clustering scheme minimizes DBI. The DBI is however not normalized, therefore its two values from different data sets are difficult to compare.

For evaluating fuzzy clustering, the Xie Beni Index is utilized (6) [11]. It minimizes the validity function S, i.e. the ratio of compactness to the separation for the fuzziness index m = 2.

3.2   Ant Clustering Algorithm (ACA)

The ACA is the discrete unsupervised learning heuristic, inspired by the behavior of the colony of ants. They are able to optimally organize the location of resources inside the hive. The method was originally designed for robotics by Deneubourga [12] and modified by Lumer and Faieta [13] for numerical data analysis. It was further utilized in graph partitioning [14] and text mining [15]. Further attempts at improving ACA were made [18], where the linear, Euclidean and cosine distance were used as measures of similarity between input data vectors (examples, patterns).

In the original version of ACA, ants are represented as agents able to traverse the two-dimensional grid, on which the data objects are randomly scattered. The grid should have more cells than the number of ants. Also, there should be more cells than data vectors [16]. While randomly moving along the grid, agents search the surrounding area of $s \times s$ cells surrounding their current position r. This forms the square neighborhood $N_{s \times s}(r)$. The agent can pick the object or drop it at any location. This depends on the similarity $f(\overrightarrow{z_i})$ and density of the objects in the vicinity of agents. Based on them, probabilities of picking up (p_p) and dropping (p_d) the data vector in the particular location are calculated.

Similarity between input feature vectors within a cluster is assessed by the dissimilarity measure, calculated as Euclidean norm. The constant α controls the scale of dissimilarity, by determining when two data vectors should be grouped together.

If the agent observes a small number of objects in its vicinity, then p_p approaches 1 and objects have a higher probability of being picked up. Alternatively, when its neighborhood is full of objects, p_p approaches 0, thus probability that an agent will pick an object is small. Similarly to the pick-up probability, if $f(\overrightarrow{z_i}) \gg k_2$, then pd approaches 1 and probability of dropping an object is high. If $f(\overrightarrow{z_i}) \ll k_2$, then pd approaches 0.

Because ACA has the tendency to create more clusters than are necessary (overfitting the data), short-term memory is introduced with respect to each agent present of the grid [17]. It is parametrized by the duration of remembering the last T actions, divided by the largest number of observable objects.

This way each agent remembers the last m data vectors and the area on the grid where he dropped it. If a data vector similar to the last m vectors is picked up, the agent will move in that direction. This method ensures, that similar data vectors are placed in the same cluster. The full summary of ACA can be found in [12].

3.3   K-Means

This algorithm, designed by MacQueen [18] and modified by Wong [19], assigns data objects to an a priori given number of k clusters. The process minimizes the sum of squares of distances between given data vectors and the corresponding cluster centroid. A proper k-means grouping minimizes the intra-cluster distance, while maximizing the inter-cluster distance.

The quality of the cluster depends on three factors: the initial values of means, the location of the initial centroids and distance metric, utilized for the purpose of data-to-centroid similarity assessment. Because the initial location of the centroid is uncertain and the input data are equally weighted, it is impossible to determine which attribute contributes more to the grouping process. This in turn can lead to suboptimal partitions. The typical solution here is to try different starting points, while normalizing each variable by its standard deviation.

3.4   Fuzzy C-Means (FCM)

The FCM method has been developed by Dunn in 1973 [20] and improved upon by Bezdek in 1981 [21]. It is frequently used in pattern recognition, being the most prominent fuzzy clustering method. This is because a single input vector can belong to more than one cluster. A fuzzy c-partition of the given data set X shows, to which extent each input vector x_i belongs to the particular (j-th) cluster, using a membership function $f_{sj}(x_i) \in (0, 1)$ [2]. To express similarity between all pairs of vectors from X, the square membership matrix U is constructed. It contains similarity measures for each vector to all clusters (initially treated as single vector groups) in such a way that the sum of $f_{sj}(x_i)$ for all j is one. The membership degree depends on the distance between vectors. A thorough look at mathematical fundamentals is in [2].

3.5   Self Organizing Map (SOM)

The Self-Organizing Map (SOM) is an unsupervised learning neural network allowing for the low-(usually two-) dimensional representation of the input space. SOM is used to cluster data objects into groups (represented by trained neurons) [22].

Computational units (neurons) in SOM are located on the two-dimensional grid. Each of M units is connected to the vector of input features (pattern) $f = \{f_1, ..., f_d\}$ through the weights $w_j = \{w_1, ..., w_d\}$ (Fig. 4). The number of units should be less than or equal to the number of patterns from the training set. Geotechnical data clustering (described by a few parameters) do not require a complex SOM. Therefore the number of units should not exceed thirty, limiting the number of categories generated.

The weight vector w is the centroid of a single cluster. Larger clusters arise by grouping similar neighboring neurons. All weights are associated with a single neuron creating a prototype.

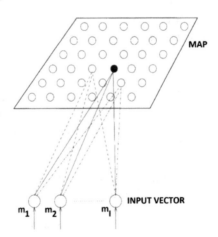

Fig. 4. Self-organizing map [22].

The training algorithm for forming a feature map is as follows [23]:

1. Initialization by random selection of initial weights w_i for all neurons
2. Repeated presentation of feature vectors to the network and adjustment of neurons' weights to them by the winning strategy based on the minimum Euclidean distance criterion between the feature vector and the vector of weights:

$$c = arg\ min_i \|f_i - w_i\|, \quad i = 1, 2, \ldots, M \tag{2}$$

The winning neuron has the minimum distance to the vector f. In the Winner-Takes-All (WTA) strategy it is the only unit adjusted to the vector. It is equivalent to k-means clustering. In the Winner-Takes-Most (WTM) strategy its neighbors are trained as well. Selection of the neighborhood function determines how many neurons are updated after the winner is selected. In the uniform neighborhood all affected neurons are trained in the same fashion (3). If the Gaussian or Mexican Hat functions are used, the change of weights is smaller for units located further from the winner.

The weights of units farther away from the winner are changed to a lesser extent than the weights of nearer neurons.

$$w_i = w_i + \eta \cdot |f_j - w_i|, \tag{3}$$

where $\eta \in (0; 1)$ is the learning-rate factor, determining the convergence and speed of learning. It should monotonically decrease in subsequent iterations. This factor is decreased with subsequent iterations i to make weights changes smaller:

$$\eta = \frac{1}{10} \cdot e^{-\frac{i}{1000}} \tag{4}$$

To scale ranges of input vector elements, normalization in used. If pattern is characterized by two parameters, the maximum values of both of them should be specified and then normalized separately, utilizing the appropriate maximum values.

The SOM quality is determined by the quantization error. Its value specifies the efficiency of the learning process, i.e. adaptation of units' weights to the input patterns. The aim of learning is the minimization of the quantization error. Also, introduction of conscience mechanism allows for considering more units during the training, which in turn leads to the smaller number of dead neurons.

4 Data Description

Clustering methods presented in Sect. 3 were evaluated on 4 datasets obtained from the DMT at the SGGW campus. Datasets consist of various numbers of patterns, depending on the location the measurements were taken: both DMT1 and DMT4 consist of 50, DMT2 of 55, while DMT3 of 45 patterns. Each dataset contains the information about the depth of the measurement (in 20 cm increments) and the values of acquired pressure parameters. The first five patterns from the DMT 2 dataset is presented in Table 1.

Table 1. Fragment of the DMT2 dataset for the clustering algorithm.

Measurement depth [cm]	p_0 [bar]	p_1 [bar]
1.0	6.500	0.300
1.2	6.300	0.280
1.4	12.000	0.200
1.6	15.000	0.533
1.8	18.000	0.600

The visualization of all these datasets, done by PCA is presented in Fig. 5. During the simulation the structural analysis of the forming clusters was evaluated in search for the optimal solution.

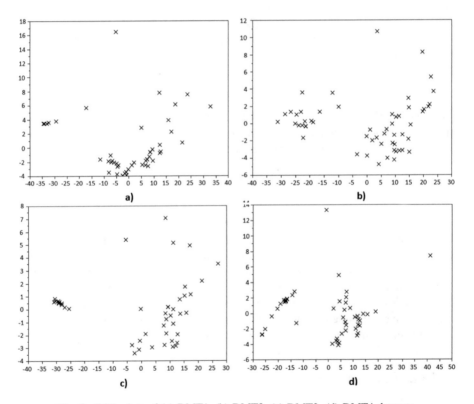

Fig. 5. PCA plots of (a) DMT1, (b) DMT2, (c) DMT3, (d) DMT4 datasets.

5   Results

This section contains experimental results for all presented clustering methods. In each case influence of the algorithm parameter on the obtained result is discussed.

The important parameters of SOM include neighborhood function for the WTM training, size of the network and the value of the conscience mechanism. All were tested during the experiments.

Gaussian and Mexican Hat neighborhood functions were used to calculate, which neurons around the winning one should be trained. Based on the acquired value of quantization error (Table 2), the former one presented more reliable clustering results. This is due to the fact that while utilizing Mexican Hat neighborhood function, the weights of the proximal neurons are subject to stronger adaptation than the weights of the more distant ones. This allows for creation of more compact clusters in the number of 5. This is the result of adapting the network to patterns to the highest possible degree, which was to be achieved in the time preceding the stop condition fulfillment (200 iterations).

Table 2. Neural network quantization error for different geotechnical datasets.

Dataset type	WTA	WTM (Gaussian)	WTM (Mexican Hat)
DMT1	0.375414648167525	0.176582663211553	0.148825981529561
DMT2	0.297117798759682	0.162445220274761	0.14153245422849
DMT3	0.225064240071421	0.202787662797606	0.0477807037876123
DMT4	0.348599906981027	0.147675405209929	0.0959529094336306

An opposite result is acquired when SOM is taught using WTA strategy. This is because neighboring neurons of the winning one are unable to adapt their weights to input patterns. Therefore more neurons cover multiple patterns and are not specialized enough. As the result, the geotechnical profile contains smaller number of soil categories. This is additionally confirmed by a substantially greater value of quantization error, which is almost threefold in comparison to WTM with Mexican Hat. The minimum potential allowing for participation in the competition was set to $p_{min} = 0.78$.

The Ant Colony grouping method was tested for the following optimal parameters: the number of ants (28), pick up gamma (0.15), drop gamma (0.15), direction probability (0.8), step size (1), neighbor size (8) and α (0.3) values. The stop condition of the ACA algorithm was 2000 iterations. In case of kmeans and its fuzzy version, the process of cluster assessment comes to assuming the number of expected clusters *a priori* and calculating the value of DBI and XBI respectfully in search for the lowest value. Figure 6 depicts the corresponding geotechnical profiles in relation to DMT2 dataset. As SOM and ACA are nondeterministic, it is possible to acquire different profiles based on the given data.

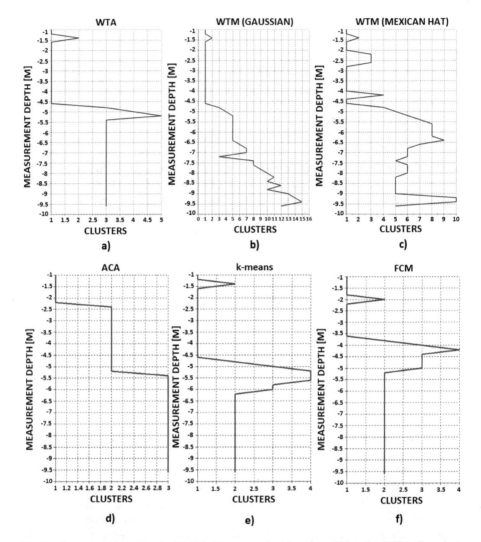

Fig. 6. Geotechnical profiles for DMT2 dataset acquired by of (a) WTA, (b) WTM (Gaussian), (c) WTM (Mexican Hat), (d) ACA, (e) k-means, (f) FCM.

The definitive geotechnical profile is acquired from the fusion of the algorithms described in Sect. 3. The final number of clusters is decided by the majority vote. When the consensus between clustering algorithms can't be achieved, the decision making process is repeated. Figure 7 presents the resulting geotechnical profile for DMT2 dataset, created based on the information from Table 3. The optimal groupings were acquired based on the DBI and XBI measures.

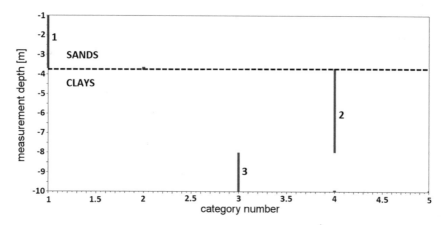

Fig. 7. Geotechnical profile for DMT2 dataset acquired from the fusion of clustering algorithms.

Table 3. Optimal grouping calculated based on DBI and XBI, acquired by kmeans, FCM, ACA and SOM for all analyzed DMT datasets.

Dataset type	kmeans	Kmeans DBI	FCM	FCM XBI	ACA	ACA DBI	SOM (Mexican Hat)	SOM DBI
DMT1	3	0.1821538	3	0.1701344	3	0.0123234	4	0.0050462
DMT2	3	0.0651066	3	0.0710239	3	0.0823233	4	0.0085305
DMT3	3	0.4238012	3	0.4140192	3	0.3545455	5	0.0624159
DMT4	3	0.3459994	3	0.2456679	3	0.1432435	5	0.0546761

6 Conclusions

Among the clustering algorithms discussed in this paper, the highest degree of elements similarity within a cluster is obtained with SOM after applying WTM learning strategy with a Mexican Hat neighborhood function. This results in the creation of many clusters, which is not always desirable. In some cases, a high accuracy of the clustering method can result in the loss of readability of the geotechnical profile.

WTA learning strategy is poorly convergent, especially when a large number of neurons are taken into consideration. The drawback of SOM is the dependency of the result on the initial random values of neuron weights. During tests different amounts of groups were obtained in the final result, often containing different elements.

The centroid-based method is a good solution if the researcher is interested in obtaining a profile with a clearly marked boundaries between different layers of soil. However, attention should be paid to the high computational costs of this solution. In case of large data analysis, the time needed to obtain the final result will be significantly extended. Regardless however of the computation time, the composition and the number of created clusters from a single set of data will be constant each time this set is processed by the centroid-based algorithm. Therefore, there is no need to conduct multiple groupings of data in order to acquire the average results. The most important feature of the ant clustering algorithm is its ability to produce solutions that are visually easily to interpret by the human expert. However provided with data described with only two features, the method fails to provide better cluster quality results than the competitive methods described in this paper.

References

1. Barry, L.: Data Mining Techniques for Marketing. Sales and Customer Support. Wiley, New York (1996)
2. Jain, A.K., Murty, M.N., Flynn, P.J.: Data clustering: a review. ACM Comput. Surv. **31**(3), 264–323 (1999)
3. Bilski, P., Rabarijoely, S.: Automated soil categorization using the CPT and DMT investigations. In: Proceedings of the 2nd International Conference on New Developments in Soil Mechanics and Geotechnical Engineering ZM2009, Nicosia, Northern Cyprus, 28–30 May 2009, pp. 368–375 (2009)
4. Shahin, M.A., Jaksa, M.B., Maier, H.R.: Neural network based stochastic design charts for settlement prediction. Can. Geotech. J. **42**, 110–120 (2005)
5. Bilski, A.: Ant clustering for the CPT and DMT-based soil profile generation. In: IDAACS 2015 (2015)
6. Marchetti, S.: In situ tests by flat dilatometer. J. Geotech. GeoEng. Div. **106**(GT3), 299–321 (1980). ASCE
7. Marchetti, S., Crapps, D.K.: Flat dilatometer manual. Internal report of GPE (1981)
8. Lechowicz Z., Rabarijoely, S.: Numerical analysis of organic soil behaviour during dilatometer test. Ann. Warsaw Agricult. Univ. Land Recl. **29**, 41–49 (2000)
9. Tan, P.N, Steinbach, M., Kumar, V.: Introduction to Data Mining. Addison-Wesley Longman Publishing Co., Inc., Boston (2005)
10. Davies, D.L., Bouldin D.W.: A cluster separation measure. IEEE Trans. Pattern Anal. Mach. Intell. **PAMI-1**(2), 224–227 (1979)
11. Xie, X.L., Beni, G.: A validity measure for fuzzy clustering. IEEE Trans. Pattern Anal. Mach. Intell. **13**(8), 841–847 (1991)
12. Deneubourg, J.-L., Goss, S., Franks, N., Sendora-Franks, A., Detrain, C., Chretien, L.: The dynamics of collective sorting. Robot-like ants and ant-like robots. In: Meyer, J.-A., Wilson, S. (eds.) Proceedings of the First International Conference on Simulation of Adaptive Behavior: From Animals to Animats 1, pp. 356–365. MIT Press, Cambridge (1991)
13. Lumer, E., Faieta, B.: Diversity and adaptation in populations of clustering ants. In: Proceedings of the Third International Conference on Simulation of Adaptive Behavior: From Animals to Animats 3, pp. 501–508. MIT Press, Cambridge (1994)
14. Kuntz, P., Snyers, D.: New results on an ant = based heuristic for highlighting the organization on of large graphs. In: Proceedings of the 1999 Congress on Evolutionary Computation, pp. 1451–1488. IEEE Press, Piscataway (1999)
15. Onan, A., Bukt, H., Korukoglu, S.: An improved ant algorithm with LDA-based representation for text document clustering. J. Inf. Sci. **43**(2) (2017)
16. Bonabeau, E., Dorigo, M., Theraulaz, G.: Swarm Intelligence: From Natural to Artificial Systems. Oxford University Press (1999)
17. Tan, S.C., Ting, K.M., Teng, S.W.: Simplifying and improving ant-based clustering. Procedia Comput. Sci. **4**, 46–55 (2011)

18. MacQueen, J.B.: Some methods for classification and analysis of multivariate observations. In: Proceedings of 5th Berkeley Symposium on Mathematical Statistics and Probability, Berkeley, vol. 1, pp. 281–297. University of California Press (1967)
19. Hartigan, J.A., Wong, M.A.: Algorithm AS 136: a k-means clustering algorithm. Appl. Stat. **28**(1), 100–108 (1979)
20. Bezdek, J.C.: Pattern Recognition With Fuzzy Objective Function Algorithms, p. 256. Plenum, New York (1981)
21. Dunn, J.C.: A fuzzy relative of the ISODATA process and its use in detecting compact well-separated clusters. J. Cybern. **3**, 32–37 (1973)
22. Kohonen, T.: Self-organizing Maps, vol. 30. Springer, Heidelberg (1995)
23. Murtagh, F.: Interpreting the Kohonen self-organizing feature map using contiguity-constrained clustering. Pattern Recognit. Lett. **16**, 399–408 (1995)

Algorithm for Optimization of Multi-spindle Drilling Machine Based on Evolution Method

Paweł Hoser[(✉)], Izabella Antoniuk[(✉)], and Dariusz Strzęciwilk[(✉)]

Faculty of Applied Informatics and Mathematics,
Warsaw University of Life Sciences,
ul. Nowoursynowska 159, 02-776 Warsaw, Poland
{pawel_hoser, izabella_antoniuk,
dariusz_strzeciwilk}@sggw.pl

Abstract. The multi-spindle drills are often used within the mass furniture production. In this case the main factor is the optimization of equipment configuration as well as the working schedule of the drill, what leads to saving time, energy and to significantly lower the manufacturing costs. The optimization problem is hard and complicated. For the equipment and working schedule optimization the specific algorithm has been suggested, incorporating a set of heuristic methods. Among those, for setting the best head equipment of the machine head, the evolution algorithm was used. The initial analysis of the algorithm duty allows to suppose, that the evolution methods may be successfully incorporated for such kind of problems.

Keywords: Multi-spindle drill · Evolution algorithms · Evolution computing

1 Introduction

During mass furniture production the multi-spindle drills are often incorporated for making series of holes in many boards. In such case using those machines is useful and profitable. It allows to save time, energy and lower the production costs significantly. It's the result of specific construction and way of duty of multi-spindle drills. The additional possibilities and profits arise from the fact, that the modern machines are computer controlled and the whole production process may be optimized with a computer program. In case of multi-spindle drills a really important notion is so called equipment of the machine head. Optimization of the equipment configuration and the duty schedule brings additional profits, especially in case of mass furniture production. It has to be stressed that the presented problem of optimization is by itself an interesting mathematical issue, very complicated one. On the other side the problem characteristics is such, that there may be different heuristic methods incorporated [6–8]. The problem has been described in details within the article [1].

For the final optimization the evolution algorithms were incorporated. It comes out, that evolutional algorithms are nearly perfect for such cases. In our case, the evolutional algorithms were used for head equipment completion, when morphological transformations were used for searching the best way of duty.

© Springer Nature Switzerland AG 2019
J. Pejaś et al. (Eds.): ACS 2018, AISC 889, pp. 34–44, 2019.
https://doi.org/10.1007/978-3-030-03314-9_3

2 Design and Duty of Multi-spindle Drill

The main feature of multi-spindle drill is the ability to make many different holes at once. A very important part of the drill is the head, where different types of bits are mounted within special nests. Drilled holes are of different types, mostly there are five kinds, signed with letters A, B, C, D, and E. The holes are made with appropriate types of bits, which are also signed with same letters. Typical head has two arms, perpendicular to each other. Within each arm there are nests settled within equal distances from each other. In the head there are in total 15 to 60 nests. We can presume, that there should always be full set of bits within the head. So it comes out, that each kind of bit should be present within the head at least once. Such presume makes that the total number of all equipment is not to be provided with a simple formula.

What more, the number of each kind of bit at the head is free. We also presume, that there are no empty nests. There comes out, that the total number of bits must be equal to the total number of nests. The whole configuration of bits within all the nests is called equipment (Fig. 1).

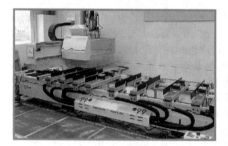 

Fig. 1. The multi-spindle drill (left) and head with bits (right).

The drilling process takes place in the following way: the head is positioned at specific position and makes holes with help of few bits at a time. If a bit is placed over a place scheduled for hole and a kind of bit refers to the kind of hole, the drilling takes place. The machine can make a few drillings at a time at one position. The more holes are drilled at once, the more effective the process is, what would be described in the further part. The aim is to perform as many holes at a time as possible. It comes out, that the placing of holes on the board should be such, that they fit the best the positions of nests within the head at different head positions. It is often presumed, that the holes coordinates (x, y) are the multiples of the distance between the two nests within the head. This allows the multiple drillings at single head positioning. The necessary condition for drilling the hole is that the kind of bit fits the kind of hole. The central point of the head is the nest at the point the two arms are crossing, this means the common nest indexed $(0,0)$ (Fig. 2).

The boards can be of different size. The length of the board side can differ from just over 10 cm up to six meters. At the mass furniture production the number of the boards can become very large, even thousands of them. The number of hole drilled in each

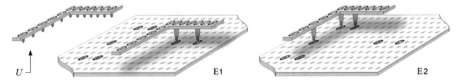

Fig. 2. Drilling process of multiple holes at a time with the multi-spindle drill.

board due to schedule can also vary from few to few dozens. It is also important that changing the head equipment during drilling is not effective. There is the equipment of the head is important to optimization effect.

3 Optimization Problem

The way of making all holes in the board depends on head equipment. With each set of equipment we have to search for the optimal working schema. Next, taking into consideration all possible equipment sets and their optimal ways of duty, we may search for the optimal equipment set for the machine. The discussed optimization problem is so complicated, as even the number of all possible ways of duty for a single equipment set is really large [9, 10]. The machine head can be set in many positions. Their number is much larger than the number of holes to be made within a board. The machine head can be positioned not directly over the hole point, it's enough that any nest gets over a point, so such a head position makes sense. So the complexity of the optimization problem is much higher than the problem of bagman [7, 8]. Additionally we have to consider a great number of equipment sets (and for each set there is a great number of working schemas). The total number of all the possibilities is gigantic.

For a fully proper description of the case we need a mathematical model. The basic notions of the model are: type of hole, type of bit, equipment, head positioning, hole configuration within the board, the optimization coefficient, way of head, duty schema, number of hits, drilling process and optimization process.

The basic notions are drill and hole types (T). Practically we sign them with letters A, B, C, D and E. Within the mathematical model it is easier to assume, that type is the natural number (from 1 to 5). The number 0 is assigned to the lack of hole in the board. Due to this the type is represented with natural number from 0 to 5.

$$T = \{0\ldots5\} \tag{1}$$

The equipment set tells us, how the configuration of the bits settled within the head nests looks like. And here is the main difference from the definition of the previous model described within [1]. Now the equipment is represented with matrix dimensioned m_x and m_y, where m_x and m_y equal to the number of nests within head arms.

Previously the equipment was represented with two vectors and within the computer program with two single dimensional tables. Using a single matrix is more comfortable, as the mathematical definitions become simpler.

$$U \in M_{m_x \times m_y(T)} \quad ; \quad U_{pq} = U(p,q) \in T \tag{2}$$

The values of the matrix are kinds of bits. We have to mark here, that within the matrix only first row and first column have non-zero indexes within the matrix. All other values are equal zero. For the first row and first column the values of indexes have to be greater than zero, as there should be no empty nest within the head. Such definition makes possible to define nests aside of the main head arms (but this occurs very rare).

The next important element of the model is the holes configuration within the board R. as the coordinates of the holes (x, y) are the multiples of some constant distance d, the matrix representation should also be used. The matrix size (M_x, M_y) depends on board size and the value d. The values of the matrix coefficients are types of holes (Fig. 3).

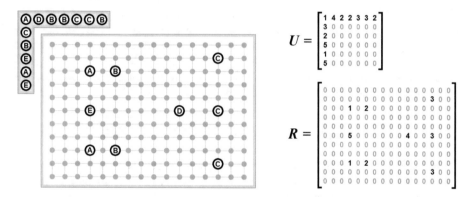

Fig. 3. Matrix representation of the machine head and holes configuration within the board.

The distance d equals to the distance between two neighbor nests within the head. It comes out, that the configuration matrix and the equipment matrix are strongly tied. If any nest within the head is positioned over a hole position, it can be stated with matrix indexes only after considering the position of the head over the board. In this way each pair of indexes (i,j) shows a point (x,y) on the board. Those points are called the knots of the net on the board. If the value of the matrix R for indexes i,j is greater than zero, it means that a hole of a kind $R_{i,j}$ should be drilled there. In such case the position of the head can be represented by two integral numbers (i, j). those numbers tell us, over which knot the head central point – the nest with the $(0,0)$ index is positioned. The matrix R can be defined as the function of two integral variables:

$$R : \{-m_x \ldots M_x\} \times \{-m_y \ldots M_y\} \rightarrow T; \tag{3}$$

The drilling process within the mathematical model represents the real drilling process of all holes in a board. Within the model the drilling process is just a series of the following positions of the head. The process P has one important parameter K.

What more, if all the holes have to be drilled, what means they should be hit at least once. Within the model it means that for each non-zero index of the matrix R there is such a head position that the hole type $(R_{i,j})$ fits the bit type mounted in the nest, which is then over the hole. Additionally we assume, that it is a $(K + 2)$ element row and first and last head position is in the upper left corner of the position matrix, so for the indexes $(-m_x, -m_y)$. In this definition, the K number is the number of steps for the whole drilling process.

$$P : \{0 \ldots (k+1)\} \rightarrow Z \times Z \quad ; \quad P(k) = \left[P_x(k), P_y(k) \right]$$

$$P(0) = P(k+1) = \left(-m_x, -m_y \right) \tag{4}$$

$$\forall i,j \left((R(i,j) > 0) \Rightarrow \exists k : R(i,j) = U \left(i - P_x(k), j - P_y(k) \right) \right)$$

In the whole optimization process there is mainly optimized the K number. In the worst case the K number equals to the total number of holes n. In such case for each head position there is only a single hole drilled. In case K number is "significantly lower" than n it means that at some positions of the head there are a few holes drilled simultaneously. But there is not only the K number that should be taken into consideration. Also the total length of the way s plays a role for the optimization process. The total way should not be allowed to become too long.

$$s = \sum_{i=0}^{k} \sqrt{\left(P_x(i+1) - P_x(i) \right)^2 + \left(P_y(i+1) - P_y(i) \right)^2} \tag{5}$$

In the total optimization process both the K number and the way s are taken into consideration. But for this both values have to be normalized, so they could get compared. Number of steps is compared with number of holes, and the length of the way with the average way. Practically, both the values k/n and s/s_r are close to 1 and should be considered only using the weighted average. There are additional parameters α and β, necessary for the aim and estimated by analyzing usage of time and energy. The optimization coefficient equals to the weighted average of normalized number of steps and normalized way. It has to be stressed that the coefficient depends on configuration of all the holes in the board (R), equipment (U) and fixed drilling process (P).

$$w_{opt}(R, U, P) = \alpha \frac{k}{n} + \beta \frac{s}{s_r} \tag{6}$$

The main cause for drill work optimization is minimization of the w_{opt} coefficient, taking into consideration all the possible equipment sets (U) and all the drilling process P. So, in result we are looking for the optimal U_{opt} and P_{opt}.

4 Suggested Optimization Algorithm

The algorithm suggested within this elaboration for optimization of equipment and drilling process is based on different heuristic methods. Choosing the best equipment *"U"* was based on the evolutional algorithm, which seems to be ideal solution for such type of issue. Searching for the best drilling process *"P"* is based, among others, on heuristic groupage of head positioning, incorporating morphological transformations and the heuristic algorithm of covering the discreet sets.

The evolutional methods in informatics arose basing on observation of biological problems [2–5]. The important common issue of evolutional method is that they are universal, but under condition it works long enough and the population is big enough. They differ, among others, with genome definitions, genome types, ways of realization mutations and crossings, selection methods and the ways of creating new generations [2, 3]. The most known methods of selection are: roulette, tournament and rank. The most important keywords are: genotype, phenotype, individual, population and adaptation function. The individual is a single problem solution. Genotype means the internal structure of the individual, what means the definition of the individual within the evolution algorithm, mostly is a countable row of numbers. Phenotype is an example of a genotype equal to solving the problem. Population is the set of the individuals. Adaptation function tells how good the solution is, what allows to choose the best individuals. Within the suggested algorithm the genotype is nearly exactly the equipment of the machine, so using the genetic algorithms is just natural. The selection method is roulette type together with inclusion of the best individual (Fig. 4).

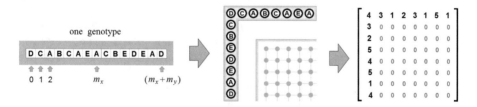

Fig. 4. One genotype, its representation as equipment set and its U matrix.

The starting population consists of random generated individuals. Generating the individual begins with pacing a single bit of each kind in random chosen nests, to avoid the situation that a kind of bit is not present. Next there are another bits randomly chosen for the empty nests. Precisely speaking, the genotype is a single dimension table with length $(m_x + m_y + 1)$. The values in the table are bit types. The table specifies the values for the equipment matrix U within first row and first column. Phenotype is the optimal drilling process for the equipment and the value for the adaptation function equals $(1 - w_{opt})$. Taking the listed definitions allows to design the whole evolution algorithm (Fig. 5).

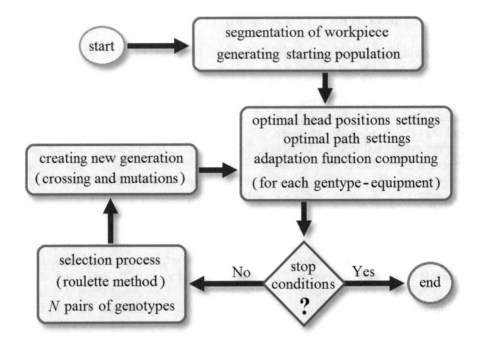

Fig. 5. Block diagram of the evolution algorithm for equipment optimization.

So, for each genome, the drilling process, being its phenotype, is optimized. We get the optimal drilling process also with the help of heuristic methods. In the beginning all the holes in the board are adequately grouped. Each group of holes is located within a specific segment. Those segments we obtain using morphological transformations with specific geometrical properties. The morphological transformations are well known from the digital picture transformations [6], but in practice they can be widely used. In the specific case the important role was taken by comfortable geometrical properties of those transformations. The transformations used are dilatation and erosion, for which the structural element is in a specific way connected with head parameters. The transformations perform their duty on the knots of the net, which represent the possible places for the holes in the board and are specified with the indexes of the matrix R. The example of such segments is shown on the picture 7. With every smaller group of holes (within a single segment) there is a subset of possible head positions connected. The number of steps K is locally optimized for each segment separately, what simplifies the complexity of the problem. It is presumed that the head operates within a single segment only. The situation with the optimization of the way s becomes similar. After choosing the head position and fixing this way the number of the steps K, we have to choose the best sequence of positions to optimize the length of the path s. this is the standard bagman problem.

The problem is then divided into few smaller problems. We presume that the head works sequentially within the segments, what means that during the drilling process all the holes within the segment are made and just then the head moves to the next segment.

It's not always such presumption is proper, but for the heuristic method we can do that. For the suggested algorithm we first check the sequence of segments by optimization of the length of the path between segments, and next we optimize path length within the segment (Fig. 6).

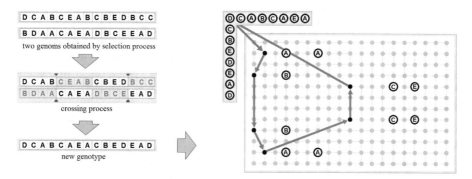

Fig. 6. Crossing operation for the pair of genomes, just after selection process.

This way from the problem of great complexity we move to the less complicated one, with some certain presumes taken. Those presumes may not always be proper, but this is typical for heuristic methods. Finding the optimal drilling process for a genotype permits to count the value of the adaptation function, which is used within the individuals selection.

The selection process is based on the roulette method. It's important to give the chance of live even for the worse cases, and roulette method permits this. To all the individuals there is a probability value p_g depends on w_{opt} value by special function. First, v_p values must be compute for each genotype.

$$v_p(i) = \frac{w_{max} - w_{opt}(i)}{w_{max} - w_{min}} \tag{7}$$

Where index i is the number of genotype in all population and w_{min}, w_{max} are minimum and maximum of w_{opt} value in whole population. Additionally, this algorithm is protected of division by zero. If $w_{min} = w_{max}$ then the value w_{max} is equal w_{max} + eps. If the same kind of hole not exist in U matrix, the value v_p is zero. Then the probability of each genotype in roulette method is compute as follow:

$$S_p = \sum_{i=1}^{N} v_p(i) \quad ; \quad p_g(i) = \frac{v_p(i)}{S_p} \tag{8}$$

This way we get the whole probability distribution for all the individuals. Next, the algorithm chooses N pairs of individuals to create next generation of individuals. Each individual can become a member of few pairs, which is not so often, as the total number of pairs is $N(N-1)/2$. The pairs of individuals are treated with crossing and mutations. In this way each pair of individuals creates a new individual (a kid).

The effect is a new population which also consists of N individuals. Such approach is similar to biology, as there the crossings and mutations occur during creation of new individuals.

Additionally, the specific exclusivity strategy, saving the best individual of previous generation, is used. In this way during each selection process the best genes are used to create next generation. The genotype with the lowest w_{opt} value is always introduced to a random chosen pair. In this case the crossing operation is really important. It's a kind of a two point crossing, designed in a way not to mix the kind of bites within the two arms of the machine head. This makes both the crossing points to apply to different arms. There is no gene exchange between two arms. It means that crossings appears separately for two parts of the table, that is separately for indexes $[0..\ m_x]$ and for $[m_x+1,.., m_x+m_y]$ (Fig. 7).

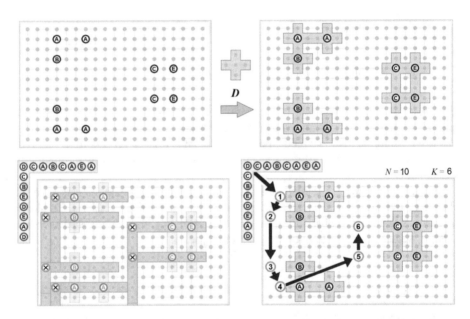

Fig. 7. The R matrix segmentation, optimal head position fixation and path optimization.

Such approach is stated with fact, that single dimension norms within the rows of R matrix and norms for the columns of R matrix can be totally different. In such situation mixing the bit types makes no sense and can even be destructive. Separately the norms within the arms may be placed at different positions. This doesn't mean anyway, that those are independent problems – on the contrary, both configurations of bit types at the machine head arms are strongly connected due to situation, when there are bits from both arms involved during a single multiple-point drilling. The stop condition consist as a conjunction of two. The whole process is stopped when the maximum permissible number of generations is exceeded or if a maximal value of the adaptation function is not getting bigger for a longer period of time.

5 The Results of the Optimization Algorithm

To test the efficiency of the new method the special computer program has been written. Program was tested with many cases. The optimization algorithm was initially tested on different kinds of input data. Algorithm was tested with different types of matrix R and sets of equipment U. The testing of this problem is not going to be easy. We have to mark, that the complexity of the problem strongly depends on the number of holes to be drilled in the board and the number of nests in the arms. In case the number of holes is big and the head has many nests, the optimization problem becomes incredibly complex. Very important, as well, is that how the holes are positioned on the board. The whole configuration of holes on the board affect on the optimization possibilities. If the configuration of holes is very regular, proper segmentation is possible and each segment reminds characteristic pattern, then effective optimization is possible. But when the configuration of holes is chaotic and irregular, the effective optimization is almost impossible. Therefore, whole experiment was divided in to two parts. One series of tests was for irregular chaotic configuration of holes and another series of tests was about regular configurations (when proper segmentation was possible and characteristic patterns was clearly visible). For each part of experiment fifty tests were done. The optimization process was tested for 10, 12, 15, 18 and 20 holes in the board. The results of experiments are presented below on the graph 9.

First row of presented results (Fig. 8) are tests of the irregular configuration of holes and second row are tests of the regular configuration of holes (where proper segmentation was done). This graph shows values: K/N, s/s_r and w_{opt} for tests results. Each bar on the graph shows the average value of ten tests (for number: 10, 12, 15, 18, 20). The R matrix had 40 columns and 30 rows in all tests.

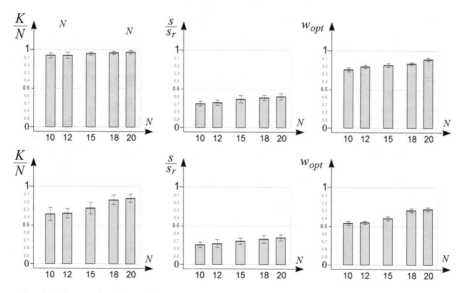

Fig. 8. The results of experiments, for irregular configuration (up) and for regular (down).

During the experiments the population number was 50 and the maximal permissible number of generations 200. The parameters α and β was ¼ and ¾. Machine head consisted 15 nests ($m_x = 7$, $m_y = 6$), so the U matrix had 7 rows and 7 column.

Increasing the population number didn't lead to essential effects, but highly increased the complexity of the calculations, as in this case we have to optimize the drilling process P for each genome. After initial tests it came out that the algorithm works well for simple cases (for example, in case there are only 10 holes in the board). The algorithm works especially well if the holes are placed within regular groups and the specific pattern can be observed. That was the aim of the method, so this is partly a success. In case of more complex situations the disturbances occur. Additionally, during experiments performed, we stated, that there is no preservation of the good norms for the head. The problem is, that with the equipment set U only a part of the bits can be set really good and so they can be used for drilling the holes within a single head positioning (large number of hits). But the crossing and mutation operations destroy these good norms and it is hard to avoid that. In this way, the protection of really good norms is not simple. It seems, that if we want to use the evolution algorithm, we have to change the definition and representation of genotype. On the other side, the tests show, that incorporating this method may be interesting and the method can be used, but for rather simple cases, for more complex cases modifications are necessary.

References

1. Hoser, P., Podziewski, P., Kurek, J., Kruk, M.: Equipment optimization problem for multi-spindle computer controlled drilling machine. In: Computing in Science and Technology, pp. 91–108. Wydawnictwo Uniwersytetu Rzeszowskiego, Rzeszów (2017)
2. Goldberg, D.E.: Genetic Algorithms in Search. Optimization, and Machine Learning. Addison-Veslay Publishing Company, Inc., Boston (1989)
3. Eiben, A.E., Smith, J.E.: Introduction to Evolutionary Computing, 2nd edn. Springer, 2003, 2015
4. Eiben, A.E., Smith, J.E.: Introduction to Evolutionary Algorithm. Springer, London (2015)
5. Michalewicz, Z.: Genetic Algorithm + Data Structure = Evolutionary Programs. Springer, Heidelberg (1996)
6. Soille, P.: Morphological Image Analysis. Springer, Heidelberg (2003)
7. Karp, R.M.: Reducibility among combinatorial problems. In: Miller, R.E., Thatcher, J.W. (eds.) Complexity of Computer Computations, pp. 85–103. Plenum, New York (1972)
8. Chvatal, V.A.: Greedy heuristic for the set-covering problem. Math. Oper. Res. 4(3), 233–235 (1979)
9. Graham, L.R., Knuth, D.E., Patashnik, O.: Concrete Mathematics. A Foundation for Computer Science, 2nd edn. (2017)
10. Cormen, T.H., Leiserson, C.E., Rivest, R.L., Stein, C.: Introduction to Algorithms, 3rd edn. Massachusetts Institute of Technology (2009)

Horizontal Fuzzy Numbers for Solving Quadratic Fuzzy Equation

Marek Landowski[(✉)]

Department of Mathematical Methods, Maritime University of Szczecin,
Waly Chrobrego 1-2, 70-500 Szczecin, Poland
m.landowski@am.szczecin.pl

Abstract. The paper presents method for solving the quadratic equation with fuzzy coefficients. Based on the horizontal fuzzy numbers the solution of fuzzy quadratic equation can be obtained. Solutions with horizontal fuzzy numbers are multidimensional. Obtained solutions are compared with results of standard fuzzy arithmetic. In examples was shown that results with standard fuzzy arithmetic are overestimated or underestimated. Method with horizontal fuzzy numbers generates the granule of information about the solution. Obtained granule gives full information about the solution. Moreover, the granule of information gives possibility to indicate the crisp quadratic equation for crisp value of the solution.

Keywords: Fuzzy quadratic equation · Fuzzy number
Horizontal fuzzy number · Fuzzy arithmetic · RDM arithmetic
Uncertainty theory · Artificial intelligence

1 Introduction

Many practical problems contain uncertain values. Operations on uncertain variables are made in artificial intelligence for example in fuzzy logic. The uncertainty can be described e.g. in the form of probability density distribution, possibility distribution or interval. In uncertainty theory [3] the fuzzy arithmetic as a calculation on fuzzy numbers is necessary to solve many problems in: granular computing [14], soft computing [5], gray systems [10], artificial intelligence [6], and computing with words [18].

The author in the paper presents the horizontal fuzzy number [9, 16, 17] as a way of uncertainty description with usage for solving fuzzy quadratic equation. To compare results obtained with horizontal fuzzy number arithmetic (HFNA) and with standard fuzzy arithmetic (SFA) the testing points method was used. The papers where researchers solved fuzzy quadratic equations are e.g. [1, 2, 4, 12, 13].

The horizontal fuzzy number arithmetic is based on the RDM interval arithmetic [7, 8, 15]. The interval $X = [\underline{x}, \bar{x}]$ in notation of RDM using RDM variable $\alpha_x \in [0, 1]$, is a set of numbers in the form of (1)

$$X = \{x : x = \underline{x} + \alpha_x(\bar{x} - \underline{x}), \alpha_x \in [0, 1]\}. \tag{1}$$

© Springer Nature Switzerland AG 2019
J. Pejaś et al. (Eds.): ACS 2018, AISC 889, pp. 45–55, 2019.
https://doi.org/10.1007/978-3-030-03314-9_4

Definition 1. [11] A parametric form of a fuzzy number u is a pair $(\underline{u}, \bar{u})$ of function $\underline{u}(r)$ and $\bar{u}(r)$, where $r \in [0,1]$, which satisfy the following requirements:

1. $\underline{u}(r)$ is a bounded monotonic increasing left continuous function,
2. $\bar{u}(r)$ is a bounded monotonic decreasing left continuous function,
3. $\underline{u}(r) \leq \bar{u}(r)$, $0 \leq r \leq 1$.

Based on the notation of RDM the fuzzy number in parametric form $U = (\underline{u}(r), \bar{u}(r))$ described as a horizontal fuzzy number is presented in (2), see [9].

$$u(r, \alpha_u) = \{u : u = \underline{u}(r) + \alpha_u(\bar{u}(r) - \underline{u}(r)), r, \alpha_u \in [0,1]\}. \tag{2}$$

On horizontal fuzzy numbers $u(r, \alpha_u) = \{u : u = \underline{u}(r) + \alpha_u(\bar{u}(r) - \underline{u}(r)), r, \alpha_u \in [0,1]\}$ and $v(r, \alpha_v) = \{v : v = \underline{v}(r) + \alpha_v(\bar{v}(r) - \underline{v}(r)), r, \alpha_v \in [0,1]\}$ basic operations $* \in \{+, -, \times, /\}$ can be made, Eq. (3), operation $/$ occurs only if $0 \notin v(r, \alpha_v)$,

$$\begin{aligned} &u(r, \alpha_u) * v(r, \alpha_v) \\ &= \{u * v : u * v = [\underline{u}(r) + \alpha_u(\bar{u}(r) - \underline{u}(r))] * [\underline{v}(r) + \alpha_v(\bar{v}(r) - \underline{v}(r))], \\ &r, \alpha_u, \alpha_v \in [0,1]\} \end{aligned} \tag{3}$$

Definition 2. Span of the result of basic operation $* \in \{+, -, \times, /\}$ on horizontal fuzzy numbers $u(r, \alpha_u)$ and $v(r, \alpha_v)$ is a fuzzy number defined as (4), operation $/$ occurs only if $0 \notin v(r, \alpha_v)$.

$$s(u * v) = \left(\min_{\substack{\alpha_u \in [0,1] \\ \alpha_v \in [0,1]}} \{u(r, \alpha_u) * v(r, \alpha_v)\}, \max_{\substack{\alpha_u \in [0,1] \\ \alpha_v \in [0,1]}} \{u(r, \alpha_u) * v(r, \alpha_v)\} \right) \tag{4}$$

The following properties for basic operations on horizontal fuzzy numbers U, V and W hold:

1. commutativity of addition and multiplication, $U + V = V + U$, $U \times V = V \times U$,
2. associativity of addition and multiplication, $U + (V + W) = (U + V) + W$, $U \times (V \times W) = (U \times V) \times W$,
3. there exist additive and multiplicative neural elements, $U + 0 = 0 + U = U$, $U \times 1 = 1 \times U = U$,
4. there exists additive invers element, $U - U = 0$,
5. there exists multiplicative invers element, $U \times 1/U = 1$, only if $0U$,
6. distributive law, $U \times (V + W) = U \times V + U \times W$,
7. cancelation law for addition, $U + W = V + WU = V$.

Shorter the fuzzy number in horizontal notation can be written as $u(r, \alpha_u) = \underline{u}(r) + \alpha_u(\bar{u}(r) - \underline{u}(r))$, where $r, \alpha_u \in [0,1]$.

Special cases of fuzzy numbers are: a triangular fuzzy number $u = (a, b, c) = (a + (b - a)r, c + (b - c)r)$ and a trapezoidal fuzzy number $v = (a, b, c, d) = (a + (b - a)r, d + (c - d)r)$. In notation of horizontal fuzzy number [9] the triangular fuzzy number is given by (5),

$$u(r, \alpha_u) = a + (b - a)r + \alpha_u(c - a + (a - c)r), \tag{5}$$

and the trapezoidal fuzzy number is presented by Eq. (6),

$$v(r, \alpha_v) = a + (b - a)r + \alpha_v(d - a + (c - d - b + a)r), \tag{6}$$

where $r, \alpha_u, \alpha_v \in [0, 1]$.

Example 1. Let us consider the fuzzy trapezoidal fuzzy number $v = (2, 4, 5, 6) = (2 + 2r, 6 - r)$, from Eq. (6) given fuzzy number in notation of horizontal fuzzy number is $v(r, \alpha_v) = 2 + 2r + \alpha_v(4 - 3r)$. Figure 1 shows trapezoidal fuzzy number in vertical form and as a horizontal fuzzy number.

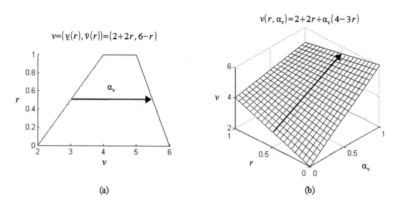

Fig. 1. Trapezoidal fuzzy number $v = (2, 4, 5, 6)$ in a standard vertical form (a) and as a horizontal fuzzy number (b).

2   Solution of Fuzzy Quadratic Equation Using Horizontal Fuzzy Number

The fuzzy quadratic equation is presented by (8)

$$Ax^2 + Bx + C = 0, \tag{8}$$

where $A = (\underline{a}(r), \bar{a}(r))$, $B = (\underline{b}(r), \bar{b}(r))$, $C = (\underline{c}(r), \bar{c}(r))$, $r \in [0, 1]$.

The coefficients of Eq. (8) in the notation of horizontal fuzzy number are given by (9)

$$
\begin{aligned}
A &= a(r, \alpha_a) = \underline{a}(r) + \alpha_a(\bar{a}(r) - \underline{a}(r)), \\
B &= b(r, \alpha_b) = \underline{b}(r) + \alpha_b(\bar{b}(r) - \underline{b}(r)), \\
C &= c(r, \alpha_c) = \underline{c}(r) + \alpha_c(\bar{c}(r) - \underline{c}(r)),
\end{aligned}
\tag{9}
$$

where $r, \alpha_a, \alpha_b, \alpha_c \in [0, 1]$.

Number of solutions of Eq. (8) depends on the Δ value, presented in Eq. (10)

$$\Delta = [\underline{b}(r) + \alpha_b(\bar{b}(r) - \underline{b}(r))]^2 \\ - 4[\underline{a}(r) + \alpha_a(\bar{a}(r) - \underline{a}(r))][\underline{c}(r) + \alpha_c(\bar{c}(r) - \underline{c}(r))] \tag{10}$$

If $\Delta > 0$ then Eq. (8) has two solutions in set of real numbers given by (11),

$$x_1 = \frac{-[\underline{b}(r) + \alpha_b(\bar{b}(r) - \underline{b}(r))] - \sqrt{\Delta}}{2[\underline{a}(r) + \alpha_a(\bar{a}(r) - \underline{a}(r))]} \text{ and } x_2 = \frac{-[\underline{b}(r) + \alpha_b(\bar{b}(r) - \underline{b}(r))] + \sqrt{\Delta}}{2[\underline{a}(r) + \alpha_a(\bar{a}(r) - \underline{a}(r))]}, \tag{11}$$

where $r, \alpha_a, \alpha_b, \alpha_c \in [0, 1]$.

If $\Delta = 0$ then Eq. (8) has one solution in set of real numbers given by (12)

$$x = \frac{-[\underline{b}(r) + \alpha_b(\bar{b}(r) - \underline{b}(r))]}{2[\underline{a}(r) + \alpha_a(\bar{a}(r) - \underline{a}(r))]} \tag{12}$$

where $r, \alpha_a, \alpha_b, \alpha_c \in [0, 1]$.

If $\Delta < 0$ then Eq. (8) has solutions in set of complex numbers.

3   Numerical Examples

Example 2. [1] Let us consider a quadratic equation with fuzzy coefficients given by (13)

$$(3, 4, 5)x^2 + (1, 2, 3)x - (1, 2, 3) = 0. \tag{13}$$

Solution with standard fuzzy arithmetic
The Eq. (13) in parametric form is given by (14).

$$(3 + r, 5 - r)x^2 + (1 + r, 3 - r)x - (1 + r, 3 - r) = 0. \tag{14}$$

where $r \in [0, 1]$.

The Δ value equals as in Eq. (15)

$$\Delta = \left(5r^2 + 18r + 13, 5r^2 - 38r + 69\right). \tag{15}$$

The results in parametric form obtained with standard fuzzy arithmetic is presented in Eq. (16), for $r \in [0, 1]$,

$$\begin{aligned} x_1 &= \left(\underline{x_1}(r), \overline{x_1}(r)\right) = \left(\frac{-(3-r) - \sqrt{5r^2 - 38r + 69}}{2(5-r)}, \frac{-(1+r) - \sqrt{5r^2 + 18r + 13}}{2(3+r)}\right), \\ x_2 &= \left(\underline{x_2}(r), \overline{x_2}(r)\right) = \left(\frac{-(3-r) + \sqrt{5r^2 + 18r + 13}}{2(5-r)}, \frac{-(1+r) + \sqrt{5r^2 - 38r + 69}}{2(3+r)}\right). \end{aligned} \tag{16}$$

Figure 2 shows the results obtained with standard fuzzy arithmetic (16).

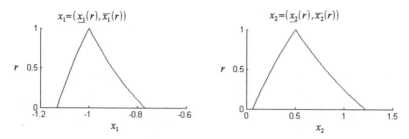

Fig. 2. Results in the form of fuzzy numbers x_1 and x_2 with standard fuzzy arithmetic.

Solution with horizontal fuzzy numbers

Given fuzzy numbers $(3, 4, 5)$ and $(1, 2, 3)$ in parametric form are $(3 + r, 5 - r)$ and $(1 + r, 3 - r)$, respectively, where $r \in [0, 1]$. Fuzzy coefficients of Eq. (13) expressed in the notation of horizontal fuzzy numbers are as in (17)

$$A = (3, 4, 5) = a(r, \alpha_a) = 3 + r + 2\alpha_a(1 - r),$$
$$B = (1, 2, 3) = b(r, \alpha_b) = 1 + r + 2\alpha_b(1 - r), \tag{17}$$
$$C = -(1, 2, 3) = -c(r, \alpha_c) = -1 - r - \alpha_c(1 - r).$$

where $r, \alpha_a, \alpha_b, \alpha_c \in [0, 1]$.

From formula (10) value Δ is calculated, Eq. (18):

$$\Delta = [1 + r + 2\alpha_b(1 - r)]^2 - 4[3 + r + 2\alpha_a(1 - r)][-1 - r - 2\alpha_c(1 - r)], \tag{18}$$

where $r, \alpha_a, \alpha_b, \alpha_c \in [0, 1]$.

The span of Δ equals (19)

$$s(\Delta) = \left(5r^2 + 18r + 13, 5r^2 - 38r + 69\right), \tag{19}$$

where minimum value of Δ is for $\alpha_a = \alpha_b = \alpha_c = 0$ and maximum for $\alpha_a = \alpha_b = \alpha_c = 1$.

For every $r, \alpha_a, \alpha_b, \alpha_c \in [0, 1]$ value of Δ from Eq. (18) is greater than 0, as was shown in Eq. (19), so Eq. (13) has two solutions in real set of numbers.

Solutions obtained with formula (11) are granules of information of solution, Eq. (20),

$$x_1(r, \alpha_a, \alpha_b, \alpha_c) = \frac{-[1 + r + 2\alpha_b(1 - r)] - \sqrt{\Delta}}{2[3 + r + 2\alpha_a(1 - r)]},$$
$$x_2(r, \alpha_a, \alpha_b, \alpha_c) = \frac{-[1 + r + 2\alpha_b(1 - r)] + \sqrt{\Delta}}{2[3 + r + 2\alpha_a(1 - r)]}, \tag{20}$$

where $r, \alpha_a, \alpha_b, \alpha_c \in [0, 1]$.

Figure 3 shows graphical representation of solution granules for $r = 0$, supports of obtained solutions.

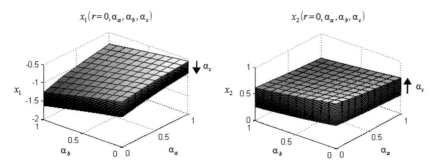

Fig. 3. Granules of solutions x_1 and x_2 of fuzzy quadratic equation for $r = 0$, $\alpha_a, \alpha_b, \alpha_c \in [0, 1]$.

Spans of the solutions x_1 and x_2 are presented in Eq. (21), $r \in [0, 1]$. The left border of the span of x_1 is for $\alpha_a = 0$ and $\alpha_b = \alpha_c = 1$, the right border of x_1 is for $\alpha_a = 1$ and $\alpha_b = \alpha_c = 0$. Analyzing the span of solution x_2 the minimum value is for $\alpha_a = \alpha_b = 1$ and $\alpha_c = 0$ and the maximum is for $\alpha_a = \alpha_b = 0$ and $\alpha_c = 1$.

$$s(x_1) = \left(\frac{-3+r-\sqrt{-3r^2-6r+45}}{2(3+r)}, \frac{-1-r-\sqrt{-3r^2+18r+21}}{10-2r} \right),$$

$$s(x_2) = \left(\frac{-3+r+\sqrt{-3r^2+10r+29}}{10-2r}, \frac{-1-r+\sqrt{-3r^2+2r+37}}{2(3+r)} \right). \tag{21}$$

Figure 4 shows the spans of solutions (20).

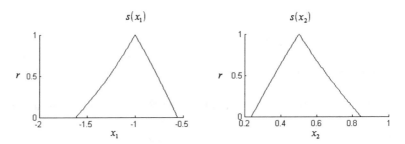

Fig. 4. Spans of solutions of fuzzy quadratic equation.

Results obtained in example 2 with SFA are presented in formula (16), supports (results for $r = 0$) of the results are: $x_1 \in [-1.1307, -0.7676]$, $x_2 \in [0.0606, 1.2178]$. Solutions with HFNA are given by Eq. (20), supports ($r = 0$) of spans of obtained solutions are: $x_1 \in [-1.6180, -0.5583]$, $x_2 \in [0.2385, 0.8471]$.

With Eq. (20) it can be generate any of solution for given quadratic fuzzy equation. Solution and coefficients of quadratic equation depend on the variable $r, \alpha_a, \alpha_b,$ $\alpha_c \in [0,1]$. For $r = \alpha_a = 0, \alpha_b = \alpha_c = 1$ value $x_1 = -1.6180$ is generated. It is the solution of equation $3x^2 + 3x - 3 = 0$, this equation is one of the cases of analyzed fuzzy quadratic equation for $r = 0$. Calculated value $x_1 = -1.6180$ do not belongs to the solution obtained with SFA, for $r = 0$. It shows that SFA do not gives full solution, result by SFA is underestimated.

Example 3. Some object moves under gravity, see Fig. 5. Find the distance x from the start point $(0,0)$ to the point (x, y) for the height $y = 4m$. Acceleration of the object is constant $g = 9.8m/s^2$. Velocities in the directions x and y are a triangular fuzzy numbers $u = (11, 12, 13)m/s$ and $v = (15, 16, 17)m/s$, respectively.

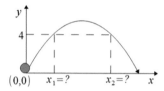

Fig. 5. Trajectory of movement of the object, the distances x_1 and x_2 for the height $y = 4$ form the start point $(0,0)$.

At the time t, the object is on the position (x, y) given by formula (22),

$$x = ut, \quad y = vt - 0,5gt^2. \tag{22}$$

From the Eq. (22) the quadratic Eq. (23) has been obtained.

$$y = \frac{v}{u}x - \frac{g}{2u^2}x^2. \tag{23}$$

Fuzzy numbers $u = (11, 12, 13)$ and $v = (15, 16, 17)$ in parametric form are: $u = (11 + r, 13 - r)$ and $v = (15 + r, 17 - r)$, $r \in [0, 1]$. To find the solution of example 3 the quadratic equation with fuzzy numbers (24) should be solved,

$$\frac{9.8}{2(11 + r, 13 - r)^2}x^2 - \frac{(15 + r, 17 - r)}{(11 + r, 13 - r)}x + 4 = 0, \tag{24}$$

Solution with standard fuzzy arithmetic
Using standard fuzzy arithmetic with fuzzy numbers in parametric form the Eq. (24) is presented as (25), $r \in [0, 1]$.

$$\left(\frac{4.9}{(13 - r)^2}, \frac{4.9}{(11 + r)^2}\right)x^2 - \left(\frac{15 + r}{13 - r}, \frac{17 - r}{11 + r}\right)x + 4 = 0, \tag{25}$$

The Δ value presents Eq. (26)

$$\Delta = \left(\frac{(15+r)^2}{(13-r)^2} - \frac{78.4}{(11+r)^2}, \frac{(17-r)^2}{(11+r)^2} - \frac{78.4}{(13-r)^2} \right), \tag{26}$$

Because $\Delta > 0$, for $r \in [0,1]$, the Eq. (25) has two solutions in the set of real numbers given by Eq. (27)

$$x_1 = \left(\underline{x_1}(r), \overline{x_1}(r) \right) = \left(\frac{\frac{15+r}{13-r} - \sqrt{\frac{(17-r)^2}{(11+r)^2} - \frac{78.4}{(13-r)^2}}}{\frac{9.8}{(11+r)^2}}, \frac{\frac{17-r}{11+r} - \sqrt{\frac{(15+r)^2}{(13-r)^2} - \frac{78.4}{(11+r)^2}}}{\frac{9.8}{(13-r)^2}} \right),$$

$$x_2 = \left(\underline{x_2}(r), \overline{x_2}(r) \right) = \left(\frac{\frac{15+r}{13-r} + \sqrt{\frac{(15+r)^2}{(13-r)^2} - \frac{78.4}{(11+r)^2}}}{\frac{9.8}{(11+r)^2}}, \frac{\frac{17-r}{11+r} + \sqrt{\frac{(17-r)^2}{(11+r)^2} - \frac{78.4}{(13-r)^2}}}{\frac{9.8}{(13-r)^2}} \right).$$

$$\tag{27}$$

Figure 6 presents results obtained with standard fuzzy arithmetic.

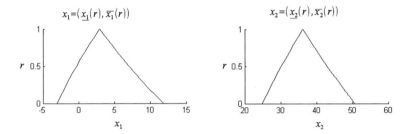

Fig. 6. Results in the form of fuzzy numbers obtained with standard fuzzy arithmetic.

Solution with horizontal fuzzy numbers

The fuzzy numbers $u = (11+r, 13-r)$ and $v = (15+r, 17-r)$ in the notation of horizontal fuzzy numbers have a form of (28), $r, \alpha_u, \alpha_v \in [0,1]$.

$$\begin{aligned} u(r, \alpha_u) &= 11+r+2\alpha_u(1-r), \\ v(r, \alpha_v) &= 15+r+2\alpha_v(1-r), \end{aligned} \tag{28}$$

The Eq. (24) with horizontal fuzzy numbers is presented by (29),

$$\frac{4.9}{[11+r+2\alpha_u(1-r)]^2}x^2 - \frac{15+r+2\alpha_v(1-r)}{11+r+2\alpha_u(1-r)}x + 4 = 0, \tag{29}$$

where $r, \alpha_u, \alpha_v \in [0,1]$.

The Δ value equals as shows Eq. (30),

$$\Delta = \frac{[15+r+2\alpha_v(1-r)]^2-78.4}{[11+r+2\alpha_u(1-r)]^2}, \tag{30}$$

For $r, \alpha_u, \alpha_v \in [0,1]$ the Δ value is greater than 0. The span of Δ is $s(\Delta) = \left(\frac{(15+r)^2-78.4}{(13-r)^2}, \frac{(17-r)^2-78.4}{(11+r)^2}\right)$, the minimum value of $s(\Delta)$ was obtained for $\alpha_u = 1$ and $\alpha_v = 0$, the maximum value of $s(\Delta)$ was achieved for $\alpha_u = 0$ and $\alpha_v = 1$.

The granules of solutions obtained with horizontal fuzzy numbers are presented in Eq. (31), $r, \alpha_u, \alpha_v \in [0,1]$.

$$
\begin{aligned}
x_1(r, \alpha_u, \alpha_v) &= \frac{[11+r+2\alpha_u(1-r)]\left[15+r+2\alpha_v(1-r)-\sqrt{[15+r+2\alpha_v(1-r)]^2-78.4}\right]}{9.8}, \\
x_2(r, \alpha_u, \alpha_v) &= \frac{[11+r+2\alpha_u(1-r)]\left[15+r+2\alpha_v(1-r)+\sqrt{[15+r+2\alpha_v(1-r)]^2-78.4}\right]}{9.8},
\end{aligned} \tag{31}
$$

Figure 7 shows the granule of information of solutions x_1 and x_2 obtained with horizontal fuzzy numbers.

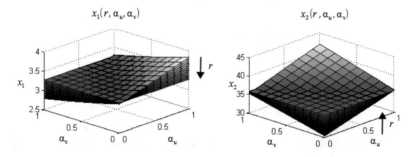

Fig. 7. Granules of solutions x_1 and x_2 of fuzzy quadratic equation for $r, \alpha_u, \alpha_v \in [0,1]$.

Spans $s(x_1)$ and $s(x_2)$ presents formula (32), $r \in [0,1]$. The left border of the span of solution $x_1(r, \alpha_u, \alpha_v)$ was obtained for $\alpha_u = 0$ and $\alpha_v = 1$, the right border for $\alpha_u = 1$ and $\alpha_v = 0$. The left border of the span of solution x_2 was obtained for $\alpha_u = \alpha_v = 0$, and the right border for $\alpha_u = \alpha_v = 1$, $r \in [0,1]$.

$$
\begin{aligned}
s(x_1) &= \left(\frac{(11+r)\left[17-r-\sqrt{(17-r)^2-78.4}\right]}{9.8}, \frac{(13-r)\left[15+r-\sqrt{(15+r)^2-78.4}\right]}{9.8}\right), \\
s(x_2) &= \left(\frac{(11+r)\left[15+r+\sqrt{(15+r)^2-78.4}\right]}{9.8}, \frac{(13-r)\left[17-r+\sqrt{(17-r)^2-78.4}\right]}{9.8}\right).
\end{aligned} \tag{32}
$$

Spans $s(x_1)$ and $s(x_2)$ of the solution of Eq. (24) presents Fig. 8.

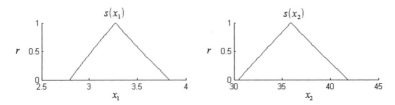

Fig. 8. Spans of solutions x_1 and x_2 of fuzzy quadratic equation obtained with use of horizontal fuzzy numbers.

Let us consider the results of example 3 with SFA and HFNA. The supports $(r = 0)$ of the results with SFA (27) are: $x_1 \in [-2.8821, 12.3949]$, $x_2 \in [24.4536, 50.5746]$. This result is the distance of the object from point $(x, y) = (0, 0)$ to the point of height equals $y = 4m$, so the solution cannot be less than 0. The SFA gives values in x_1 that are less than 0. Results of SFA are overestimated, this arithmetic gives the results that are not solutions. Solutions with HFNA are presented in formula (31), the support of spans $(r = 0)$ of solutions are: $x_1 \in [2.7926, 3.865]$, $x_2 \in [30.4272, 41.8017]$. By granule of solution (31) we can generate all solutions of analyzed equation. Each solution depend on the value $r, \alpha_u, \alpha_v \in [0, 1]$. For example, for $r = \alpha_u = \alpha_v = 0$, from Eq. (31) the crisp solutions are $x_1 = 3.2463$ and $x_2 = 30.4272$, also from Eq. (29) the crisp quadratic equation can be generated: $0.0405x^2 - 1.3636 + 4 = 0$.

4   Conclusion

The paper presents the application of horizontal fuzzy numbers for solving the quadratic fuzzy equation. Solutions were compared with results obtained with standard fuzzy arithmetic. It was shown that solutions obtained with HFNA are multidimensional. Results with standard fuzzy arithmetic are only the spans and can be underestimated or overestimated. The span is only the indicator of the full solution. The HFNA gives the granule of information about solution that generates the full set of solutions. By obtained granule can be calculated the crisp solution for fuzzy quadratic equation. Moreover, also crisp quadratic equation from fuzzy equation can be obtained. Single value of solution is strictly correlated with crisp values of the coefficients of fuzzy quadratic equation. Presented RDM method with RDM variables using horizontal fuzzy numbers gives possibility to describe correlation between solution and the problem (equation) by formula. On the base of granule of solution not only span can be calculated, also other indicators of solution can be obtained, like the center of gravity or cardinality distribution of all possible solutions.

References

1. Abbasbandy, A.: Homotopy method for solving fuzzy nonlinear equations. Appl. Sci. **8**, 1–7 (2006)
2. Allahviranloo, T., Moazam, L.G.: The solution of fully fuzzy quadratic equation based on optimization theory. Sci. World J. **2014**, Article ID 156203 (2014). http://dx.doi.org/10. 1155/2014/156203. Hindawi Publishing Corporation
3. Baoding, L.: Uncertainty Theory, 4th edn. Springer, Berlin (2015)
4. Banerjee, S., Roy, T.K.: Intuitionistic fuzzy linear and quadratic equations. J. Inf. Comput. Sci. **10**(4), 291–310 (2015)
5. Dymova, L.: Soft Computing in Economics and Finance. Springer, Heidelberg (2011)
6. Konar, A.: Artificial Intelligence and Soft Computing. CRC Press, London, New York (2000)
7. Landowski, M.: Differences between Moore and RDM interval arithmetic. In: Angelov, P., et al. (eds.) Intelligent Systems 2014. Advances in Intelligent Systems and Computing, vol. 322, pp. 331–340. Springer, Cham (2015)
8. Landowski, M.: Comparison of RDM complex interval arithmetic and rectangular complex arithmetic. In: Kobayashi, S., et al. (eds.) Hard and Soft Computing for Artificial Intelligence, Multimedia and Security, ACS 2016. Advances in Intelligent Systems and Computing, vol. 534, pp. 49–57. Springer, Cham (2017)
9. Landowski, M.: Method with horizontal fuzzy numbers for solving real fuzzy linear systems. Soft Comput. 1–13 (2018). https://doi.org/10.1007/s00500-018-3290-y
10. Li, Q.X., Liu, S.F.: The foundation of the grey matrix and the grey input-output analysis. Appl. Math. Model. **32**, 267–291 (2008)
11. Ming, M., Friedman, M., Kandel, A.: A new fuzzy arithmetic. Fuzzy Sets Syst. **108**, 83–90 (1999)
12. Mashinchi, M.H., Mashinchi, M.R., Shamsuddin, S.M.: A genetic algorithm approach for solving fuzzy linear and quadratic equations. Int. J. Math. Comput. Phys. Electr. Comput. Eng. **1**(4), 215–219 (2007)
13. Nurhakimah, A.R., Lazim, A.: Multi-solutions of fuzzy polynomials: a review. Int. J. Eng. Appl. Sci. **3**(3), 8–16 (2013)
14. Pedrycz, W., Skowron, A., Kreinovich, V.: Handbook of Granular Computing. Wiley, Chichester (2008)
15. Piegat, A., Landowski, M.: Two interpretations of multidimensional RDM interval arithmetic - multiplication and division. Int. J. Fuzzy Syst. **15**(4), 488–496 (2013)
16. Piegat, A., Landowski, M.: Horizontal membership function and examples of its applications. Int. J. Fuzzy Syst. **17**(1), 22–30 (2015)
17. Piegat, A., Landowski, M.: Fuzzy Arithmetic Type 1 with Horizontal Membership Functions. In: Kreinovich, V. (ed.) Uncertainty Modeling. Studies in Computational Intelligence, vol. 683, pp. 233–250. Springer, Cham (2017)
18. Zadeh, L.A.: From computing with numbers to computing with words - from manipulation of measurements to manipulation of perceptions. Int. J. Appl. Math. Comput. Sci. **12**(3), 307–324 (2002)

Regression Technique for Electricity Load Modeling and Outlined Data Points Explanation

Krzysztof Karpio$^{(\boxtimes)}$ ⓘ, Piotr Łukasiewicz$^{(\boxtimes)}$ ⓘ,
and Rafik Nafkha$^{(\boxtimes)}$ ⓘ

Department of Informatics, SGGW,
Nowoursynowska 159 02-787 Warsaw, Poland
{krzysztof_karpio,piotr_lukasiewicz,
rafik_nafkha}@sggw.pl

Abstract. Hundreds of explanatory variables such as historical consumption data, climate variables, socioeconomic and demographics parameters, etc. are used to forecast major forms of commercial energy consumptions. The scientists increasingly face the problems of big data: huge amount of data which grows in time. This article presents the analysis of the relationships between electricity amount to deliver and factors having more or less significant impact on electricity consumptions. The linear regression algorithm is used to reduce the set of explanatory variables and to evaluate their importance. A reduction of number of variables without significant loss of accuracy of the model is presented. Next, regression decision trees are used both to evaluate the quality of the modeled energy consumption, as well as to further explanatory variables reduction. The last part of the article deals with the outlining data points and explains reasons they come off model predictions.

Keywords: Electricity load modeling · Regression decision tree
Electricity short-term forecasting

1 Introduction

Specialists staying with modeling development point out that one of the key problems of any formalized forecasting process is the so-called calibration of the model, i.e. adjusting its results to the balancing values for the base (starting) year of the calculations. Typically, the base year is the one for which the required set of data can be obtained, which in some case is a complex and/or time-consuming task. Electricity demand forecasting models are an important issue and at the same time a challenge due to the high variability of load characteristics over time, which is difficult to model. Variabilities may be regular or random. Regular load changes are caused by astronomical factors, user category (industrial or municipal), type of day (working or non-working) or other causes that are easily predictable such as macroeconomic indicators. Random changes occur irregularly and are caused by random factors, such as: outdoor temperature, cloud cover, failures and shortage in the power system, etc. The application

© Springer Nature Switzerland AG 2019
J. Pejaś et al. (Eds.): ACS 2018, AISC 889, pp. 56–67, 2019.
https://doi.org/10.1007/978-3-030-03314-9_5

of regular and irregular factors and the choice of appropriate modelling techniques depends on the forecasting horizon.

Short-term forecasting (an hour to a week) models, the mostly used for scheduling and analyses of the distribution network, parameters such as daily electricity consumption, average and peak load influenced by the rhythm and people life activities are very essential. Mid-term forecasting (a month to 5 years) models, mostly applied for planning the power production resources and tariffs. Besides historical consumptions data, climate variables, socioeconomic and demographics parameters state important input of the models. Long-term forecasting (5 to 20 years) models, usually applied for global electricity demand planning and development investments, introduce indexes related to electricity generation efficiency, energy supplies security and environmental impact of the power industry. The time periods adopted above are arbitrary and there are often slightly different divisions in the literature.

Each of the above models may have different modeling techniques and calculation methods, which can be grouped broadly in three major groups [1]: Traditional Forecasting Technique, Modified Traditional Forecasting and Soft Computing Technique. In the early days, the traditional forecasting techniques used traditional and conventional mathematical techniques and contain regression, multiple regression, exponential smoothing and Iterative reweighted least-squares techniques. The modified versions of these traditional techniques are adaptive load forecasting, stochastic time series and support vector machine based techniques. Soft computing techniques are computer-based intelligent systems with a remarkable ability of a human mind to reason and learn, contain genetic algorithms, fuzzy networks or neural networks.

The review of the literature presents an overview of electricity demand models correlating the energy demand with particular features. The most important features that have significant impact on electricity consumption are selected as primary data for our tree decision short term forecasting model.

2   Literature Review

The different relationship between electricity consumption and the number of independent economic variables are used in different countries [2–6]. Parameswara Sharma et al., have generated econometric demand models to forecast three major forms of commercial energy (electricity, petroleum products and coal) in the state of Kerala [7]. Economic explanatories like state domestic product and its sectoral components, per capita income, indexes of industrial production, etc. have been considered in the models. Similar explanatory variables including GDP, per capita income, agricultural production output, industrial production output, and capital investment are used to determine sectoral energy demand for six provinces in China [8] and electricity consumptions forecast of industry, households and services in Italy [9]. Hsiao-Tien Pao, use linear and nonlinear models like artificial neural networks to analyze the economic impact of the national income, population, GPD and customer tariff rates on electricity consumption in Taiwan [10]. His study led to the finding that expected GPD and customer tariff rate have less impact on forecasting electricity demand than the population and national income.

While economic features are useful for long- and medium-term electricity fore-casting, they have little practical use estimating day-to-day load flows, which are vital to utility short-term planning. Electricity consumption depends on a number of factors. The impact of customers number, their nature (industry or households), the type of day (working or holiday), and the value of the installed capacity of individual entities are most often examined. Furthermore, many forecasting methods search for additional supplementary information that may improve the prediction accuracy. A number of studies have demonstrated for example, that weather influences energy consumption patterns. Hor et al. [11] investigated the impact of weather variables on monthly electricity demand in England and Wales. Initially, the electricity demand model to forecast monthly electricity demand was built as a function of cooling and heating days, further gross domestic product, and population growth variables are added. The inclusion of degree-days and relative humidity in the model has improved the demand forecast especially during the summer months. A year-to-year comparison of the electricity demand characteristics for London and Athens and its relationships with both climate and non-climate related factors are investigated by Psiloglou et al. [12]. Several other effects such as weekends and holidays effects, unrelated to weather conditions were examined for the two cities. The influence of temperatures feature appeared significantly on electricity demand during summer for Athens, and the rela-tionship sensitivity was greater during the cold period of the year for both cities. Most papers [13–15], have indeed pointed out the non-linearity between outdoor temperature and electricity consumptions. Any increase or decrease of temperature over or below given threshold increases the demand of electricity. Gajowniczek et al. remarked that although humidity has no effect on real temperature it can intensify the severity of hot climate. Therefore, for the prediction of daily load at domestic consumers it is rec-ommended to consider apparent temperature instead of real temperature [16]. In addition to classical consumption and weather features, Weron et al. reminded, in day-ahead electricity forecasting the daily and weekly seasonality must be always taken into account [17].

To understand users demand drivers we analyzed first the relationship between the electricity to deliver and factors having less or more significant impact on users' electricity consumptions. We used regression methods to model the relationship of load demand and features such as weather, day type and historical consumption. Features having a significant impact on the amount of electricity consumed have been then selected. The F-ratio was used to assess features significance. The followed regression tree technique was utilized to further limit the number of features and to evaluate the quality of energy consumption modeling. The model's outlining points were studied in order to explain reasons of their origin.

3   Dataset Characteristics and Processing

This study was performed based on the historical data representing total energy con-sumption in Polish power system [18]. The consumption is denoted on the hourly basis covering time span between January 1st 2008 and June 24th 2015. The power system load is presented in time series form, showing annual, weekly and daily seasonal cycles as shown in Fig. 1.

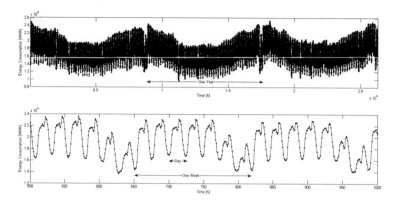

Fig. 1. Analyzed energy consumption vs time. Top plot presents: first three years, white line indicating the trend and one year (2009) period. Bottom plot contains three weeks period, one week from Monday to Sunday and one day period–Wednesday

The top plot contains first three years. The bottom plot contains 500 data points, what corresponds to about three weeks. The one week period reflects period of time between Sunday/Monday nights, the day period indicates load on Wednesday. Changes in the daily load shape and load level during the year could be influenced by weather conditions including temperature, humidity and daylight hours. The weekly cycles seem to be mostly determined by weekdays and weekends, what is visible in Fig. 1. The energy consumption on Sunday is significantly lower than during weekdays and consumption on Saturdays seems to be partly influenced too. The long range linear trend indicated by white line in Fig. 1 has been subtracted using the formula (1)

$$E = E_0 - 0,0181 * t, \tag{1}$$

where t – time [h], E/E_0–energy with/without long range trend, *the number*–directional coefficient of the linear trend.

The coefficient has been calculated by fitting the linear function to the whole data series. Then we leave the first data point intact and correct the remaining data points. The transformation makes all energies smaller except the first one. Thanks to the transformation every year can be treated equally and a year number will not be included in the analysis making results more universal. However, at the first stage of the analysis we include the time as one of the variables to show that it is of the small or none importance. In order to model the energy consumption at i-th data point we start with the set of variables described in Table 1. They precede the modeled data point up to a week and three hours what corresponds to 170 h. That's why we excluded from modeling data points for $t \leq 170$. Thus, each data point has 43 features associated with it. The set of all variables is presented in Table 1 containing their symbols, types and descriptions.

Table 1. List of all features/variables used in the model

Symbol	Type	Description
En	Continuous	Energy at n-th data point; the modeled value
t	Discrete	Ordinal number of the data point – equivalent to time
Month	Discrete	Month's number
Day	Discrete	Day of the month
Hour	Discrete	The hour
nDay	Discrete	Day of the week
nWeek	Discrete	Week of the year
Holiday	Boolean	Holiday or working day
Night_Day	Boolean	Night or day
Tem	Continuous	Temperature
Hum	Continuous	Humidity
En_01	Continuous	Energy consumption one hour before
En_02	Continuous	Energy consumption two hours before
En_03	Continuous	Energy consumption three hours before
Tem_01	Continuous	Temperature one hour before
Tem_02	Continuous	Temperature one two hours before
Tem_03	Continuous	Temperature one three hours before
Hum_01	Continuous	Humidity one hour before
Hum_02	Continuous	Humidity two hours before
Hum_03	Continuous	Humidity three hours before
En_10	Continuous	Energy consumption one day before
En_11	Continuous	Energy consumption one day and one hour before
En_12	Continuous	Energy consumption one day and two hours before
En_13	Continuous	Energy consumption one day and three hours before
Tem_10	Continuous	Temperature one day before
Tem_11	Continuous	Temperature one day and one hour before
Tem_12	Continuous	Temperature one day and two hours before
Tem_13	Continuous	Temperature one day and three hours before
Hum_10	Continuous	Humidity one day before
Hum_11	Continuous	Humidity one day and one hour before
Hum_12	Continuous	Humidity one day and two hours before
Hum_13	Continuous	Humidity one day and three hours before
En_70	Continuous	Energy consumption one week before
En_71	Continuous	Energy consumption one week and one hour before
En_72	Continuous	Energy consumption one week and two hours before
En_73	Continuous	Energy consumption one week and three hours before
Tem_70	Continuous	Temperature one week before
Tem_71	Continuous	Temperature one week and one hour before
Tem_72	Continuous	Temperature one week and two hours before
Tem_73	Continuous	Temperature one week and three hours before

(*continued*)

Table 1. (*continued*)

Symbol	Type	Description
Hum_70	Continuous	Humidity one week before
Hum_71	Continuous	Humidity one week and one hour before
Hum_72	Continuous	Humidity one week and two hours before
Hum_73	Continuous	Humidity one week and three hours before

In the Table 1 the meanings of day and week are 24 and 168 h respectively. The *En* is described variable and is not counted toward 43 descriptive variables. Our aim is to describe the influence of features onto the modeled value of energy consumption. That's why we deal with the *Tem*, *Hum* despite they are not known in advance and they must be predicted in real scenario.

4   Numerical Experiment

The analysis starts by modeling energy consumption by means of linear regression taking into account all the features. Then, the backward selection [19] to limit number of features is used. This method is based on the comparison of the given model with the reduced model. The comparison is performed by calculating the *F*-ratio measure. The procedure is performed in steps. At each step a variable with the smallest *F*-ratio is eliminated while the *F*-ratio does not exceed a fixed threshold value. Precisely, variable i is deleted from the p-term equation if:

$$F_i = \min_i \left(\frac{SSE_{p-i} - SSE_p}{\sigma_p^2} \right) < F_{out}, \qquad (2)$$

where $\sigma_p^2 = SSE_p/(n - p - 1)$ is residual mean squares and SSE_{p-i} denotes the residual sum of squares obtained when variable i is deleted from the current p-term model. The value F_{out} has been set experimentally to 15. This way, we started with all 43 variables but only 19 of them were statistically significant. In the first step we excluded a variable with the smallest importance *Hum_10* and repeated the procedure. A number of statistically significant variables for each step are presented in Fig. 2. Finally, after 25 steps we got 18 variables. Quality of models measured using *F*-ratio was systematically raising (Fig. 2). The number of important variables was rising at first and after the 14 steps started to drop. Among variables was t – time. This variable should not be important and indeed its importance is one of the lowest because it has been eliminated in the fourth step (Table 2).

In the second step of the analysis we used a regression decision tree to further model the energy consumption using the final 18 features. Regression trees are decision trees that deal with a continuous dependent variable. This method is a combination of decision tree and linear regression. It is used to forecast continuous target attribute based on a set of input attributes [20, 21]. A linear regression is fit to all the points in

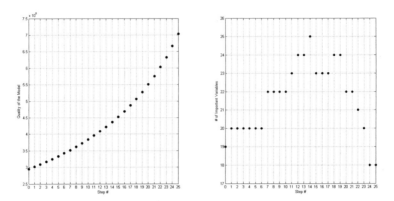

Fig. 2. The stages of backward selection. Left plot: the quality of the model (model *F*-ratio) vs step number. Right plot: number of statistically significant variables vs step number

Table 2. Eliminated and remaining features

Eliminated 25 variables in the order of elimination	Hum_10, Tem_02, Hum_72, t, Hum_01, Tem_70, Tem_11, Tem_73, Hum_71, Hum_70, Tem_01, Tem_03, Tem_72, Hum_02, Hum_13, Hum_12, Tem_12, Tem_13, Night_Day, Month, nWeek, Day, Hum, Hum_11, Tem_10
Final 18 statistically important variables	Hour, nDay, Holiday, Tem, En_01, En_02, En_03, Hum_03, En_10, En_11, En_12, En_13, En_70, En_71, En_72, En_73, Tem_71, Tem_73

each tree's node (all the attributes must be quantitative). Then the possibility of splitting the fitting region into ranges which improves the fit quality is checked. The choice of the best split at each node of the tree is based on a least squares error criterion. At each step, heterogeneity is measured by the within-node sum of squares of the response

$$V_c = \sum_{i=1}^{n_c} (y_i - \widehat{y}_i)^2,$$ (3)

Where, $y_i, \widehat{y}_i$ are empirical and theoretical values of dependent variable and n_c is the number of node objects. The heterogeneity for each potential split is the sum of the variances for the nodes that would result:

$$S = \sum_{j=1}^{k} n_{c_j} V_{c_j},$$ (4)

where c_j (two or more) are the descendants of the node c. The sum of squares V_c of the parent node is compared to the combined sums of squares S from each potential split into descendants nodes. The split is chosen that reduces most this within-nodes sum of squares.

The algorithm of the regression tree was implemented in SQL Server 2012 Analysis Services? The implemented algorithm has three options: *binary* (nodes are split into two subsets only, binary tree), *complete* (nodes are split into maximum number of subsets based on the all possible values of the attribute) and, *both* (during each split of a node a decision is being made, based on an effectiveness, which of the previous options to use). We chose the *both* option to do not impose any limitations on the number of splits. In order to limit a growth of the tree we used 40 as minimum support of nodes and penalty parameter $\alpha = 0.9$. The regression tree was built based on the 30% random sample. The remaining part of the data was used for validation. One obtained the tree consisting of 3 to 11 levels depending on branch. At the first level the *Hour* attribute was taken into account generating 24 splits of the root node. For another splits various variables were taken into account.

Partial regression models evaluated in leaves allows calculating predicted energy consumption. The results are presented in Fig. 3.

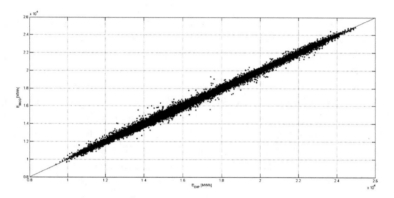

Fig. 3. Predicted vs experimental energy consumption

The tree model is characterized by the residual sum of squares $SSE = 2.87 \cdot 10^9$, while the total variance of energy consumption $SST = 5.56 \cdot 10^{11}$. These numbers yield to the coefficient of determination equal to 0.9948. Mean squared error is equal to 209.5 MWh. The real energy consumption within an hour is between ~ 9000 MWh and ~ 25000 MWh, so the error can be regarded as small. Most of the relative differences are smaller than 1% of the minimal experimental energy (Fig. 4).

Their size seems to be independent on the energy consumption. The visible small differences in width in the left plot are due to the distribution of the energy. Points in regions with bigger statistics look like they were distributed more widely. The relative errors are presented in Fig. 4 right plot together with the line representing Gaussian shape. The empirical distribution follows a Gaussian shape but exhibits fat tails. The standard deviation of errors is 0.8%. We extracted 2243 data points where the distortions between model predictions and experimental data were greater than 2%. This

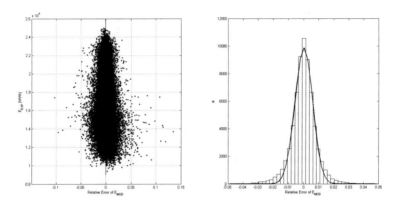

Fig. 4. Results of the decision tree algorithm. Left plot: experimental energy consumption vs relative error of the prediction. Right plot: distribution of relative errors of the prediction

value is related to 2.3 of the standard deviation. There are 1058 data points with over predicted consumption and 1185 points where model predicted too small consumption. They correspond to 1.61% and 1.81% respectively, so the 96.58% of points are within the 2% error.

In the third step we adopted classification decision tree to track the predictions outside the 2% boundaries. We would like to distinguish such points based on their features. That information could shed light on the origin of such points, the reasons of irregular energy consumption. The regular tree was based on the 18 features and flag indicating type of point was the prediction features: 0 – point lays within the 2% boundaries, 1 – protruding point (3.42%). We obtained the decision tree build up from 59 nodes (42 in the leaf level) arranged on 4 levels. The Fig. 5 contains the Lift chart presenting the percentage of protruding points vs sample size in per cents. There are also two models on the chart: the ideal (dotted-dashed line), which reaches 100% at sample of 3.42% and the random one (dashed line). Our model, when comparing to the random one is more than 4 times better at 10% of the population, more than 3 times better at 20%, even at 40% it is more than 2 times better. We listed leaves with shares of 1-type points at least 10% in Table 3.

The best leaf has 46.22% of protruding points, the next one 31.65%. The presented leaves contain 778 out of total 2243 protruding points – more than one third of them. The features (attributes) ordered from the most to the least important are: *Hour*, *En_02*, *Tem_71*, *En_01*, *Holiday*, *nDay*, *En_13*, *En_10*, *En_70*, *En_11*, *En_03*, and *En_12*. The most important feature *Hour* has the values: $6 \div 8$ and $17 \div 21$. They correspond to the rush hours.

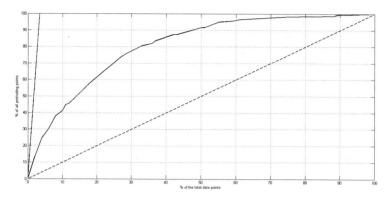

Fig. 5. The lift plot for the decision tree–solid line, ideal model–dotted-dashed line random model–dashed line

Table 3. The leaves with shares of 1-type points at least 10%

Condition	Percentage	Number of points
Hour = 1 & *En_10* < 14062.24	13.30%	63
Hour = 6 & *En_03* < 14062.24	14.65%	144
Hour = 7 & *En_01* < 15619.87	24.29%	351
Hour = 8 & *En_10* < 15619.87	10.64%	56
Hour = 17 & *Tem_71* < −5	31.65%	31
Hour = 19 & *nDay* = 7 & *En_12* < 18735.13	17.04%	38
Hour = 20 & *En_02* < 18735.13 & *En_70* ≥ 18814.22	18.11%	40
Hour = 21 & *nDay* ≠ 7 & *En_02* < 15619.87	46.22%	55

5 Summary and Concluding Remarks

Most of the studies use very large number of descriptive features in order to model and predict power consumption. Some of these features are not statistically significant. That leads to the idea that it could be possible to limit the number of analyzed variables without the significant loss of model precision. In this paper, we studied the quality of the electricity usage models versus the number of explanatory variables. At the beginning, the simple, commonly known and very fast backward linear regression algorithm was used. That allowed for precisely evaluation of the importance of every variable and the model quality as a whole. After the reduction of the total set of variables we then used regression tree algorithm in order to model the energy consumption and to evaluate its precision. Next, protruding points are pointed in order to further examine reasons of their disagreement with the model.

Basically, one can say that, the resulting regression tree model in this study is perfectly suited to empirical data. While the backward selection algorithms based on electricity consumption theory, provided good results, able to reduce explanatory variables from 43 to 18, a few theoretical outlining data points have been identified using the decision tree method. The results of the model can be used to predict energy consumption values and the forecast error in turn.

It is worth considering the efficiency of the resulting model in this paper by comparing algorithms for predictors reducing and electricity consumption modeling e.g. including forward selection algorithms and neuron network method respectively. The other problem arises from the so called 'big data' problems. There is a lot of data already but their amount constantly grows. The process of selecting variable can easily performed based on the sample of data, because data exhibit strong perpendicularity (daily, weekly, monthly, yearly). Other approach which is already under investigation is the search for correlations between energy consumptions during various hours of the same day. It seems that this approach could successfully reduce number of studied data to about 30% of the original volume.

References

1. Singh, A.K, Ibraheem, S.K., Muazzam, Md.: An overview of electricity demand forecasting techniques. In: Proceedings of National Conference on Emerging Trends in Electrical, Instrumentation & Communication Engineering, vol. 3, No. 3, pp. 38–48 (2013)
2. Rajan, M., Jain, V.K.: Modelling of electrical energy consumption in Delhi. Energy **24**, 351–361 (1999)
3. Soytas, U., Sari, R.: Energy consumption and GDP: causality relationship in G–7 countries and emerging markets. Energy Econ. **3**, 33–37 (2003)
4. Yan, Y.Y.: Climate and residential electricity consumption in Hong Kong. Energy **23**(1), 17–20 (1998)
5. Shiu, A., Lam, P.L.: Electricity consumption and economic growth in China. Energy Policy **32**, 47–54 (2004)
6. Egelioglu, F., Mohamad, A.A., Guven, H.: Economic variables and electricity consumption in Northern Cyprus. Energy **26**(4), 355–362 (2001)
7. Parameswara Sharma, D., Chandramohanan Nair, P.S., Balasubramanian, R.: Demand for commercial energy in the state of Kerala, India: an econometric analysis with medium-range projections. Energy Policy **30**, 781–791 (2002)
8. Yang, M., Yu, X.: China's rural electricity market—a quantitative analysis. Energy **29**(7), 961–977 (2004)
9. Gori, F., Takanen, C.: Forecast of energy consumption of industry and household and services in Italy. Heat Technol. **22**(2), 115–121 (2004)
10. Pao, H.T.: Comparing linear and nonlinear forecasts for Taiwan's electricity consumption. Energy **31**, 2129–2141 (2006)
11. Hor, C.L., Watson, S.J., Majithia, S.: Analyzing the impact of weather variables on monthly electricity demand. IEEE Trans. Power Syst. **20**, 2078–2085 (2005)
12. Giannakopoulos, Ch., Psiloglou, B.E.: Trends in energy load demand for Athens, Greece: weather and non-weather related factors. Clim. Res. **31**, 97–108 (2006)
13. Pardo, A., Meneu, V., Valor, E.: Temperature and seasonality influences on Spanish electricity load. Energy Econ. **24**, 55–70 (2002)
14. Sailor, D.J.: Relating residential and commercial sector electricity loads to climate – evaluating state level sensitivities and vulnerabilities. Energy **26**, 645–657 (2001)
15. Valor, E., Meneu, V., Caselles, V.: Daily air temperature and electricity load in Spain. J. Appl. Meteor. **40**, 1413–1421 (2001)

16. Gajowniczek, K., Nafkha, R., Ząbkowski, T.: Electricity peak demand classification with artificial neural networks. Comput. Sci. Inf. Syst., ACSIS **11**, 307–315 (2017)
17. Marcjasz, G., Uniejewski, B., Weron, R.: Importance of the long-term seasonal component in day-ahead electricity price forecasting revisited: Neural network models. HSC Research Reports HSC/17/03 (2017)
18. Polish power system dataset. http://www.pse.pl/index.php?dzid=77. Accessed 12 Aug 2017
19. Hocking, R.R.: The analysis and selection of variables in linear regression. Biometrics **32**(1), 1–49 (1976)
20. Breiman, L., Friedman, J.H., Olshen, R.A., Stone, C.J.: Classification and Regression Trees. CRC Press, Boca Raton (1984)
21. Rokach, L., Maimon, O.: Data mining with decision trees. Theory and applications. World Scientific Pub Co Inc., Singapore (2008)

Correct Solution of Fuzzy Linear System Based on Interval Theory

Andrzej Piegat and Marcin Pietrzykowski[✉]

Faculty of Computer Science and Information Technology,
West Pomeranian University of Technology, Żołnierska 49, 71-210 Szczecin, Poland
{apiegat,mpietrzykowski}@wi.zut.edu.pl

Abstract. In this paper authors would like to make a critical review of the low-dimensional method proposed in [3] by Allahviranloo and Gandavi for solving fuzzy linear system with crisp square matrix and a fuzzy right-hand side vector. The solution presented in the mentioned work is, in general incorrect. The authors prove that given method is incorrect and propose an alternative solution based on multidimensional Relative-Distance-Measure (RDM) interval arithmetic and fuzzy RDM interval arithmetic that gives correct results.

Keywords: Fuzzy linear system · Interval linear system
Multidimensional Fuzzy RDM arithmetic
Multidimensional Interval RDM arithmetic

1 Introduction

Many real systems with uncertain parameters and variables are described with use of interval or fuzzy linear equation systems [4,9,17]. Most authors of papers on fuzzy linear systems use to solving them methods directly or indirectly (α-cuts) based on the standard interval arithmetic. However, this arithmetic is in the general case incorrect, though it correctly solves some simple and not complicated problems (e.g. basic arithmetic operations on intervals). This fact for many scientists creates appearance of its correctness. And it is the main reason why the standard interval arithmetic is continually used in the world. Meanwhile, many problems are solved by this method with great and sometimes even with not acceptably great errors. Hence standard interval arithmetic should not be used, and especially not for solving fuzzy linear systems. Further on few important disadvantages of this arithmetic will be given. But earlier a simple and easily understandable example of application will be presented that expressively shows its incorrectness. Let us imagine that a truck driver has to transport from one to another city load of a commodity. In the start city A the load had been weighed on a balance with measurement error of 1 ton. From this measurement we know that the start weight $a \in [32, 34]$ tons. In the course of the journey the driver stole a part x of the commodity, which was seen by

© Springer Nature Switzerland AG 2019
J. Pejaś et al. (Eds.): ACS 2018, AISC 889, pp. 68–75, 2019.
https://doi.org/10.1007/978-3-030-03314-9_6

witnesses. After arriving to the destination city C the load had been weighed also by a balance with the same error and the measurement gave the same result as in the start city $c \in [32, 34]$, where c means weight in the destination city. The task is to calculate how much load x could be stolen by the driver? In terms of standard interval arithmetic [7] the problem can be expressed as $A - X = C$ or more exactly as $[a, a] - [x, x] = [c, c], x > 0$. Inserting numerical values we get $[32, 34] - [x, x] = [32, 34]$. Solving the interval equation according to rules of the standard interval arithmetic we get: $32 - x = 32$ and $34 - x = 34$ which results in $x = 0$ and $x = 0$ or in interval of the stolen load $X = [0, 0]$. The result means that with full certainty no part of the load has been stolen. However, a simple common sense analysis shows that the stolen amount lies in the interval $[0, 2]$ tons. It corresponds to situation $a = 34$ (start load) and $c = 32$ (destination load). This example distinctly shows what errors can be made by standard interval arithmetic. Because most of uncertainty analysis methods is based on this arithmetic, hence results delivered by them are less or more, depending on the case, incorrect and imprecise. In the paper [3] T. Allahviranloo and M. Ganbari (shortly TA&MG have presented "a new approach to solve fuzzy linear systems (FLS) involving crisp square matrix and a fuzzy right-hand side vector". This approach is based on interval inclusion linear system (IILS) and standard interval arithmetic [7,9]. According to TA&MG the method allows for obtaining the unique algebraic solution. In the paper numerical examples are given to illustrate the proposed method. Investigation of the TA&MG method shows, that, in general, it is incorrect. Below, few comments concerning the method are given.

1. TA&MG try to determine the algebraic solution of a FLS. However, in the case of uncertain equations not algebraic but universal algebraic solutions are to be determined. TA&MG consider FLS of the form $A\tilde{X} = \tilde{Y}$ where $\tilde{X} = (\tilde{x}_1, \tilde{x}_2, \ldots, \tilde{x}_n)^T$, $\tilde{Y} = (\tilde{y}_1, \tilde{y}_2, \ldots, \tilde{y}_n)^T$ are fuzzy number vectors and A is square matrix with elements being crisp numbers. In their opinion a fuzzy number vector $\tilde{X}$ is an algebraic solution of the FLS $A\tilde{X} = \tilde{Y}$. However, the equation $A\tilde{X} = \tilde{Y}$ is only one of possible model forms of a real system. Other equivalent forms are $A\tilde{X} - \tilde{Y} = 0, \tilde{X} = A^{-1}\tilde{Y}$ [4,6,12,13,17]. The universal algebraic (UA) solution $\tilde{X}$ has to satisfy not only the model form $A\tilde{X} = \tilde{Y}$ but also other possible, equivalent model forms. Otherwise, the "algebraic" solution causes unnatural behaviour in modeling the system (UBM phenomenon) and various paradoxes [4,6,10,13,17].
2. Correct UA-solution of the FLS $A\tilde{X} = \tilde{Y}$ is not the vector $\tilde{X} = (\tilde{x}_1, \tilde{x}_2, \ldots, \tilde{x}_n)^T$ consisting of 2D fuzzy numbers $\tilde{x}_i$ but a vector consisting of multidimensional fuzzy granules.
3. The method proposed by TA&MG does not take into account dependences existing between uncertain variables and parameters in a FLS. It increases error of solutions.
4. The notation $\tilde{X} = (\tilde{x}_1, \tilde{x}_2, \ldots, \tilde{x}_n)^T$ used in the discussed paper is incorrect [5, 12,13], though it can be met in many papers. It can be used only as symbolic one. This notation causes incorrect understanding of uncertain equations.

5. Solutions of numerical examples provided by the TA&MG method are in general incomplete and imprecise. It can be seen on examples given in the discussed paper.

2 Comparison of the Discussed TA&MG Method and of the Multidimensional Fuzzy Arithmetic

The equation $A\tilde{X} = \tilde{Y}$ is a mathematical model of a real system, that, in the case of dimensionality $n = 2$ is ruled by (1).

$$\begin{bmatrix} a_{11} & a_{12} \\ a_{21} & a_{22} \end{bmatrix} \begin{bmatrix} \tilde{x}_1 \\ \tilde{x}_2 \end{bmatrix} = \begin{bmatrix} \tilde{y}_1 \\ \tilde{y}_2 \end{bmatrix}, \quad A\tilde{X} = \tilde{Y} \tag{1}$$

In a real, stationary system values of coefficients a_{ij}, $i,j \in \{1,2\}$, are constant and have crisp values. Similarly, values of the system variables. Apart from the first model form $AX = Y$ there exist also few other equivalent crisp models of the system, as e.g. given by (2). X and Y are here crisp vectors.

$$AX - Y = 0, \quad X = A^{-1}Y \tag{2}$$

If all coefficients of the crisp matrix A and of the crisp vector Y would be known precisely then all possible model forms would deliver the same crisp solution $X = [x_1, x_2]^T$. This solution, substituted in all equivalent model forms would satisfy them. However, if only coefficients of A are known precisely but the vector Y not, if it is only known approximately in form of fuzzy numbers $\tilde{Y} = [\tilde{y}_1, \tilde{y}_2]^T$ then to each of possible crisp model forms corresponds one fuzzy model extension [4,17]. Fuzzy extensions of crisp models (2) are given by (3).

$$A\tilde{X} = \tilde{Y}, \quad A\tilde{X} - \tilde{Y} = 0, \quad \tilde{X} = A^{-1}\tilde{Y} \tag{3}$$

The universal algebraic solution $\tilde{X}$ is such solution, which satisfies all possible fuzzy extensions [6,12,13]. The method of solving the FLS $A\tilde{X} = \tilde{Y}$ proposed by TA&MG is based on their method of solving the interval linear system (ILS) $A[X] = [Y]$, where $[X] = ([x_1], [x_2], \ldots, [x_n])^T$ and $[Y] = ([y_1], [y_2], \ldots, [y_n])^T$ are interval vectors. TA&MG define an algebraic solution of the ILS in Definition 2.15 as the interval number vector $[X] = ([x_1], \ldots, [x_n])^T$ which satisfies system of linear Eq. (4).

$$\sum_{j=1}^{n} a_{ij}[x_j] = [y_i], \quad i = 1, 2, \ldots, n \tag{4}$$

However, they do not consider all equivalent forms of the ILS $A[X] = [Y]$, as they are given by (3). They also assume that solution of an ILS is interval vector and not vector of multidimensional granules. Correctness of the solution of the fuzzy linear system fully depends on correctness of the ILS solution. But the ILS-solution method given by TA&MG is, in general, incorrect. Proof of this opinion can be solution given in Example 3.10 presented in the paper. In this example the ILS given by (5) is to be solved.

$$x_1^* + 2x_2^* = z_1, \quad z_1 \in [-2, 5]$$
$$x_1^* - x_2^* = z_2, \quad z_2 \in [-2, 2] \tag{5}$$

Solution achieved by TA&MG is given by (6).

$$[x_1^*] = [-2, 3], [x_2^*] = \left[-\frac{4}{3}, \frac{7}{3}\right] \tag{6}$$

It is easy to check that this solution does not satisfies the ILS (5). After substituting it in (5) results shown in (7) are achieved.

$$[x_1^*] + 2[x_2^*] = \left[-4\frac{2}{3}, 7\frac{2}{3}\right] \quad \neq \quad [z_1] = [-2, 5]$$
$$[x_1^*] - [x_2^*] = \left[-4\frac{1}{3}, 4\frac{1}{3}\right] \quad \neq \quad [z_2] = [-2, 2] \tag{7}$$

The solution (6) does not satisfies also other equivalent forms of Eq. (5).

The universal, algebraic solution $[X] = ([x_1], [x_2], \ldots, [x_n])^T$ of the ILS $A[X] = [Y]$ can be determined with use of the multidimensional RDM interval arithmetic (RDM-IA), [11]. In this case model of z_1 in (5) has form $z_1 = -2 + 7\alpha_{z_1}, \alpha_{z_1} \in [0, 1]$ and of z_2 has form $z_2 = -2 + 4\alpha_{z_2}, \alpha_{z_2} \in [0, 1]$, where RDM means Relative-Distance-Measure. Then Eq. (5) can be written in new form (8).

$$x_1^* + 2x_2^* = -2 + 7\alpha_{z_1}, \quad \alpha_{z_1} \in [0, 1]$$
$$x_1^* - x_2^* = -2 + 4\alpha_{z_2}, \quad \alpha_{z_2} \in [0, 1] \tag{8}$$

Solving Eqs. (8) delivers solutions given by (9).

$$x_1^* = -2 + \frac{7}{3}\alpha_{z_1} + \frac{8}{3}\alpha_{z_2}$$
$$x_2^* = \frac{7}{3}\alpha_{z_1} - \frac{4}{3}\alpha_{z_2}, \quad \alpha_{z_1}, \alpha_{z_2} \in [0, 1] \tag{9}$$

One can easily check that the multidimensional solution (9) is the universal algebraic solution of (5), i.e. it satisfies not only ILS in the form presented by (8) but also all equivalent forms of (8). In Chap. 4 of the discussed paper TA&MG described their method of solving Fuzzy Linear System (FLS). However, this method is based on the incorrect method of solving Interval Linear Systems (ILS) described in Chap. 3, hence it also is incorrect. The best verification of a method correctness are numerical experiments showing how the method performs in concrete examples. On the end of Chap. 4 TA&MG give Example 4.9 of their method applied to solve the FLS (10), where $\tilde{y}_j$ are known fuzzy numbers and values of $\tilde{x}_i$ should be determined.

$$2\tilde{x}_1 - \tilde{x}_2 + \tilde{x}_3 = \tilde{y}_1, \quad [\tilde{y}_1]_r = [r - 2, 2 - 3r],$$
$$-\tilde{x}_1 + \tilde{x}_2 - 2\tilde{x}_3 = \tilde{y}_2, \quad [\tilde{y}_2]_r = [1 + 2r, 7 - 4r], \tag{10}$$
$$\tilde{x}_1 - 3\tilde{x}_2 + \tilde{x}_3 = \tilde{y}_3, \quad [\tilde{y}_3]_r = [r - 3, -2r].$$

According to TA&MG the unique algebraic solution of FLS (10) is given by (11), where r means membership level, $r \in [0, 1]$, and $[x_i]_r$ are triangular fuzzy numbers determined in L-R notation.

$$[x_1]_r = [r - 2, 2 - 3r],$$
$$[x_2]_r = [1 + 2r, 7 - 4r], \tag{11}$$
$$[x_3]_r = [r - 3, -2r].$$

However, substituting solutions (11) in (10) shows that they are not the unique solutions, because they do not give equality of left-hand and right-hand sides of equations, see (12).

$$
\begin{aligned}
2\tilde{x}_1 - \tilde{x}_2 + \tilde{x}_3 &= [-r, 5 - 6r] &\neq&& \tilde{y}_1 &= [r - 2, 2 - 3r], \\
-\tilde{x}_1 + \tilde{x}_2 - 2\tilde{x}_3 &= [3 - 3r, -1 - r] &\neq&& \tilde{y}_2 &= [1 + 2r, 7 - 4r], \\
\tilde{x}_1 - 3\tilde{x}_2 + \tilde{x}_3 &= [16 - 10r, 5 + r] &\neq&& \tilde{y}_3 &= [r - 3, -2r].
\end{aligned} \tag{12}
$$

The main reason of incompatibility of the results presented in (12) is the authors assumption that the main, original and direct results of operations on fuzzy numbers are also fuzzy numbers (the same mathematical objects), what is not true. The direct results are multidimensional information granules. The correct and verifiable universal algebraic solutions of FLS can be achieved with use of multidimensional fuzzy RDM arithmetic which uses special horizontal membership functions (MFs), [10,11,14–16]. This arithmetic has been successfully applied by scientists in solving various problems, see e.g. [1,2,6,8,18]. In the case of the triangle fuzzy number $X = (a, b, c)$ the horizontal MF is given by (13).

$$x = [a + (b - a)\mu] + (c - a)(1 - \mu)\alpha_x, \qquad \alpha_x \ in[0, 1] \tag{13}$$

Values of a and c mean borders of the support and b means the position of the core of FN. Formulas (14) present the horizontal form of FNs $\tilde{y}_1, \tilde{y}_2, \tilde{y}_3$ that occur in (10). They are RDM models of the true values of variables $\tilde{y}_1, \tilde{y}_2, \tilde{y}_3$.

$$
\begin{aligned}
\tilde{y}_1 &= (-2 + \mu) + 4(1 - \mu)\alpha_{y_1}, \quad \alpha_{y_1} \in [0, 1], \\
\tilde{y}_2 &= (1 + 2\mu) + 6(1 - \mu)\alpha_{y_2}, \quad \alpha_{y_2} \in [0, 1], \\
\tilde{y}_3 &= (-3 + \mu) + 3(1 - \mu)\alpha_{y_3}, \quad \alpha_{y_3} \in [0, 1].
\end{aligned} \tag{14}
$$

With use of known Cramer formulas or with the method of variables cancellation the FLS (12) can be solved. Its solutions given by (15).

$$
\begin{aligned}
x_1 &= \frac{1}{7}[(-5 + 8\mu) + (1 - \mu)(20\alpha_{y_1} + 12\alpha_{y_2} - 3\alpha_{y_3})] \\
x_2 &= \frac{1}{7}[(6 - 4\mu) + (1 - \mu)(4\alpha_{y_1} - 6\alpha_{y_2} - 9\alpha_{y_3})], \qquad \alpha_{y_1}, \alpha_{y_2}, \alpha_{y_3} \in [0, 1] \\
x_3 &= \frac{1}{7}[(2 - 13\mu) + (1 - \mu)(-8\alpha_{y_1} - 30\alpha_{y_2} - 3\alpha_{y_3})]
\end{aligned}
$$

$$\tag{15}$$

Substituting solutions (15) in the FLS (12) gives equality of left- and right-hand sides of equations. The same result is achieved in the case of all alternative, equivalent forms of the FLS (12). It means that solutions (15) are universal algebraic solutions of FLS (12). It should be noted that solutions (15) are not usual fuzzy numbers defined in 2D-space, i.e., $\mu_1 = f_1(x_1), \ldots, \mu_3 = f_3(x_3)$ as TA&MG have assumed. Solutions of the FLS (12) are functions existing in 5D-space, because $x_1 = g_1(\mu, \alpha_{y_1}, \alpha_{y_2}, \alpha_{y_3})$, similarly as x_2 and x_3. Only multi-dimensional granules can be solutions of FLSs. Such granules cannot be visualized in 2D-space. However, their low-dimensional indicators as span, cardinality distribution, center of gravity can be determined and visualized [10,11,13]. The span $s(x_i)$ can be determined from (16). The span $s(x_i)$ informs about the maximal uncertainty of the multidimensional solution x_i that cannot be seen. It gives us some low-dimensional imagination about x_i. In low-dimensional arithmetic types the span is assumed as direct result of calculation. However, it is not true.

$$s(x_i) = \left[\min_{\alpha_{y_1}, \alpha_{y_2}, \alpha_{y_3}} x_i(\mu, \alpha_{y_1}, \alpha_{y_2}, \alpha_{y_3}), \max_{\alpha_{y_1}, \alpha_{y_2}, \alpha_{y_3}} x_i(\mu, \alpha_{y_1}, \alpha_{y_2}, \alpha_{y_3}) \right], \quad (16)$$

$$\mu, \alpha_{y_1}, \alpha_{y_2}, \alpha_{y_3} \in [0, 1].$$

Spans $s(x_i)$ of particular solutions are in the case of the FLS (12) are triangular fuzzy numbers given by (17) and presented in Fig. 1.

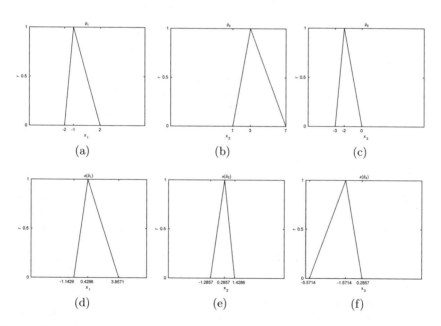

Fig. 1. Comparison of low-dimensional solution according to TA&MG (figure a, b, c) (11) and spans of multidimensional solutions $s(x_i)$ (17) (figure d, e, f).

The correctness of both methods can be checked with the method of point (crisp) solutions. E.g. for $\mu = 0, \alpha_{y_1} = \alpha_{y_2} = 1, \alpha_{y_3} = 0$ MFAr gives solution

$x_1 = 3.857, x_2 = 0.571, x_3 = 5.714$. By inserting it in FLS (10) one can check that this solutions satisfies the FLS. However, according to the TA&MG method (11) this solutions are impossible, see also Fig. 1. It shows lack of precision of the TA&MG method.

$$s(x_1) = \left[-\frac{8}{7} + \frac{11}{7}\mu, \frac{27}{7} - \frac{24}{7}\mu \right],$$

$$s(x_2) = \left[-\frac{9}{7} + \frac{11}{7}\mu, \frac{10}{7} - \frac{8}{7}\mu \right], \tag{17}$$

$$s(x_3) = \left[-\frac{39}{7} + \frac{28}{7}\mu, \frac{2}{7} - \frac{13}{7}\mu \right],$$

The spans $s(x_i)$ (17), in any case, are not solutions of FLS (10) or (12). They are only simplified 2D information pieces (indicators) about multidimensional solution granules $x_i = g_i(\mu, \alpha_{y_1}, \alpha_{y_2}, \alpha_{y_3})$. Because of this fact they should not be used in possible next calculations and formulas. These spans can also be presented in forms of triples $s(x_1) = \left(-\frac{8}{7}, \frac{3}{7}, \frac{27}{7}\right), s(x_2) = \left(-\frac{9}{7}, \frac{2}{7}, \frac{10}{7}\right), s(x_3) = \left(-\frac{39}{7}, -\frac{11}{7}, \frac{2}{7}\right)$ representing triangle fuzzy numbers.

3   Conclusion

The paper shows comparative results of application of the low-dimensional method of solving FLSs proposed by TA&MG in [3] and of multidimensional fuzzy arithmetic. Comparison of both methods has been made on concrete FLSs. It has shown that sometimes low-dimensional methods of fuzzy arithmetic deliver imprecise of fully incorrect results. Instead, multidimensional fuzzy arithmetic delivers precise result. It can be checked by point verification method or with computer simulation of possible results.

References

1. Aliev, R.: Operations on z-numbers with acceptable degree of specificity. Procedia Comput. Sci. **120**, 9–15 (2017). 9th International Conference on Theory and Application of Soft Computing, Computing with Words and Perception, ICSCCW 2017, 22–23 August 2017, Budapest, Hungary
2. Aliev, R., Huseynov, O., Aliyev, R.: A sum of a large number of z-numbers. Procedia Comput. Sci. **120**, 16–22 (2017). 9th International Conference on Theory and Application of Soft Computing, Computing with Words and Perception, ICSCCW 2017, 22–23 August 2017, Budapest, Hungary
3. Allahviranloo, T., Ghanbari, M.: On the algebraic solution of fuzzy linear systems based on interval theory. Appl. Math. Model. **36**, 5360–5379 (2012)
4. Dymova, L.: Soft Computing in Economics and Finance. Springer, Heidelberg (2011)
5. Lodwick, W.A., Dubois, D.: Interval linear systems as a necessary step in fuzzy linear systems. Fuzzy Sets Syst. **281**, 227–251 (2015). Special Issue Celebrating the 50th Anniversary of Fuzzy Sets

6. Mazandarani, M., Pariz, N., Kamyad, A.V.: Granular differentiability of fuzzy-number-valued functions. IEEE Trans. Fuzzy Syst. **26**(1), 310–323 (2018)
7. Moore, R.E., Kearfott, R.B., Cloud, M.J.: Introduction to Interval Analysis. Society for Industrial and Applied Mathematics, Philadelphia, PA, USA (2009)
8. Najariyan, M., Zhao, Y.: Fuzzy fractional quadratic regulator problem under granular fuzzy fractional derivatives. IEEE Trans. Fuzzy Syst. **PP**(99), 1–15 (2017)
9. Pedrycz, W., Skowron, A., Kreinovich, V.: Handbook of Granular Computing. Wiley-Interscience, New York (2008)
10. Piegat, A., Landowski, M.: Horizontal membership function and examples of its applications. Int. J. Fuzzy Syst. **17**(1), 22–30 (2015)
11. Piegat, A., Landowski, M.: Fuzzy arithmetic type 1 with horizontal membership functions. In: Kreinovich, V. (ed.) Uncertainty Modeling, pp. 233–250. Springer International Publishing, Cham (2017). Dedicated to Professor Boris Kovalerchuk on his Anniversary
12. Piegat, A., Landowski, M.: Is an interval the right result of arithmetic operations on intervals? Int. J. Appl. Math. Comput. Sci. **27**(3), 575–590 (2017)
13. Piegat, A., Landowski, M.: Is fuzzy number the right result of arithmetic operations on fuzzy numbers? In: Kacprzyk, J., Szmidt, E., Zadrożny, S., Atanassov, K.T., Krawczak, M. (eds.) Advances in Fuzzy Logic and Technology 2017, pp. 181–194. Springer International Publishing, Cham (2018)
14. Piegat, A., Pluciński, M.: Computing with words with the use of inverse RDM models of membership functions. Int. J. Appl. Math. Comput. Sci. **25**(3), 675–688 (2015)
15. Piegat, A., Pluciński, M.: Fuzzy number addition with the application of horizontal membership functions. Sci. World J. **2015**, 1–16 (2015)
16. Piegat, A., Pluciński, M.: Fuzzy number division and the multi-granularity phenomenon. Bull. Pol. Acad. Sci. Tech. Sci. **65**(4), 497–511 (2017)
17. Sevastjanov, P., Dymova, L.: A new method for solving interval and fuzzy equations: linear case. Inf. Sci. **179**(7), 925–937 (2009)
18. Zeinalova, M.L.: Application of RDM interval arithmetic in decision making problem under uncertainty. Procedia Comput. Sci. **120**, 788–796 (2017). 9th International Conference on Theory and Application of Soft Computing, Computing with Words and Perception, ICSCCW 2017, 22–23 August 2017, Budapest, Hungary

Processing of Z^+-numbers Using the k Nearest Neighbors Method

Marcin Pluciński$^{(\boxtimes)}$ iD

Faculty of Computer Science and Information Technology,
West Pomeranian University of Technology, Żołnierska 49, 71-210 Szczecin, Poland
mplucinski@wi.zut.edu.pl

Abstract. The paper presents that with the application of Z^+-numbers arithmetic, the k nearest neighbors method can be adapted to various types of data. Both, the learning data and the input data may be in the form of the crisp number, interval, fuzzy or Z^+-number. The paper discusses the methods of performing arithmetic operations on uncertain data of various types and explains how to use them in the kNN method. Experiments show that the method works correctly and gives credible results.

Keywords: Z^+ numbers arithmetic · Fuzzy numbers arithmetic
k nearest neighbors method

1 Introduction

In today's world, we perceive and process huge amounts of information of various types. A part of it is determined with absolute precision. However, most of it is information that is uncertain, imprecise or incomplete. Humans have a great capability to make rational decisions based on such information [1]. For this reason, there is a need to develop such data processing methods that will cope with uncertainty of various types. An example of such solution may be the k nearest neighbors method. It can be adapted to work with information that has various levels of uncertainty as: intervals (level 1), fuzzy or random numbers (level 2) and Z or Z^+-numbers (level 3).

The k-nearest neighbors (kNN) method belongs to the memory based approximation methods. It is one of the most important between them and probably one of the best described in many versions [2–4], but what is significant it is still the subject of new researches [5–8]. Other popular memory based techniques are methods based on locally weighted learning [2,3] which use various ways of samples weighting.

Thanks to the different kinds of arithmetics (interval arithmetic, fuzzy number arithmetic, random numbers arithmetic, Z and Z^+-numbers arithmetic) described further on, the kNN method can be applied to various and mixed types of data. Both, the learning data and the input data may be in the form of the crisp number or uncertain (interval, fuzzy, Z or Z^+) number. Exemplary results of work with such data are presented in subsequent sections.

© Springer Nature Switzerland AG 2019
J. Pejaś et al. (Eds.): ACS 2018, AISC 889, pp. 76–85, 2019.
https://doi.org/10.1007/978-3-030-03314-9_7

2 Z-numbers and Z^+-numbers

A Z-number can be defined as an ordered pair: $Z = (A, B)$, where A is a fuzzy number playing a role of a fuzzy restriction on values that a random variable X may take (X is A) and B is a fuzzy number playing a role of a fuzzy restriction on the probability measure of A ($P(A)$ is B) [1,9].

With help of Z-numbers, sentences expressed in natural language can be described in a convenient, structured way. For example the sentence: 'the probability that the unemployment rate will be small next year is high' can be represented in the form:

X = 'the unemployment rate next year' is Z = ('small', 'high').

A and B are possibilistic restrictions applied to the variable X and its probability.

A Z^+-number is a pair consisting of a fuzzy number A and a random number p_X: $Z^+ = (A, p_X)$ where A plays the same role as above and p_x is the probability distribution of random variable X. By definition, the Z^+-number carries more information than the Z-number [1,9]. First of all, the exact probability value $P(A)$ can be calculated as:

$$P(A) = \int_{\text{supp}(A)} \mu_A(u) \cdot p_X(u) du, \qquad (1)$$

where: $\mu_A(u)$ – is a membership function of the fuzzy number A and supp(A) means its support.

2.1 Z^+-numbers Arithmetic

Let's assume that $*$ is a binary operation and its operands are Z^+-numbers: $Z_X^+ = (A_X, p_X)$ and $Z_Y^+ = (A_Y, p_Y)$. By definition [9]:

$$Z_X^+ * Z_Y^+ = (A_X * A_Y, p_X * p_Y), \qquad (2)$$

and of course the operation is realized in different way for fuzzy numbers: $A_X * A_Y$ and probability distributions: $p_X * p_Y$.

Fuzzy Numbers Arithmetic. The main concepts connected with fuzzy numbers (FN) are well described in many literature positions, e.g. in [10–12]. Let's recall some basic definitions.

The fuzzy subset of the real numbers set $\mathbb{R}$, with the membership function $\mu : \mathbb{R} \to [0, 1]$, is a fuzzy number if:

(a) A is normal, i.e. there exists an element $x_0 \in \mathbb{R}$ such that $\mu(x_0) = 1$;
(b) A is convex, i.e. $\mu(\lambda x + (1 - \lambda)y) \geq \mu(x) \wedge \mu(y)$, $\forall x, y \in \mathbb{R}$ and $\forall 0 \leq \lambda \leq 1$;
(c) μ is upper semicontinuous;
(d) supp(μ) is bounded.

Each fuzzy number can be described as:

$$\mu(x) = \begin{cases} 0 & \text{for } x < a_1 \\ f(x) & \text{for } a_1 \le x < a_2 \\ 1 & \text{for } a_2 \le x < a_3 \\ g(x) & \text{for } a_3 \le x < a_4 \\ 0 & \text{for } x \ge a_4 \end{cases} \qquad (3)$$

where: $a_1, a_2, a_3, a_4 \in \mathbb{R}$. f is a nondecreasing function and is called the left side of the fuzzy number. g is a nonincreasing function and is called the right side of the fuzzy number.

The next important concept are α-levels of the fuzzy set. The α-level set A_α of the fuzzy number A is a nonfuzzy set defined by:

$$A_\alpha = \{x \in \mathbb{R} : \mu(x) \ge \alpha\}. \qquad (4)$$

The family $\{A_\alpha : \alpha \in (0,1]\}$ can be a representation of the fuzzy number.

From the definition of the fuzzy number results that α-level set is compact for each $\alpha > 0$. As a consequence, each A_α can be represented by an interval:

$$A_\alpha = [f^{-1}(\alpha), g^{-1}(\alpha)], \qquad (5)$$

where: $f^{-1} = \inf\{x : \mu(x) \ge \alpha\}$ and $g^{-1} = \sup\{x : \mu(x) \ge \alpha\}$.

If A_α is the α-level set of the fuzzy number A, then it can be represented in the form:

$$A = \bigcup_{\alpha \in [0,1]} \alpha, A_\alpha. \qquad (6)$$

Each α-level set is an interval, so rules of interval arithmetic [13] can be applied in formulation of basic arithmetic operations of fuzzy numbers. If we have two interval numbers $[a_1, a_2]$ and $[b_1, b_2]$ then:

$$[a_1, a_2] \oplus [b_1, b_2] = [a_1 \oplus b_1, a_2 \oplus b_2], \qquad (7)$$

$$[a_1, a_2] \otimes [b_1, b_2] = [\min(a_1 \otimes b_1, a_1 \otimes b_2, a_2 \otimes b_1, a_2 \otimes b_2), \\ \max(a_1 \otimes b_1, a_1 \otimes b_2, a_2 \otimes b_1, a_2 \otimes b_2)], \qquad (8)$$

where: $\oplus \in \{+, -\}$, $\otimes \in \{\times, \div\}$ and $0 \notin [b_1, b_2]$ if $\otimes = \div$.

Above interval operations can be extended to fuzzy numbers [10, 14–16]. Let:

$$A = \bigcup_{\alpha \in [0,1]} \alpha, [a_1^\alpha, a_2^\alpha] \quad \text{and} \quad B = \bigcup_{\alpha \in [0,1]} \alpha, [b_1^\alpha, b_2^\alpha],$$

be two fuzzy numbers, then:

$$A \circ B = \bigcup_{\alpha \in [0,1]} \alpha, ([a_1^\alpha, a_2^\alpha] \circ [b_1^\alpha, b_2^\alpha]), \qquad (9)$$

where: $\circ = \{+, -, \times, \div\}$.

Random Numbers Arithmetic. Let p_X and p_Y be probability density functions of two independent random variables. Distributions resulting from arithmetic operations on such variables can be calculated as [17,18]:

$$p_{X+Y}(u) = \int_{-\infty}^{\infty} p_X(v) \cdot p_Y(u-v)\, dv\,,$$

$$p_{X-Y}(u) = \int_{-\infty}^{\infty} p_X(v) \cdot p_Y(v-u)\, dv\,,$$

$$p_{X \cdot Y}(u) = \int_{-\infty}^{\infty} p_X(v) \cdot p_Y(u/v) \cdot \frac{1}{|v|}\, dv\,,$$

$$p_{X/Y}(u) = \int_{-\infty}^{\infty} p_X(u \cdot v) \cdot p_Y(v) \cdot |v|\, dv\,. \tag{10}$$

2.2 Distance Between Z^+-numbers

A distance between Z^+-numbers can be calculated as [9]:

$$d(Z_1^+, Z_2^+) = d_{FN}(A_1, A_2) + d_P(p_1, p_2)\,, \tag{11}$$

where: $d_{FN}(A_1, A_2)$ – is the distance between fuzzy numbers A_1 and A_2, $d_P(p_1, p_2)$ – is the distance between random numbers described by their distributions p_1 and p_2.

Fuzzy numbers do not form a natural linear order, like e.g. real numbers, so different approaches are necessary for calculating the distance between them. Many methods have been described in the literature [11,19,20]. Each one has its own advantages and disadvantages, so it is hard to decide which one is the best. In this paper, methods proposed in [11] will be applied.

The distance, indexed by parameters $p \in [1, \infty)$, $q \in [0, 1]$, between fuzzy numbers A and B can be calculated as:

$$d_{FN}(A, B) = \begin{cases} \sqrt[p]{(1-q) \int_0^1 |f_B^{-1}(\alpha) - f_A^{-1}(\alpha)|^p\, d\alpha + q \int_0^1 |g_B^{-1}(\alpha) - g_A^{-1}(\alpha)|^p\, d\alpha} \\ \hspace{8cm} \text{for } 1 \le p < \infty \\[2ex] (1-q) \sup_{0<\alpha\le1} (|f_B^{-1}(\alpha) - f_A^{-1}(\alpha)|) + q \sup_{0<\alpha\le1} (|g_B^{-1}(\alpha) - g_A^{-1}(\alpha)|) \\ \hspace{8cm} \text{for } p = \infty \end{cases} \tag{12}$$

where: $A_\alpha = [f_A^{-1}(\alpha), g_A^{-1}(\alpha)]$ and $B_\alpha = [f_B^{-1}(\alpha), g_B^{-1}(\alpha)]$. The parameter q characterizes the weights connected with sides of fuzzy numbers. If there is no reason to distinguish any side, $q = 0.5$ is recommended.

If we assume that all fuzzy numbers are elements of the space $F(\mathbb{R})$ then it can be proved that $(F(\mathbb{R}), d_{FN})$ is a metric space [11].

The distance between random numbers can be also calculated in many ways [21]. One of them base on engineer's metric (EM). Such distance can be calculated as:

$$d_p(p_1, p_2) = |E(p_1) - E(p_2)|, \tag{13}$$

where: $E(p)$ – is expected value of the distribution p. EM belongs to the group of primary metrics (identity property is fulfilled only for certain features of distributions – here expected values).

3   The k-nearest Neighbors Method

The kNN method realizes a local regression. It means that the output for the considered input point $\mathbf{x}^*$ is calculated on the base of a local model created only for k samples nearest (in a meaning of an applied metric) to $\mathbf{x}^*$ [8].

In the classic kNN method, the model output is calculated as a mean value of target values of k neighbor samples. It can be also calculated as the weighted mean value and in such case, weight values usually depend on a distance $d(\mathbf{x}^*, \mathbf{x})$ between the input point $\mathbf{x}^*$ and analyzed neighbors $\mathbf{x}$, for example:

$$w_{\mathbf{x}^*, \mathbf{x}} = \frac{1}{1 + m \cdot d(\mathbf{x}^*, \mathbf{x})/k^2}, \tag{14}$$

where: the m parameter is determined empirically.

The main parameter of the kNN method is the number of neighbors k that are used in calculations. It can be constant for entire data set, but it can be also dynamically varied – according to the input point location in the input space. The most popular techniques of k evaluation are applying crossvalidation or applying two distinct data sets: training data – that are memorized by the model, and testing data – to evaluate the real model error. The best k value is the value that gives the lowest test or crossvalidation error and in this way it guarantees the lowest real error of the model and the best generalization.

Thanks to the arithmetics described in previous section, the kNN method can be adapted to various types of data. Both, the learning data and the input data may be in the form of the crisp number, interval, fuzzy number or Z^+-number. However, to be able to calculate the distance between the input vector and the samples and to calculate the output, it is necessary to unify all data. Each attribute in the training data set and in the input vector must be represented by the Z^+number. Crisp numbers can be replaced by fuzzy numbers with a singleton membership function and interval numbers – by fuzzy numbers with a rectangular membership function. Because for such numbers, the probability distribution is not explicitly specified, it will be always assumed that it is uniformly distributed on the range from minimum to maximum of each input data attribute.

4 Results of Experiments

All experiments were realized with use of a specially prepared Python library that allows calculations on interval, fuzzy and Z^+-numbers. Calculations on random numbers (probability distributions) were realized with use of PaCAL library [18], freely available for download at: http://pacal.sourceforge.net/.

4.1 Experiment 1 – SISO Model

The objective of first experiments was to determine whether the method works correctly and its results are credible. For this reason, the research was carried out on training data with only one input and one output attribute. Data were prepared to be diverse as much as possible, thus both attributes were Z^+-numbers and the value restriction had a form of crisp numbers, interval and fuzzy numbers with uniform probability distribution. Data are presented in Table 1 and in Fig. 1.

Table 1. Learning data used in experiment 1

Input x with uniform distribution $U(0, 10)$	Output y (target value) with uniform distribution $U(-4, 4)$
0	0
1	1
[2, 2.5]	[2.5, 4]
3	Triangle FN [1, 1.5, 2]
Triangle FN [3.5, 4, 4.5]	Triangle FN [−0.5, 0, 0.5]
Triangle FN [5, 5.5, 6]	[−0.5, 0]
6	−1
[7, 7.5]	Triangle FN [−2.5, −2, −1.5]
8	Trapezoid FN [−4, −4, −3.5, −3]
Trapezoid FN [8, 8.5, 9, 9.5]	[−1.5, −1]
10	0

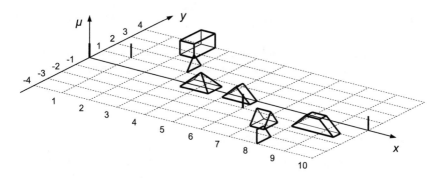

Fig. 1. View of the learning data used in experiment 1

In experiments, a distance measure d_{FN} described by formula (12) was applied with parameters $p = 2$ and $q = 0.5$. Calculations were carried out for input vectors in the form of Z^+-numbers with the value restriction in the form of a crisp number, interval or fuzzy number and with the uniform probability distribution $U(0, 10)$.

Some exemplary results are presented in Figs. 2 and 3. All calculations were performed for the number of neighbors $k = 3$. Results obtained for the kNN method and its weighted version are similar, but the weighted kNN method is more sensitive to the input value $\mathbf{x}^*$.

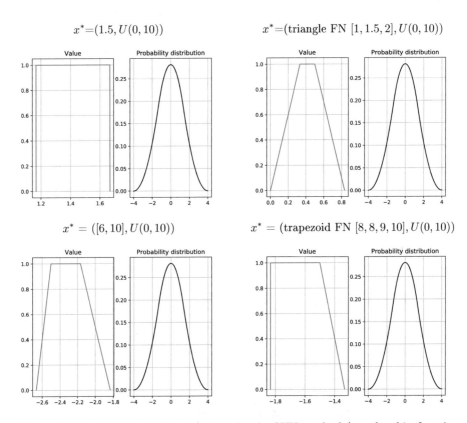

Fig. 2. Exemplary results of calculations for the kNN method (membership function of the value restriction of the output and its probability distribution)

4.2   Experiment 2 – Crisp Learning Data

The purpose of the second experiment was to show that it is possible to process Z^+-numbers, also in a situation in which learning data has a form of crisp

$x^*=(1.5, U(0, 10))$ $x^*=(\text{triangle FN } [1, 1.5, 2], U(0, 10))$

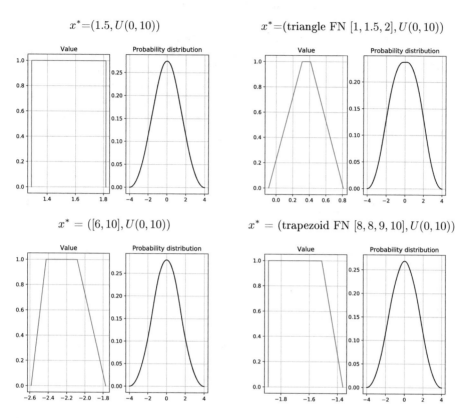

$x^* = ([6, 10], U(0, 10))$ $x^* = (\text{trapezoid FN } [8, 8, 9, 10], U(0, 10))$

Fig. 3. Exemplary results of calculations for the weighted kNN method (membership function of the value restriction of the output and its probability distribution)

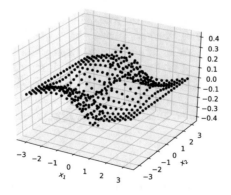

Fig. 4. View of the learning data used in experiment 2

numbers. Learning data are presented in Fig. 4 – each sample has 2 inputs and one output (target value).

Because we determined the model's response for inputs in the form of Z^+-numbers, it was necessary to unify the data. All attributes of the learning data (both input and output) was treated as a Z^+-number with the value restriction in the form of singleton number and the probability uniformly distributed.

Figure 5 presents results of some exemplary calculations. Because the model's response is determined as a weighted mean of target values of k nearest neighbors it has the form of singleton fuzzy number (the same as the learning data).

$$x_1^* = (\text{triangle FN } [0.7, 0.8, 0.9], U(-\pi, \pi))$$
$$x_2^* = (\text{triangle FN } [0.1, 0.2, 0.3], U(-\pi, \pi))$$

$$x_1^* = ([-1.5, -1], U(-\pi, \pi))$$
$$x_2^* = (\text{trapezoid FN } [-0.5, -0.2, 0, 0], U(-\pi, \pi))$$

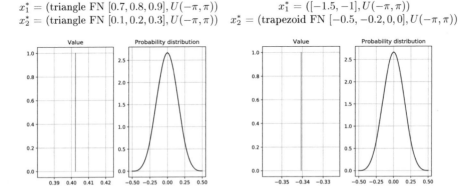

Fig. 5. Exemplary results of calculations for the weighted kNN method ($k = 4$) in experiment 2

5   Conclusions

One of the greatest advantages of the kNN method (as well as other local modeling methods) is its 'flexibility' in data processing. This is largely due to the simplicity of the calculations performed during the determination of the response. Thanks to the use of different kinds of arithmetic, the method can be easily adapted to the processing of various data: crisp or imprecise (interval, fuzzy and Z^+-numbers). Experiments described in the paper showed that the method gives correct and credible results. So in the case of data uncertainty, memory based methods can be an attractive approach for the approximation task.

References

1. Zadeh, L.A.: A note on Z-numbers. Inf. Sci. **181**, 2923–2932 (2011)
2. Atkeson, C.G., Moore, A.W., Schaal, S.A.: Locally weighted learning. Artif. Intell. Rev. **11**, 11–73 (1997)
3. Cichosz, P.: Learning Systems. WNT Publishing House, Warsaw (2000). [in Polish]

4. Hand, D., Mannila, H., Smyth, P.: Principles of Data Mining. The MIT Press, Cambridge (2001)
5. Kordos, M., Blachnik, M., Strzempa, D.: Do we need whatever more than k-NN? In: Rutkowski, L., Scherer, R., Tadeusiewicz, R., Zadeh, L.A., Zurada, J.M. (eds.) ICAISC 2010. LNCS, vol. 6113, pp. 414–421. Springer, Heidelberg (2010)
6. Korzeń, M., Klęsk, P.: Sets of approximating functions with finite Vapnik-Czervonenkis dimension for nearest-neighbours algorithm. Pattern Recogn. Lett. **32**, 1882–1893 (2011)
7. Pluciński, M.: Application of the information-gap theory for evaluation of nearest neighbours method robustness to data uncertainty. Przegląd Elektrotechniczny **88**(10b), 272–275 (2012)
8. Pluciński, M., Pietrzykowski, M.: Application of the k nearest neighbors method to fuzzy data processing. Przegląd Elektrotechniczny **93**(1), 77–81 (2017)
9. Aliev, R.A., Huseynov, O.H., Aliyev, R.R., Alizadeh, A.A.: The Arithmetic of Z-Numbers: Theory and Applications. World Scientific, Singapore (2015)
10. Dubois, D., Prade, H.: Operations on fuzzy numbers. Int. J. Syst. Sci. **9**(6), 613–626 (1978)
11. Grzegorzewski, P.: Metrics and orders in space of fuzzy numbers. Fuzzy Sets Syst. **97**, 83–94 (1998)
12. Piegat, A.: Fuzzy Modeling and Control. Physica, Heidelberg (2001)
13. Moore, R.E., Kearfott, R.B., Cloud, M.J.: Introduction to Interval Analysis. Society for Industrial and Applied Mathematics, Philadelphia (2009)
14. Dutta, P., Boruah, H., Ali, T.: Fuzzy arithmetic with and without using α-cut method: a comparative study. Int. J. Latest Trends Comput. **2**(1), 99–107 (2011)
15. Hanss, M.: Applied Fuzzy Arithmetic. Springer, Heidelberg (2005)
16. Kaufmann, A., Gupta, M.M.: Introduction to Fuzzy Arithmetic. Van Nostrand Reinhold, New York (1991)
17. Springer, M.D.: The Algebra of Random Variables. John Wiley & Sons, New York (1979)
18. Jaroszewicz, S., Korzeń, M.: Arithmetic operations on independent random variables: a numerical approach. SIAM J. Sci. Comput. **34**(3), 1251–1265 (2012)
19. Diamond, P., Rosenfeld, A.: Metric spaces of fuzzy sets. Fuzzy Sets Syst. **35**, 241–249 (1990)
20. Tang, W., Li, X., Zhao, R.: Metric spaces of fuzzy variables. Comput. Ind. Eng. **57**, 1268–1273 (2009)
21. Rachev S.T., Klebanov L., Stoyanov S.V., Fabozzi F.: The Methods of Distances in the Theory of Probability and Statistics. Springer Science and Business Media (2013)

Fingerprint Feature Extraction with Artificial Neural Network and Image Processing Methods

Maciej Szymkowski$^{(\boxtimes)}$ and Khalid Saeed

Faculty of Computer Science,
Bialystok University of Technology, Bialystok, Poland
szymkowskimack@gmail.com, k.saeed@pb.edu.pl

Abstract. Fingerprints are often claimed as the safest measurable human trait. As one can observe they are commonly used in many different solutions. Nowadays they are applied for instance in financial institutions where clients can confirm their identities with fingerprints. These biometrics are the kind of the password that no one can lost or forget. An approach to extract fingerprints features with artificial neural network and image processing algorithms is presented in this work. Soft computing methods are becoming more popular. It leads to their usage in the human recognition procedures. In the case of the algorithm presented in this paper, the results from neural network are confirmed by Crossing-Number (CN) algorithm. The two-step confirmation ensures more precise results than obtained with each of them separately. The final stage of the algorithm is minutiae extraction and marking them in the analyzed image.

Keywords: Fingerprint · Biometrics · Physiological biometrics
Artificial neural network · Minutiae detection

1 Introduction

In the ancient times, comparison of the fingerprints was one of the ways in which human identity can be confirmed. Historians point out that this measurable human trait was used during trades on the market. What is more the first signs of using fingerprints for trading come from ancient Babylon and ancient China. During the transaction, fingerprints were imprinted on the clay tablets. It is how the trade was confirmed by the both sides. We can notice that fingerprints were used as a unique signature.

Nowadays they are also used in the human identification. Of course methods that are practiced today are much more advanced. We have different scanners, algorithms and their implementations that enable us to simplify the human recognition process. One of the trends in the fingerprint recognition solutions is soft computing algorithms usage. In the literature there are multiple different methods like: evolutionary algorithms, artificial neural networks or deep learning that are applied in the biometrics solutions.

© Springer Nature Switzerland AG 2019
J. Pejaś et al. (Eds.): ACS 2018, AISC 889, pp. 86–97, 2019.
https://doi.org/10.1007/978-3-030-03314-9_8

Fingerprints like other measurable human traits have their advantages and drawbacks. Their main benefits are connected with ease of retrieval and uniqueness. On the other hand disadvantages are mostly fused with the simplicity of their spoofing. In the different technical articles we can find descriptions of how to reproduce one's fingerprint. Moreover authors of the mentioned articles also provided results of their experiments that were set to deceive biometrics system. In the most of the cases hacker was identified as a man who is registered in the system database. Logical conclusion connected with this fact is that improvement of fingerprint software and hardware is needed. These changes are not needed in the case of fingerprint quality nonetheless they are to provide higher accuracy of the biometrics system.

In the program that is the significant part of this paper, an approach to describe fingerprint with artificial neural network and image processing algorithms was implemented. Authors have used soft computing method due to the fact that it can assure high accuracy level and is not computationally demanding. Moreover it makes the algorithm more sensitize to different little changes. This paper focuses on fingerprints because their advantages are still overbalance drawbacks. It should be noticed that every man have different fingerprints. It is truth also in the case of twins. The goal is to prepare a method that despite difficulties connected with spoofing will make the right decision only in the case of the users registered in the database.

2 Literature Review

In the literature there are multiple different approaches connected with the fingerprints. To present them, the authors of this paper decided to describe some of the recent articles.

In [1] the Authors claimed that different pieces of information about a fingerprint image, such as ridge orientation and frequency must be considered. Moreover their solution is based on numerous papers about fingerprint processing and feature extraction. Feature vector that is a result of the Authors algorithm contains information about minutiae-type, location and its orientation. In the first step ("Image analysis") of the proposed approach a fingerprint image is analyzed. In this stage, initial segmentation based on simple statistical functions: mean value and variation is performed. What is more an orientation matrix is computed using Sobel operator. After both these operations core and deltas are detected. As the next block of algorithms, image filtering with Gabor filter is done. When it is finished the minutiae can be localized. It is fulfilled with threshold-based binarization, thinning algorithm that is KMM [2] and Crossing-Number (CN) algorithm. At the end of the whole procedure "Final processing" is performed. Segmentation, spurious minutiae removing and feature vector generation are performed in this stage. In the paper only fingerprint processing and feature vector generation algorithms are presented. The Authors do not describe any information about classification and its accuracy.

The another interesting approach was presented in [3]. In the case of this solution the image is represented in bit-plane technique. The Authors claimed that this image type requires less storage space than a grayscale image. At the beginning the Region of Interest (ROI) is extracted using blob analysis. Then the ROI core point is located for dimension reduction process. The next step is image enhancement with Fourier transform. After this operation bit-plane images are extracted. When all images are available fingerprint matching with phase-only correlation (POC) function is performed. In the article information about implemented algorithm accuracy is also presented. The Authors claimed that their solution achieved 81.16% recognition rate on FVC2002-Db1a database and 89.78% on FingerDOS database. This approach can be classified as an interesting idea for fingerprint feature extraction because it does not base on minutiae that is a kind of a standard in fingerprint feature vectors.

In [4] simple survey about fingerprints was presented. The Authors focused on the different characteristics connected with this traits algorithms. They concluded that the most commonly used in fingerprint matching are minutiae-based feature vectors. Moreover their study provided the main blocks of fingerprint processing algorithms. As the first of them, the Authors pointed a binarization. In their work, the whole process that take place in this step was described although there are not included any information about advantages of this operation. Then as the next block thresholding is considered. At the end of the whole procedure, thinning is performed. The Authors observed that after this operation not only real but also false minutiae are still in the picture. Nonetheless in the paper there is no information how to remove spurious elements.

The Authors of [5] took into consideration not only fingerprint processing software but also hardware implementation of the available algorithms. In this paper, fingerprint minutiae extraction algorithm based-on Crossing Number (CN) method was implemented with FPGA device. At the beginning the Authors prepared fingerprint extraction and matching algorithms as a simple software that can be run on every PC then when it worked as expected, implementation for FPGA device was prepared. In the paper two main blocks of the algorithms were studied. The first of them was preprocessing consists of image enhancement with Gabor filters, binarization and thinning. The second one was connected with feature extraction. In this stage following procedures were implemented and accomplished: minutiae points detection with Crossing Number algorithm and minutiae points parameters calculation. The parameters were calculated on the basis of the minutiae angles. In the article FPGA hardware implementation is also described. The system consists of a Fujitsu MBF200 fingerprint sensor connected to the FPGA board. The Authors also claimed that their solution has 14.05% of EER with the computational time that is 18 s.

In the literature one can easily find diversified articles connected with fingerprint but do not present any novel approach. These papers refer to the review of the existing methods. During our study we also found a few worth-reading examples [6–9].

It also should be pointed out that nowadays a fingerprint can also be used to identify a user in the mobile devices, for example smartphones. The most popular technique is Apple Touch ID [10]. An interesting approach was presented in [11]. In the case of the solution presented in this paper, the Authors implemented their own program for Android operation system. They used simple fingerprint matching algorithm for user validation. In the paper one can read full program description even with screenshots and pieces of the code implementation although there is no information about the accuracy. Similar approach can be find in [12]. In this paper the Authors also prepared fingerprint validation program for Android operating system and they also do not present any information about the accuracy of their approach.

Sometimes a fingerprint is not the only one human trait that is taken into consideration in the biometrics system. The systems that take more than one feature as an input are known under "multimodal systems" [13] name. In [14] the Authors proposed to identify a man on the basis of his fingerprint and face. For fingerprints analysis, simple processing algorithm consists of six steps was created. The Authors used also classification technique that is k-Nearest Neighbors algorithm. Their approach has about 60% accuracy level when fingerprint was analyzed alone. When it was combined with face, recognized by Eigenfaces algorithm, the solution accuracy has grown to 81%. It shows that combination of two traits can also provide satisfactory accuracy level for humans recognition.

The another articles type that can be find in the literature refers to feature vector construction. An interesting approach was presented in [15]. In this paper, the Authors created novel solution connected with the fingerprint description. At the beginning the image was enhanced with manual thresholding binarization algorithm. When it was finished median filtering was performed. As the third step of the solution, thinning performed with KMM algorithm was done. At the end of the whole procedure minutiae detection with Crossing Number (CN) algorithm was done. The feature vector was constructed on the basis of image division into separate blocks. In each block the number of the detected minutiae was calculated. The n-th number in the feature vector consists of sum of $n–1$ value and the number of the detected minutiae in n-th block. The Authors claimed that their solution has 100% accuracy level for specific algorithm parameters.

3   Methodology

The authors have already published articles in the field of the image processing [14–17]. In this chapter we present the complete algorithm that was worked out. The proposed approach is based on experiences gained during previous works and it is a continuation of our work introduced in [15].

The first step of the proposed system is fingerprint image preprocessing that starts with an image cut. The borders by which sample is cut are designated by the first pixels of fingerprint edges. The original image and its form after cut are presented in Fig. 1.

The second step of the proposed approach is binarization. In the case of this system, the authors have considered different algorithms: Manual thresholding, Otsu algorithm and Niblack method. A comparison between the obtained results is presented in Fig. 2. It should be pointed out that in this case binarization is an image segmentation method. It means that black pixels will represent fingerprint and white ones correspond to the background. The best results were obtained with Otsu method. By this algorithm all elements of the fingerprint are clearly visible. Selected method allows to avoid inaccurate representation of a fingerprint. However, a few elements that are not a part of a fingerprint have black color.

The third step is connected with the image enhancement. It means that all spurious elements visible after binarization have to be removed in this step. The best results were obtained with multiple median filtering. The results are presented in Fig. 3.

(a)   (b)

Fig. 1. Original fingerprint image (a) and its form after cut (b)

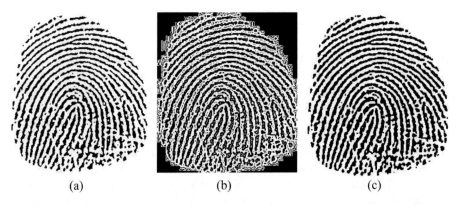

(a)   (b)   (c)

Fig. 2. Fingerprint image binarized with: manual thresholding (a), Niblack algorithm (b) and Otsu method (c).

In Fig. 2 manual binarization was done with threshold set to 68. It could be claimed that manual thresholding can be better option because it can be customized to the proper purpose although it also have to be pointed out that Otsu algorithm is an automatic method that gives better results and it is self-adaptive to the input image. The self-adaptation cannot be observe in manual binarization method.

(a)   (b)

Fig. 3. Image after binarization (a) and after median filtering (b)

The another step is connected with a data reduction. It is performed due to the information redundancy caused by too wide fingerprint edges. The operation that reduce fingerprint to 1-pixel wide lines is called thinning. The authors have considered two algorithms: KMM and K3M [18, 19]. K3M is an extension of KMM method. Both of them were analyzed because of low computational complexity and low processing time. The comparison between results obtained with both methods is presented in Fig. 4. The authors have decided to use K3M algorithm because it returned much better results. In this case it means that fingerprint lines were not connected and all details are clearly separable.

3.1 Feature Extraction

The second part of the algorithm is connected with fingerprint feature extraction and feature vector creation. The authors decided to use a neural network combined with Crossing Number (CN) algorithm. This two-step extractor provides more accurate results than each of the methods separately. The first step of the algorithm bases on a neural network. The image is divided into 3 × 3 squares and then each of the separated elements is an input for the neural network. When information is given, it is processed and decision about minutiae type can be obtained (000 – means that there is no minutiae, 010 – ridge ending, 001 – ridge bifurcation). Selected soft computing method was configured to detect minutiae although it also returned additional points that should not be selected. When this operation is finished, Crossing Number algorithm is performed for each point indicated by the neural network. It enables to remove spurious minutiae with low computational complexity. The detection final result is presented in Fig. 5.

(a)

(b)

(c)

Fig. 4. Enhanced image (a) and image after thinning: KMM (b), K3M (c)

The crucial part of the algorithm was the configuration of an artificial neural network. In the case of this algorithm the authors created 4-layers neural network consists of one input, one output and two hidden layers. Each of them has 9, 25, 25 and 3 neurons respectively. The authors considered two main minutiae types: ridge endings and ridge bifurcations. The output can return three values because it is significant to know whether an analyzed point is a minutiae or not.

The significant decision had also been made in the case of activation function for each layer. The authors tested different possible configurations although the most accurate results were obtained when input and output layer was activated with sigmoid function (1) as well as tanh (2) function for hidden layers. The neural network configuration is presented in Fig. 6. In the proposed neural network $x_1 \ldots x_9$ – are each pixel value from the input square and y_1, y_2, y_3 are the outputs that indicate minutiae type.

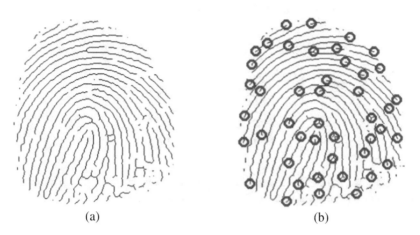

(a)   (b)

Fig. 5. Image after thinning with K3M algorithm (a) and detected minutiae (b)

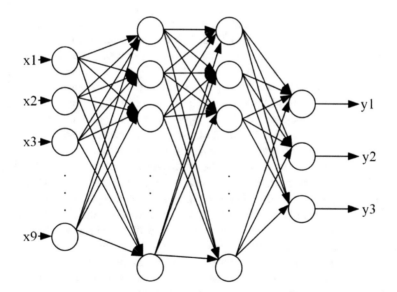

Fig. 6. Configuration of the used neural network for minutiae detection

After the results were returned by the neural network, Crossing Number algorithm was used to confirm minutiae. This value is calculated as in (3). The computed CN number enables easy and fast classification of each pixel. When CN = 0 it means that the analyzed point is not a minutiae, CN = 1 pixel is a ridge terminal, CN = 2 means that the pixel is a ridge continuation, when CN = 3 the pixel is a ridge bifurcation.

$$f(x) = \frac{1}{1 + e^{-x}} \tag{1}$$

$$g(x) = \frac{e^x - e^{-x}}{e^x + e^{-x}} \tag{2}$$

$$CN = \frac{1}{2} \cdot \sum_{i=1}^{8} |P_i - P_{i-1}| \tag{3}$$

As the last step of the approach, false minutiae removal was performed. It was based on the distance between found minutiae described in [1].

4   Experiments

The significant part of the work is connected with the performed experiments. The first of them was connected with the configuration of the neural network. The authors took into consideration different numbers of hidden layers and neurons in each of them. The most precise results were obtained when the neural network was configured as in Fig. 6. The authors also observed that when a neural network consists of one hidden layer only, too much points were accepted as a minutiae, in the case of more hidden layers than two only a part of the real minutiae were selected. During experiments also different numbers of neurons in input layer were taken into consideration. The experiments were performed for 9 and 25 input neurons. Nonetheless when the neural network has 25 inputs, too much false minutiae were selected.

During our study we used two different fingerprint scanners: U.Are.U 5160 and Futronic FS80. The first of them can provide 1000 dpi images whilst the second one returns only images in 500 dpi. The comparison between input images of the same fingerprint are present in Fig. 7. The authors used exactly the same algorithm described in this paper to process obtained images. We would like to check whether one algorithm can be used for images from different scanners also whether image resolution can have crucial impact on the processed image quality. The results of the comparison are presented in Fig. 8. The conclusion the from conducted experiments is that image quality has huge impact on the next algorithm processing steps. In the case of Futronic FS80 special version of the algorithm has to be created. The authors algorithm is a good way to preprocess high quality images.

Fig. 7. Fingerprint image obtained with: U.Are.U 5160 (a), Futronic FS80 (b)

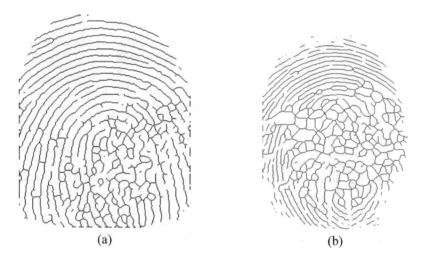

Fig. 8. Results of processing algorithm: U.Are.U 5160 (a), Futronic FS80 (b)

5 Conclusions and Future Work

In the last few years it can be observed that fingerprint is one of the most popular trait and it is often used in the commercially available projects. It is widely implemented due to the ease of retrieval and high reliability. Another advantage of this measurable human trait is that it can be fast obtained by proper scanners. Nowadays fingerprint devices are used not only in the financial institutions but also in everyday use devices like smartphones.

In this work a novel approach to fingerprint processing was presented. It was implemented in Java Programming Language and tested in the real development environment. The whole algorithm was tested on more than 50 samples. Fingerprint images were obtained with two different scanners: U.Are.U 5160 and Futronic FS80. It should be claimed that image quality has crucial impact on the image processing algorithm. If an image is in lower quality then the algorithm needs additional steps that can enhance sample. The implemented solution gives user a possibility to display each stage of the image processing algorithm. The final result of the proposed solution is an image with the marked minutiae. The feature vector can be easily created from information presented on an image.

In the article the authors do not present any results connected with the proposed solution accuracy. This value will be measured in the future when the authors database will be expanded with additional samples. The soft computing method used in the presented algorithm can provide satisfactory results although two-step extractor assures much more accurate minutiae detection. It also shows that artificial neural network can be effectively used for image processing and feature extraction.

The authors' current work is to improve the image quality for much better final output image. What is more we are working under fingerprint classification with convolutional neural network.

Acknowledgment. This work was supported by grant S/WI/3/2018 from Białystok University of Technology and funded with resources for research by the Ministry of Science and Higher Education in Poland.

References

1. Surmacz, K., Saeed, K., Rapta, P.: An improved algorithm for feature extraction from a fingerprint fuzzy image. Optica Applicata **43**(3), 515–527 (2013)
2. Saeed, K., Rybnik, M., Tabędzki, M.: Implementation and Advanced Results on the Non-interrupted Skeletonization Algorithm, Computer Analysis of Images and Patterns, LNCS, vol. 2124, pp. 601–609 (2001)
3. Francis-Lothai, F., Bong, D.B.L.: A fingerprint matching algorithm using bit-plane extraction method with phase-only correlation. Int. J. Biometrics **9**(1), 44–66 (2017)
4. Singh, I., Sharma, R.: A survey on fingerprint minutiae extraction. Int. J. Adv. Res. Ideas Innovations Technol. **3**(3), 264–267 (2017)
5. Arief Sudiro, S., Trisno Yuwono, R.: Adaptable fingerprint minutiae extraction algorithm based-on crossing number method for hardware implementation using FPGA device. Int. J. Comput. Sci. Eng. Inf. Technol. **2**(3), 1–30 (2012)
6. Ali, M.M.H., Mahale, V.H., Yannawar, P., Gaikwad, A.: Overview of fingerprint recognition system. In: Proceedings of ICEEOT 2016 – International Conference on Electrical, Electronics and Optimization Techniques
7. Kaushal, N., Kaushal, P.: Human identification and fingerprints: a review. J. Biometrics Biostatics **2**(4) (2011)
8. Bose, P.K., Kabir, M.J.: Fingerprint: a unique and reliable method for identification. J. Enam Med. Coll. **7**(1), 29–34 (2017)

9. Kanbar, A.B.: Fingerprint identification for forensic crime scene investigation. Int. J. Comput. Sci. Mobile Comput. **5**(8), 60–65 (2016)
10. https://support.apple.com/en-us/HT204587. Accessed 9 Apr 2018
11. Dospinescu, O., Lisii, I.: The recognition of fingerprints on mobile applications – an android case study. J. East. Europe Res. Bus. Econ. **2016**, 1–11 (2016)
12. Jo, Y.-H., Jeon, S.-Y., Im, J.-H., Lee, M.-K.: Security Analysis and improvement of fingerprint authentication for smartphones. Mobile Inf. Syst. **2016**, 1–12 (2016)
13. Gavrilova, M., Monwar, M.: Multimodal Biometrics and Intelligent Image Processing for Security Systems. IGI Publishing Hershey, PA (2013)
14. Szymkowski, M., Saeed, K.: A multimodal face and fingerprint recognition biometrics system. In: Proceedings of 16th IFIP TC8 International Conference on Computer Information Systems and Industrial Management CISIM 2017, Białystok, Poland, pp. 131–140, 16–18 June 2016
15. Szymkowski, M., Saeed, K.: A novel approach to fingerprint identification using method of sectoralization. In: IEEE Proceedings of International Conference on Biometrics and Kansei Engineering ICBAKE 2017, Kyoto, Japan, 15–17 September 2017
16. Szymkowski, M., Saeed, E., Saeed, K.: Retina tomography and optical coherence tomography in eye diagnostic system. In: Chaki, R., Cortesi, A., Saeed, K., Chaki, N. (ed.) Advanced Computing and Systems for Security, vol. 5, pp. 31–42 (2018)
17. Szymkowski, M., Saeed, E.: A Novel Approach of Retinal Disorder Diagnosing using Optical Coherence Tomography Scanners, Transactions on Computational Science XXXI, pp. 31–40 (2018)
18. Saeed, K., Tabędzki, M., Rybnik, M., Adamski, M.: K3M: a universal algorithm for image skeletonization and a review of thinning techniques. Int. J. Appl. Math. Comput. Sci. **20**(2), 317–335 (2010)
19. Tabędzki, M., Saeed, K., Szczepański, A.: A modified K3M thinning algorithm. Int. J. Appl. Math. Comput. Sci. **26**(2), 439–450 (2016)
20. http://www.elsi.es/get:2febdd776466b8b4b029cdd43de4edea. Accessed 11 Apr 2018
21. http://www.futronic-tech.com/product_fs80.html. Accessed 11 Apr 2018

An Investment Strategy Using Temporary Changes in the Behavior of the Observed Group of Investors

Antoni Wilinski$^{(\boxtimes)}$ and Patryk Matuszak

Faculty of Computer Science and Information Technology, Westpomeranian University of Technology, Zolnierska 49, 71-210 Szczecin, Poland
{awilinski,patryk_matuszak}@zut.edu.pl

Abstract. The article considers an investment strategy based on observing the behavior of a certain organized group of investors belonging to the *oanda.com* platform. This platform provides data on the distribution of open positions between long and short for many different financial instruments. A relatively simple and quite effective investment strategy was developed, which was tested for various time ranges and various currency pairs. This data was generated artificially trying to keep statistical similarity to data published by *Oanda*. The basic observed variable was the share of long positions in the total number of open positions. The basic input variable of the strategy was the first derivative of the number of these open long positions. The investment risk was controlled by means of mechanisms typical of the brokerage platform. The tests were carried out on the selected fixed data set both in the *Matlab* environment and using the *MetaTrader* platform tester.

Keywords: Behavioral finance · Internet investment platform
Investment strategies · Forecasting · Financial markets

1 Introduction

Behavioral finances are of interest to world-class economists, psychologists and sociologists, including several recent Nobel prize winners like Richard Thaler (Nobel Prize in 2017), Shiller (the prize in 2013) or more than a decade earlier, Kahneman (2003). These outstanding researchers were interested in and are interested in various aspects of human behavior in investment decision-making situations. Traditional economics referring to Eugen Fama's hypotheses about market efficiency assumed a series of simplifications - all investors know everything about the price, which includes past and future factors affecting it, price has cycles (history repeats itself), there are patterns in time series, etc. (Fama 1998; Shiller 2003, Krutsinger 1997). Behavioral finances are an open, still-formed field in which, in addition to the factors already taken into account, unexpected new discoveries arising from human emotions and weaknesses.

Richard Thaler in his last book mocks the traditional approach to markets, which assume that the knowledge and competencies of investors are equal, that they have equal time and that they have comparable means. Meanwhile, they have a completely

© Springer Nature Switzerland AG 2019
J. Pejaś et al. (Eds.): ACS 2018, AISC 889, pp. 98–106, 2019.
https://doi.org/10.1007/978-3-030-03314-9_9

different, very subjective reality (Thaler 2015). Real investor's behavior also depends on such hard-to-measure factors as overconfidence, decision perspective, aversion to losses, cognitive dissonance, anchoring, intuition and emotions (Ricciardi and Simon 2000). Also many others are still defined or forming other aspects of those mentioned. Behind Nicholas Barberis, behavioral finance can be defined as any attempt to explain and predict the behavior of markets containing anomalies, which are man-made, both as an individual investor and as a group or larger community (Shiller 1999; Barberis 2001; Thaler 2005).

From the point of view of an ordinary daily work of a trader seeking a concept for building a system of rules allowing for systematic trade these new discoveries are troublesome. As a rule, there is a lack of data in the form of patterns or derivatives from the base time series on which such a trader tests his ideas. Therefore, one should look for indirect proofs of man's weaknesses and errors, and possibly use them, for example, to avoid them.

In addition to commonly available data constituting base financial instruments such as exchange rates or commodity prices, there are derivatives of them. The word "derivative" of the financial world has a strict narrow meaning (contracts, options), it is rather a broader context here. It is about the time series associated and accompanying the underlying series. These will not only be the series created by the underlying series, such as prices, exchange rates and indices. These will also be time series created directly by market participants. For such series, for example, investor activity measured by the volume of transactions, the number of open positions, the number of contracts, the number of pending orders, number of occurrences of keywords in social media, chats, etc. can be classified. The time series are also interesting and useful for assessing the direction of the changes. So, we should consider also "derivatives of these derivatives" e.g. the first derivatives of these series.

An interesting property of such data is the difficulty of assessing their place in the cause-and-effect chain. Each given data can be considered both as the reason for the next change and as a result of the previous one. It is not entirely clear to what extent the observed values of these variables create a market and in what is the level they mirror it. This work is a small experiment using such series accompanying the base series.

As an excuse to present a strategy based on such data, the observations published by the *oanda.com* brokerage platform [www.oanda.com/forex-trading/analysis/historical-positions] will be used.

Oanda is one of the oldest online brokers in the world co-founded in New York in 1996 by Dr Michael Stumm and Dr Richard Olsen, a computer scientist and economist, respectively [http://www.oandareviews.com/oanda-wiki/]. Today is a global leader among brokerage platforms. A great social contribution of the platform to the development of education of investing is the creation of the virtual laboratory *Oanda* [https://www.oanda.com/forex-trading/analysis/labs/]. There, with the participation of users, innovations supporting new investment strategies are presented, mainly in the area of forex (currency pairs). This laboratory has, first of all, a lot of interesting statistical data, processed for the needs of possible strategies and inspirations.

A very interesting web page in these resources is Historical Position Ratios containing graphs with the same content as in Fig. 1. In the upper part of such graphs, the number of open long positions is displayed graphically in the percentage of the number of all open positions (the field filled by "+") according to the formula $D = (D/(D+S))$ *100*; where D is the percentage of long positions, S - short positions. In the bottom part, below the division axis, the number S is presented. The graph also contains the second important variable - the value of the exchange rate pair named P. We remind you that the long position is opened by the investor when he thinks that the price will increase, short position, when he thinks that the price will fall.

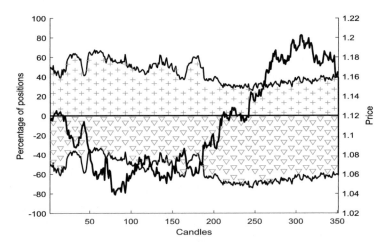

Fig. 1. Proportions between open long and short positions for the EURUSD currency pair over a period of about one year (from Oct. 2016 to Sept. 2017–350 daily candles).

On the example chart which is shown in Fig. 1, it can be noticed that investors are not always right. One can go further with this thesis since the majority lose (Barber and Odean 2002, Barber et al. 2008), they are probably making more mistakes than they who are right. Figure 1 was generated by the authors artificially in the form of charts from the *oanda.com* laboratory.

In Fig. 1, covering a period of 350 candles, and thus a period of about one year, one can notice a period of several months, approximately from about 80 candles to about 310 candles with an almost monotonous price increase. In the same period, the number of open long positions was less than the number of open short positions for the majority of this period. This means that all investors with open short positions increased their losses or closed positions after reaching the Stop Loss level, and thus also at a loss. Shorter incidents illustrating the illogical behavior of investors can also be found on charts of other currency pairs. In turn, from the candle, around 310 to 350, the fall in prices is accompanied by the advantage of short and thus winning positions.

It is worth noting that the platform *oanda.com* recommends investment caution when using charts as in Fig. 1:

This is for general information purposes only - Examples shown are for illustrative purposes and may not reflect current prices from OANDA. It is not investment advice or an inducement to trade. Past history is not an indication of future performance.

The platform itself does not make any suggestions regarding this data. He simply reveals them kindly, leaving the way they are used to the creativity of their users.

Observed results of a collective, in a sense, investment are the effects, not the reasons, of investors' behavior. Their investment decisions may be based on both irrational and compatible with traditional trading. For example, an investor with knowledge about the growing number of short positions with the simultaneous price increase and so opens the short position counting on the reversal of the trend. The other will open a long position waiting for the trend to continue. Both attitudes can be considered as normal, classic behavior. The choice of one or the other decision is most often in the case of observing the principles of rational behavior, the result of a technical analysis.

Observing the graph such as in Fig. 1 and similar, authors decided to study the following hypothesis derived from the area of behavioral finance.

The data obtained is obtained/generated with the frequency of one instance per day. So if a single investor has a chance to observe these changes, let his next reaction expressed by the change in the number of open positions depend on the change in the price P. So, the first derivative of P will affect the effectiveness of the strategy in which the investment decision, taken every day, will be the result of the observed change P?

The article will present the concept of the initially described strategy and then present the results of tests performed on static, permanent data in the *Matlab* environment. The results of these simulations will then be confronted with the results of simulations carried out on the *MetaTrader* brokerage platform. This environment reflects actual changes in the market and imposes to investigations its specific limitations. The discussion of results will end the work.

2 An Investment Strategy Based on the First Derivative of the Number of Open Long Positions

Let's consider any of the above charts, for any currency pair and for any period (with an indication for one year, during this period we will focus in the research) and define the variables appearing on it as:

$D(i), i = 1, 2, \ldots, N_i$ - the percentage share of open long positions at time t_i in the total number of all open positions at the moment;

$S(i), i = 1, 2, \ldots, N_i$ - the percentage share of open short positions at time t_i in the total number of all open positions at the moment;

$$\text{There is } D(i) + S(i) = 100, \text{ for } i = 1, 2, \ldots, N_i \qquad (1)$$

where N_i - the number of elements of the considered time series, e.g. in the charts, the period of one year sampled every one day was considered.

Another variable appearing on the chart (in Fig. 1 it is a bold line running through the fields denoting the percentage of long and short positions) is:

$P(i), i = 1, 2, \ldots, N_i$ - price of the variable observed, in this case the EURUSD currency pair sampled once a day.

The number of open positions is a characteristic of the derivatives market and an attempt to use it for the prediction of changes in currency pairs is, in the authors' opinion, an interesting challenge. That's why the *oanda.com* broker's initiative is noteworthy and praiseworthy.

The very idea of using the first derivative of the number of open positions is to check whether the relative number of open positions (e.g. long) at a given moment $D(i)$ is greater or smaller than this relative number in the previous period $D(i-1)$.

Let's assume that the strategy will be defined as follows:

$$\text{if } D(i) > D(i-1) \text{ then open short; for } i = 1, 2, \ldots, N_i \tag{2}$$

$$\text{if } D(i) < D(i-1) \text{ then open long; for } i = 1, 2, \ldots, N_i \tag{3}$$

Therefore, we assume that when the number of open long positions increases, we will open a short position. When it drops, we will open a long one.

Such investor behavior cannot be explained rationally. If the number of open positions increased in the observed moment, it should be assumed that there is a temporary advantage in the market of investors who believe in the growth of the value of the instrument, not in a decline. Why do we open short positions instead of long ones? The simplest explanation can be the assumption that this temporary majority is wrong.

The only confirmation of this hypothesis can be its empirical verification. Of course, we can also expect that if the strategy proves to be right, it will probably be right for a moment. In the next episode of the time series, it may happen that the majority will be right. Such dilemmas are typical for behavioral finance.

To verify the correctness of this approach (2)–(3) let us be guided by the approach called computational intelligence, according to the principle, until we calculate, we do not know if we are right. This principle adheres to the inductive approach to science, in which the correct rule is assumed which is statistically verified. However, predicting that strategy (2)–(3) will not always give a positive result, it will be complemented by two Stop Loss mechanisms, adequate to the direction of opening of the position.

The return after one sampling period for a short position will be

$$Z(i) = -P(i+1) + P(i); \text{for } i = 1, 2, \ldots, N_i \tag{4}$$

And after taking into account the security in the form of Stop Loss:

$$Z(i) = -SL \text{ and } H(i+1) - P(i) > SL, \tag{5}$$

where $H(i+1)$ - the highest value of the variable P in $(i+1)$ -th candle

While, the return after one sampling period for a long position will be

$$Z(i) = P(i+1) - P(i); \text{for } i = 1, 2, \ldots, N_i \tag{6}$$

And after taking into account the security in the form of Stop Loss:

$$Z(i) = -SL \text{ if } P(i) - L(i+1) > SL, \qquad (7)$$

where $L(i+1)$ - the lowest value of the variable P in $(i+1)$ -th candle;

A strategy prepared in this way basically has only one parameter, whose changes may influence the result. This is the level of acceptance of the SL loss.

3 Data for the Simulation

Preparation of data for simulation of the investment strategy considered here requires obtaining data from two sources:

1. From a random data generator generating data similar to those from the *oanda.com* laboratory [https://www.oanda.com/forex-trading/analysis/historical-positions#] in terms of average and first statistical moments. The data from the laboratory can be downloaded as .csv files.
2. From the *bossaFX* brokerage platform, price data (values of the course) and OHLC candle vectors (*Open, High, Low and Close* values in the period of the candle).

Assuming that the model will consider making decisions once a day, *OandaLab* can give historical data from one year back from the current date. The vector of processed data of the percentage share of long positions in the total open positions on the *Oanda* brokerage platform, let us denote further $D(i)$.

The first major EURUSD pair was chosen for the first research, as in Fig. 1 sampled every one day.

From the *MetaTrader* platform belonging to *bossaFX* [bossa.pl] after opening the investment account, in the toolbar Tools/HistoricalData, you can get data for various values, including of course for dozens of currency pairs, sampled at standard frequencies: every 5, 15, 30 min, every 1 and 4 h and every 1 day, 1 week and 1 month. The data is contained in the columns {Date Time Open High Low Volume}. Time for data downloaded every day is recorded at midnight, i.e. at 0000 CET. The number of instances (rows of data matrix) is definitely higher than for *Oanda* and ranges from several to several dozen thousand lines depending on the sampling period.

The preparation of data should start with generating about 350 values of the time series D symbolizing the number of open long positions as in *oanda.com* and then supplementing this series with data from the *bossaFX*.

4 Investigation of the Strategy

The strategy described in conditions (2) and (3) together with the formula for calculating the return after each candle (4) and (6) and the Stop Loss (5) and (7) mechanisms have been implemented in the *Matlab* environment.

The final result of the cumulative profit was obtained as $Zs = 0.2702(2702 \text{ pips})$.

An additional assessment of the results should be an estimating of a risk. The Calmar Ratio (CR) index was used, which is the quotient of the final profit to the largest capital drawdown (Young 1991).

$$CR = Z_i(N_i)/MDD \tag{8}$$

Where $Z_i(N_i)$ - the final value of the cumulative strategy's profit;

MDD - maximum drawdown on the cumulative gain curve;

A very good Calmar Ratio index of $CR = 9.51$ was obtained (Fig. 2). The profit achieved on average every day (profit per candle) was $ppc = 5.5$ pips, which should also be considered a very good result, especially in the context of low risk expressed by CR. A fairly typical Stop Loss of 20 pips was used. A pip (price interest point) is the smallest price movement made by currency if it is quoted to four decimal places.

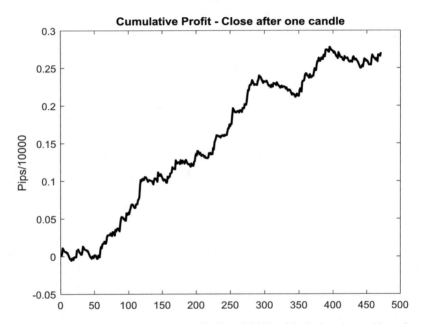

Fig. 2. Cumulative profit for the strategy with $SL = 0.0020$ with closing the position after the period of one candle

In the simulation, the positions were evenly distributed - 269 long positions and 201 short positions were opened, and 129 positions were closed by Stop Loss.

For the same data of the EURUSD currency pair, simulation tests were carried out on the *MetaTrader4* platform. This is a different programming environment that requires implementation in mql4, and the initial assumptions for which research is carried out are completely different. It is necessary to introduce initial capital and possibly a spread characteristic for a given broker.

A cumulative profit diagram was presented in Fig. 3. A contractual initial capital of USD 10,000 was assumed. This is the value that was not necessary for the *Matlab* simulation. The simulation was carried out for $SL = 0.0020$, as in *Matlab*. Positions were closed after one candle. A profit of US $ 9392 was achieved (93.92% in a year) with a Calmar of 1.54.

These results compared to those obtained in *Matlab* are worse, especially in the aspect of risk assessment

Fig. 3. Cumulative profit obtained in the *MetaTrader* strategy tester with SL = 0.0020 with closing the position after the period of one candle

5   Discussion

The results obtained can be considered satisfactory by confronting them, for example, with assessments we can find on the Internet (Pasche 2014). In the above description, only one value of the SL parameter was considered. The final results depend on this parameter, to the extent that it is reasonable to adapt this size to the market through machine learning. The same approach is suggested by analyzing the changes in cumulative profit in Fig. 3 made in *MetaTrader*. It can be seen periodically, in three different periods, stabilization of achieved profits or losses incurred. The use of machine learning, which is not considered here, would probably allow the choice of the parameter or parameters that improve the liquidity of capital growth. The parameter could be *SL*. Parameters which are not considered here could be, for example, the number of candles after which the positions would be closed or calculating the derivative, not for one candle as in (2) or (3), but after several such periods.

According to the authors, the article argues that there is still little-explored potential for support in the behavioral data for different decision-making systems, including forecasts in the financial markets.

At work, the authors did not use any design or institutional support.

References

Barber, B.M., Lee, Y.T., Liu, Y.J., Odean, T.: Just how much do individual investors lose by trading? Rev. Financ. Stud. **22**(2), 609–632 (2008)

Barber, B.M., Odean, T.: Online investors: do the slow die first? Rev. Financ. Stud. **15**(2), 455–488 (2002)

Barberis, N., Huang M.: Mental accounting, loss aversion, and individual stock returns. J. Finance **56**(4), 1247–1292 (2001)

Fama, E.F.: Market efficiency, long-term returns, and behavioral finance. J. Financ. Econ. **49**(3), 283–306 (1998)

Kahneman, D.: Maps of bounded rationality: Psychology for behavioral economics. Am. Econ. Rev. **93**(5), 1449–1475 (2003)

Krutsinger, J.: Trading Systems: Secrets of the Masters, p. 242. McGraw-Hill, New York (1997)

Pasche, R.: How Many Pips Should We Target Per Day? DailyFX, 8 July 2014. www.dailyfx.com. Accessed Aug 2017

Ricciardi, V., Simon, H.K.: What is behavioral finance? Bus. Educ. Technol. J. **2**(2), 1–9, Fall 2000 (2000)

Shiller, R.J.: From efficient markets theory to behavioral finance. J. Econ. Perspect. **17**(1), 83–104 (2003)

Shiller, R.J.: Human behavior and the efficiency of the financial system. Handb. macroecon. **1**, 1305–1340 (1999)

Thaler, R.H. (Ed.).: Advances in Behavioral Finance, vol. 2. Princeton University Press (2005)

Thaler, R.H.: Misbehaving: The Making of Behavioral Economics. WW Norton & Company (2015)

Young, W.T.: Calmar Ratio: A Smoother Tool, Futures (magazine), October 1991

Software Technology

Measuring Gender Equality
in Universities

Tindara Addabbo[1], Claudia Canali[2], Gisella Facchinetti[3],
and Tommaso Pirotti[3(✉)]

[1] Department of Economics Marco Biagi,
University of Modena and Reggio Emilia, Modena, Italy
{tindara.addabbo,claudia.canali}@unimore.it
[2] Department of Engineering Enzo Ferrari,
University of Modena and Reggio Emilia, Modena, Italy
[3] Economics Department, University of Salento, Lecce, Italy
{gisella.facchinetti,tommaso.pirotti}@unisalento.it

Abstract. The paper proposes a fuzzy expert system for gender equality evaluation in tertiary education that has been experimented in 6 European universities in Italy, Lithuania, Finland, Germany, Portugal, Ukraine within the EQUAL-IST Horizon 2020 project with the goal to design and implement Gender Equality Plans (GEPs) for IST Research Institutions. We propose a Fuzzy Expert System (FES), a cognitive model that, by replicating the expert way of learning and thinking, allows to formalize qualitative concepts and to reach a synthetic measure of the institution's gender equality (ranging from 0 to 1 increasing with gender equality achievements), that can be then disentangled in different dimensions. The dimensions included in the model relate to gender equality in the structure of employment (academic and non academic) and in the governance of the universities, to the equal opportunity machinery and to the work-life balance policies promoted by the institutions. The rules and weights in the system are the results of a mixed strategy composed by gender equality experts and by a participatory approach that has been promoted within the EQUAL-IST project. The results show heterogeneity in the final index of gender equality and allow to detect the more critical areas where new policies should be implemented to achieve an improvement in gender equality. The value of the final gender equality index resulting from the application of the FES is then compared to the gender equality perceived by each institution involved in the project and will be used to improve also the awareness in gender gap in important dimensions in tertiary education setting.

Keywords: Fuzzy expert · Gender equality · Fuzzy logic
Tertiary education · Work-life balance

1 Introduction

Gender inequality in research and innovation is well documented by [26], and tools to measure and monitor it are presented in [13,17,28] and within EU

© Springer Nature Switzerland AG 2019
J. Pejaś et al. (Eds.): ACS 2018, AISC 889, pp. 109–121, 2019.
https://doi.org/10.1007/978-3-030-03314-9_10

funded projects as GenderTime [5], Effective Gender Equality in Research and Academia [9] and Gender-Net [16].

The reasons to achieve gender equality in academic and research institutions are very well described in the Norwegian Committee for Gender Balance and Diversity Web site [22] in terms of fairness, democracy and credibility, national research objectives, research relevance and quality, competitive advantage and by EIGE [13] in terms of compliance with national and European legislations, creation of better working environments, attraction and retention of talents, improved success probability in accessing funds as (gender equality is referred as an additional criterion to access to public funding), capability of reaching new target audiences, beneficiaries or final users/customers of services produced by the institutions thanks to the change induced by adopting a gender perspective and more gender diverse teams, enhancement of sense of community and ownership within the institution developed by trying to pursue the objective of gender equality. Moreover:

"The quest for excellence and quality has become a major issue for research organizations and higher education institutions. It is driven by intense competition for skills, funding and innovations. Bringing a gender dimension in research and innovation content improves the overall quality of research design, hypotheses, protocols and outputs in an ample variety of fields. It not only allows addressing gender bias and building more evidence-based and robust research, but also contributes to pluri-disciplinarity. As science and innovation are increasingly framed as working for/with society, reflecting the diversity of final users from the early research stage has become a must. "Gender blindness" (understood as the lack of consideration for gender-related aspects) often goes with neglecting other relevant social or experiential parameters. Challenging this blindness, on the contrary, creates awareness for a broader set of variables than just sex and/or gender." EIGE [13, p. 12]

This contribution presents a new proposal in the measuring field and has been developed within the European Union's Horizon 2020 research and innovation programme funded project "EQUAL-IST" (Gender Equality Plans for Information Sciences and Technology Research Institutions). This project aims at introducing structural changes in research organizations to enhance gender equality within Information System and Technology Institutions. The project supports the seven European Research Performing Organizations (RPOs) of the EQUAL-IST consortium in developing and implementing Gender Equality Plans (GEPs) focusing on four main levels: human-resources practices and management, research design and delivery, student services and institutional communication. It is in this framework that the idea to get a synthetic index of gender equality to be used for comparative analyses within the network of institutions of the consortium has been translated in an operational system. Following what is indicated in the literature on gender equality in research and academic institutions, the research group has detected a set of variables that define the employment structure of a given institution, distinguishing by academic and administrative and technical roles within the institution, and evaluating the degree of gender equality at different levels (administrative employment, academic composition by level and gender, students enrolled in different areas within the institution) and governance. The three universities chosen are located in different countries: Italy (Unimore), Germany (WWU) and Lithuania (KTU), all characterized by under-representation of women in higher academic positions [27] but by different levels of gender equality index [15]. The latter is on average 66.2 in EU 28 countries, 65.5 in Germany, 62.1 in Italy and 56.8 in Lithuania. The University of Münster (WWU) is a public university located in the city of Münster, North Rhine-Westphalia in Germany. The WWU is part of the Deutsche Forschungsgemeinschaft, a society of Germany's leading research universities. The WWU has also been successful in the German government's Excellence Initiative. Kaunas University of Technology (KTU) is the largest Technical University in Lithuania. The University of Modena and Reggio Emilia (UniMORE) has a longstanding tradition (founded in 1175) and currently offers 80 bachelor and master classes through its ten departments. In compliance with the national laws, Unimore recently established a Unified Committee for the promotion of equal opportunities (CUG) that aims to promote gender equality and fight any kind of discrimination within employees and students population. We have collected administrative data on the structure of employment by gender and analyzed the system of governance within each institution involved. Our purpose has led us to see everything in a multi-criteria context and then to draw a bottom up decision tree that sees in the final part (its root) the aggregate evaluation and in each leaf the individual indicators that will produce the evaluation. Each node of the tree will provide macro-indicators useful for intermediate evaluations. Given the overall problem in this perspective, we just have to choose which mathematical or statistical tool to use. At this point, we have to choose which is the aggregative tool that best suits the presented problem. This choice must be made among all those available in the context of multi-criteria analysis. Literature shows that the problem can be approached with many techniques, usually divided in two conceptual branches: *data-based* methods and *knowledge based systems*. The first ones usually need big amount of data to learn and forecast; the second ones instead rely on experts and just a really small amount of data are required. Since every University is represented by a single record of data, our case is close to the second branch of techniques. Due to the small database dimension, even the simplest statistical aggregations, such as averages, cannot be applied. The small database dimension has suggested to go in other directions. We propose a fuzzy multi-criteria analysis method as a fuzzy expert system (FES) [21, 25, 30] in which the linguistic rules take the place of weights and averages and experts inject their knowledge during the construction of the decision tree. This is a cognitive model that, by replicating the expert way of learning and thinking, allows to formalize qualitative concepts and to reach a synthetic measure of the institution's gender equality (ranging from 0 to 1 increasing with gender equality achievements), that can be then disentangled in its different dimensions. The latter characteristic of the model that we propose can be fruitfully used by Policy Makers and Equal Opportunity

Officers in order to detect and address the critical elements in the organization and carry out changes to improve gender equality. Other papers involving the same methodology have been published [2–4, 10–12]. A more recent application of the model has been experimented within the EQUAL-IST project and is available for other universities and research institutions wishing to obtain an assessment of their organization in terms of gender equality [1]. Further developments of the model, together with its wider implementation, include the assessment by using fuzzy logic of gender equality policies and institutional factors affecting gender equality within the institution. In Sect. 2, we present the proposed methodology, in Sect. 3 we present the systems we have produced, in Sect. 4 we present a subsystem in more details, in Sect. 5 we comment upon the results of the system application to 3 of the universities belonging to the EQUAL-IST project while concluding remarks and further developments can be found in the last Section of the paper (Sect. 6).

2   FES Methodology

The main inspiration behind the introduction of fuzzy sets theory was the necessity for modeling real-world phenomena, which are inherently vague and ambiguous. Starting from the idea that human knowledge about complex problems can be successfully represented using the imprecise terms of natural language, the theories of fuzzy sets and fuzzy logic provide formal tools for mathematical representation and efficient processing of such idea. Here we want to deal with a fuzzy system, that is a system with the addition of a fuzzy opportunity [21,25,30]. The term "System" is connected with the idea of inputs-outputs process. The addition of term "Fuzzy" to systems lets to face towards information processing, where the usage of classical sets theory and binary logic is impossible or difficult. The inputs-outputs procedure shows that these systems are based on aggregation methods. The idea of aggregation functions is rather simple: they aim to summarize the information contained in an n-tuple of input values by means of a single representative value (output). Starting from the simplest example, the arithmetic mean, many other kinds of means were applied in numerous applications in various areas. Several other kinds of aggregation functions, such as the conjunctive and the disjunctive ones, are an indispensable mathematical model not only of logical operations in the area of many-valued logics, but also of many other theoretical and applied fields [18–20, 23]. In the literature, terms such as fuzzy system, system based on fuzzy rules, fuzzy expert systems are used depending on which are the techniques used to define the "weights" of the "averaging procedure" we consider. In a FES, the weights are made by experts, not in a numerical way but using "If-Then Rules" involving symbolic knowledge representation.

The typical structure of a fuzzy system consists of four functional blocks: the fuzzifier, the fuzzy inference engine, the knowledge base, and the defuzzifier. Both linguistic values (defined by fuzzy sets) and crisp (numerical) data can be used as inputs for a fuzzy system. If crisp data are applied, then the inference process is preceded by fuzzification, which assigns the appropriate fuzzy set to the crisp input. The values of input variables are mapped into linguistic values of the output variable by means of the appropriate method of approximate reasoning (inference engine) using expert knowledge, which is represented as a collection of fuzzy conditional rules (knowledge base). The fuzzy system output needs to be translated in a crisp number by defuzzification methods. Several are the advantages to use this aggregation methods. First of all, we have to notice that the data set is not so wide to use machine learning techniques and so we are obliged to use a sort of average. The choice of this type of average is connected with the "transparency" of the decision maker choice. Everything is written in the rules-blocks and in the inference method he chooses. Anything is changeable if he is not satisfied of outputs replay. Another advantage is given by its "modularity" form. The initial inputs (if their number is sufficiently high) don't need to be aggregated directly into the final output, but other intermediate variables, perhaps not present in our data set, are created to be able to develop the system in a modular way. This step simplifies the several rules-blocks buildings, that reduces their dimensions giving origin to a set of sub-systems that have their separate meaning in the description of system in its global aspect. For the system final tuning, we may start from the output backwards to the inputs following the single sub-system and better understand where changes have to be done.

3   The Fuzzy Expert System "Total Equality Index"

In Fig. 1 we present the main system (Table 1).

Fig. 1. Total Equality Index

Table 1. List of abbreviations for Total Equality Index

Equal_Opportunity_Policy	Equal Opportunity Policy Index
Gender_Equity	Gender Equity Index
WL_Balance	Work-Life Balance Index

We have called Total Equality Index the final output of the system. The experts involved in the project established that the final evaluation of equality reached in academic institutions can be declined through three macro-indicators: Work Life Balance, Gender Equity and Equal Opportunity machineries. Starting from this structure, the experts, acting backwards, established what could be the inputs or the sub-macro indicators that could produce each of those three initial variables as an aggregate response. In this way, each of the variables of the system in Fig. 1 became the output of another FES.

In this paper we will go into the detail of the sub-macro indicator "gender equity index". Before starting, let's give some explanation of what information are provided by the other two sub-macro indicators.

"Work-life balance". As the literature shows, work-life balance is a crucial dimension in the achievement of gender equality in private and public institutions: this is related to the different gender roles in social reproduction paid by men and women. Its importance within research and university institutions can be detected by the high diffusion of the need of including work-life balance policies within Gender Equality Plan as results in the survey carried out by EIGE [14] and also within the experience collected during the research project EQUAL-IST, as reported in the project Deliverable "D2.5 Gender Equality Assessment Report" [29]. The importance of taking into account work-life balance policies within universities is confirmed by applied studies [6–8,24].

This sub-macro indicator is the result of the aggregation of several dimensions concerning to the direct or indirect (in terms of available external services at a special price) availability of child care services and summer camps, and the time flexibility in terms of bank hours, part-time work, teleworking, flexible entry or exit time.

The last component of the model deals with Equal Opportunities Machineries. Here we take into account whether the university has a GEP and its quality, as well as the presence of gender budgeting and of equal opportunities machineries. Surveys on the diffusion of gender auditing and gender budgeting in EU universities and research institutions show that they are quite underdeveloped and limited to a few institutions [14] (Table 2).

4   The Gender Equity Index

Figure 2 shows the gender equity index FES.

Also in this case "gender equity index" (GE) is obtained by the aggregation of three macro-sub indicators as Academia, Public Technical Administration (PTA) and Governance.

"Accademia" carries information on gender equity in the academic positions like Full Professors, Associated Professors and Researchers positions. PTA is connected with gender equity in administrative employment with five levels of classifications. Governance presents gender equity in governance position starting from the highest position at the university level (that is, the Rector) down to the executive board, that means the dean of a faculty or the department director.

In the above system (Fig. 2) we have given more importance to the presence of gender equity in the governance of the system, followed by the academic part of the Institution employment, and by the administrative staff, taking into account the different degree of power in the structure. Analysing the sub-macro

Table 2. List of abbreviations for gender equity index

Full_Professors_WoT	Full professors - ratio women over total
Associate_Professors_WoT	Associate professors - ratio women over total
Researchers_WoT	Researchers - ratio women over total
First_Level	First level employees - ratio women over total
Second_Level	Second level employees - ratio women over total
Third_Level	Third level employees - ratio women over total
Fourth_Level	Fourth level employees - ratio women over total
Fifth_Level	Fifth level employees - ratio women over total
Students_WoT	Students - ratio women over total
Rector	Rector's gender
ViceRector	Vice Rector's gender
Academic_Senate_WoT	Percentage of women in Academic Senate
Exec_Board_WoT	Percentage of women in Executive Board

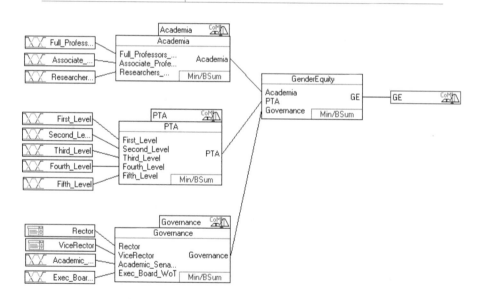

Fig. 2. Gender equity index

indicator Governance obviously the presence of women in the positions of rector or vice-rector are more important than in the other lower positions.

In Academia a higher importance is given to the presence of women in the Full professor level followed by associate professor and researcher. Turning to the administrative staff, the maximum weight has been given to the presence in the first level with a decreasing importance to the gender equity in the lower levels. All this information, corresponding to the aggregation weights, will be transferred through the rules-blocks building.

In this section we discuss more in details the sub-macro indicators "Academia".

The three inputs are connected with the average of gender presence in the several university roles. The experts decided to attribute to this variable three linguistic distinctions: low, medium and high. The threshold are represented in percentage. If the presence of women is below 5%, we assume the assessment as "Low". If it exceeds 50%, it is assumed "High" while the "Medium" variable has been positioned with a maximum value of 25% and then standardized. (Figure 3)

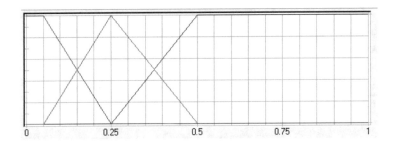

Fig. 3. Associate professors - ratio of women over total

To obtain the data inputs we have performed some descriptive statistics on the components of the different part of the system (Tables 3, 4 and 5)

The results shown below (Sect. 5) are rather in line with the structure of employment in academic institutions as found in [27].

To translate the "weights" the experts have fixed we present the rule-block that connects the three variables with the "academia" macro indicator (Table 6).

Table 3. Gender equity in academia

University	Full_Professors_WoT	Associate_Professors_WoT	Researchers_WoT
Unimore	0.25	0.38	0.48
KTU	0.27	0.45	0.43
WWU	0.2	0.19	0.4

Table 4. Gender equity in the administrative staff

University	First_Level	Second_Level	Third_Level	Fourth_Level	Fifth_Level
Unimore	0.4	0.72	0.68	0.73	0.38
KTU	0.33	0.43	0.28	1	0.78
WWU	0.35	0.74	0.77	0.81	0.5

Table 5. Gender equity in governance

University	Exec_Board_WoT	Academic_Senate_WoT	ViceRector	Rector
Unimore	0.36	0.28	0	0
KTU	0	0.27	1	0
WWU	0.5	0.43	0.75	0

Table 6. Rules of the rule block "Academia"

IF			THEN
Full_Professors	Associate_Professors	Researchers	Academia
low	low	low	Low
low	low	medium	Low
low	low	high	Low
low	medium	low	Low
low	medium	medium	Medium
low	medium	high	Medium
low	high	low	Medium
low	high	medium	Medium
low	high	high	Medium
medium	low	low	Low
medium	low	medium	Medium
medium	low	high	Medium
medium	medium	low	Medium
medium	medium	medium	Medium
medium	medium	high	Medium
medium	high	low	Medium
medium	high	medium	High
medium	high	high	High
high	low	low	Medium
high	low	medium	Medium
high	low	high	Medium
high	medium	low	Medium
high	medium	medium	High
high	medium	high	High
high	high	low	High
high	high	medium	High
high	high	high	High

5   Results

The first results of the model application can be analysed.

By putting together the different results of the subsystems that compose the Total Equity Index we obtain the results available in Tables 7 and 8.

Table 7. Gender equity

University	Academia	PTA	Governance	GE
Unimore	0.39	0.5	0.25	0.5
KTU	0.43	0.38	0.5	0.6
WWU	0.17	0.5	0.5	0.5

Table 8. Total equity index

	Equal opportunity policy	Gender equity	WL balance	Total equality index
KTU	0,00	0,51	0,25	0,38
UNIMORE	0,49	0,58	0,53	0,54
WWU	0,54	0,50	0,69	0,52

As we can see, Unimore has reached the higher score 0.54 that is however well lower than 1, followed by WWU (0.52) and by KTU. The latter has a particularly low performance in two out of 3 items (WLB and Eq. Opp.). We may notice that, if the two dimensions (WLB and EO) were not included in the model, the results would have looked more similar; in this case, however, two important dimensions of gender equity (work life balance and the presence and quality of the Equal Opportunity Machineries) would not have been taken into account. It is worth to notice that KTU is located in Lithuania where, according to the comparative analysis on the legislations on equal opportunities with special reference to academic and research institutions carried out by EIGE [14] there isn't a legal provision on GEP, while it is present in Italy and in Germany (at federal level for research institutions; at Länder for Universities). This should call for an extension to all EU countries of GEP legal provisions and also for a more developed control of the functioning of the active machineries.

6   Conclusions

There are different advantages in achieving gender equality in research and academic institutions, as outlined in the literature and summarized in the introduction of this paper.

The aim of our work is to provide a tool that can be used across research and academic institutions for their self-evaluation of the status in terms of gender equality. Moreover, our proposal provides a system that could allow to detect the areas where gender equality is visible in the employment structure and governance composition, and to measure key factors in gender equality, such as work-life balance and equal opportunities machineries within the institutions.

Collecting data in these different fields and sharing the rules to assign values to the observed variables has been possible thanks to the Horizon 2020 project EQUAL-IST.

Differently from other indicators of gender equity of institutions, the construction of this Index implies a Fuzzy Expert System modelization that expands the evaluation of gender equity within the institution by taking into account the structure of governance, academic and non-academic employment in terms of gender equity attending different levels of degree and areas within the University. As shown in Sect. 3 the Total Equality Index puts together the gender equity with the degree of work life balance in the structure as well as with a judgment on equal opportunity machineries consistently with the importance that the literature has assigned to these two factors in affecting gender equality. This, as the results of our application show (Table 8), produces different results in terms of the evaluation of the three institutions in the Project that we have analysed.

The extension of the application of our model to other institutions in EU countries that takes into account the diversity in the policies developed within each EU countries, as referred in [14], can provide a useful policy instrument in order to detect by multivariate econometric analysis the very effect of different institutional settings on the obtained outcome. The analysis will also include the evaluation of the country gender equality as provided by EIGE Gender Equality Index [15].

Acknowledgments. This paper has benefited from funding from the EQUAL-IST Horizon 2020 project: Gender Equality Plans for Information Sciences and Technology Research Institutions - Grant Agreement N. 710549.

We thank Dr Alessandro Grandi for having provided the data used to implement the system.

References

1. Addabbo, T., Canali, C., Facchinetti, G., Grandi, A., Pirotti, T.: Gender equality in tertiary education and research institutions: an evaluation proposal. In: Azevedo, A., Mesquita, A. (eds.) Proceedings of the International Conference on Gender Research, pp. 1–9. ISCAP, Academic Conferences and Publishing International Limited, Reading, UK. ISBN: 978-1-911218-77-7, Porto, Portugal, 12–13 April 2018
2. Addabbo, T., Facchinetti, G., Mastroleo, G.: Child well being and parents' work: the evaluation of firm's compliance to work-life balance. Pol. J. Environ. Stud. 18(4A), 18–26 (2009)
3. Addabbo, T., Facchinetti, G., Mastroleo, G., Lang, T.: "Pink seal" a certification for firms' gender equity. In: WIRN, pp. 169–176 (2009)

4. Addabbo, T., Facchinetti, G., Mastroleo, G., Solinas, G.: A fuzzy way to measure quality of work in a multidimensional perspective. In: Advances in Information Processing and Protection, pp. 13–23. Springer (2007)
5. Badaloni, S., Perini, L.: A Model for Building a Gender Equality Index for Academic Institutions. Padova University Press, Padova (2016)
6. Baker, M.: Choices or constraints? Family responsibilities, gender and academic career. J. Comp. Fam. Stud. **41**(1), 1–18 (2010)
7. Baker, M.: Academic Careers and the Gender Gap. UBC Press (2012)
8. Baker, M.: Women graduates and the workplace: continuing challenges for academic women. Stud. High. Educ. **41**(5), 887–900 (2016)
9. EGERA. http://www.egera.eu/. Accessed 17 May 2018
10. Facchinetti, G., Addabbo, T., Pirotti, T., Mastroleo, G.: A fuzzy approach to face the multidimensional aspects of well-being. In: 2012 Annual Meeting of the North American Fuzzy Information Processing Society (NAFIPS), pp. 1–6. IEEE (2012)
11. Facchinetti, G., Mastroleo, G., Pirotti, T.: Quality of daily work & life index– a definition and its evaluation in a fuzzy way. In: Advances in Fuzzy Logic and Technology 2017, pp. 36–47. Springer (2017)
12. Facchinetti, G., Solinas, G., Pirotti, T.: Quality of work and elderly care– preliminary experiments. In: 2013 Joint IFSA World Congress and NAFIPS Annual Meeting (IFSA/NAFIPS), pp. 484–489. IEEE (2013)
13. European Institute of Gender Equality, E.: Gender Equality in Academia and Research. GEAR tool. Publications Office of the European Union, Luxembourg (2016)
14. European Institute of Gender Equality, E.: Integrating gender equality into academia and research organisations. Analytical paper. Publications Office of the European Union, Luxembourg (2016)
15. European Institute of Gender Equality, E.: Gender Equality Index 2017: Measuring gender equality in the European Union 2005-2015 - Report. Publications Office of the European Union, Luxembourg (2017)
16. Gender Net. http://www.egera.eu/. Accessed 17 May 2018
17. Goltz, S.M., Hietapelto, A.B.: Translating the social watch gender equity index for university use. Chang. Mag. High. Learn. **45**(3), 66–73 (2013)
18. Grabisch, M., Marichal, J.L., Mesiar, R., Pap, E.: Aggregation functions, volume 127 of encyclopedia of mathematics and its applications (2009)
19. Kahraman, C.: Fuzzy Multi-Criteria Decision Making: Theory and Applications with Recent Developments, vol. 16. Springer, Berlin (2008)
20. Kahraman, C., Onar, S.C., Oztaysi, B.: Fuzzy multicriteria decision-making: a literature review. Int. J. Comput. Intell. Syst. **8**(4), 637–666 (2015)
21. Kasabov, N.K.: Foundations of neural networks, fuzzy systems, and knowledge engineering. Marcel Alencar (1996)
22. Norwegian Committee for Gender Balance and Diversity. http://www.kifinfo.no/. Accessed 17 May 2018
23. Mardani, A., Jusoh, A., MD Nor, K., Khalifah, Z., Zakwan, N., Valipour, A.: Multiple criteria decision-making techniques and their applications–a review of the literature from 2000 to 2014. Econ. Res. Ekonomska Istraživanja **28**(1), 516–571 (2015)
24. Picardi, I.: La dimensione di genere nelle carriere accademiche: Riflessività e cambiamento nel progetto pilota GENOVATE@UNINA, vol. 2. FedOA-Federico II University Press (2017)
25. Piegat, A.: Fuzzy modeling and control (studies in fuzziness and soft computing). Physica, 742 (2001)

26. for Research, E.C.D.G., Development, I.D.A.P., Internal, C.U.A., external commu-
    nication: Open Innovation. Open Science. Open to the world. A vision for Europe.
    Publications Office of the European Union, Luxembourg (2016)
27. for Research, E.C.D.G., Innovation, D.B..O.I., Open Science, U.B.S.w., for Society:
    She figures 2015. Publications Office of the European Union, Luxembourg (2016)
28. Ruest-Archambault, E., von Tunzelmann, N., Iammarino, S., Jagger, N., Miller,
    L., Kutlaca, D., Semencenko, S., Popvic-Pantic, S., Mosurovic, M.: Benchmarking
    Policy Measures for Gender Equality on Science. Office for Official Publications of
    the European Communities of the European Communities (2008)
29. Sangiuliano, M., Grandi, A.: EQUAL-IST project deliverable D2.5 gender equality
    assessment report (2017)
30. Von Altrock, C.: Fuzzy Logic and Neurofuzzy Applications in Business and
    Finance. Prentice-Hall Inc., Upper Saddle River (1996)

Transitive Closure Based Schedule of Loop Nest Statement Instances

Wlodzimierz Bielecki[(✉)] and Marek Palkowski

Faculty of Computer Science and Information Systems,
West Pomeranian University of Technology in Szczecin,
Zolnierska 49, 71210 Szczecin, Poland
{wbielecki,mpalkowski}@wi.zut.edu.pl
http://www.wi.zut.edu.pl

Abstract. A novel algorithm of loop nest statement instance scheduling is presented. It is based on the transitive closure of dependence graphs. The algorithm is implemented in the publicly available optimizing TRACO compiler, which allows for automatic parallelization of program loop nests and automatic generation of parallel compilable code in the OpenMP standard. Results of an experimental study demonstrate that the algorithm is able to generate parallel code for popular benchmarks and this code achieves satisfactory speed-up on modern machines. The computational complexity of the approach is low. Future algorithm improvements are discussed.

Keywords: Schedule · Transitive closure · Dependence graphs
Automatic loop nest parallelization · OpenMP

1 Introduction

In this paper, we deal with parallelization of loop nests that usually comprise most calculations in real-life codes. To parallelize a loop nest, first we need to form a legal schedule — a function which maps each loop nest statement instance to a discrete time when this instance should be executed.

Loop nest statement instance schedules can be formed by means of affine transformations [5,6] or the power k of a dependence relation [3]. However, it is well-known that for some classes of loop nests, affine transformations lack to find schedules [7], while the calculation of the power k of a dependence relation may be a computationally heavy task that in practice does not allow us to produce any schedule at short time. So there still exists necessity for development of new loop nest statement instance schedule algorithms of reduced computational complexity.

In this paper, we present a new approach to find schedules based on the transitive closure of dependence graphs. We demonstrate that the computational complexity of calculating transitive closure for dependence graphs, associated with real-life programs, is low and it is possible to form legal schedules for those programs at short time.

© Springer Nature Switzerland AG 2019
J. Pejaś et al. (Eds.): ACS 2018, AISC 889, pp. 122–131, 2019.
https://doi.org/10.1007/978-3-030-03314-9_11

2   Background

The algorithm presented in this paper uses relations to represent data dependences. A dependence relation is a tuple relation of the form $[input\ list] \rightarrow [output\ list]$: *formula*, where *input list* and *output list* are the lists of variables and/or expressions used to describe input and output tuples, and *formula* describes the constraints imposed upon input and output lists. It is a Presburger formula built of constraints represented by algebraic expressions and using logical and existential operators [10].

In the presented algorithm, standard operations on relations and sets are used, such as intersection ($\cap$), union ($\cup$), difference ($-$), domain (dom R), range (ran R), relation application ($S' = R(S) : e' \in S'$ iff exists e s.t. $e \rightarrow e' \in R$, $e \in S$). In detail, the description of these operations is presented by Verdoolaege [11].

The positive transitive closure of a given lexicographically forward relation R, named R^+, is defined as follows [11]: $R^+ = \{e \rightarrow e' : e \rightarrow e' \in R \vee \exists e''\ s.t.\ e \rightarrow e'' \in R \wedge e'' \rightarrow e' \in R^+\}$.

It describes which vertices e' in a dependence graph (represented by relation R) are connected directly or transitively with vertex e.

A weakly connected component (WCC) is a maximal subgraph in which all components are connected by some path, ignoring direction. All WCCs of a dependence graph are independent and can be executed in parallel.

A loop nest domain, LD, is the union of the iteration domains of all statements.

A *schedule* is a function $\sigma : LD \rightarrow \mathbb{Z}$ that assigns a discrete time of execution to each loop nest statement instance. A schedule is *valid* if for each pair of dependent statement instances, $s_1(I)$ and $s_2(J)$, satisfying the condition $s_1(I) \prec s_2(J)$, the condition $\sigma(s_1(I)) < \sigma(s_2(J))$ holds true, i.e., the dependences are preserved when statement instances are executed in an increasing order of schedule times.

3   Basic Concept

Main idea of the approach discussed in this paper is to split a loop nest dependence graph into weakly connected components and then for each of them find a legal schedule. Let us consider the following working loop nest.

```
for ( i =1; i <4; i++){
s1:    b[i] = b[i−1]+a[i−1][1];
    for ( j =1; j <4; j++){
s2:      a[i][j] = a[i−1][j] + b[i];
}}
```

The normalized relation[1] below represents dependences in this loop nest.

---

[1] To allow for computing the union of all dependence relations for imperfectly nested loops as well as valid code generation, we have to preprocess sets and relations. Preprocessing makes the sizes of tuples to be the same by means of scattering [1] and inserts identifiers of loop nest statements in the last position of tuples.

$$R := \{[i, -1, 1] \rightarrow [i, j, 2] : 1 \leq i \leq 3 \ \wedge \ 1 \leq j \leq 3; \ [i, j, 2] \rightarrow [1 + i, j, 2] :$$
$$1 \leq i \leq 2 \ \wedge \ 1 \leq j \leq 3; \ [i, 1, 2] \rightarrow [1 + i, -1, 1] : 1 \leq i \leq 2; [i, -1, 1] \rightarrow$$
$$[1 + i, -1, 1] : 1 \leq i \leq 2\}.$$

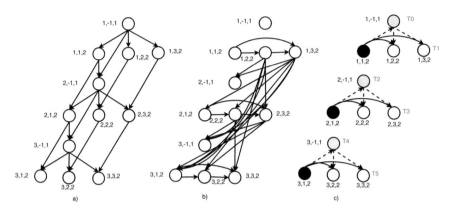

Fig. 1. (a) dependence graph, (b) graph described with relation R_IND, (c) graph formed according to relation R_T and schedule time for statement instances of the working example.

Figure 1(a) shows the dependence graph for the working example. There exists only one weakly connected component in this graph.

The next step is to form a relation, R_IND, representing independent statement instances within a given weakly connected component, WCC, as follows,

$R\_IND = \{[e] \rightarrow [e'] : e, e' \in WCC \ \wedge \ e' \notin R^+(e) \ \wedge \ e' \succ e\}$, where $R^+(e)$ denotes the application operation of transitive closure R^+ to element e. For each statement instance e within a WCC, this relation exposes all lexicographically greater statement instances e' within this WCC that are independent of e.

For the working example, relation R_IND is as follows.

$$R\_IND := \{[i, j, 2] \rightarrow [i', j', 2] : 1 + i \ \leq i' \leq 3 \ \wedge \ i \geq 1 \ \wedge \ j \leq 3 \ \wedge \ 1 \leq j'$$
$$\leq -1 + j; [i, j, 2] \rightarrow [i, j', 2] : j \geq 1 \ \wedge \ 1 + j \leq j' \leq 3 \ \wedge \ 1 \leq i \leq 3; [i, j, 2] \rightarrow [i', -1, 1] :$$
$$1 + i \leq i' \leq 3 \ \wedge \ i \geq 1 \ \wedge \ 2 \leq j \ \leq 3; [i, 2, 2] \rightarrow [i', 3, 2] : 1 + i \ \leq i' \leq 3 \ \wedge \ i \geq 1 \ \}.$$

Figure 1(b) represents the graph formed according to relation R_IND above. As we can see from this graph, independent statement instances can belong to different times of a schedule.

We assign to the first schedule time nodes contained in set $UDS = domain\ R - range\ R$ including ultimate dependence sources, i.e., they are the statement instances that are not destinations of any dependences. Then we assign to the same schedule time the nodes that are independent and have a common parent or common grandparent, or common great-grandparent,..., or common great-great-...-grandparent. Such a concept yields a valid schedule if for the nodes assigned to the same time partition, the maximal length of the paths connecting

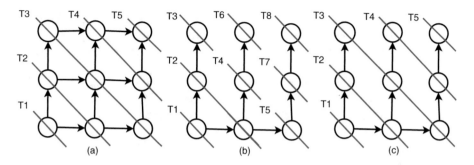

Fig. 2. Illustration of time partitions generated for dependence graphs with (a) k = 1; (b) k = 1; (c) k = 2.

them with ultimate dependence sources is the same. In Sect. 5, we demonstrate that this condition is true for all loop nests from the Polybench benchmark suite being examined by us. If this condition is not true, then the algorithm presented in Sect. 4 discovers that fact and does not return any schedule.

Let us consider the dependence graph in Fig. 1(a). The ultimate dependence source in that graph is node $(1, -1, 1)$. We can assign nodes $(2, 1, 2)$, $(2, 2, 2)$, and $(2, 3, 2)$ to the same time partition because the length of the maximal paths connecting them with the ultimate dependence source $(1, -1, 1)$ is the same.

We form relation R_T, which implements the above concept, it connects nodes belonging to the same time partition:

$$R\_T = \{[i1]-> [i2] : i2 \succ i1 \wedge i2 \in R\_IND(i1) \wedge \exists i3 : ((i1 \in R(i3) \wedge i2 \in R(i3)) \vee (i1 \in R^2(i3) \wedge i2 \in R^2(i3)) \vee ... \vee (i1 \in R^k(i3) \wedge i2 \in R^k(i3))\}.$$

The relation above connects nodes that are (i) independent $(i2 \in R\_IND(i1)))$ and (ii) have a common parent or common grandparent, or a common great-grandparent,..., or a common great-great-...-grandparent $(\exists i3 : ((i1 \in R(i3) \wedge i2 \in R(i3)) \vee (i1 \in R^2(i3) \wedge i2 \in R^2(i3)) \vee ... \vee (i1 \in R^k(i3) \wedge i2 \in R^k(i3))$.

The value of k defines the maximal length of a path connecting a node in the dependence graph with its descendants. This length impacts the number of time partitions represented with a generated schedule. Let us consider Fig. 2. For the dependence graph in Fig. 2(a), the minimal number of time partitions is equal to 5, those partitions are generated with choosing $k = 1$. For the dependence graph in Fig. 2(b), choosing $k = 1$ leads to 8 time partitions, while choosing $k = 2$, we obtain 5 time partitions.

Figure 1(c) shows the graph built according to relation R_T for the working example. Markers $Ti, i = 0, 1, ..5$ show the schedule time for statement instances.

For each time partition, including two or more nodes, we calculate a representative – the lexicographically minimal node belonging to this partition – as follows.

$REPR1 = domain\ R\_T - range\ R\_T.$

For the working example, set $REPR1$ is equal to $\{(1, 1, 2), (2, 1, 2), (3, 1, 2)\}$. In Fig. 1(c), representatives are shown in black. From Fig. 1(c), we can see that some schedule partitions include only one node, for example, time partition $T0$ includes a single node $(1, -1, 1)$.

To form set, $REPR2$, including such representatives, we apply the following formula $REPR2 = WWC - (domain\ R\_T \cup range\ R\_T)$, where WCC is the set including all the nodes of a given weakly connected component. For the working example, applying this formula yields the following set $\{(1, -1, 1), (2, -1, 1), (3, -1, 1)\}$. In Fig. 1(c), the nodes belonging to set $REPR2$ are shown in gray.

Next, we form relation, R_SCHED, describing a loop nest statement instance schedule as follows

$$R\_SCHED := \{[I] - > [I'] : \{I \in REPR1\ AND\ I' \in R\_T^+(I)\} \cup \{[I] - > [I] : I \in REPR2\}.$$

This relation connects a time partition representative with all the other nodes belonging to the same time partition. Let us note that for a node $p \in REPR2$, $R\_SCHED(p) = p$. For the working example, relation R_SCHED is as follows: $R\_SCHED := \{[i, 1, 2] \rightarrow [i, j', 2] : 1 \le i \le 3 \ \wedge \ j' \le 3 \ \wedge \ j' \le 2; [i, -1, 1] \rightarrow [i, -1, 1] : 1 \le i \le 3\}.$

Let us note that, in general, a weakly connected component can include only one node, i.e., such nodes are independent statement instances in the iteration space. To find all such nodes included in set IND, we apply the following formula $IND = IS - (domain\ R \cup range\ R)$, where IS is the iteration space set. For the working example set IND is empty.

4   Formal Algorithm

The formal algorithm is presented below. It includes four steps. The first one splits the iteration space into WCCs. The second one forms a schedule and checks whether it is valid. For each component i, this step includes the following sub-steps. First, relation R_IND_i is formed, which represents independent nodes. Then a set of ultimate dependence sources within a component is calculated and relation R_T_i is generated. This relation connects nodes to be map to the same time partition. The next sub-step calculates representatives of time partitions – the lexicographically minimal nodes. The algorithm recognizes two types of representatives: the first type ($REPR1$) is associated with time partitions, which include at least two nodes, while the second ($REPR2$) defines time partitions including a single node. Relation R_SCHED, which represents a schedule, maps each representative of the first type to all the nodes that belongs to the same time partition that a given representative does, while each representative of the second type is mapped to itself. The last sub-step is to check whether a schedule obtained is valid. For this purpose, relation $VALIDITY$ is formed. It checks whether all original dependences are respected under a generated schedule. If relation $VALIDITY$ is not empty, this means that the schedule obtained is invalid: it cannot be used for code generation. Step three calculates all independent instances and finally the last step generates target parallel code.

Algorithm 1. Loop nest statement instance scheduling

Input: Relation R, describing all the dependences in a loop nest; iteration space, IS, the maximal length of a path connecting a node in the dependence graph with its descendants, k.

Output: A schedule.

Method:

1. Applying the algorithm presented in paper [2], split the iteration space into weakly connected components WCC_i, where r is the number of components,
$1 \leq i \leq r$
WCC_i is the set including all the statement instances of component i.

2. **for each** WCC_i **do**
$1 \leq i \leq r$

   2.1. Form relation, R_IND_i, describing independent statement instances:
   $R\_IND_i = \{[e] \rightarrow [e'] : e, e' \in WCC_i \land e' \notin R^+(e) \land e' \succ e\}$.

   2.2. Form set UDS_i including ultimate dependence sources within WCC_i:
   $UDS_i = (domain\ R - range\ R) \cap WCC_i$.

   2.3. Form relation R_T_i describing statement instances belonging to the same time partition:
   $R\_T_i = \{[i1] -> [i2] : (i1 = lexmin(UDS) \land i2 \in UDS \land i2 \succ i1) \lor i1, i2 \in WCC_i \land$
   $i2 \in R\_IND(i1)) \land \exists i3 : ((i1 \in R(i3) \land i2 \in R(i3)) \lor (i1 \in R^2(i3) \land i2 \in R^2(i3)) \lor ... \lor (i1 \in R^k(i3) \land i2 \in R^k(i3))\}$.

   2.4. Form sets including time partition representatives as follows
   $REPR1_i = (domain\ R\_T_i - range\ R\_T_i)$,
   $REPR2_i = WWC_i - (domain\ R\_T_i \cup range\ R\_T_i)$.

   2.5. Form relation R_SCHED_i, representing a schedule for WCC_i
   $R\_SCHEDi := \{[I] -> [I'] : I \in REPR1_i \land I' \in R\_T_i^+(I)\} \cup \{[I] -> [I] : I \in REPR2_i\}$.

   2.6. Calculate the following relation
   $VALIDITY_i = \{[i1] \rightarrow [i2] : i1 \in domain\ R \land i2 \in R(i1) \land R\_SCHED_i^{-1}(i1) \succeq R\_SCHED_i^{-1}(i2)\}$
   and check whether it is empty; if not, then the end, the schedule obtained is invalid.

   end for

3. Calculate set, IND, describing all independent statement instances
$IND = IS - (domain\ R \cup range\ R)$.

4. Generate final code of the following structure

```
parfor enumerating WCCi, i=1 to r
    for enumerating time partitions T represented with the union of
        all sets REPR1i and REPR2i
        parfor enumerating nodes of each time partition contained in
            the union of all sets ( R_SCHEDi(T) union T)
    parfor  enumerating nodes belonging to set IND
```

Code generated for each weakly connected component enumerates time partition representatives in lexicographical order, which defines the order of the execution of time partitions according to a schedule generated. For each such representative, code enumerates all statement instances to be execute at the same schedule time.

All WCCs are independent, so if a schedule produced for each WCC is valid, this means that it is valid for the whole dependence graph.

For the working example, target code generated by means of isl AST [8] is the following.

```
for (t1 = 1; t1 <= 3; t1 += 1) {
  b[t1]=b[t1-1]+a[t1-1][1];
  #pragma omp parallel for
  for (c3 = 1; c3 <= 3; c3 += 1)
    a[t1][c3]=a[t1-1][c3]+b[t1];
}
```

5   Experiments

We chose four applications of numerical algorithms for experimental study: three typical programs from compute-intensive sequences of algebra operations *seidel*, *lu*, *jacobi2d*, and one simulation stencil *heat3d*. To carry out experiments, we have used a machine with a processor Intel Xeon E5-2699 v3 (3.6 Ghz, 32 cores, 45 MB Cache) and 128 GB RAM. All programs were compiled by means of the Intel C++ Compiler (*icc* 15.0.2) with the -O3 flag of optimization. For all examined programs, choosing $k = 1$ (about k see Sect. 3) leads to generating a valid

Table 1. Execution times of the studied programs, in seconds

Program/code gen. time	Problem size	Number of CPUs					
		1	2	4	8	16	32
LU	2500	3.826	1.937	0.971	0.614	0.548	0.474
	3500	12.101	7.089	4.119	2.543	2.223	2.103
1.743	5000	36.842	26.323	13.549	11.622	10.166	9.568
JACOBI-2D	1000	1.564	0.808	0.438	0.246	0.165	0.156
	2000	23.528	14.243	8.065	3.747	2.901	2.588
0.932	3000	80.053	47.615	25.485	23.363	20.883	16.793
HEAT-3D	100	0.438	0.249	0.127	0.096	0.087	0.077
	200	10.747	5.653	3.031	2.332	2.001	1.876
1.252	300	39.408	21.388	15.655	10.733	9.714	8.197
SEIDEL	500	1.890	0.955	0.481	0.250	0.145	0.098
	1000	13.783	7.096	3.699	1.975	1.073	0.582
1.522	1500	46.858	23.982	13.256	6.483	3.410	1.829

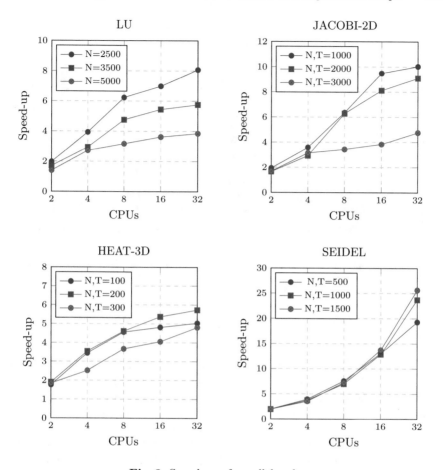

Fig. 3. Speed-up of parallel codes.

schedule and shedule generation time is low (less than 1 s). The performance of generated codes was studied to compare it with that of the original one.[2]

The results in Table 1 shows that parallel codes are significantly faster than sequential ones for various problem sizes. Table 1 presents also parallel code generation time of the examined programs, which are acceptably short (the first column). Speed-up is presented also graphically in Fig. 3.

6   Related Work

The presented approach is within the Index Set Splitting (ISS) framework [7]. It allows for automatic splitting of a loop nest's iteration space into independent

---

[2] The original programs are available at the webpage http://repo.or.cz/w/pluto.git/ tree/HEAD:/examples. The generated codes are available at the repository https:// sourceforge.net/p/traco/code/HEAD/tree/trunk/examples/lu.

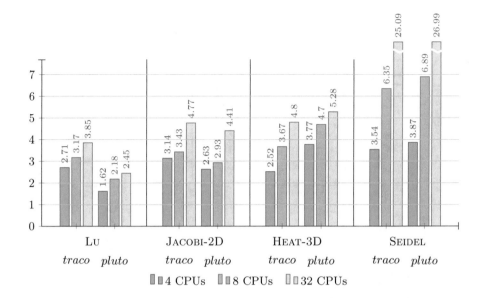

Fig. 4. Speed-up of parallel codes generated by TRACO and PLUTO.

WCCs, and forming a schedule for each WCC, independently. For both splitting and scheduling, the transitive closure of dependence graphs is applied.

For affine dependences, the most powerful algorithm for building schedules is Feautriers one, which is based on multi-dimensional affine schedules [5,6]. But as mentioned by Feautrier, it is not optimal for all codes with affine dependences. The approaches published in [4,9] present different ways of building affine schedules, but none of them guarantees producing scheduling for the general case of loops with affine dependences.

Tools such as the powerful state-of-the-art compiler, PLuTo [4], have provided empirical confirmation of the success of polyhedral-based optimization. The Polyhedral model has proven to be a valuable tool for improving memory locality and extracting parallelism for optimizing dense array codes. Figure 4 shows that the speed-up of generated codes is comparable to that of parallel codes generated by means of PLuTo.

7 Conclusion

This paper introduces a new approach to schedule loop nest statement instances based on the transitive closure of a dependence graph. Its main merit is low computational complexity, it does nor require forming, linearizing, and solving big time partition constraints, as it is required with the affine transformation framework. We are going to extend the discussed approach to allow for code locality improvement, in particular by means of loop nest tiling and compare code generated with optimized code generated with related tools. In general, scheduling of generated tiles is not a trivial problem. Dependences along tiles can be much more irregular than those among statement instances, hence the extraction of parallelism in tiled code becomes a more complex task. We plan to address these issues in our forthcoming research.

Acknowledgements. Thanks to the Miclab Team (miclab.pl) from the Technical University of Czestochowa (Poland) that provided access to high performance multi-core machines for the experimental study presented in this paper.

References

1. Bastoul, C., Cohen, A., Girbal, S., Sharma, S., Temam, O.: Putting polyhedral loop transformations to work. In: International Workshop on Languages and Compilers for Parallel Computers, LCPC 2016. LNCS, Texas, vol. 2958, pp. 209–225 (2003)
2. Beletska, A., et al.: Coarse-grained loop parallelization: iteration space slicing vs affine transformations. Parallel Comput. **37**, 479–497 (2011)
3. Bielecki, W., Palkowski, M., Klimek, T.: Free scheduling for statement instances of parameterized arbitrarily nested affine loops. Parallel Comput. **38**(9), 518–532 (2012)
4. Bondhugula, U., et al.: A practical automatic polyhedral parallelizer and locality optimizer. SIGPLAN Not. **43**(6), 101–113 (2008)
5. Feautrier, P.: Some efficient solutions to the affine scheduling problem: I. one-dimensional time. Int. J. Parallel Program. **21**(5), 313–348 (1992)
6. Feautrier, P.: Some efficient solutions to the affine scheduling problem. Part II. multidimensional time. Int. J. of Parallel Program. **21**(6), 389–420 (1992)
7. Griebl, M., Feautrier, P., Lengauer, C.: Index set splitting. Int. J. Paralleling Program. **28**, 607–631 (2000)
8. Grosser, T., Verdoolaege, S., Cohen, A.: Polyhedral AST generation is more than scanning polyhedra. ACM Trans. Program. Lang. Syst. **37**(4), 12:1–12:50 (2015)
9. Lim, A., Cheong, G.I., Lam, M.S.: An affine partitioning algorithm to maximize parallelism and minimize communication. In: Proceedings of the 13th ACM SIGARCH International Conference on Supercomputing, pp. 228–237. ACM Press (1999)
10. Pugh, W., Wonnacott, D.: An exact method for analysis of value-based array data dependences. In: Sixth Annual Workshop on Programming Languages and Compilers for Parallel Computing. Springer (1993)
11. Verdoolaege, S.: isl: an integer set library for the polyhedral model. In: Mathematical Software, ICMS 2010. LNCS, vol. 6327, pp. 299–302. Springer, Heidelberg (2010)

Design of the BLINDS System for Processing and Analysis of Big Data - A Pre-processing Data Analysis Module

Janusz Bobulski[(⊠)] and Mariusz Kubanek

Institute of Computer and Information Science,
Czestochowa University of Technology, 73 Dabrowskiego Str.,
Czestochowa, Poland
{januszb,mariusz.kubanek}@icis.pcz.pl

Abstract. Big Data is one of the most important challenges of the modern digital world. The possibilities of processing large amounts of data of various types and complexity, coming from various information sources, are used in many areas. The use of Big Data systems will take place in practical areas of all life. The article proposed the system BLINDS, its characteristics and assumptions of the data pre-processing module.

Keywords: Big data · Intelligent systems · Data pre-processing
Multi-data processing

1 Introduction

Every day, millions of GB of new digital data are generated from various sources of information, such as computer systems, portals, databases or IoT (Internet of Things). It is a global phenomenon in all areas of life that led to a situation in which traditional techniques of processing and storing information have become insufficient. As a result, the trend of managing Big Data resources has emerged [1, 2]. The term Big Data refers to technology for collecting and analysing data with high volume and complexity. The data comes from both traditional databases, e.g. those operating in enterprises that contain so-called internal data as well as external sources (documents, e-mails, blogs, social media, various types of electronic sensors, location devices). The data has both a definite and indefinite structure, which makes it difficult to distribute and process them using available IT infrastructure (analytical architecture and tools) and calculation methods.

Semi-structured data (XML) and unstructured data (text files, PDFs, images, videos) dominate. This complexity and diversity of data creates a new challenge for data analysts.

One of the Big Data definitions is based on the concept of three attributes: volume (volume), variety (variety), processing speed (velocity) [3]. In 2012 Gartner introduced an additional two dimensions that relate to large data: variability and complexity. Big Data is a large number of data that requires the use of new technologies and architectures, so that it is possible to extract the value flowing from these data by capturing and analysing the process, the operative part expressed by the authors of the publication

© Springer Nature Switzerland AG 2019
J. Pejaś et al. (Eds.): ACS 2018, AISC 889, pp. 132–139, 2019.
https://doi.org/10.1007/978-3-030-03314-9_12

[4, 6]. In 2013, IBM defines Big Data as a variety of data generated from various sources, at high speed and in large quantities. IBM characterized Big Data with four attributes: volume (volume), processing speed (velocity), variety (variety) and reliability (veracity). SAS defines Big Data as a tendency to seek and use business value hidden in available, ever larger volumes of data, which are characterized by high volatility and complexity. SAS, describing Big Data, draws attention to the additional two attributes: variability and complexity.

To sum up, Big Data is a term used for such data sets, which at the same time are characterized by high volume, diversity, real-time stream inflow, variability, complexity, as well as require the use of innovative technologies, tools and methods in order to extracting new and useful knowledge from them.

For Big Data should be used methods that techniques combining solutions characterized by: [4–8]:

- The volume is characterized by a significant dynamics of data growth, for which new database technologies are required.
- Speed - data flowing quickly, stream, which requires additional processing power to analyse them in real time. Data which due to the limited network capacity should be collected in portions and only those that have a significant informative value should be selected.
- Diversity - data comes from many sources and often appears in a variety of formats and is saved using various models and expressed in any form, e.g. numerically, in text, image, sound, and generated in a variety of ways.
- Volatility - data whose intensity is variable over time, and data flows are subject to periodic cycles and trends, as well as to peaks, which is also related to the dynamics of economic and political processes and events.
- Complexity - the complexity of the data is closely related to diversity. It is characterized by different data ordering. These include data of a specific structure, having a specific type and format, mixed-structured data, partially ordered (semi-structured, "quasi" structured), having some organizational properties, and data not having a natural structure (unstructured), which should be integrated to discover unknown relationships, relationships and hierarchies.
- Value - a unique information value hidden in large and complex data structures, giving the opportunity to draw new conclusions, which then contribute to the increase in the effectiveness of the organization on various levels. The value is an important Big Data attribute. It can be understood as unique knowledge from a scientific point of view, as well as information value being a business benefit, having an impact on reducing the costs of an organization's activities or on improving business relationships and profits.

Big Data is a new challenge and information possibilities [5]. Correct interpretation of data can play a key role in the global and local economy, social policy and enterprises. In management, greater data availability is more accurate analysis and better decisions leading to greater operational efficiency, lower costs and lower risk.

Together with the general challenges related to Big Data, we identify challenges related to data, their sources, quantity, multidimensionality, quality, information that can be obtained from them, and the business value that can be translated into the specific purpose of the organization. Data of the Big Data class come from various sources, contain various types of data and appear in various formats. In part, these data are irrelevant from the point of view of the organization's business purpose and contain incorrect or false information. It is therefore a serious challenge to assess their quality.

A particular problem of Big Data management is the reduction of the dimensionality of data in the context of their analysis and visualization - visual data reduction (Visual Data Reduction) [7]. This is one of the most demanding challenges of data analysis, because data records are multidimensional, often multi-billion. Visualization of such large data sets is often impossible and usually unreadable to the user. In addition, computer systems and technologies limit the ability to act quickly when creating visualizations [9, 10].

2   Big Data Architectures

2.1   Lambda Architecture

One of the two architectures used in Big Data systems is the Lambda architecture, which provides parallel processing of large data sets and the possibility of continuous access to them in real time. The idea behind this architecture is to create two separate flows, where one is responsible for data processing in batch mode, while the other one for accessing them in real mode [11]. A steady stream of data is directed to both layers (Fig. 1).

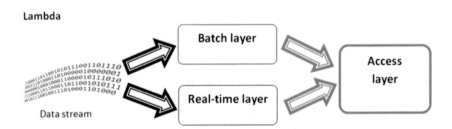

Fig. 1. Idea of Lambda architecture.

In the batch layer, calculations are performed on the entire data set. It happens at the expense of time, but the data received in return contain the full history and high quality. It is assumed that the data set in the batch layer has an undivided form, which should only be expanded rather than delete data from it. In this way, you can ensure data consistency and access to historical data.

The real-time layer processes incoming data in real mode. The short time of access to data in this layer translates into the possibility of faster information retrieval. Unfortunately, the lack of access to historical data means that not all calculations are possible. Often the quality and faith-dignity of data from the real time layer is not as high as from the batch layer. The latter should be considered more reliable, but due to the longer time needed to load them, the real-time layer proves to be an invaluable help in order to guarantee the possibility of processing data in real mode.

The access layer is responsible for creating views based on the batch and real-time layers. The data is aggregated in such a way that the end-user sees it as a single, coherent whole. Views should be prepared in such a way as to ensure the possibility of performing all kinds of ad-hoc analyses, while enjoying fast access to data.

The Lambda architecture concept provides many advantages, primarily a perfect compromise between batch and real-time processing. The biggest and most often mentioned disadvantage is the need to maintain two independent applications - one for powering the batch layer, and the other for the real-time layer. The tools used in each layer are different, so it's hard to choose one that can be used for two purposes. Unfortunately, this architecture is more complicated and more expensive to maintain, so for our system we chose the second of the popular architectures used in Big Data - Kappa.

2.2   Architektura Kappa

Kappa's architecture appeared as a response to criticism related to implantation and maintenance of systems based on Lambda architecture [11]. The new architecture was based on four main assumptions:

1. Everything is a stream - the stream is an infinite number of completed data packages (batches), so that each data source can be a data stream generator.
2. Data is immutable - raw data is persisted in the original form and does not change, so that you can use them again at any time.
3. The KISS rule - Keep is short and simple. In this case, using only one engine for data analysis instead of several, as in the case of Lambda architecture.
4. The ability to restore data state - calculations and their results can be refreshed by retrieving historical and current data directly from the same data stream at any time.

It is critical for the above rules to ensure that the data in the stream remains unchanged and original. Without this condition being met, it is not possible to obtain consistent (deterministic) calculation results.

The idea of Kappa architecture is shown in Fig. 2, and we may see a simplified structure in relation to Lambda architecture.

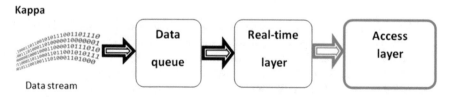

Fig. 2. Idea of Kappa architecture.

Just like in the Lambda architecture, we have a Real-Time layer and an Access layer that performs the same functions here. In contrast, there is no layer of the Batch, which has become redundant because history can be reconstructed at any time from the data stream at the Access layer using identical data processing engine.

3   BLINDS System Project

The name of the BLINDS system is an abbreviation of the names of the component system technologies that is: Bigdata, Learning, Intelligent, Neural network, Deep learning, Self-learning [11–13].

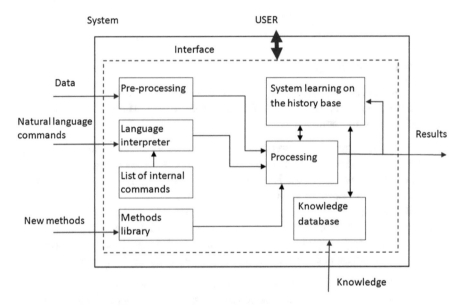

Fig. 3. Idea of BLINDS system

Elements of BLINDS system

The general scheme of the system is shown in the Fig. 3. Main parts of the systems are:

1. System: A system for processing any type of data that is able to handle the system, having its own language and grammar, can learn, is intelligent and able to find solutions to new problems; universal for tasks requiring reasoning in a general sense, using the knowledge base; a system capable of learning and gathering knowledge, as well as data mining and extracting new knowledge from the knowledge possessed; the goal of the system is to gather knowledge, not data.
2. User: a person, device or system having the ability to communicate with the system, having the ability to handle system input/output streams. This applies to

formulating questions, characterizing the analysed problem, and interpreting the results of the system's operation.

3. Data: a set of information in the form of a data stream as well as a file. The system should process data in both known formats and new types [14, 15].

4. External commands: queries from the user in natural language, possibly as simple as possible, constituting correct sentences of a subset of natural language defined in accordance with syntax - grammar, i.e. alphabet, and construction rules.

5. Methods library: a set of functions that will work on a set of external data provided to the system, based on algorithms of artificial intelligence, deep learning, application algorithms, data processing and analysis; this library will be used by the processing unit.

6. Pre-processing: recognition and interpretation of input streams or streams in order to select optimal methods of information processing; recognition of data forms and their processing to the necessary internal form with the analysis of correctness and error signalling.

7. Language interpreter: program that verifies the correctness of external commands, intelligent, interactive parser and generator of executable code or an internal language suitable for the system to perform in the processing; the interpreter can be in the form of an agent or bot using preliminary analysis and algorithms based on fuzzy sets and artificial neural networks, speech recognition [12, 13].

8. List of internal commands: a set of internal system instructions that are not available to an external user; using a library of methods, its own internal data format and its own language and grammar.

9. System history: a collection of information collected by the system during its operation. This knowledge is not generally available and is used only by the system to improve its work; a set of data related to the system operation history stored in accordance with specific rules; record of internal commands issued based on external commands based on preliminary analysis of activities and historical results.

10. Knowledge database: Knowledge of the system stored in accordance with specific rules; objects and relations between them introduced from outside or deduced by the system and stored in the system in a specific way (human and system-readable) enabling quick access to them and their use in taking partial actions in solving problems and in requesting; knowledge base will use elements 3–9.

11. Processing: transformation of input streams into output streams based on elements 3–10; it is a part of the system that enables the data to be processed using all of its isolated elements (3–10), mainly inference; execution of the code or internal language generated by the language interpreter when using the knowledge base and library of methods.

12. Results: streams of output data that meet the expectations formulated by the user and presented in a way that is understandable to him; they also supply the knowledge base; this is every effect/result of the system operation.

13. System input: there are data on the system input - see point 3.

14. Additional data and information entered into the system: the system has the option of downloading additional external data in the form of knowledge or new methods and information about new data formats.

15. System output: a data stream obtained in the results of data processing by the system.
16. Communication and system connections: The system can communicate with other systems and use their knowledge in accordance with the principle of 6 degrees of separation in search of a solution according to the principle: I do not know how to solve it but → I know who can know [14].

4 Pre-processing Unit

Data will not be collected. They will be used to extract knowledge. The "raw" data being the basis of the knowledge discovery process is very often characterized by the following features: measurement errors, missing values in data, disruption during sampling, human mistakes. The system should be resistant to such errors. Therefore, replacing such data with empty values is more justified than with null values, because "zero" is also a value and can lead to incorrect conclusions being drawn.

Preliminary analysis of data should lead to the formulation of only such statistical statements regarding data for which the value of "true" is the invariant of correct transformations, e.g. median, mean or range. Therefore, the pre-processing unit should control the quality of data, because according to the GIGO principle (garbage in - garbage out) - entering erroneous data results in erroneous results - conclusions. Poor quality data make it difficult to draw correct conclusions and, as a result, knowledge exploration and rational decision making. Loaded data and dependencies derived from them can have serious consequences when it comes to formulating laws and rules.

Pre-processing should include data cleaning and transformation to prepare for exploration. It is estimated that the initial data processing is 70–80% of the knowledge discovery process. Data after verification of, for example, the range, will be converted to the internal format of meta-based tags. Thanks to this we will obtain a unified data structure. An additional benefit of this stage of data processing will be their standardization. A detailed description of the data format will be developed in further works and parallel processing of multiple data streams is also planned [16, 17].

5 Conclusion

Big Data technology creates new opportunities for researchers. A large amount of diverse data from many sources is a factor that has a decisive impact on the quality of results in scientific research. Better, in the qualitative sense, knowledge bases determine the extraction of unique, for scientific reasons, conclusions. In addition, new data collection and processing technologies will enable interdisciplinary scientific research. For entrepreneurs, Big Data also means new perspectives for the company's development. Future large IT systems will be based on techniques that use Big Data, so new technologies that can process large amounts of data and extract useful knowledge from them should be developed.

Further work on the system proposed in the article will concern subsequent modules and own data format. It will also be necessary to develop your own language and grammar based on the ideas of Zadeh and fuzzy sets.

References

1. Buhl, H., Röglinger, M., Moser, F., Heidemann, J.: Big data. Bus. Inf. Syst. Eng. **5**(2), 65–69 (2013)
2. Zikopoulos, P., Eaton, C., deRoos, D., Deutsch, T., Lapis, G.: Understanding Big Data: Analytics for Enterprise Class Hadoop and Streaming Data. McGraw Hill, New York (2012)
3. Doug, L.: Data Management: Controlling Data Volume, Velocity, and Variety, Application Delivery Strategies. META Group, Gartner, Stamford (2011)
4. Chang, C., Kayed, M., Girgis, M.R., Shaalan, K.F.: A survey of web information extraction systems. IEEE Trans. Knowl. Data Eng. **18**(10), 1411–1428 (2006)
5. Jinchuan, C., Yueguo, C., Xiaoyong, D., Cuiping, L., Jiaheng, L., Suyun, Z., Xuan, Z.: Big data challenge: a data management perspective. Front. Comput. Sci. **7**(2), 157–164 (2013). SP Higher Education Press
6. Katal, A., Wazid, M., Goudar, R.H.: Big data: issues, challenges, tools and good practices. In: 2013 Sixth International Conference on Contemporary Computing (IC3), pp. 404–409. IEEE, Noida (2013)
7. Keim, D., Kohlhammer, J., Ellis, G., Mansmann, F.: Mastering The Information Age – Solving Problems with Visual Analytics. Eurographics Association, Goslar (2010)
8. Labrinidis, A., Jagadish, H.: Challenges and opportunities with big data. Proc. VLDB Endow. **5**(12), 2032–2033 (2012)
9. Tabakow, M., Korczak, J., Franczyk, B.: Big data–definition, challenges and information technologies. Bus. Inform. **1**(31), 138–153 (2014)
10. Maslankowski, J.: Analysis of the quality of data obtained from websites using Big Data solutions. Ann. Collegium Econ. Anal. **38**, 167–177, (2015)
11. Marz N., Warren J.: Big Data Principles and Best Practices. Manning Publications Co., Greenwich (2015)
12. Zadeh, L.A.: Computing with Words: Principal Concepts and Ideas. Springer Publishing Company Incorporated, Heidelberg (2012)
13. Schank, R.C.: Conceptual Information Processing. Yale University, New Haven (1975)
14. BBC: Connecting with people in six steps. http://news.bbc.co.uk/2/hi/programmes/more_or_less/5176698.stm
15. Bobulski, J.: Multimodal face recognition method with two-dimensional hidden Markov model. Bull. Pol. Acad. Sci. Tech. Sci. **65**(1), 121–128 (2017)
16. Kubanek, M., Bobulski, J., Adrjanowicz, L.: Characteristics of the use of coupled hidden Markov models for audio-visual Polish speech recognition. Bull. Pol. Acad. Sci. Tech. Sci. **60**(2), 307–316 (2012)
17. Bobulski,, J.: Parallel facial recognition system based on 2DHMM. In: Advances in Intelligent Systems and Computing, vol. 534, pp. 258–265. Springer (2017)

QoS and Energy Efficiency Improving in Virtualized Mobile Network EPC Based on Load Balancing

Larysa Globa$^{(\boxtimes)}$, Nataliia Gvozdetska$^{(\boxtimes)}$, Volodymyr Prokopets$^{(\boxtimes)}$, and Oleksandr Stryzhak$^{(\boxtimes)}$

National Technical University of Ukraine "Igor Sikorsky Kyiv Polytechnic Institute", Kyiv, Ukraine
lgloba@its.kpi.ua, n.gvozdetska@gmail.com,
vprokopets95@gmail.com, sae953@gmail.com

Abstract. Virtualization is the modern trend of networks development. Due to the SDN and NFV technologies, network equipment is being replaced by distributed software installed on the common servers. Thus, such challenges as load balancing and resource allocation become relevant for virtualized networks. In this paper three approaches to load balancing in EPC network are proposed: static approach, adaptive approach and energy aware approach. These approaches' efficiency is analyzed using Matlab modeling. The results showed that proposed approaches allow to avoid QoS degradation in case of overloads in some parts of the network and energy efficiency of traffic processing is improved at the same time.

Keywords: Load balancing · EPC · Virtualization · Energy efficiency · QoS

1 Introduction

Nowadays, more and more Internet services are appearing every day. It causes the rapid growth of network traffic, particularly mobile data traffic. According the Ericson forecast, mobile data traffic is going to reach 110 EB per month till 2023 [1]. Rapid growth of mobile data traffic produces challenges for mobile network operators. One of the biggest challenges is how to make network flexible enough to meet the requirements of fast growing data traffic processing? To deal with this challenge, such concepts as SDN (Software defined network) and NFV (Network Function Virtualization) were proposed. According to [2], NFV is the concept aimed to reduce the dependency on hardware, where mobile network functions are deployed as software virtual network functions on commodity servers at cloud infrastructure, i.e., data centers, and SDN is the concept that provides a programmable and flexible network control by decoupling the mobile network functions into control plane and data plane functions.

Due to the SDN and NFV technologies using specialized network hardware can be replaced by common servers with special software. Thus, it becomes possible to manage virtual networks resources as it is done for server clusters' and distributed data

© Springer Nature Switzerland AG 2019
J. Pejaś et al. (Eds.): ACS 2018, AISC 889, pp. 140–149, 2019.
https://doi.org/10.1007/978-3-030-03314-9_13

centers' resources. In this regard, one of the problems that become significant for virtualized network management is load balancing – in terms of load balancing between hardware servers that are used for virtualized functions placement and among the geographically distributed parts of the virtualized network as well. The process of load balancing should ensure all the traffic is processed with the required QoS (Quality of Service) and with minimal costs.

At the same time, as soon as network equipment is being replaced by common servers in data centers, the problem of data centers' energy efficiency improving becomes more and more relevant. According this, it is a need to find optimal load balancing approaches for virtualized mobile networks to provide subscribers with high QoS, save mobile network operators' costs and ensure energy efficiency as well. According this, three approaches to load balancing in SDN networks are proposed and analyzed. These approaches are: static, adaptive and energy aware approach.

The paper is structured as follows: Sect. 2 contains state of the art and analysis of load balancing problem for SDN networks. Section 3 explains the problem to be solved by proposed approach. Section 4 introduces proposed approaches. Section 4 presents a model for considered approaches analysis and comparison. Section 5 includes the summary.

2 State of the Art and Background

The problem of load balancing in virtualized networks is being widely investigated due to its high relevance. In [3] an algorithm of load balancing in modified LTE core network was proposed. As the base, LTE architecture modified in [4] was used, where OpenFlow (OF) switches are used instead of SGW (Serving Gateway) traditional LTE network elements. According to the algorithm proposed in [3], the network load statistics is collected by the OF switches and sent periodically to the controller, which keeps the general state of traffic load in the network. The load is balanced on the base of the statistics that is kept by the controller. The efficiency of the algorithm was proven by simulation. The algorithm in its application to the described network allowed to balance the load while reducing latency. However, the proposed approach is only aimed to process successfully all traffic but does not consider the computing and energy resources efficient usage as a goal.

The problem of efficiency improvement of data center resources using is investigated in [5]. Authors propose three optimization models that aim at finding the optimal dimensioning and planning for a mobile core network based on SDN and NFV, in terms of network load cost and data center resources cost. The balance between cost factors of network and data center was achieved by using Pareto optimal multi-objective model. However, authors do not pay enough attention to the energy efficiency of data processing in different data centers. Energy efficiency can vary from data center to data center due to different types of hardware use and different data center design. Thus, it is a need to pay additional attention to the energy efficiency problem while choosing the data center.

In [6] the load balancing scheme for data center based on SDN is proposed. The scheme means dynamic scheduling according to the server status. It balances applications using the feature of programmability in OpenFlow network.

However, proposed methods do not take into account the energy efficiency of mobile traffic processing, but this issue is very important nowadays. Within this paper, three approaches to load balancing in SDN networks are proposed including energy aware one, that bases on the approach that we previously proposed in [7]. The approaches named static and adaptive are considered in paper too. Their efficiency is analyzed with the help of Matlab modeling.

3 Problem Definition

Consider that the virtualization is applied in mobile network EPC of the national mobile network operator. EPC is a part of typical architecture of LTE mobile network. LTE architecture is presented in Fig. 1, where EPC is highlighted. Each of EPC network elements can be virtualized according to NFV approach. The functions of each EPC element are described in details in the LTE standard [8].

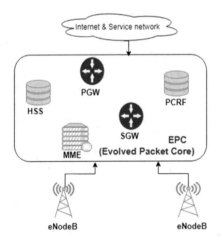

Fig. 1. Evolved Packet Core network within LTE architecture

Assume that the servers with virtualized network functions placed, are located in different regions around the country and local traffic is sent to the local EPC. Consider these local network subsystems are called clusters of the virtualized EPC network. They mean the physical datacenters that are used for distributed virtualized infrastructure implementation.

In case of some cluster is overloaded (for example when some public event or holiday is carried out in a region), traffic processing in the local EPC cluster can lead to poor service quality and subscribers' dissatisfaction.

Thus, the problem is:

- to provide sufficient quality of service for subscribers in overloaded areas (through load balancing among remote EPC clusters);
- to provide energy efficiency of load balancing process.

4 Proposed Approaches to Load Balancing in Virtualized EPC Network

In this paper three approaches to load balancing in virtualized EPC network are analyzed. These approaches include:

1. Static load balancing
2. Adaptive load balancing
3. Energy aware load balancing

These approaches are analyzed using Matlab modeling.

4.1 Static Load Balancing

According to this approach, that is the simplest one, in case of a lack of resources for one of the virtualized network functions, resources are provided in the equal amount *res* from the total resource pool of each cluster. The amount of resources that can be allocated from each cluster is defined on the base of statistical data ensuring sufficient local resources amount saving to meet QoS requirements. The load is balanced between the clusters not taking into account their current state.

4.2 Adaptive Load Balancing

In case of a lack of resources for one of the virtualized EPC network clusters, the system works as follows:

1. The amount of resources *lack* that lacks for ensuring required QoS for the overloaded cluster is defined
2. The amount of free resources res_free_i for each of the operator's EPC clusters is defined
3. All available free resources form a pool of standby resources: $pool = \Sigma_i \ res\_free_i$
4. Overloaded cluster routes traffic to the other clusters in an amount proportional to the amount of standby resources of each cluster, that are included into the common pool.

4.3 Energy Aware Load Balancing

This approach means that the power consumption of each cluster is measured periodically and the statistics for power consumption is available for each cluster. Average power consumption should be normalized to take into account the scale of each local cluster. Normalized values of power consumption can vary due to the different physical servers that can be used and different local data centers' design. As the tariffs for electricity can also vary from city to city around the country, operator should evaluate each cluster's energy efficiency not only by the amount of consumed power, but also by cost for this amount.

Thus, according to the energy aware load balancing resources are allocated in proportion to the amount of free resources and the energy efficiency of each cluster. For load balancing an approach that we proposed in [7] is used. It is proposed to modify this approach in order to improve efficiency of SDN networks.

The modification consists of the following steps:

1. To evaluate energy efficiency of each cluster i on the base of statistics of power consumption P_i
2. To sort clusters by their energy efficiency $\rightarrow$ {List_P_i}
3. To sort clusters by the amount of available resources that can be used for data processing $\rightarrow$ {List_res_i}
4. To put mark to each cluster for energy efficiency $mark_P$ and resources reserve $mark_{res}$
5. To find the sums of the marks for each cluster i: $mark_i = mark_{P\_i} + mark_{res\_i}$
6. Allocate the load in proportion to the summary marks $mark_i$ of each cluster

To evaluate the efficiency of modified energy aware approach Matlab modeling was used.

5   Modeling

In order to evaluate the approaches' efficiency, the EPC network of Ukrainian LTE network was modeled with some assumptions. Assume that the LTE network EPC clusters are deployed in 5 biggest cities in different parts of Ukraine: Kyiv, Kharkiv, Odesa, Dnipro, Lviv. The assumed network topology is shown in the Fig. 2.

Fig. 2. EPC network topology in Ukraine

5.1   Static Load Balancing

The load balancing in the virtualized EPC network according to the static approach is realized by simulating the overload of one cluster. Within the model the case of "Atlas Weekend" annual mass event holding in Kyiv was used. During this event the load exceeds the capability of an existing pool of resources in Kyiv. In case of network virtualization implementation, additional resources can be allocated from other local EPC clusters.

A simple static load balancing approach is implemented by allocating a static number of resources from each cluster, which is calculated based on the statistics of the load of each data center during the year, and statistics of each data center during mass events. In the model, the values are taken so that the overloaded node continues to function properly, and the other nodes did not risk being overloaded during the hour-peak of the full-time workload of the working day. The implementation of the load balancing is shown in Fig. 3.

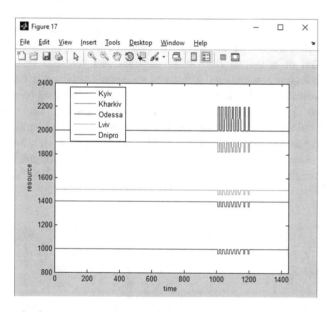

Fig. 3. The graph of resource pools changes for each city with the usage of static load balancing approach

As it is possible to see in Fig. 3 load redistribution was carried out in the time interval of 1000… 1200 m, which corresponds to the busy hour.

In the Fig. 4 it is possible to see the amount of resources of Kyiv EPC cluster in discrete resource units for different time moments.

As can be seen from Fig. 4 at the moment of 1159 h, an assessment of the Kyiv local EPC cluster necessary resources took place and the decision was made to allocate additional 200 resource units for this cluster. After the load returned to normal, the next

Variables - Resource_Pool_Kyiv

| Resource_Pool_Kyiv ✕ | Resourses_Kyiv ✕ | ResoursesFreeKyiv ✕ |

1x1440 double

	1153	1154	1155	1156	1157	1158	1159	1160	1161
1	2000	2000	2000	2000	2000	2000	2200	2200	2200
2									

Fig. 4. Resource pool of Kyiv city changing in time

assessment of the resources was carried out and all allocated resources were returned for processing local traffic, which can be seen in Fig. 3.

As the result, the traffic which was excessive for Kyiv cluster was processed by other clusters using their redundant resources and avoiding QoS degradation.

5.2   Adaptive Load Balancing

The use of static load balancing is appropriate only in the presence of a large amount of statistical data, in which a clear pattern of change in daily load can be found. In this case, it is expedient to use a static load balancing that is simple and proves its efficiency in particular cases. However, at the initial stages of system deployment, when the required amount of statistical data is not accumulated yet, and in situations with extremely high load peaks that may go beyond statistical load fluctuations, the statistic-based model will not be able to solve the problem of efficient load balancing.

In this case, it is advisable to use adaptive load balancing. The mechanisms of adaptive load balancing analyze the amount of applications that cannot be serviced without a queue at a given time in considered EPC network cluster, and additionally analyzes the load of other network clusters. The result of such calculations is the table of routing for redundant traffic to other unloaded data centers (clusters). Resource pools of each considered network clusters changing is shown in Fig. 5.

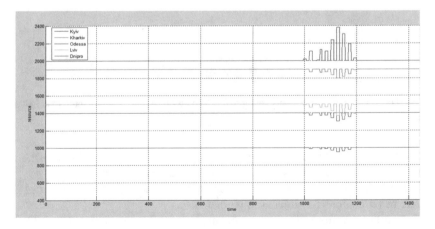

Fig. 5. The graph of resource pools changes for each city with the usage of adaptive load balancing approach

As can be seen from Fig. 5, each cluster state in terms of its load is evaluated and load balancing is performed according to the volume of free resources in other network clusters. The resources are allocated in proportion to the amounts of free capabilities for each cluster. This can be seen from the example of the Dnipro resource pool (purple line): the amount of resources of this cluster that were used for traffic from Kyiv processing is much smaller than for Kharkiv (green line), in which standard resource pool is greater than 25%.

As the result, Kyiv got the additional amount of resources sufficient for its traffic processing and there is no lack of capabilities in other clusters as well. This is a consequence of taking into account the actual load on each network node. Therefore, adaptive network load balancing should be used at the initial stage of deployment of the network, when a sufficient amount of statistics has not yet been collected. First of all, this applies to cases of mass public events: concerts, festivals, etc., which have the character of a local load peak what is not accounted by statistics. Secondly, not only geographical peaks of loading, but also calendar peaks: holidays, seasonal migration, etc. are taken into account.

5.3   Energy Aware Load Balancing

According to the energy aware load balancing approach, the clusters are evaluated by their amount of free resources and energy efficiency. Assume that the statistic of power consumptions of each considered cluster is known. It is shown in Table 1 (assume that the scale of each local cluster is the same, no need to normalize power consumptions).

Table 1. Results of energy aware load balancing

City	Average power cons., kW/day	Mark for power cons.	Amount of free resources, %	Mark for reserve	Total mark	% of resources to be allocated
Odesa	343,2	4	15%	3	7	35%
Kharkiv	367,5	1	25%	4	5	25%
Lviv	345,2	3	10%	2	5	25%
Dnipro	355,7	2	5%	1	3	15%

The difference in power consumptions can be caused by difference in physical servers used and data centers' construction. The Energy aware load balancing was modeled according to the following algorithm [9]:

1. To evaluate energy efficiency of each cluster on the base of statistics of power consumption
2. To sort clusters by their energy efficiency
3. To sort clusters by the amount of available resources that can be used for data processing
4. To put mark to each cluster for energy efficiency and resources reserve

5. To find the sums of the marks for each cluster
6. To allocate the load in proportion to the summary marks of each cluster.

Assume Kyiv cluster is overloaded and needs 10% of its traffic to be reallocated to other clusters to ensure QoS. Clusters evaluation is carried out according to data in Table 1. The power consumption data is chosen according to the statistics from [10], where it is stated that one server that can be used for SDN system approximately consumes 0,4 kW per hour. Thus, one server consumes around 9,6 kW per day. Assuming that for all virtualized network functions placement 35–50 servers are needed, the power consumptions for each of regional clusters per day are listed in Table 1.

Resulting data shows, that the traffic from Kyiv cluster that is unable to be processed locally should be distributed among the clusters of Ukraine in such proportion: 35% of excessive traffic is rational to process in Odessa cluster, 25% - in Kharkiv cluster, 25% - in Lviv cluster and 15% - in Dnipro cluster.

As the result, excessive traffic from Kyiv cluster was re-distributed among other clusters, that helped to avoid QoS degradation and improve energy efficiency of the whole system. Energy aware approach is the most expedient to use in case when clusters in different cities differ significantly by their energy consumption, that can be caused by different hardware and data centers' design.

6   Summary

The problem of load balancing among virtualized EPC network clusters was investigated. Within the research three approaches were proposed and analyzed. According to static approach, load is redistributed in equal proportion between other clusters in case of some cluster is overloaded. Adaptive approach additionally takes into account the state of all available clusters and balances load among most unloaded clusters. Energy aware approach takes into account not only servers' availability but also average power consumption of each cluster. These approaches are analyzed using the Matlab modeling. Modeling showed, that due to the approaches usage it is possible to avoid QoS degradation caused by network overloads in some areas and improve energy efficiency due to load allocation to the most energy efficient clusters. Proposed approaches will allow network operators to save costs through saving energy and ensure required QoS for the subscribers.

References

1. Ericson, Mobile data traffic growth outlook (2017). https://www.ericsson.com/en/mobility-report/reports/november-2017/mobile-data-traffic-growth-outlook
2. Basta, A., Blenk, A., Hoffmann, K., Morper, H.J., Hoffmann, M., Kellerer, W.: Design for a 5G mobile core network based on SDN and NFV. IEEE Trans. Netw. Serv. Manag. **14**(4), 1061–1075 (2017)

3. Adalian, N., Ajaeiya, G., Dawy, Z., Elhajj, I.H., Kayssi, A., Cheha, A.: Load balancing in LTE core networks using SDN. In: IEEE International Multidisciplinary Conference on Engineering Technology (IMCET) (2016)
4. Said, S.B.H., Sama, M.R., Guillouard, K., Suciu, L., Simon, G., Lagrange, X., Bonnin, J.M.: New control plane in 3GPP LTE/EPC architecture for on-demand connectivity service. In: 2013 IEEE 2nd International Conference on Cloud Networking (CloudNet), pp. 205–209. IEEE (2013)
5. Basta, A., Blenk, A., Hoffmann, K., Morper, H.J., Hoffmann, M., Kellerer, W.: Towards a cost optimal design for a 5G mobile core network based on SDN and NFV. IEEE Trans. Netw. Serv. Manag. **14**(4), 1061–1075 (2017)
6. Cui, H., Yang, L., Yu, T., Fang, Y., Zhang, H., Xia, Z.: Load balancing scheme for data center based on SDN. In: The 20th International Symposium on Wireless Personal Multimedia Communications (WPMC 2017) (2017)
7. Schill, A., Globa, L., Stepurin, O., Gvozdetska, N., Prokopets, V.: Power consumption and performance balance (PCPB) scheduling algorithm for computer cluster. In: proceedings of the UkrMiCo 2017, Odesa, Ukraine, pp. 1–8 (2017). http://ieeexplore.ieee.org/document/8095365/
8. Signals Research Group The LTE Standard, April 2014
9. Gvozdetska, N.A., Stepurin, O.V., Globa, L.S.: Experimental analysis of PCPB scheduling algorithm. In: CADSM 2017, 21–25 February, 2017, Polyana-Svalyava (Zakarpattya), Ukraine (2017)
10. Lefergy, C., Rajahmani, K.: Freeman Roson: Commercial servers powerconsumption management. https://www.osp.ru/os/2004/02/183912/

The Approach to Users Tasks Simplification on Engineering Knowledge Portals

Larysa Globa$^{(\boxtimes)}$, Rina Novogrudska, and O. Koval

National Technical University of Ukraine "Igor Sikorsky Kyiv Polytechnic Institute", Peremoga Ave. 37, Kyiv 03056, Ukraine
lgloba@its.kpi.ua, rinan@ukr.net, avkovalgm@gmail.com

Abstract. The paper the approach to computer-aided workflow designing for engineering tasks (engineering web-services) on knowledge portals that can be used to increase the efficiency of engineering tasks performance. The method of engineering tasks simplification is proposed that allows to form the minimized set of engineering tasks elements used for such tasks execution. Specific algebraic system of engineering tasks is described that form the basis for method of engineering tasks simplification. Algebraic system involves formal contextually independent structures for engineering tasks elements representation. The example of the approach usage for real engineering tasks is depicted, the quantitative evaluation of the efficiency increasing for engineering tasks of "Strength of materials" problem domain is given.

Keywords: Engineering tasks · Computer-aided workflow designing
Algebraic system · Web services execution

1 Introduction

Nowadays, different types of web services are widely used for different purposes. Usually the such services usage is oriented on performance of some problem oriented tasks on the end user request. Usually to perform such user's task it is in need to realize the process of web services selection, composition, optimization, etc. The performance of such user's task should be held in most optimal way aiming to increase their execution efficiency.

Usually web services are used for the performance of different processes, as:

- general user's tasks,
- calculation and computational tasks of different subject domain,
- business processes.

This research deals with performance, execution and realization of engineering tasks – specific user's tasks that represent calculation and computational tasks of different engineering subject domain.

The efficiency of engineering tasks performance depends on their execution time. The execution time depends on several factors: time of web services composition (web services are to be composed in dynamic way to form the specific sequence that correlates to certain engineering task), the amount of web services included in the

© Springer Nature Switzerland AG 2019
J. Pejaś et al. (Eds.): ACS 2018, AISC 889, pp. 150–158, 2019.
https://doi.org/10.1007/978-3-030-03314-9_14

engineering task sequence; the time for such sequence execution. The time of web services composition depends on the composition method that is used and the complexity of certain engineering tasks, while the time of engineering task sequence execution can be optimized by the amount of elements minimization (web services, stages) for such engineering task.

The paper describes the approach to computer-aided workflow designing for engineering tasks (engineering web-services) on knowledge portals. This approach can be used to increase the efficiency of engineering tasks performance based on a novel method of engineering tasks simplification. The method of engineering tasks simplification uses specific algebraic system for engineering tasks composition that allows to form the minimized set of web services that uses for such tasks execution.

The structure of the paper is: Sect. 2 gives analyses of related works and introduces basic notions. Section 3 describes specific algebraic system for engineering tasks representation. In Sect. 4 the method of engineering tasks simplification is given. Section 5 presents the estimation and verification of the proposed approach for the application of real engineering task from "Strength of materials" subject domain. Section 6 depicts conclusions and plans for future work.

2   Related Works and Basic Notions

The efficiency of web services composition was examined in various research. In [1, 2, 21–25] different types of web services composition approaches are described. Some approaches present the composition of web services based on service discovery phases and service selection stage [3, 4]. Other research proposes to use semantic for web services composition [5–7, 21–23, 25], or extended SOA architecture [8, 9]. The paper [10] presents the approach to web services composition for user's tasks represented on knowledge portals, the paper [24] deals with the graph evolution technique for web service composition.

Thus, efficiency of engineering tasks performance (time of their execution) depends not only from time needed for web services composition. The time of engineering tasks execution can be minimized by the web services sequence optimization that was formed after their composition. Unfortunately, described researches do not deal with this thesis. To realize such minimization it is urgent to use some formal system that will set some specific formal structures for engineering tasks (or web services) description. That it will be possible to handle some formal simplification method on the elements of such system. There are two types of formal systems that can be used for such purposes: mathematical logic and algebraic system [11–13].

Let us describe basic notions the research deals with:

Engineering knowledge portals – knowledge portals containing information, knowledge and services in engineering subject domain.
User's tasks – specific tasks that are performed by the request of the end user.
Engineering tasks – user's calculation tasks, the implementation of which is regulated by the norms of engineering subject domain.

Web services – specific program module that executes certain stage of engineering task (or engineering task in total).

Metha descriptions – a specific set of descriptions that describes engineering tasks characteristics and features.

This paper deals with engineering tasks that are represented on engineering knowledge portal. The described research operates with formal system and method of simplification used for such engineering tasks.

3   Algebraic System of Engineering Tasks

Algebraic system of engineering tasks allows to obtain engineering task as set of its parameters and elements by introducing specific formal structures for their description. As the result such formal description will be used for correct and fast composition of engineering tasks elements (or web services).

Algebraic system of engineering tasks – A_{ET} is universal algebra (algebra) [12] and is characterized by set of objects (engineering tasks - *ET*), set of operation (*O*) and empty set of relations:

$$A_{ET} = \langle ET, O \rangle \tag{1}$$

Let us introduce the description of algebraic system of engineering tasks elements

Objects – are basic elements of the algebraic system, all operations are conducted on them. For algebraic system of engineering tasks objects are represented by the set of engineering tasks.

Data – is the set of elements that are given to the input and output of the system [14]. For algebraic system of engineering tasks the data can include: different constants, variables, results of operations, etc. At the physical level, the data is represented by the values of various parameters and characteristics of the subject domain, formulas, values boundaries, as well as partial services and computational procedures that represent user's calculation tasks.

Operations. Operations of algebraic system of engineering tasks are divided into several groups according their characteristics. Let us defined such groups of operations:

Simple operations include:

1. Elementary operations (operations of elementary algebra according to the definition):

   + – summation,
   * – multiplication,

2. Set operations (operation on sets given in the set theory):

   $\subseteq$ – set inclusion,
   $\subset$ – proper set inclusion,
   $\not\subset$ – negation of inclusion,

$\in$ – affiliation,
$\notin$ – non affiliation.

3. Logical operations (logical operations of algebra of relations):

d – supplement,
-1 – inversion,
$\vee$ – disjunction,
$\wedge$ – conjunction,
$\backslash$ – difference,
$\sim$ – equivalence,
$\circ$ – composition

Complex operations are:

O^{pc} – the operation of parallel connection,
O^{sc} – the operation of serial connection,
O^{c} – operation of logical composition,
O^{i} – operation of inversion,
O^{m} – matching operation.

The group of simple operations includes standard operations [15–17] and in the research it is demonstrated and proved the possibility of their usage for manipulating with elements of engineering tasks algebraic system. The group of complex operations includes novel operations that are used for the combination of the stages and elements of engineering tasks in integral complex engineering task on end user demand.

As well under the research properties of all given operations were detected. The research had shown that

– for simple operation – their properties are set and proved by appropriate formal theory (elementary algebra, set theory, theory of relations, etc.),
– for complex operations: all operations have the associativity property; all operations except logical composition operation have such properties as commutativity, neutrality, universality and borders idempotence; operation of logical composition has property of distributivity.

More detailed description of proposed algebraic system is given in the research described in [18]. In this paper only main operations are listed and properties of operations are mentioned as they are used as the bases for engineering tasks simplification.

4   The Method of Engineering Tasks Simplification

The algebraic approach, methods of composing tasks into a single workflow, methods of combining and selecting task sets based on ontologies are presented in detail in the papers [10, 18]. The method of engineering tasks simplification that allows forming the minimized set of engineering tasks elements used for such tasks execution can be presented below.

According to the algebraic system of engineering tasks each engineering task can be presented as

$$ET_k = \langle M_k, A_k \rangle \qquad (2)$$

where
M_k –   the set of metha description of k-th engineering task,
A_k –   logical formula of k-th engineering task.

Logical formula of engineering task – $F(ET)$ is the formula that includes elements of engineering tasks – $El(ET)$ (engineering tasks solving stages, data of algebraic system for sequence of engineering tasks design, etc.) connected by operations – O of algebraic system for sequence of engineering tasks design. The logical formula of engineering task is the mathematical representation of computer-aided workflow of the end user's task execution.

To increase the efficiency of engineering tasks execution it is in need to reduce the amount of elements in engineering task logical formula (Fig. 1). For this reason the method of engineering tasks sequence simplification is proposed. The method of engineering tasks sequence simplification is specific mathematical method that is used to represent the logical formula of engineering task in minimal form.

The method of engineering tasks simplification is based on simplification criterion, operations properties and simplification rules. The criterion – C is the amount of elements in engineering task logical formula. The aim of optimization process is to minimize the criterion:

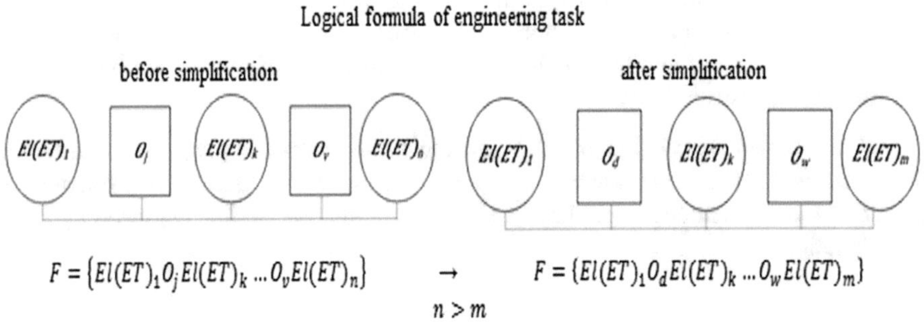

Fig. 1. Engineering task logical formula during the process of simplification

$$F = \{El(ET)_1 O_j El(ET)_k \ldots O_v El(ET)_n\},$$
$$n \to min$$

(3)

The properties of engineering tasks algebraic system operations allow to held equivalent conversion of engineering tasks formulas. Simplification regulations are introduced based on operation properties. Simplification regulations set the rules of formulas transformations while proposed criterion allows to give preference to one transformation over the other.

Some additional rules are involved in the method that allow to choose from the set of invariants one formula which is optimal for engineering task representation. For this purpose the additional definitions, rules and theorems were described under the research. Let us depict some of such rules:

1. The principle of substitution in the formula is:
   if $ET_i = ET_j$, then in any formula containing ET_i instead of ET_i can substitute ET_j and an equivalent formula will be obtained.
2. All equivalent conversion of formulas are carried out in accordance with the basic laws of Algebraic system of engineering tasks.
3. The simplification phase is performed if and only if the formula of engineering tasks after the application of the phase contains a smaller number of elements than before simplify.
4. The choice of engineering task from engineering tasks set (taking into account the transformation on the basis of properties) which must be included in the formula of the general engineering tasks is carried out on the basis of the optimality criterion.
5. The advantage of unary operations over the binary operations by priority of execution is determined.
6. The advantage of simple operations simplification over the complex operations simplification by the priority of execution is determined.
7. The advantage of complex operations conversion over the simple operations simplification by the priority of execution is determined.

In the paper [18] more detailed mathematical description of logical rules is given. The proposed engineering tasks simplification method allows to reduce the time for their execution by minimization of elements amount in their formula. This method uses the set of logical rules from the engineering domain that allows to minimize the amount of the elements (web-services) in end users' workflow.

5 The Estimation of Engineering Tasks Execution Efficiency

For the estimation and verification of the proposed approach applicability the test group of engineering tasks was chosen. For this purpose, a computer experiment was conducted. The tasks present engineering calculations from subject domain "Strength of materials" [19, 20]:

1. Calculation of the strength for the power components of magnetic systems.
2. Calculation of the strength for equipment and pipelines of nuclear power systems.

3. ITER Structural Design Criteria for magnetic components.
4. Magnet DDD 1.1–1.3. Magnet System Design Criteria.
5. Calculation of the strength for the elements of equipment and pipelines of ship nuclear steam generating systems with water reactors.

All elements of such engineering task were represented using algebraic system of engineering tasks. The results of the computer experiment that used proposed method show the possibility to rise the engineering tasks (from test group) execution efficiency.

The time for engineering tasks execution depends on process of its formation and composition of all elements that take part in its execution. During such process the method of engineering tasks simplification is used. As the criteria of simplification the amount of elements in engineering task logical formula $F(ET)$ was used. The amount of elements n in engineering task logical formula was minimized that correspond to the decrease of the amount of requests to storage of services that are used for engineering tasks execution.

Let us analyze the example in Table 1. One of the engineering task from the test group is shown and it is described that before simplification its formula consists of $n = 157$ elements and after the usage of proposed method the amount of its formula elements $n = 131$.

Table 1. The amount of elements *in* engineering calculation formula

The title of the complex engineering calculation	The amount of elements of engineering calculation formula	
	Before simplification	After simplification
Calculation of the strength for the elements of equipment and pipelines of ship nuclear steam generating systems with a water reactors	157	131

Thus, the usage of proposed method allows to reduce the amount of elements in engineering task formula (for the engineering tasks from the test group) in 1,2 times. This led to the decreasing of the amount of requests to services storage that are used for engineering tasks in 1,2 times, which in turn led to the minimization of the time for engineering tasks execution on 17%.

6   Conclusions

The paper presents the approach to computer-aided workflow designing for engineering tasks (engineering web-services) on knowledge portals that can be used to increase the efficiency of engineering tasks performance. The approach bases on the method of engineering tasks simplification using specific algebraic system of engineering tasks aiming to form the minimized set of web services used for such tasks execution.

This method allows to rise the efficiency of various computational tasks execution represented on knowledge portals in different engineering subject domain.

Future works are oriented on implementation of suggested method to user's tasks of various subject domains that allows to show method efficiency and applicability on different real world scenarios.

References

1. Pukhkaiev, D., Kot, T., Globa, L., Schill, A.: A novel SLA-aware approach for web service composition. In: IEEE EUROCON, pp. 327–334 (2013)
2. Moghaddam, M., Davis, J.G.: Service selection in web service composition: a comparative review of existing approaches. Web Services Foundations, pp. 321–346. Springer, New York (2014)
3. Shehu, U., Epiphaniou, G., Safdar, G.A.: A survey of QoS-aware web service composition techniques. Int. J. Comput. Appl. (2014)
4. Martin, D., Paolucci, M., McIlraith, S., Burstein, M., McDermott, D., McGuinness, D., Parsia, B., Payne, T., Sabou, M., Solanki, M., Srinivasan, N.: Bringing semantics to web services: the OWL-S approach. In: Semantic Web Services and Web Process Composition, pp. 26–42. Springer, Heidelberg (2004)
5. Moghaddam, M., Davis, J.G.: Service selection in web service composition: a comparative review of existing approaches. In: Web Services Foundations, pp. 321–346. Springer, New York (2014)
6. Kolb, D.G.: Web-oriented realization of semantic models for intellectual systems. In: Proceedings of scientific conference "Open Semantic Technologies for Intelligent Systems (OSTIS-2012)", 111–122 pp., Minsk (2012)
7. Maximilien, E.M., Singh, M.P.: A framework and ontology for dynamic web services selection. IEEE Internet Comput. **8**, 84–93 (2004)
8. Ngan, L.D., Kanagasabai, R.: Semantic Web service discovery: state-of-the-art and research challenges. Pers. Ubiquitous Comput. **17**(8), 1741–1752 (2013)
9. Hatzi, O., Vrakas, D., Bassiliades, N., Anagnostopoulos, D., Vlahavas, I.: The PORSCE II framework: using AI planning for automated semantic web service composition. Knowl. Eng. Rev. **28**, 137 (2011)
10. Koval, A., Globa, L., Novogrudska, R.: The approach to web services composition. In: Hard and Soft Computing for Artificial Intelligence, Multimedia and Security. Advances in Intelligent Systems and Computing, vol. 534, pp. 293–304. Springer (2017)
11. Barwise, J.: Handbook of Mathematical Logic. North Holland, Studies in Logic and the Foundations of Mathematics (1989)
12. Burris, S.N., Sankappanavar, H.P.: A Course in Universal Algebra. Springer, New York (1981)
13. Glushkov, V.M., Ceitlin, E.L., Yushenko, E.L.: Algebra. Languages. Programing, 376 p. Naukova Dumka, Kyiv (1989)
14. Shahovska, N.B.: Data space in sphere of scientific research. In: Modeling and Information Technologies, vol. 45, pp. 132–140 (2008)
15. Rosen, K.H.: Discrete Mathematics: And Its Applications. McGraw-Hill College (2007). ISBN 978-0-07-288008-3
16. Dwyer, J.: An Introduction to Discrete Mathematics for Business & Computing (2010). ISBN 978-1-907934-00-1

17. Telenik, S.F.: Logic of computation processes representation in the intellectual system SmartBase. In: System Technologies. System Modeling of Technological Processes, pp. 131–139, Kyiv (1999)
18. Globa, L.S., Novogrudska, R.L.: An approach to formal system for knowledge portals development. Ontol. Des. **2**(11), 40–59 (2014) ISSN 2223-9537
19. Norms for the strength calculation for the power components of magnetic systems, 73 p. ISP, Kyiv (1984)
20. Norms for the strength calculation for equipment and pipelines of nuclear power systems. Energoatomizdat, 525 p. (1989)
21. Bansal, S., Bansal, A., Gupta, G., Blake, M.B.: Generalized semantic web service composition. Serv. Oriented Comput. Appl. **10**(2), 111–133 (2016)
22. Rodriguez-Mier, P., Pedrinaci, C., Lama, M., Mucientes, M.: An integrated semantic web service discovery and composition framework. IEEE Trans. Serv. Comput. **9**(4), 537–550 (2016). https://doi.org/10.1109/tsc.2015.2402679. ISSN 1939-1374
23. Petrie, C.J.: Web Service Composition. Springer, Heidelberg (2016)
24. da Silva, A., Ma, H., Zhang, M.: GraphEvol: a graph evolution technique for web service composition. In: Chen, Q., Hameurlain, A., Toumani, F., Wagner, R., Decker, H. (eds.) DEXA 2015. LNCS, vol. 9262, pp. 134–142. Springer, Heidelberg (2015). https://doi.org/10.1007/978-3-319-22852-5_12
25. Wang, C., Ma, H., Chen, A., Hartmann, S.: Comprehensive quality-aware automated semantic web service composition. In: Peng, W., Alahakoon, D., Li, X. (eds.) AI 2017: Advances in Artificial Intelligence, AI 2017. LNCS, vol. 10400. Springer, Cham (2017)

Repository Model for Didactic Resources

Andrzej Jodłowski[1]([envelope]), Ewa Stemposz[2]([envelope]), and Alina Stasiecka[2]([envelope])

[1] Warsaw University of Life Sciences – SGGW, Warsaw, Poland
andrzej_jodlowski@sggw.pl
[2] Polish Japanese Academy of Information Technology – PJATK,
Warsaw, Poland
{ewag,alas}@pjwstk.edu.pl

Abstract. The paper concerns the teaching methodology that defines a learning path as a sequence of teaching steps, where all important subjects and notions are related to adequate didactic resources. For a certain course of object-oriented analysis and software engineering, a characteristics of an exemplary didactic resource is shown with the correspondence to subject groups, subjects and notions. The work presents a proposal of a teaching resources repository introducing a structure model (class diagram) of didactic materials and an adequate model of learning path organization.

Keywords: Teaching methodology · Learning · Learning path
Didactic resource · Didactic resource repository

1 Introduction

The experience in teaching courses in object-oriented analysis and software engineering has enabled us to create various didactic resources. We have observed that courses are significantly better performed when are defined by teaching steps that act on adequate didactic materials. In the authors' research, several teaching approaches have been considered to identify the required characteristics of teaching materials that support students and teachers [1, 2, 6, 7]. A such structure for teaching materials has been sought that would allow their usage in both traditional and distance learning, where the learner's activities mainly rely on independent learning with a limited teacher's support. As a result, the model of effective learning was chosen. One of the features distinguishing this model from the others is that it systematizes the teaching process by specifying the stages and defines for each of them the activities that should be performed in the next steps. Moreover, the model of effective learning strongly emphasizes the necessity to use various types of didactic materials.

In previous works [3–5, 8, 9], we have introduced a proposal of methodology that defines a teaching/learning path as a process and encompasses for a given course all its important subjects and notions in a proper order and with the use of adequate didactic resources. According to the proposal, for each didactic resource, one has to define its correspondence to adequate didactic paths, subjects groups, subjects, notions and its usability to the effective learning model [1].

© Springer Nature Switzerland AG 2019
J. Pejaś et al. (Eds.): ACS 2018, AISC 889, pp. 159–169, 2019.
https://doi.org/10.1007/978-3-030-03314-9_15

In this paper we present a proposal for a description (a structural model) of didactic resources repository that utilizes courses, learning paths, subject groups, subjects and notions.

The paper is organized as follows: Sect. 2 describes the general concepts of the approach, identifies subject groups for a given course and shows a characteristics of an exemplary didactic resource, Sect. 3 presents a proposal for a teaching resources repository that includes a didactic material structure and a model for learning path organization. Section 4 concludes the paper.

2   Teaching Methodology

Initially, the methodology was developed for courses in object-oriented analysis and software engineering, but we assume that it can be applied for other courses (with some assumptions being fulfilled).

An important principle is that the teaching process ought to be precisely specified. For each course one has to define a teaching/learning path using the terms of the methodology. A teaching path consists of teaching steps related to didactic resources used by teachers or students. The approach supports the model of effective learning [1]. Therefore for each didactic resource used in a course, its reference and usability to effective learning model should be specified - see Fig. 1 (in the UML notation [10]).

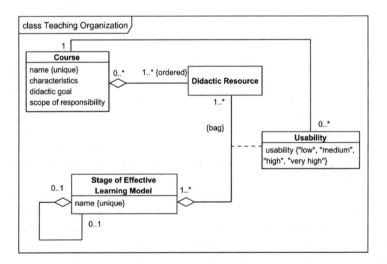

Fig. 1. Teaching organization

The following types of didactic resources were identified: courses, learning paths, subject groups, subjects, notions and didactic materials. The didactic resource is described with such features as: unique name, characteristics, didactic goal and scope of responsibility (some features can be optional, e.g. concepts). The resource may be both an aggregate and a component of another resources. For each didactic resource, its usability and/or difficulty level can be described in relation to other resources on a four-level scale: low, medium, high, and very high. A resource may be extended by information included in other resources. For any extension, it is possible to determine the degree of usability on a scale as before (see Fig. 2).

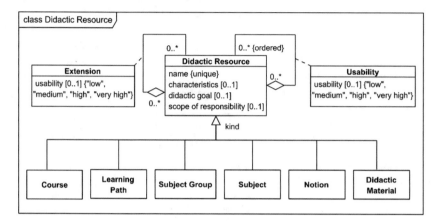

Fig. 2. Didactic resource structure

The didactic materials are divided into: simple tasks, design problem tasks, fragments of lectures, metrics (sets of measures), explanations of common errors, presentations of exemplary solutions, project documentation templates, test and examination kits, etc. According to the authors' experience, design problem tasks are most useful in teaching from both the teacher's and the learner's points of view. Therefore it was decided to incorporate them as a first type of didactic materials into the repository model.

Learning paths can be simple or complex. It is known which of the paths is the essential one. The essential teaching path of a course specifies its core, that is, the minimal or basic portion of knowledge that a learning person has to learn. In addition, a teacher can define for an individual person a so-called individual path – it is a modified version of the basic path where the learning person's interests, abilities, and skills are taken into account.

In order to organize a teaching course, one has to define subject groups and subjects that can be, or ought to be, included in learning paths for a given course. Each of the defined paths must be associated with at least one subject group; similarly, each subject group must have at least one subject related to it. Both subject groups and subjects can be repeated within a path. Subjects can be further refined into more detailed terms, called notions. A subject may be attached to a learning path if it belongs to a subject group within that path. The order of subjects within a given path should be conformant both to the order of subject groups within that path and to the order of subjects within each subject group; similar rules applies to the notions (see Fig. 3).

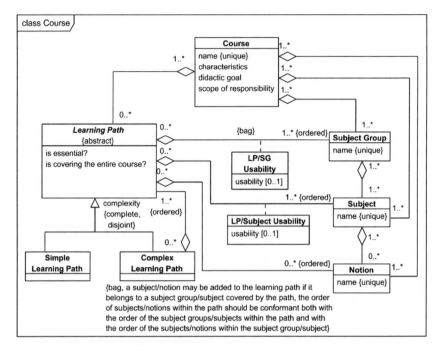

Fig. 3. Teaching course organization

2.1   Sample Lists of Subject Groups, Subjects and Notions

A proposal of the organization of subject groups, subjects and notions concerns a course that lectured at Polish-Japanese Academy of Information Technology in Warsaw. The aim of the course is to introduce general concepts of object-oriented analysis and software engineering using the UML notation [10]. For this course, four subject groups were distinguished, which cover all related subjects: *Course Introduction, Functional Analysis, Structural Analysis,* and *Dynamic Analysis* [9]:

- *Course Introduction* identifies basic concepts of the course and common subjects of all subject groups (stages and main activities characterizing the software development process, model and diagram types used in software development with focusing on package diagrams, basic notions used for building models and diagrams, extensibility mechanisms, name rules of diagram elements concerning the improvement of visibility, diagram structures for the improvement of readability);
- *Functional analysis* encompasses system requirements analysis based on a given requirements text (defining system requirements, building use case models, creating notion vocabularies, constructing use case scenarios, the decomposition of models/diagrams, creating iteration plan for the use case model);
- *Structural analysis* encompasses the construction of models/diagrams that represent structures based on a given requirements text and is conformant to a certain problem domain (building a conceptual schema, transforming a conceptual model into a design model and considering limitations of implementation environments,

transforming a conceptual model into a relational schema, decomposing models/diagrams, building component diagrams and interface specifications);

- *Dynamic analysis* encompasses dynamic analysis and further changes to structural models/diagrams (performing dynamic analysis for use cases, building activity diagrams, state machines diagrams, interaction diagrams realized both on a conceptual and a design level, interaction overview diagrams and timing diagrams, incorporating changes to structural models/diagrams as an effect of dynamic analysis).

The detail description of subject groups, subjects and covered by them notions is presented in the exercise book [4, 9]. The book consists of 11 design problem tasks corresponding to information systems of size from 10 to 15 business classes.

Each design problem task is related to an individual problem domain and includes: user requirements, explained solutions of functional, structural, partial dynamic analysis, fragments of design model, and a partial transformation of conceptual models into entity-relationships diagrams. Table 1 presents an exemplary list of subjects and corresponding notions for the *Structural Analysis* subject group.

Table 1. The *Structural Analysis* subject group: subjects and notions

Subject	Scope of responsibility; Notions
Class vs. object	Class as classifier, object as instance, class invariants
Class attributes	Simple, complex, optional, multi-valued, instance-scoped, class-scoped, derived attribute
Class methods	Method vs. procedure, method signature, method body, abstract method, implemented method, class-scope method, instance-scoped method, method overriding, method overloading, method polymorphism, operation vs. method vs. message
Generalization/specialization structures	Class inheritance, dynamic inheritance, complete inheritance, incomplete inheritance, overlapping inheritance, ellipsis inheritance, multiple aspect inheritance
Abstract class vs. concrete class	Direct class instance, indirect class instance, abstract class, concrete class
Associations	Association as classifier, association hierarchy, link as association instance, association name vs. association end, association cardinality (multiplicity), association end, association class, directed association, aggregation, composition, recursive aggregation, qualified association
Value analysis	Initial (default) value, derived value, constraint threshold value
Composite structure diagrams	Modeling complex attributes
Transformation into design model	Limitations to implementation environment, conceptual (abstract) schema, design (logical) schema, bypassing class inheritance
Transformation into relational schema	Table vs. class vs. class extension, instance vs. table row, primary keys, foreign keys, entity hierarchy, dictionary entity/table, changes to attributes and methods, bypassing class inheritance, changes to associations
Component diagram	Building component diagrams with interface specifications

2.2   An Example of Design Problem Task Characteristics

A general characteristics of each design problem task presents its name, description of a problem domain, goal, scope of responsibility, difficulty level, size, usability for every subject group separately, and usability for every subject group with the conformance to (four) stages of effective learning model. For each subject group, a detail task characteristics describes its difficulty level for a given group, covered subjects with difficulty levels and covered notions.

The difficulty levels for particular tasks treated as a whole, as well as the difficulty levels of subjects (considered within a given subject group for each of the tasks separately) should be defined from the perspective of a person who is starting to learn. A similar principle has been adopted for determining the usability of a task, separately for each of the subject groups within a given task, separately for each stage of learning, and in accordance to the model of effective learning.

The general characteristics of an exemplary design problem task (The "Old Cars" Auction House) is shown in Tables 2 and 3. Table 4 presents the detail characteristics of that resource for the *Structural Analysis* subject group [5].

Table 2. The general characteristics of the exemplary design problem task (the basic description)

General characteristics of task	
Name	The "Old Cars" Auction House
Problem domain	The system can be applied to any domain where there is a need for managing different types of auctions
Goal	The main system's purpose is to support the activity of the "Old cars" Auction House by using internet and to increase the number of customers and offered auctions
Scope of responsibility	The evidence of persons as sellers and buyers, the evidence of auctions and cars, creating new auctions, bidding, generating simple reports (e.g., lists of unsold cars), the management of historical data for auctions and cars
Difficulty level	medium
Size	10 classes

Table 3. The general characteristics of the exemplary design problem task (the usability for subject groups)

Subject group	Usability for a subject group	Conformance to stages of effective learning model
Course introduction	Low	*Introduction:* low *Main content:* low *Summary:* low *Evaluation:* low
Functional analysis	High	*Introduction:* low *Main content:* high *Summary:* high Evaluation: high
Structural analysis	Medium	*Introduction:* low *Main content:* medium *Summary:* high *Evaluation:* high
Dynamic analysis	Medium	*Introduction:* medium *Main content:* medium *Summary:* medium *Evaluation:* high

Table 4. The detail characteristics of exemplary design problem task for the *Structural Analysis* subject group

Detail characteristics of task		
Subject group	Structural Analysis	
Difficulty level of a subject group	Easy	
Covered subjects	Class attributes, class methods, generalization/specialization structures, abstract class vs. concrete class, associations, value analysis, transformation into design model, transformation into relational schema, extensibility mechanisms	
Covered subject	Difficulty level related to a subject for a subject group	Covered notions
Class attributes	Easy	Optional attribute, multi-valued attribute, derived attribute, instance-scoped attribute, class-scoped attribute
Class methods	Easy	Implemented method, class-scope method, instance-scoped method, method overriding, method polymorphism
Generalization and specialization structures	Medium	Overlapping inheritance, dynamic inheritance, complete inheritance, ellipsis inheritance, multiple aspect inheritance
Abstract class vs. concrete class	Easy	Abstract class, concrete class
Associations	Easy	Aggregation, qualified association
Value analysis	Easy	Derived value, initial (default) value, constraint threshold value
Transformation into design model	Easy	Bypassing overlapping inheritance
Transformation into relational schema	Easy	Bypassing overlapping inheritance
Extensibility mechanisms	Easy	«dynamic» stereotype, constraint, comment

3 The Model of Teaching Resources Repository

At present, the structural model of a resource repository encompasses only problem design tasks (which are, in authors' opinion, amongst the most useful type of didactic materials for teaching and learning), see Fig. 4. A general characteristics of a task is expressed using the following elements:

- the *Design Problem Task* class;
- the aggregation between the classes *Design Problem Task* and *Subject Group* (with the features *usability* and *difficulty level*);

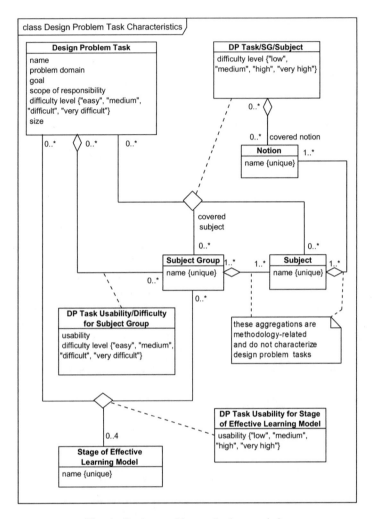

Fig. 4. Design problem task characteristics

- and the ternary association among the classes: *Design Problem Task, Subject Group* and *Stage of Effective Learning Model* (with the *usability* feature).

A description of a detail characteristics of a task is realized using:

- the aggregation between the classes *Design Problem Task* and the *Subject Group* (with the *difficulty level* attribute);
- the ternary association among the classes: *Design Problem Task, Subject Group* and *Subject* that describes subjects covered by a task and with a reference to a certain subject group (with the *difficulty level* feature);
- and the aggregation between the *Task/Subject Group/Subject* association class and the *Notion* class. When a certain subject is covered for a given problem domain task and within a given subject group, it is possible to indicate notions that are covered.

It is worth noting that there are two associations identifying relationships among some terms which are methodology-related (for a certain course) and do not describe design problem tasks. This includes the following aggregations: between the *Subject Group* class and the *Subject* class, and between the *Subject* class and the *Notion* class.

The structure of design problem tasks presented above enables the realization of a wide range of search and browsing operations within repository, such as:

- finding resources that encompass certain subject groups, regarding their usability or difficulty level;
- finding resources that cover certain subjects, regarding their difficulty level;
- finding resources that include certain notions;
- finding resources that are conformant to a given stage of effective learning model, regarding their usability;
- finding paths that cover certain terms (subject groups, subjects or notions);
- and complex explorations that include all of the above.

The organization of a learning path (see Fig. 5) starts with the selection of subject groups, which are to be included at a given learning step (the aggregation between the

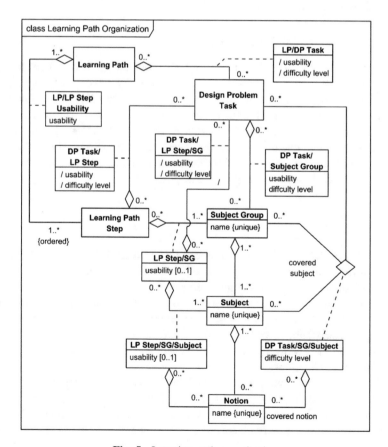

Fig. 5. Learning path organization

classes *Learning Path Step* and *Subject Group*). For each subject group addressed, one specifies subjects to be covered at this step (the aggregation between the classes *LP Step/SG* and *Subject*) and may optionally indicate related notions (the aggregation between the classes *LP Step/SG/Subject* and *Notion*). For a better readability, the relation (the ternary association) between design problem tasks, subject groups and stages of the effective learning model has been omitted from the diagram.

The choice of a resource, that could be assigned to a certain learning step, depends on its usability and difficulty level regarding a given subject group (the aggregation between classes *Design Problem Task* and *LP Step/SG*). These features are derived and can be calculated using the general and detail task characteristics (see Fig. 4). The model allows that several tasks may be associated to a certain learning path step and, in consequence, to a given learning path. For the two relationships: the aggregations between the classes *Learning Path* and *Design Problem Task*, and between the classes *Learning Path Step* and *Design Problem Task*, the attributes *usability* and *difficulty* can be computed in a similar way. One can search for resources with the appropriate (usually maximal) usability and matching the difficulty level to a given one as needed, i.e. with respect to a given subject group, a learning path step or a whole learning path. Such an organization enables the creation of learning paths, consisting of steps associated with resources in a static or a dynamic manner.

3.1   The Repository Implementation

Initially, the implementation of the repository has been addressed to systems that are an institutional repository, used both at universities and by government organizations, available at no charge, active, still developed, and with a sufficient documentation, such as DSpace [11], Islandora [12], Samvera (formerly Hydra) [13]. The first attempt on building a prototype has been based on DSpace. The chosen platform has turned out to be not suitable enough due to limitations of metadata structures and, consequently, also to search and sorting operations. In this situation, there have been started preparations for building the repository as a standalone application.

4   Conclusions

The presented teaching methodology imposes that each teaching path should utilize all covered subjects and notions with the use of adequate didactic resources. For a certain course of object-oriented analysis and software engineering, we present a characteristics of an exemplary didactic resource. Basing on it, we introduce a proposal of a teaching resources repository for specifying resource characteristics and organizing learning path using subject groups, subjects and notions that are related to a given course. The model of repository includes the object models (class diagrams) of resource and learning path structures.

In the further works, we plan to extend the repository to other types of didactic materials (e.g. simple tasks, exemplary solutions, explanations of common problems/errors, project documentation templates, test and examination kits). The repository will allow to classify didactic materials. It will also help in searching various teaching paths for a given course: a general (basic) or individual ones with the conformance to teaching interests or learning person's abilities and skills.

References

1. Allesi, S.M., Trollip, S.R.: Multimedia for Learning: Methods and Development. Allyn and Bacon, Needham Heights (2001)
2. Driscoll, M.P.: Psychology of Learning for Instruction, 3rd edn. Pearson, Boston (2005)
3. Habela, P., Płodzień, J., Stasiecka, A., Stemposz, E.: A teaching methodology proposal for courses in software engineering. In: Proceedings of the 4th Polish National Conference on Software Engineering, Tarnowo Podgórne (2002, in Polish)
4. Jodłowski, A., Stemposz, E., Stasiecka, A.: Organizing teaching paths, computing in science and technology. In: Monographs in Applied Informatics, Warsaw, pp. 258–268 (2015/2016)
5. Jodłowski, A., Stemposz, E., Stasiecka, A.: Proposal for a didactic resources repository, computing in science and technology. In: Monographs in Applied Informatics, Rzeszów, pp. 109–118 (2017)
6. Joyce, B., Calhoun, E., Hopkins, D.: Models of Learning, Tools for Teaching, 3rd edn. McGraw-Hill Open University Press (2010)
7. Smith, L.P., Ragan, T.J.: Instructional Design, 3rd edn. Wiley (2004)
8. Stasiecka, A., Płodzień, J., Stemposz, E.: ObAn – an application supporting e-learning. In: The WSEAS International Conference on Applied Mathematics, WSEAS Transactions on Computers, Malta, vol. 2, no. 2, pp. 305–310 (2003)
9. Stemposz, E., Jodłowski, A., Stasiecka, A.: Zarys metodyki wspierającej naukę projektowania systemów informacyjnych (in polish). Wydawnictwo PJWSTK, Warsaw (2013)
10. UML Modeling Language. http://www.omg.org/spec/UML/
11. DSpace. http://duraspace.org/dspace/
12. Islandora. https://islandora.ca/
13. Samvera (formerly Hydra). http://samvera.org/

SLMA and Novel Software Technologies for Industry 4.0

Andriy Luntovskyy[(⊠)] 🆔

BA Dresden University of Cooperative Education, Hans-Grundig-Street 25,
01307 Dresden, Germany
Andriy.Luntovskyy@ba-dresden.de

Abstract. The paper discusses how active deployment of modern networks in Industry 4.0 stimulates the further development of the novel software technologies for IoT, Fog Computing and Robotics. Such novel applications and mobile apps operate nowadays on advanced hardware and software platforms. The approaches to their development are often aimed on productive and energy-efficient Cloud and Fog co-operation scenarios. The important paradigms and relevant approaches for development of so-called novel SLMA (Server-Less Mobile Apps) have been discussed. An overview of the up-to-date robotics applications and platforms was given. The existing platforms and frameworks for their development are considered as well as analysis of the properties of the future robot platforms is offered. The specials kind of the advanced web apps, so-called PWA (Progressive Web Apps) are suitable for the purpose of spared data use, autonomous access and participation on clouds, increasing of performance, fault tolerance, as well as of online activity of the users.

Keywords: Novel software technologies · Industry 4.0 · Robotics
Progressive web apps · Server-Less Mobile Apps · Bio-inspired engineering
Fog-Cloud platforms

1 Backgrounds: "Industry 4.0" and Intelligent Networking

The algorithmically and energy-efficient intelligent networking becomes step-by-step one of the most important components of the 4th-industrial revolution called "Industry 4.0". The paper discusses how active deployment of modern networks stimulates the further development of mobile apps, IoT and fog computing (Fig. 1). In the future "green" technologies will be wide-spread. The services will be often shifted from the clouds on "the network edges". The energy autarky can be realized via self-sufficient and low-energy nodes. Such interoperable software solutions and platforms help to the providers with energy saving and enhancing of resource efficiency. Together with the client-server and cloud-centric applications and apps the novel applications are permanently developed. These apps must also provide autarky in sense to be only loose coupled to the servers. One of the further goals of the paper was to introduce the readers the basic knowledge about the modern, self-organizing networks and robots. These contents will be an important part of the teaching at the BA Dresden as well as TU Dresden in the future. Modern intelligent and energy-efficient networks, IoT and

© Springer Nature Switzerland AG 2019
J. Pejaś et al. (Eds.): ACS 2018, AISC 889, pp. 170–184, 2019.
https://doi.org/10.1007/978-3-030-03314-9_16

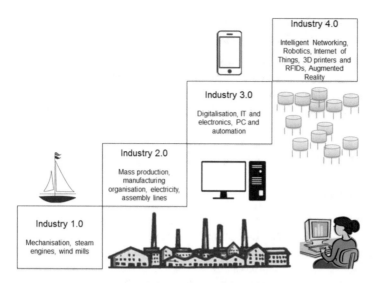

Fig. 1. Context "Industry 4.0" or the four industrial revolutions (own presentation)

fog stimulate the further development of "Industry 4.0". The further works are focused on programming of new scenarios on bio-inspired engineering, deployment of the robots to pedagogical aims as well as behavioral human modeling. The "Industry 4.0" (2011) is one of the future important aims in the German Federal Government's high-tech strategy. The driving force [1–3] for the new generation of the industry (new industrial revolution) is the further automation and computerization of production processes via IoT, fog and cloud computing based on intelligent networking. The goal is as follows: smart factories have to be characterized by adaptiveness, resource efficiency and ergonomic working conditions as well as the integration of customers and business partners into the business value chain [1]. Within Industry 4.0, information and communication technologies as well as automation and production technologies become increasingly and more than ever interfaced to each other. The political ambition is to defend and extend the traditional core of the German industry with its internationally outstanding positions. The manufacturing processes are not only auto-mated (like in Industry 3.0) but also the processed components obtain their additional intelligence via the equipment with not expensive chips (processors, sensors, wireless mini-senders and RFIDs). As further important aim the integration of customers and business partners in an optimized value chain has to be reached through their extensive networking and under taking account of aspects of data security, privacy and anon-ymity [1–5] in these chains. Therefore, we discuss the novel software solutions in this scope: mostly mobile and interfaced to the PHY-world; compact, energy- and algorithmic-efficient; fog and cloud based; progressive, loosely-coupled to the server part, often server-less. The next sections of the paper are organized as follows:

1. A periodization for the novel software.
2. Techniques and trends to PWA, mobile OS and SLMA.
3. Robotics SLMAs and platforms.
4. Case studies for fog and robotics.

2 Trend to the Server-Less Mobile Apps

2.1 A Periodization for the Novel Software

Within the history of the development of the server-based and server-less applications there were permanently the fluctuations between fat and thin client pattern. The both possess some cons and pros in dependence from the requirements, network QoS and use context. E.g. in 1994 Sun Java terminal as thin clients were favourited, but the lack of network performance forced to change to Web Services in 2003. Then P2P systems as server-less structures gained on popularity, then the Cloud-based solutions became a trend (2011) under predominant use of the load-balanced thin clients with functionality delegation to the clouds. A periodization of development of the server-based and server-less apps is depicted in Fig. 2.

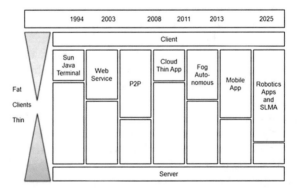

Fig. 2. Periodization of the server-based and server-less applications (own presentation)

2.2 Distribution Techniques and a Trend to the Server-Less Mobile Apps

The so called Server-less Mobile Apps (SLMA) are examined below. The algorithmic efficiency and the performance are the mostly important conditions for so-called server-less mobile apps. The illustration to the performance optimization for client-server systems is given in Fig. 3. The following main approaches are generally used to provide the increasing of the QoS in the novel software systems, i.a. for Server-less Mobile Apps, i.e. SLMA [1, 4, 6]:

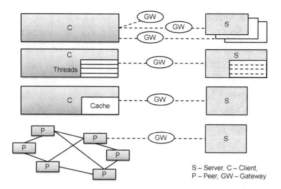

Fig. 3. Performance optimization methods for the server-based and server-less applications

1. Server replication (S) via the available gateways (GW) for more availability and reaction time minimization.
2. Multithreading on the client (C) for fine parallelization with the goal of the speedup of the run-time.
3. Caching of URLs, objects, components, data on the client often used aimed for shorten of the network access.
4. P2P systems without fixed ordering of servers (P - peer) and with so-called Autonomous Participation in Cloud Services.

SLMAs can be divided as follows: (1) conventional or traditional CPS, i.e. the applications embedded in machines; (2) cloud-centric IoT; (3) server-less Fog computing as well as (4) the robots. The mostly used client-server processing paradigms are as follows: fat and thin clients. The last mentioned are nowadays frequently in use in context of the cloud technology. The modern cloud services provide high-performance for the distributed computing and support multimedia data transfer from hidden clusters and storage media to the outside. By leveraging of virtualized computing and storage resources as well as of Web technologies (Web clients, Web services, file system Web-DAV etc.), the clouds permit scalable, network-centric abstracted IT-infrastructures, platforms and applications "on-demand". The billing of these services is user-dependent, while the access is realized by the thin clients. As a disadvantage for cloud technology, the following must be noted: non-uniformity of the data security and safety aspects [1–6].

2.3   Progressive Web Apps

Use of so-called "Progressive Web Apps" (PWA) can now become an efficient alternative to native mobile apps. Instead of the complex development and installation of native apps, the web browser can be used together with optimized web apps. PWAs are also more efficient than plain web apps [7]. PWA is a website with multiple characteristic functions of a native application. Such an app is normally implemented using the standard techniques for RWD (Responsive Web Design) like HTML5, CSS3, JavaScript and can be used across all platforms. The so called "Service Workers" are herewith the core of a PWA. The Service Workers [7] are usually used as proxies as well as OS containers simultaneously with different caching concepts to allow offline functionality for a web app or web application (Fig. 4). In contrast to normal web apps, the PWAs use HTTPS, a cryptographically secured protocol. The development of such client applications for mobile devices for existing versions of the web browsers on the basis of so called "Service Workers" is an up-to-date technical problem. The PWAs are deployed with the purpose of the increasing of performance, fault tolerance, online activity by the users. Generally, the so-called Progressive Web Apps can be distinguished from native apps and regular web apps by the following quality criteria:

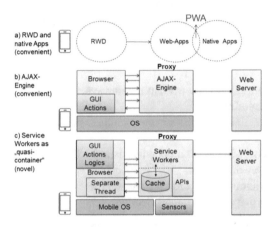

Fig. 4. Comparison and demarcation of PWA to the conventional technologies

- Reduced conversion by installation and update Process
- Findability of online content
- Link-ability of the available information
- Access restriction and optimized memory usage.

The PWAs are available via the following web browsers: (1) mobile PWA: Android Browser, Chrome for Android; (2) desktop PWA: Opera, Firefox, Chrome. Figure 4 depicts a comparison and demarcation of PWA from regular RWD, AJAX concepts as well as from container technology (sandboxes). The complexity of the installation process for novel PWA is significantly reduced for a pair clicks in comparison to the native apps. The advantages of PWAs under use of Service Workers are as follows:

- Web App in form PWA behaves like a native app and can be stored on the smartphone's home screen.
- Service Workers enable the offline use of a PWA even by partial network decoupling (3G, 4G, Wi-Fi).
- Data volume and traffic consumption can be reduced.

Additionally it will profit by the commercial and marketing effects. The users are better connected to the advertised product through efficient push notifications which are sent via PWA. The positive user experiences are improving the user loyalty to the advertised products by increasing of the conversion rate via spared data rates and volumes. The development tools for PWAs with Service Workers can be supported by Ionic Lab framework as well as by the specialized Google Lighthouse and Chrome Canary. The mentioned frameworks offer many interfaces (APIs), which are necessary for the use of PWA.

2.4 Co-operation Architectures and Technical Platforms "Fog-Cloud"

The last times the new paradigm, called "Fog Computing" (also "Edge Computing") obtains more and more meaning. Fog Computing is also a distributed computing environment that supports the IoT approach and brings processing closer to periphery with the augmentation of the cloud functionality [1–6]. The fog energy autarky can be realized via the wireless self-sufficient low-energy nodes on "the edge" e.g. RFID, NFC, ZigBee, EnOcean, 6LoWPAN which are interoperable with 5G nets. The autarky fog nodes provide the monitoring of technological processes, of hospital areas, of environmental contamination, and facility management as well as can cooperate with the clouds and deliver the acquired data. The services and pre-processing are shifted on the "the network edge", i.e. closer to the users (partially or completely). The communication between a cloud and the fog nodes is performed via the three planes (refer the figure) aimed to flexible data pre-processing, VM and functionality migration as well as load balancing. The first Fog-Cloud co-operation platforms exist already [1, 4].

The demarcation of the Fog to other systems like traditional CPS (cyber-PHY-systems) and cloud-centric IoT with their inter-operability aspects are shown in Fig. 5. Traditional CPS (cyber-PHY-systems) are compared with cloud-centric IoT and the loosely-coupled Fog solutions based on light-weighting protocols like REST, MQTT etc. The distinguishing between the PHY-world, heterogenic interfaces, analytic blocks, coordination agents, adaptive role-based interfaces as well as a flexible data flows to the fog and cloud demarcation are used within the presented architectural

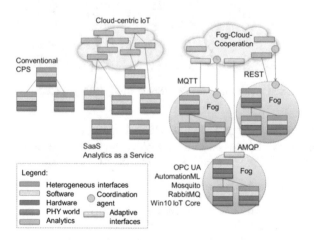

Fig. 5. Fog demarcation and inter-operability aspects [6, 10]

solutions for the novel SLMA. The used "Fog-Cloud" platforms (refer Figs. 5, 6) provide cloud and fog computing inter-operability together with efficient desktop and mobile access via smartphones and tablets. The small intelligent "stuff" cooperates with "big clouds" and provides optimal load balancing as well as costs expenditures [1–6]. For deployment of such kind of co-operation platforms is supported via the industrial IoT gateways from Siemens, Intel, Bosch etc. Fog applications enable optimize service latencies, network bandwidth and available QoS. The security, privacy and anonymity in fog solutions are an open common research topic nowadays [1–6], and, therefore, require more detailed discourse. Concrete technical platforms and software solutions for Fog Computing are rare. They remain mostly a vague technical concept to be fully realized within the next years. Still, a few preliminary architectures exist. One such implementation platform to cloud and fog computing interoperability is offered in [1–6]. Such platform (Fig. 6) possesses the following structure: five layers (0–4), inclusive SDN and hypervisors, as well as multi-layered stack (3–4 layers) for digital appliances and sensor nodes. The platform is oriented to heterogeneous applications. The cloud contains primarily all necessary elements: CPUs, RAM, HDD, SSD, DAS as well as interfaces and combines the advantages of image-streaming with template-based management. The configuration have to be built on-the-fly, based on the configuration profiles, just, like e.g. by HPE [1, 4, 5, 15, 16]. The usable server profile templates are as follows:

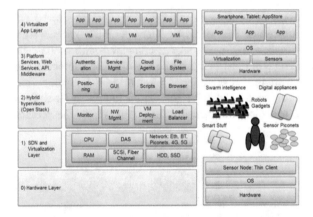

Fig. 6. Co-operation architecture for Fog (Smart Stuff, IoT) and Clouds [1, 4, 5, 15, 16]

- shared hypervisors, VM and physical OS and containers
- computing nodes, storages as well as DAS (Direct Attached Storage)
- coupling interfaces like Ethernet, SCSI, FC, iSCSI, FCoE etc.

The fog part can contain as a rule

- thin clients on the sensor nodes under tiny OS (ARM).
- thin clients for small robots (swarm intelligence).
- the mobile apps offered via digital appliances and so called Smart Stuff.

By the co-operation between Clouds and Fog systems, the influence of transfer times and delays as well as service failure probabilities should be taken into account. The transfer of a computing task to a cloud can cause real-time problems because the execution can be delayed. One of the important construction principles for Fog Computing is namely real-time capability. In the face of the latencies within mobile networks which were significantly reduced down to several microseconds: e.g. 5 ms for 4G mobile radio as well as 1 ms for the 5G, which will be deployed by year 2020.

Furthermore, the Fog paradigm requires the development of novel mobile app architectures and middleware platforms. The mostly useful protocols, exchange formats and selected interfaces for the IoT and Fog solutions are as follows [1, 4, 5, 15, 16]:

- MQTT (Message Queue Telemetry Transport)
- AMQP (Advanced Message Queuing Protocol)
- OPC UA (Open Platform Communications Unified Architecture) or IEC 62541 Specification Google Protocol Buffers (analogous to XML, JSON)
- Markup Languages, e.g. a XML-based AutomationML as well as further frameworks
- APIs and IoT-oriented services (e.g. Mosquito-API).

They are today available on the market as freeware for the practitioners. Furthermore, there is a diversity of implementation techniques for the application and apps: containers in C++, applications in Open62541, Java, MS. NET, JavaScript, Node-OPCUA and Node. JS. Some of them, e.g. Node-RED, can provide automated software design. Furthermore, a very interesting topic will be providing of a SOA/web services oriented to IoT, fog computing and app development automation. Multiple available API can be used: web services over OASIS, Eclipse IDE, JMS, Apache servers, RabbitMQ, Windows 10 IoT Core etc.

2.5 New Paradigms with Google Fuchsia

Nowadays, along with the existing OS from Google called Google Android and Google Chrome OS, a completely new, next generation real-time OS is created to supporting apps from such domains like desktop, mobile, Smart Stuff, CPS and IoT. Google Fuchsia OS promises a significant contribution to the development paradigms for mobile apps in midterm [17]. One of the prompting factors for the development of the Fuchsia project was the clear desire of Google to get rid of the dependence on Java as soon as possible. Fuchsia OS is built as a real-time OS on an efficient micro-kernel Zircon but not a Linux-kernel (cp. Google Android and Google Chrome OS). A hybrid architectural model (Fig. 7) based on micro-kernel and quasi-consistent layers enable efficient memory management, communication between the expanding server and application processes, as well as network communication.

Fuchsia OS provides more data security based on the concept of micro-kernel with surrounding server processes and extending quasi-consistent layers. The application processes (A) order to the server processes (R) the requests (Req) for the app execution. The execution follows only through the secured micro-kernel. The micro-kernel delivers the responses (Rsp) with the resulting data and visualizations itself. In addition, the HAL (HW Abstract Layer) was aimed to supporting of hardware and software

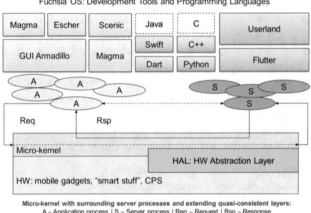

Fuchsia OS: Development Tools and Programming Languages

Fig. 7. Fuchsia architecture based on micro-kernel [17]

drivers for sensors and OS components. The apps development is based on BSD, MIT, Apache licenses and mainly oriented to Dart (proprietary client-sided language for native apps within Fuchsia), Swift (Apple's for compatibility with iOS), Python and some further programming languages. GPL Java will be used no more within Fuchsia as well as some further PL like e.g. "plain old C". This offers more freedom in the use of the development tools and frameworks as well as affects positively the end-price of apps to be developed. The current state of the Fuchsia OS development is as follows: voluminous documentation and an existing working prototype are already available.

3 Server-Less Apps for Robotics

3.1 Motivation

A so-called SLMA (Server-Less Mobile Apps) is per definition a composite application executed in parallel across physical and virtual spaces [4, 6]. The SLMA are nowadays only partially cloud-based and based on the CPS (cyber-PHY-systems). SLMAs are divided as follows: conventional or traditional CPS, i.e. the applications embedded in machines, cloud-centric IoT, server-less Fog computing, as well as the robots. The following Fig. 8 depicts the SLMA-related terms and architecture components for robotics [4, 6, 15].

3.2 Taxonomies and SLMA Robot Platforms. Robotics Standards

On this place, I would like to cite the following opinion of B. Roth [8]: "I had a strong feeling that a meaningful robotics scientific and ethic laws could be developed, and that it would be best to think in terms of general concepts rather than concentrate exclusively on particular devices" (refer Table 1).

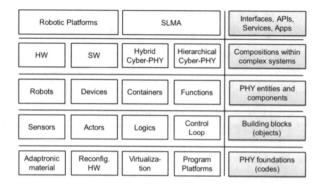

Fig. 8. SLMA and robotics: composite applications across physical and virtual spaces [6–15]

Table 1. The codex of the robot ethic laws

The "Three Laws of Robotics" (1942) by Isaac Asimov from his story "Runaround" as appropriate inspiration for starting the section can be cited as follows:	In 1974–1986 the additional laws 4–6 of robotics were introduced by L. Dilov, N. Kesarovski, V. Ivanov, H. Harrison:
(1) A robot may not injure a human being or, through inaction, allow a human being to come to harm	(4) A robot must establish its identity as a robot in all cases
(2) A robot must obey the orders given it by human beings except where such orders would conflict with the First Law	(5) A robot must know it is a robot
(3) A robot must protect its own existence as long as such protection does not conflict with the First or Second Laws	(6) A robot must reproduce. As long as such reproduction does not interfere with the First or Second or Third Law

Based on the available statistics [9] a fastest growth of professional service robot sales till 2020–2025 can be recognized, i. a. mobile service robots with SLMAs. There are already ca. 5 million personal/domestic robots and ca. 50,000 service robots and drones in professional domains have to appear till 2020. Due to their diversity let's to give a shorten classification. The following professional domains are recognized with the approximated robotics quota: defense (45%), agriculture (21%), logistics (7%), medical (5%), as well as industrial (22%). The robot architectures and platforms possess [8–12] the following specifics:

- Client-centric, loosely-coupled to the server
- Dominant "Sense-Plan-Act"-paradigm with mobility
- Standing in interaction, often asynchronous, and in real-time under uncertainty.

Furthermore, the mostly used robots taxonomies are as follows (cp. Fig. 8)

- programmable, multi-function devices
- domain of use: industrial, service, entertainment, medicine, household

- appearance: hard vs. soft hardware
- movement: walking, driving, flying, stationary (refer Fig. 8).

Due to the shifted accent on the performance, energy and algorithmic efficiency for the SLMAs the following important algorithms have to be optimized [8–15]

- Robot movement and trilateration in 3D-space (unrestricted movement: communication – computation; stationary and mobile robots)
- Robot body Collision Detection with Deadlock Prevention (sensing and routing; autonomous relative positioning)
- Path planning autonomous driving.

The specific methods there are too: (1) no planning, only reaction or (2) mapping and localization + path planning. The responses are waiting in varying temporal scopes: ms for feedback (deployment by 5G mobile is good possible), hours for complex tasks (often to clouds delegated). Commonly, the used nowadays robotics platforms can by classified [8–14] to the four following categories (refer Fig. 8):

- Hardware platforms (for walking, driving, flying, stationary programmable and multi-function robots)
- Coordination platforms (e.g. for swarm robots)
- Development platforms (language-specific, oriented to SLMA for all types: industrial as well as service robots)
- Runtime-platforms (SLMA, all types).

SLMAs are supported via multiple robotic platforms, which are so much important for the further development. Some examples of robot kinds and maintaining platforms are as follows [11–15]:

- Harvard Swarm Robots supported via Amtel WinAVR and Amtel AvrStudio tools
- Robotics BIOLOID robots with RoboPlus Motion platform
- Aldebaran Robotics with NAO Emotion robots
- Webots development platform
- RobotOS runtime architecture (ROS with so-called Autonomous Participation in Cloud Services)
- extended Texas Instruments Architecture for robotics
- as well as the recent development standards for robotics (s. below).

To the recent official and industrial development standards for robotics belong as follows: Texas Instruments Architecture, ISO 8373/TC 184. ISO 8373 contains a definition for industrial robots: automatically controlled and reprogrammable multipurpose manipulators in three or more axes. ISO TC 184 is a movement-focused standard under development. The future requirements are as follows: robot swarms; code offloading from robots; virtual robots and machine learning [8, 16].

4   Case Studies

4.1   Case Study 1: A Freeware Solution for Fog

This case study is based on the materials from the workshop on IBH IT-Service GmbH Dresden on 16.3.2017 [16]. A typical deployment scenario within computing centers is considered [16] and co-operation "fog - private cloud" is examined. The usage scenario is targeted to middle-range cloud or computing center parameter polling (air temperature, humidity, voltage analysis). Furthermore, the scenario can be extended via physical entry control and access detection within the middle-range cloud centers [16]. The measured data from the cloud (computing, networking, storage, environmental) are acquired and analyzed. Let us to discuss how small sensors detect overheat in provider racks. The provider hosts are managed via SNMP and Nagios-tools. The sensors are managed via MQTT, AMQP under use of queue-based "Publisher-Broker-Subscriber-Model" (PBSM). Multiple API are available, i.a. web services over OASIS, Eclipse, JMS, Apache, RabbitMQ, Windows 10 IoT Core etc. The used PBSM allows to the MQTT clients to communicate in P2P as well as broadcast modes (Fig. 9). The typical workflow consists of the following steps:

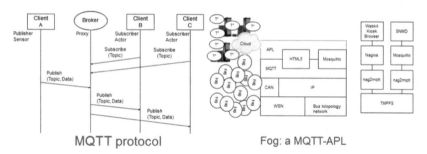

Fig. 9. Case study 1: MQTT-solution for IoT and Fog (own representation based on [7, 8])

- All three clients (so called subscribers) open TCP connections with the Broker.
- Clients B and C subscribe to the topic "Temperature T" and wait for the messages.
- At a later stage Client A publishes a value to the topic T = 22.5 °C.
- The Broker redirects the message to the topic to all Subscribers.

The corresponding sequence diagram for the publishing and subscripting is depicted. All the subscribers are with the actual temperature delivered. A freeware MQTT-based solution for IoT and Fog is shown below (cp. Fig. 9). The important components of the examined architecture are as follows: Mosquitto, Nagios (SNMP-based tool), TMPFS as well as interfaces to Web and Nagios. The Mosquitto tool is an IoT-project (iot.eclipse.org) and provides a broker for MQTT and other protocols and tools. SNMD (Sensor Node Management Device) is SNMP-based tool as well as an efficient interface to Nagios and Moscuitto. The TMPFS (temporary filesystem) is a file system which is used in many Unix-like operating systems to create a RAM disk.

With TMPFS some parts of the real working memory can be integrated and described as a hard disk [15, 16]. The examined scenario shows the energy-efficient co-operation "fog - private cloud provider" as well as provides PW- factor increasing.

4.2   Case Study 2: Kilobots and "Bio-Inspired Engineering"

"Bio-Inspired Engineering" investigates how to use the biological methods and models in industrial processes generally as well in "industry 4.0". Inter alia the swarm intelligence of socialized animals has been taken into account. The scientists from Harvard University (Cambridge, Massachusetts) discovered how to increase the efficiency of a swarm of 1024 = 1 K small robots (so called "kilobots"). The kilobots [5] are the small collective operating robots which can model a kind of "swarm intelligence" that is the property of socialized animals like fishes, ants, cranes, bees etc. Under use of specialized programming tools "the machine armada" can arrange itself different 2D-forms (Fig. 10). The kilobots can be programmed to imitate the forms of fish and bird swarms as well as other socialized animals and biological forms. In their construction the kilobots are not expensive microprocessor boards with three twiggy legs, two vibrating motors which help them go forward or rotate. The both motors can be configured using the 255 available power values and give the robots a differential drive. Each kilobot possesses few advanced sensors to be coordinated, e.g. a RGB LEDs and IR-sensors, as well as a small CPU ATmega-328 (8-Bit/8 MHz clock) [11–15]. The kilobots negotiate and communicate with their neighbors on distances till to 7 cm via IR-reflection on the ground. Via receiving a message per sensors, the distance to the kilobot that it sent can be determined by their signal power. The ambient light can also be detected by a kilobot. As IDE in C and open source-GUI the software Amtel WinAVR and Amtel AvrStudio is available (cp. Fig. 10). A swarm of kilobots is programmed together as a wireless cluster. The kilobots can interact, avoiding each other, playing follow-the-leader, or even performing simulations of nutritional processes. The deployment aims for kilobots are quite different:

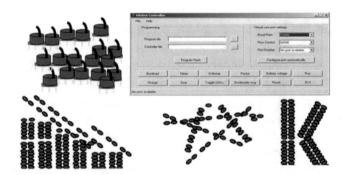

Fig. 10. Case study 2: swarm robots 2D-configurations and programming tools [9–14]

- The modeled 2D-structures allow comprehending the behavior of human masses in order to avoid the emergences or in the mass panic situations as well as help to handle easier such situations.
- Furthermore, the behavioral aspects of the colonies of the microorganisms or animals can be modeled.
- As further deployment examples certain urban as well as suburb construction and development scenarios can be considered as well as some historical scenarios of the propagation of the human populations too.
- The collaboration in the clean-up actions for environmental protection or in the periods after disasters and after environment contaminations can be reconstructed under use of the kilobots.
- And last but not least, in mid-term the modeling of the millions of self-driving cars (Google's Waymo or Tesla autopilots) on multiple roads will be possible via the discussed techniques.

The main goal of such experiments with small und inexpensive robot consists of modeling of technical processes and engineering objects ("Biologically Inspired Engineering" by Rubenstein) [11–15]. The robots help to obtain an inexpensive and adequate perception about real of technical processes and engineering objects. The modeling of socialized behavior of some animals or even humans in a crisis or catastrophe situation can be reached via providing of large groups of robots working together to complete a formulated task that they can't do on their own (cp. Fig. 10).

5 Conclusions

The important paradigms and relevant approaches for development of so-called novel SLMA have been discussed. An overview of the up-to-date robotics applications and platforms was given. The existing platforms and frameworks for their development are considered as well as analysis of the properties of the future robot platforms is offered.

The special kind of the advanced web apps, so-called PWAs are suitable for the purpose of spared data use, autonomous access and participation on clouds, increasing of performance, fault tolerance, as well as of online activity of the users. Novel secured real time OS Fuchsia with advanced programming languages promises new software development paradigms for mobile apps in midterm.

The co-operation architectures for Fog (Smart Stuff, IoT) and Clouds are discussed. The efficient access to the platform must be provided via mobile apps. Then it was concentrated on the case studies.

In the future "green" technologies will be widespread. The services will be often shifted from the clouds on "the network edges". The energy autarky can be realized via self-sufficient and low-energy nodes. Such interoperable software solutions and platforms help to the providers with energy saving and enhancing of resource efficiency. Hence such platforms are often heavy-weighing.

The following case study 1 was examined: a light-weighing solution for fog and cloud cooperation was represented. One of the further goals of the paper was to introduce the readers the basic knowledge about the modern, self-organizing networks and robots. The following case study 2 regarding to the kilobots and "bio-inspired engineering" was examined. The further works are focused on programming of new scenarios on bio-inspired engineering, deployment of the robots to pedagogical aims as well as behavioral human modeling.

Acknowledgment. Author's acknowledgements to the BA Dresden colleagues Prof. Dr. L. Zipfel, Prof. Dr. D. Gembris and the IT-students of BA Dresden, as well as to the colleagues from Springer Nature London, also to Prof. Dr. F. Fitzek, Dr. D. Guetter (the both TU Dresden) und Dr.habil. J. Spillner (ZHAW Zurich) for technical support, inspiration and challenges by fulfilling of this work.

References

1. Luntovskyy, A., Spillner, J.: Architectural Transformations in Network Services and Distributed Systems: Service Vision. Springer, Wiesbaden (2017). ISBN 9-783-6581-484-09
2. Luntovskyy, A., Guetter, D., Melnyk, I.: Planung und Optimierung von Rechnernetzen: Methoden, Modelle, Tools für Entwurf, Diagnose und Management im Lebenszyklus von drahtgebundenen und drahtlosen Rechnernetzen. Springer Vieweg, Wiesbaden 411 S. (2011). ISBN 978-3-8348-1458-6.(in German)
3. Luntovskyy, A., et al.: New Architectures in Distributed Systems, Monograph. Kolo, Lviv-Drogobych, 328 p. (2015). ISBN 978-617-642-185-6. (in Ukrainian)
4. Luntovskyy, A., Guetter, D., Klymash, M.: Up-to-date paradigms for distributed computing. In: 2nd International IEEE Conference AICT-2017, July 2017, pp. 113–119. IEEE Xplore (2017). https://doi.org/10.1109/aiact.2017.8020078
5. Luntovskyy, A., Globa, L., Stepurin, O.: Performance-energy tradeoff models for distributed computing. In: 13th International IEEE Conference TCSET-2016, February 2016, pp. 613–617. IEEE Xplore (2016). https://doi.org/10.1109/blackseacom.2016.7901554
6. Spillner, J.: Autonomous participation in cloud services, Dresden (2012)
7. PWA. http://t3n.de/news/wahre-konversionskiller-lange-594340/
8. Siciliano, B., Khatib, O.: Handbook of Robotics, p. 1624. Springer, Heidelberg (2008). ISBN 978-3-540-30301-5
9. Khatib, O.: Stanford Univ. CA, USA, International Foundation of Robotics Research. http://www.ifrr.org/
10. Spillner, J.: Current and Future Platforms for Robotics, Manuscript to a probe talk for Habilitation, Dresden, 33 p. (2015)
11. Cornejo, A., Nagpal, R.: Kilobots: Distributed Range-Based Relative Localization of Robot Swarms. Algorithmic Foundations of Robotics XI, April 2015
12. Spillner, J.: ROS and Cloud Robotics, Teaching Lecture, Zurich (2017)
13. Rubenstein, M.: A self-organizing thousand-robot swarm. Biologically Inspired Engineering (2017). https://wyss.harvard.edu/a-self-organizing-thousand-robot-swarm/
14. NoDNA.de. https://nodna.de/Bioloid-Premium-Robot-Kit-V2
15. Luntovskyy, A., et al.: Intelligent networking and bio-inspired engineering. In: 2nd International IEEE Conference UkrMiCo 2017, September 2017, Odessa, Ukraine, 4 p. IEEE Xplore (2017). http://ieeexplore.ieee.org/document/8095421/
16. IBH IT-Service GmbH Dresden Workshop on 16 March 2017
17. Google Fuchsia OS GitHub. https://github.com/fuchsia-mirror/

Applications of Multilingual Thesauri for the Texts Indexing in the Field of Agriculture

Waldemar Karwowski$^{(\boxtimes)}$ [ID], Arkadiusz Orłowski [ID], and Marian Rusek [ID]

Faculty of Applied Informatics and Mathematics,
Warsaw University of Life Sciences, Nowoursynowska 166,
02–787 Warsaw, Poland
waldemar_karwowski@sggw.pl

Abstract. The problem of simultaneous automatic indexing of texts related to agriculture in different languages is discussed. A short survey of multilingual dictionaries of the vocabulary content related to agriculture is presented. Dictionaries containing Polish vocabulary are discussed in more details. Available thesauri are assessed and advantages of AGROVOC thesaurus are underlined. Available indexing tools based on AGROVOC and, at the same time, the author's indexing system in Polish are shortly described. Eventually parallel text indexing with the use of summaries of articles in Polish and English is tested.

Keywords: Multilingual dictionary · Thesaurus · SKOS · Text indexing

1 Introduction

The spoken language is far from being precise. Every day during the conversation we use specific phrases understood by a small circle of listeners only. Even in the different regions of the same country colloquial vocabulary can vary greatly. This lack of precision has impact on the written language. However in many fields such as law, medicine, engineering, or agriculture it is necessary to use the more precisely defined vocabulary. Acceptable words, synonyms, idioms, etc., must be carefully determined. Even higher precision requirements apply to scientific texts. The precision of the used terms is particularly important when we are dealing simultaneously with texts in different languages. There is often a need to translate the text into other language or make the comparison of texts in different languages in order to determine whether their subjects are similar. Such a situation appeared in the project eFarmer (INTERREG IIIC East zone) [1], in which the main idea was the stimulation of entrepreneurship, competitiveness, and regional development in rural areas of Finland, Poland, and the Czech Republic. To ensure the appropriate cooperation among partners, it was necessary to prepare a multilingual dictionary in English, Finnish, Swedish and Polish (Czech version has not been implemented). Electronic dictionary of terms for agribusiness has been implemented as a Web application on the Moodle platform and contained about 1000 terms in four languages. Design and implementation process of the dictionary is

© Springer Nature Switzerland AG 2019
J. Pejaś et al. (Eds.): ACS 2018, AISC 889, pp. 185–195, 2019.
https://doi.org/10.1007/978-3-030-03314-9_17

described in [2]. Dictionary was based on Moodle Glossary module which uses its own XML format. XML format allows using the dictionary not only by the persons through the Web interface but also by software tools. The electronic dictionary project showed that the precise multilingual dictionary, in a well-defined format, is necessary for analysis of the professional texts in different languages. In the mentioned project terms were developed by experts, but in the field of agriculture such an elaborate dictionary exists, it is multilingual thesaurus AGROVOC.

The main goal of this paper is to check relevance of AGROVOC thesaurus to the parallel texts indexing in the field of agriculture. At the same time, we will compare the quality of abstracts translation and we make comparison of three AGROVOC-based text indexing systems. We first discuss methods of automatic indexing. After then we present available multi-language thesauri, especially AGROVOC, and standards on which they are based. In section four annotation tools are described. Next we make an overview of experiment with parallel text indexing and present its results. At the end the most important conclusions are formulated.

2   Methods of Automatic Indexing

Text indexing is one of important natural language processing (NLP) and information retrieval (IR) tasks. It is the process of labeling a document by assigning set of terms (tags, keywords, key phrases). Generally terms are extracted from the text. Text indexing is connected with other NLP tasks like text summarization, text clustering, or checking texts similarity. A first indexing method, simple but still important, was formulated by Luhn [3]: the weight of a term that occurs in a document is simply proportional to the term frequency. It means that documents might be indexed by their most frequent terms. Of course the principle is true when we remove so called stop words. Many times, indexing is performed on big collections of texts; in such situations the most popular is TF-IDF (term frequency–inverse document frequency) method. It is generalization of Luhn statistic method intended to reflect how important a word is to a particular document in the context of the whole collection [4]. In TF-IDF, words present in each and every document in the collection are not essential. We have to note that ignoring most common words apart from stop words sometimes is not proper. For example in hot topic extraction, topics that appear frequently over a period of time are the most important. It means that there are two important factors: how often a term appears in a document and the number of documents that contain the term [5]. Other known and popular methods are Naive Bayes Classifier or Latent Semantic Analysis. Nowadays, machine learning methods, like support vector machines, are very effective with big collections. Exhausting review of key phrase extraction methods and their evaluations is presented in [6]. A key phrase extraction system typically operates in two steps: first extracting a list of words/phrases that serve as candidate key phrases using some heuristics; and next determining which of these candidate key phrases are correct. Sometimes, to perform the second step, external resource-based features like knowledge bases (domain ontologies, Wikipedia taxonomy, etc.) are very useful. Many times set of keywords from ontology is more important than other keywords gathered from indexed documents. Especially, such approach is proper for texts connected to specific domain. There are many examples of this approach: ontology-based indexing method for engineering documents in aerospace industry is presented in [7], ontology based annotation of text segments in agriculture is discussed in [8]. This method is also very important in bioinformatics [9].

Our task is to analyze rather small corpus of documents and for this purpose methods based on machine learning are not proper. Because we analyze documents with similar subjects in agriculture domain also methods like TF-IDF are not proper. Similarly like in hot topic extraction, words connected to a domain and present in every document are important keywords candidates. Moreover for our task, taxonomy or ontology based indexing is desired. Because we analyze texts in Polish and English, multilingual thesauri are needed.

3 Multilingual Thesauri

When we want to preserve the precision in texts indexing, we need dictionary with organized lists of words and phrases. Moreover we need notation systems, that are used to initially tag content, and then to find it through navigation or search [10]. To precisely describe a domain, for example agriculture, we have to select a set of words related to it, with a well-defined meaning. Controlled vocabulary is a kind of dictionary that meets such conditions. It consists only predefined, authorized, carefully selected terms and phrases that are used for the determination of units of information. For example controlling synonyms can be very effective in minimizing the variants introduced by natural language [10]. For our purposes the most useful are thesauri. They use controlled vocabulary, create the hierarchy of terms and organize the issues such as synonyms, antonyms or homographs by specifying the relationships and dependencies between them. The best known standard is ISO 25964 [11] "Thesauri for information retrieval" (based on earlier ISO 2788 and ISO 5964). Simple Knowledge Organization System (SKOS) [12], being recommended by W3C, is close to ISO standards [13–15]. It is a part of the Semantic Web family of standards built upon RDF and RDFS. Many knowledge organization systems share a similar structure. SKOS captures much of this similarity and makes it explicit, to enable data and technology sharing across diverse applications.

The Food and Agriculture Organization of the United Nations (FAO) offers, from our point of view, the most interesting dictionary. AGROVOC is multilingual thesaurus which covers all subject areas related to FAO interests [16]. AGROVOC is releasing more than 32,000 connected concepts that return standards-based term metadata. AGROVOC indexes content of a FAO multilingual bibliographic database for agricultural science – AGRIS, and links it to bibliographic records managed by external repositories. AGRO-VOC has long history. It was first published, in printed form, at the beginning of the 1980s in English, French and Spanish. Today, AGROVOC is available in 29 languages as an SKOS-XL concept scheme and published as a Linked Open Data (LOD) set aligned to 16 other data sets related to agriculture. With the publication of AGROVOC as LOD dataset, FAO has not only exposed its first data set in the linked data world, but the largest data set about agriculture that is now out there for public use. It is possible to download the entire thesaurus as XML file, using SPARQL endpoint or web service. In Polish AGROVOC provides near 20,000 terms, and we can conclude that it is a very good tool for indexing text in agricultural domain. We have to note that there are other interesting dictionaries. The most famous a large lexical database of English language, widely used in scientific research of natural language processing, is WordNet. In the WordNet, nouns, verbs, adjectives, and adverbs are grouped into sets of cognitive synonyms (synsets), each expressing a distinct concept [17]. Synsets are interlinked by means of conceptual-semantic and lexical relations. WordNet was originally prepared for the English language and is not multilingual. However, in the meantime, different versions of WordNet in several languages were created. Among them Polish version, called plWordNet, was developed at Wrocław University of Technology [18]. Unfortunately there is no mapping between terms in English and Polish versions, it means that it is not useful for us. The United Nations clearly needs to use many languages. This means that it must use multi-lingual dictionaries. The UNESCO thesaurus is a controlled and structured list of terms used in subject analysis and retrieval of documents and publications [19]. It is available in SKOS format via SPARQL endpoint. Therefore, the use of the thesaurus in computer programs is easy, but in the context of our purposes the drawback is the lack of Polish language. Thesaurus is available only in English, French, Spanish, and Russian. The other dictionaries were created in the European Union institutions. The EU has three "proce-dural" languages: English, French, and German but formally has 24 official languages and all documents are prepared in all of them. GEMET (GEneral Multilingual Environmental Thesaurus) is a multilingual thesaurus that aims to define a core terminology for the environmental domain [20]. The dictionary contains many themes, agriculture being one of them. GEMET supports more than 30 languages among them Polish, additionally GEMET is available in SKOS format. Moreover thesaurus has Web interface, and in addition to this REST service is available. These features indicate that GEMET is a useful for text indexing. Unfortunately, the number of terms related to agriculture is small because the thesaurus focuses on issues of environmental protection, so for our purposes it is not enough. The next EU vocabulary is EuroVoc [21]. It is a multilingual, multidisci-plinary thesaurus which contains terms in 23 EU languages. The thesaurus is covering fields which are sufficiently wide-ranging to encompass both Community and national points of view, with a certain emphasis on parliamentary activities. The aim of the the-saurus is to provide the information management and dissemination services with a coherent indexing tool for the effective management of their documentary resources and to enable users to carry out documentary searches using controlled vocabulary. Agriculture, forestry, and fisheries are examples of 21 domains, inside of which we have subdomains: agricultural policy, agricultural structures and production, farming systems, cultivation of agricultural land, means of agricultural production, and agricultural activity. Also EuroVoc is in SKOS format. Apart from the Web interface there is possible download of RDF file as well as the PDF documents. Additionally, for programmers, there is a web service but unfortunately with very limited functionality. All these features indicate that EuroVoc may be useful for text indexing although like GEMET it has limited number of concepts devoted to agriculture.

In order to verify the possibilities of multilingual dictionaries it was necessary to choose the most suitable dictionary. To perform our experiments we selected the AGROVOC. It fulfilled important criteria such as vocabulary in the field of agriculture, multilingualism (including Polish) and availability in SKOS format.

4 Annotation Tools

We noted in Sect. 2 that for the text analysis some initial processing is necessary. The first step is to remove stop words. After then, problems with inflexion and various form of words, must be solved. To reduce inflectional and derivationally related forms of a word to a common base form, stemming algorithms are applied. The literature review about stemming can be found in the second chapter of [4]. Most known stemming methods are good for the English language. Stemming algorithms are developed for many western languages, but they are not perfect for languages with a very extensive inflection like Polish language. The second important issue during indexing is to identify the part of speech, i.e. to classify words in a text as nouns, verbs, etc. Nowadays in English texts part of speech tagging is fairly accurate [22].

Selected text indexation tools are described in [23]. One of the most interesting for us is Agrotagger [24], a FAO initiative, which for keyword extraction uses AGRO-VOC. Second indexation tool that uses AGROVOC is Agroportal Annotator [25]. Both these tools use stemming algorithms and part of speech tagging. Current version of Agrotagger uses the Maui packages [26]. It means that it uses machine learning and for training, bigger collection of tagged texts is needed. Agrotagger is able to index also texts in French, German, and Spanish because Maui has implemented such stemmers. Annotator indexes only English texts. It does not index in other languages, even in French, although it was created by a French institution (but work on other languages is in progress) [27]. The reason is that it is connected with BioPortal Annotator prepared in English [28]. Despite the fact that Agroportal Annotator algorithms are not described we can conclude that they are similar to BioPortal Annotator algorithms. Algorithms for BioPortal Annotator are described in [29]. First, the user's free text is given as input to a concept recognition tool along with a dictionary (or lexicon), i.e., a list of strings that identifies ontology concepts. This primary set of annotations serves as input for the semantic expansion components, which expand the annotations extracted from the first step using the structure and/or semantics of one or more ontologies. Annotator is much easier to use than Agrotagger because it has user friendly web interface. We have to mention that there were few Agrotagger versions prepared by FAO with web interface [23], but those services are not active now. Maui-based Agrotagger version is running locally, has only command line interface, and requires configuration of the environment. Unfortunately both applications cannot index Polish texts. To index text in the Polish language a special application was created [23, 30]. Our system in Polish works as a Windows application, and is based on database of words with inflected forms from open-source dictionary of the Polish language (http://www.sjp.pl) and full version of AGROVOC. AGROVOC thesaurus is accessed through Web Service. In our application we created a huge structure containing the whole inflection dictionary together with a part of speech tags. In the structure every word has a pointer to the base form tagged by part of speech. In this way, we solved the stemming and part of speech tagging issues. The details of creating such a structure have been presented in [31].

5 Experiment

The three systems, mentioned in Sect. 4, formed the basis of the experiment. Polish indexer has been slightly improved with respect to [30]. To perform analysis of words we have improved checking synonyms as well as, broader and related terms in semantic distance [31]. Although we index Polish text, the selected keywords are in English. This is possible thanks to the fact that AGROVOC is multilingual. Because Maui package has machine learning background, we used in Agrotagger the model trained on 780 texts from the FAO text resources in agriculture area. This model is given with Agrotagger but we have to note that those texts are not connected with specific agriculture domain.

According to [32] most of a scientific paper's key phrases should appear in the abstract and the introduction. Consequently we analyzed only titles and abstracts of selected texts. In this way, we had analogous texts in Polish and English. Of course, translations of abstracts, made by authors, do not need to be accurate; however, they are a good material to compare indexers parallel for different languages. To compare the search results of indexing we selected two groups of publications in Polish with English summary from Agricultural Engineering Journal (Inżynieria Rolnicza). The first group consisted of seven papers (A–G) and was generally connected with maize. We have to note that the same papers were used in the analysis performed in [30 - Sect. 4] on whole texts but only in Polish. The second group was taken from the same journal. There are twelve papers (H-S) generally connected with potatoes.

Table 1 shows the result indexes extracted from abstracts. Annotator extracts only nouns or noun phrases; Polish indexer extracts nouns and verbs that are present in the AGROVOC. Polish indexer, in current version, does not choose phrases but it extracts

Table 1. Indexes (keywords) extracted from texts.

Text	Polish indexer	Annotator	Agrotagger
A	product, kernel, Zea (maize), model, temperature, water, labour, to apply, to forecast	information, data, geometry, agricultural products, processes, elements, nodes, temperature, moisture content	agricultural products, data collection, economic distribution, software development, moisture content, processing, processes, models, mechanics, maize
B	labour, quality, evaluation, humidity, enterprises, Zea (maize), grain, dryers, drying, cleaning	assessment, quality, seeds, processing, cleaning, threshing, weight, bagging, units, leaves, drying, selection	processing, quality, sowing, bags (storage), bagging (pest control), bagging, processes, cleaning, weight, Zea mays
C	Zea (maize), firmness, kernel, time, *breeds*, labour, methods, to exploit	maize, firmness, seed, measurement, paper, stages, tissue, properties, fruit, research, duration, time, labour	testa, seed pelleting, sowing, fruits, fruiting, fruit, measurement, mestes, methods, maize
D	Zea (maize), grain, uses (usage), wheels, evaluation, profitability, fields, labour, quality, summer, soil, to irrigate	irrigation, maize, grain, evaluation, quality, light, soil, subsoil, research, yield increases, profitability, needs, plants	grain, field experimentation, plants, planting, sandy soils, yield increases, yields, maize, irrigation, quality

(continued)

Table 1. (*continued*)

Text	Polish indexer	Annotator	Agrotagger
E	quality, scapula, uses (usage), Zea (maize), choppers, grain, *sawnwood*, labour, to shred	quality, shredding, forage, cutting, improvement, tests, research, elements, grain	maize, labour, greening, cutting back, mowing, cuttings, cross cutting, cutting, methods, Zea mays
F	grain, Zea (maize), methods, quality, *breeds*, evaluation, nitrogen, labour, raw material, to froze	threshing, grain, sugar, quality, freezing, varieties, methods, separating, tests, assessment, nitrogen	grain, maize, calcium, methods, Zea mays, nitrogen, foods, cutting, separating, threshing
G	technology, sowing, Zea (maize), silage, models, plants, evaluation, precocity, *breeds*, computers, internet, farms, to use, to fertilize	decision support, production, maize, silage, technology, models, seed drilling, assessment, plant protection, dates, fields, selection, development, application, internet, weather, precocity, uses	decision support, protective plants, protected species, technology, plant protection, seed drilling, seed drills, silage making, weather data, decision support systems
H	technology, solanum tuberosum (potato), seed potatoes, methods, *breeds*, reproduction, seed production, plants	technology, potatoes, production, tillage, paper, quality, cultivation, weeds, infection, viroses, diseases, seed, seed production, reproduction, varieties, fields	seed potatoes, viroses, yields, seed production, cultivators, sowing, ridging, tillage, reproduction, reproductive performance
I	tubers, *breeds*, solanum tuberosum (potato), evaluation, separating, friction, sand,	evaluation, chemicophysical properties, bulbs, harvesting, crops, species, efficiency, separating, friction coefficient, light, sand	chemicophysical properties, friction coefficient, *crop (bird)*, crops, potatoes, processing, processes, selection, properties, separating
J	soil, tubers, species, friction, solanum tuberosum (potato), plate towers, mass, to apply	application, image processing, chemicophysical properties, tubers, separating, separators, components, mass, soil, friction coefficient, experimentation, light	chemicophysical properties, soil, soil types, tubers, *tuber (truffles)*, separating, potatoes, processing, processes, friction coefficient
K	density, solanum tuberosum (potato), tubers, models, *breeds*, labour, to store, to shrink, to fertilize	density, growth, properties, storage, paper, tubers, losses, shrinkage, varieties, peel, models, tests	tubers, *tuber (truffles)*, storage, properties, property, potatoes, fractionation, peel, growth, density
L	*bean* sprouts, solanum tuberosum (potato), seed potatoes, tubers, stimuli, labour, *breeds*, radiation	irradiation, growth, dynamics, paper, research, seed, radiation, energy sources, time, state	tubers, *tuber (truffles)*, growth, microwave radiation, seed potatoes, sowing, radiation, processing, processes, seeds
M	models, solanum tuberosum (potato), evaluation, time, surface area, leaves, larvae, foraging, damage	assessment, damage, *Colorado*, leaves, analysis, area, defects, larvae, imago, feeding, regression analysis, guidelines	models, regression analysis, potatoes, damage, mathematics, self feeding, feeds, feeding, statistics, plants
N	solanum tuberosum (potato), losses, larvae, time, leaves, foraging, damage, to use	losses, leaves, predation, larvae, *Colorado*, individuals	potatoes, available days, damage, losses, adults, coleoptera, *Colorado*, larvae, predation, assessment

(*continued*)

Table 1. (*continued*)

Text	Polish indexer	Annotator	Agrotagger
O	plate towers, solanum tuberosum (potato), tubers, separating, to clean	design, cleaning, cleaning, separating, harvesting, systems, *Australian Capital Territory*, rollers	legislation, separators, tubers, *tuber (truffles)*, separating, equipment, cleaning, potatoes, advertising, processes
P	tubers, solanum tuberosum (potato), separating, mass, velocity, farms, *breeds*, soil, to irrigate, to sort	irrigation, tubers, separating, grading, compensation, processing, research, fields, farms, scientists, varieties, soil, weight, dimensions, velocity, filling, rubber	tubers, *tuber (truffles)*, *crop (bird)*, crops, processing, processes, clay soils, plantations, irrigation, selection
Q	tubers, solanum tuberosum (potato), soil, plants, height, mass, *breeds*, uses, technology, seed potatoes, to fertilize	evaluation, chemical composition, quality, factors, tests, tubers, varieties, maintenance, tillage, soil, cultivation, methods, research, measurement, volume, seed, potatoes, analysis, mass, technology	tillage, maintenance, arable soils, tubers, *tuber (truffles)*, *crop (bird)*, potatoes, Andean region, measurement, methods
R	solanum tuberosum (potato), plants, history, sawn-wood, production, time, technology,	history, research, production, management	production controls, crops, cultivators, cultivation, potatoes, technology, potato products, history, plants, planting
S	tubers, *breeds*, starch, solanum tuberosum (potato), quality, surface area, size, proteins, evaluation, length, mass, to evaluate	evaluation, chemical composition, quality, tubers, starch, varieties, field experimentation, length, vegetation, re-search, components, rain, area, weight	tubers, *tuber (truffles)*, chemical composition, yields, potatoes, potato starch, vegetation, quality, selection, starch

only single words, in spite of AGROVOC which contains many phrases like "moisture content", "seed pelleting", etc. First observation is that results are not identical, even for Annotator and Agrotagger although both tools use the same English text version. It should be noted that in AGROVOC we have many semantic relations. Most fundamental relation is preferred label. In translations of abstracts prepared by authors some problems appeared. They, for example, translate Polish word "kukurydza" into "maize". "Kukurydza" has as preferred label Latin name "Zea". In English AGROVOC "Zea" is used on higher level in taxonomy hierarchy than "maize": "Zea"–(narrower)–> "Zea mays"–(produces)–> "maize". Analogous relation in Polish is "Zea"–(narrower)–> "Zea mays"–(produces)–> "Kukurydza (ziarno)". We have to note that "Zea mays" has in Polish alternative label "kukurydza zwyczajna", but in English the alternative label is "corn (zea)". We can conclude that between "kukurydza" and "maize" the semantic distance is 2. The second example is Polish word "odmiana", used for plants is translated by authors as "variety". Unfortunately in AGROVOC English term for "odmiana" is "breed", but only for animals. Polish term for "variety" in AGROVOC is "odmiana roślin uprawnych". Because authors used for short "odmiana" it caused bad index in Polish indexer. A similar mistake appears with the Polish word "listwa". In AGROVOC in English it is "sawnwood" but authors mean "part of cutting machine". Polish word "ocena" is in AGROVOC "evaluation" but authors sometimes translated it as "assessment" (in AGROVOC there is no Polish term for "assessment"). Moreover in English phrase "Colorado beetle" was not recognized as AGROVOC term "Colorado potato beetle" and in consequence alone name Colorado appeared. Another mistake in English is that the verb "act" was recognized by Annotator as Australian Capital Territory (ACT).

First, after reading texts, we can conclude that Polish indexer works well and generally keywords in English are proper besides the faults listed above. Second conclusion is that if the authors inconsistently use AGROVOC terminology, the quality of translation and consequently indexing is at the medium level. Third conclusion is that surprisingly in Annotator, the main subject is often not completely included. In texts A–G maize appears only in C, D, and G. In texts H–S potato appears only in H. It seems that Annotator has bad preprocessing method, especially stemming. In AGROVOC, English terms are generally in plural form, i.e., potatoes. Annotator evidently ignores this. Some final conclusions are connected with Agrotagger. Agrotagger was trained on texts not only associated with the maize and potato cultivation and processing; hence the results may be different than in Annotator. Moreover in Agrotagger may appear keywords that are not at all in the analyzed text like "Andean region" in text Q. Additional mistake in Agrotagger is that it extracts some homonym terms like "tuber (truffles)" or "crop (bird)" evidently not connected with texts. Finally, it should be added that abstracts are short and Agrotagger based on machine learning methods may work worse than on longer texts.

It was decided to compare extracted indexes pairwise i.e. Polish indexer with Annotator, Polish indexer with Agrotagger, and Annotator with Agrotagger. Because a term occurrences number is not produced by Agrotagger, the Jaccard measure was selected (the number of common terms divided by the number of all distinct terms) to compare results. Moreover before evaluation some manual corrections especially to Agrotagger results were performed. E.g. words such as "processing" and "process" were treated as the same word. Also we removed from Agrotagger results evident mistakes (duplications) like "tuber" (truffles) and "crop" (kind of bird). Finally we treated as the same term alternative labels like "Zea mays" and "maize". After manual corrections, average Jaccard similarity for Polish indexer and Annotator was about 0.31, it means that roughly half of terms in every pair were common. The best result was for paper F - 0.5, the worst for paper L - 0.19. Similarity for Polish indexer and Agrotagger was about 0.25, the best for paper R - 0.45, the worst for papers C and E - 0.14. Similarity between Annotator and Agrotagger was about 0.27, the best for paper D - 0.54, the worst for E only 0.07.

6 Conclusions and Future Work

Analysis of thesauri, in the context of standards, agriculture vocabulary, and availability of terms in the English and Polish language, showed that the AGROVOC fulfills formulated demands. Presented indexers demonstrated that it is possible to integrate

AGROVOC with indexing applications. An initial experiment showed that parallel text indexing for Polish and English is fairly compatible. Some differences are due to a not-too-precise translation of the texts. Similarity level between Polish indexer and Annotator would probably be better if Annotator had a proper text preprocessing. Indexing the same English text by Annotator and Agrotagger turned out to be worse than expected. The reason is that the Agrotagger training set was apparently too small. One step in the future research is obvious. It is necessary to prepare a suitable text preprocessor for Annotator, which would convert nouns to the plural form, it is also desirable to modify Polish indexer to allow indexing of the phrases contained in the thesaurus. There is also need to increase the semantic distance of analyzed terms (broader and narrower terms etc.). This should solve the problem of imprecise translation. Moreover, in a longer perspective, further research requires the preparation of the corpus of texts both in Polish and English with similar subjects.

References

1. INTERREG IIIC Operations. http://www.interreg4c.eu/list-of-interreg-iiic-operations
2. Rusek, M., Karwowski, W., Orłowski, A.: Internet dictionary of agricultural terms: a practical example of extreme programming. Studies & Proceedings of Polish Association for Knowledge Management, vol. 15, pp. 91–97 (2008)
3. Luhn, H.P.: A statistical approach to mechanized encoding and searching of literary information. IBM J. Res. Develop. **1**(4), 307–319 (1957)
4. Manning, C.D., Raghavan, P., Schütze, H.: Introduction to Information Retrieval. Cambridge University Press, Cambridge (2008)
5. Hot topic extraction apparatus. U.S. Patent US 7,359,891 B2, April 15 2008
6. Hasan, K.S., Ng, V.: Automatic keyphrase extraction: a survey of the state of the art. In: Proceedings of the 52nd Annual Meeting of the Association for Computational Linguistics, pp. 1262–1273 (2014)
7. Fang, W., Guo, Y., Liao, W.: Ontology-based indexing method for engineering documents retrieval. In: IEEE International Conference on Knowledge Engineering and Applications (ICKEA), pp. 172–176 (2016)
8. El-Beltagy, S., Hazman, M., Rafea, A.: Ontology based annotation of text segments. In: Proceedings of the 2007 ACM Symposium on Applied Computing (SAC), pp. 1362–1367 (2007)
9. Shah, N.H., Jonquet, C., Chiang, A.P., Butte, A.J., Chen, R., Musen, M.A.: Ontology-driven indexing of public datasets for translational bioinformatics. BMC Bioinform. **10**(Suppl 2), S1 (2009)
10. Warner, A.J.: A taxonomy primer. https://www.ischool.utexas.edu/~i385e/readings/Warner-aTaxonomyPrimer.html
11. ISO 25964-1:2011 - Thesauri and interoperability with other vocabularies - Part 1: Thesauri for information retrieval
12. SKOS Recommendation, 18 August 2009. http://www.w3.org/TR/skos-reference
13. SKOS Primer Note 18 August 2009. http://www.w3.org/TR/skos-primer
14. ISO 25964. http://www.niso.org/schemas/iso25964
15. Correspondence between ISO 25964 and SKOS/SKOS-XL Models. http://www.niso.org/apps/group_public/download.php/12351/CorrespondenceISO25964-SKOSXL-MADS-2013-12-11.pdf

16. AGROVOC thesaurus. http://aims.fao.org/vest-registry/vocabularies/agrovoc-multilingual-agricultural-thesaurus
17. WordNet https://wordnet.princeton.edu
18. Słowosieć. http://plwordnet.pwr.wroc.pl/wordnet
19. UNESCO thesaurus. http://vocabularies.unesco.org/browser/thesaurus/en
20. GEMET thesaurus. http://www.eionet.europa.eu/gemet
21. EuroVoc thesaurus. http://eurovoc.europa.eu/drupal
22. Manning, C.D.: Part-of-speech tagging from 97% to 100%: is it time for some linguistics? In: Proceedings of 12th International Conference on Computational Linguistics and Intelligent Text Processing, CICLing 2011, Part I (2011)
23. Karwowski, W., Wrzeciono, P.: Methods of automatic topic mining in publications in agriculture domain. Inf. Syst. Manag. **6**(3), 192–202 (2017)
24. Agrotagger. http://aims.fao.org/agrotagger
25. Annotator. http://agroportal.lirmm.fr/annotator
26. Maui package. https://github.com/zelandiya/maui
27. Jonquet, C., Toulet, A., Arnaud, E., Aubin, S., Yeumo, E.D., Emonet, V., Graybeal, J., Laporte, M., Musen, M.A., Pesce, V., Larmande, P.: AgroPortal: a vocabulary and ontology repository for agronomy. Comput. Electron. Agricult. **144**, 126–143 (2018)
28. Bioportal Annotator. http://bioportal.bioontology.org/annotator
29. Jonquet, C., Shah, N.H., Musen, M.A.: The Open Biomedical Annotator. AMIA Summit on Translational Bioinformatics, March 2009, San Francisco, CA, United States, pp. 56–60 (2009)
30. Karwowski, W., Wrzeciono, P.: Automatic indexer for Polish agricultural texts. Inf. Syst. Manag. **3**(4), 229–238 (2014)
31. Wrzeciono, P., Karwowski, W.: Automatic indexing and creating semantic networks for agricultural science papers in the polish language. In: 2013 IEEE 37th Annual Computer Software and Applications Conference Workshops (COMPSACW), Kyoto (2013)
32. Lancaster, F.W.: Indexing and Abstracting in Theory and Practice. Library Association, London (2003)

On Code Refactoring for Decision Making Component Combined with the Open-Source Medical Information System

Vasyl Martsenyuk[1]([⊠])(iD) and Andriy Semenets[2](iD)

[1] Department of Computer Science and Automatics,
University of Bielsko-Biala, Bielsko-Biała, Poland
vmartsenyuk@ath.bielsko.pl
[2] Department of Medical Informatics,
Ternopil State Medical University, Ternopil, Ukraine
semteacher@tdmu.edu.ua

Abstract. The work is devoted to the facility of decision making for the open-source medical information systems. Our approach is based on the code refactoring of the dialog subsystem of platform of the clinical decision support system. The structure of the information model of database of the clinical decision support subsystem should be updated according to the medical information system requirements. The Model - View - Controller (MVC) based approach has to be implemented for dialog subsystem of the clinical decision support system.

As an example we consider OpenMRS developer tools and corresponding software APIs. For this purpose we have developed a specialized module. When updating database structure, we have used Liquibase framework. For the implementation of MVC approach Spring and Hybernate frameworks were applied. The data exchanging formats and methods for the interaction of the OpenMRS dialog subsystem module and the Google App Engine (GAE) Decision Tree service are implemented with the help of AJAX technology through the jQuery library.

Experimental research use the data of pregnant and it is aimed to the decision making about the gestational age of the birth. Prediction errors and attribute usage were analyzed.

Keywords: Medical information systems
Electronic medical records · Decision support systems · Decision tree
Open-source software · MIS · EMR · OpenMRS · CDSS · Java
Spring · Hibernate · Google App Engine

1 Introduction

The importance of wide application of the Medical Information Systems (MIS) as a key element of informatization of healthcare, especially in Ukraine, is shown

© Springer Nature Switzerland AG 2019
J. Pejaś et al. (Eds.): ACS 2018, AISC 889, pp. 196–208, 2019.
https://doi.org/10.1007/978-3-030-03314-9_18

in [2,23]. The development of information technologies makes it possible to improve the quality of medical care by providing medical personnel with hardware and software tools for the efficient processing of clinical information [2,3]. A conceptual direction of modern information technologies adoption in hospitals pass through patient's Electronic Medical Record (EMR) formation and support [1,2,23].

An overview of approaches of implementation into as well as brief list of the leading MIS developers is given in [23]. MIS global market has stable positive dynamics as it is shown in [4]. A few high-quality MIS has been created by Ukrainian software development companies too, for example, "Doctor Elex" (http://www.doctor.eleks.com), "EMSiMED" (http://www.mcmed.ua), etc. In fact, all they are commercial software with a high cost [2].

An open-source-based software solutions for healthcare has been actively developing for the last decade along with the commercial software applications [1,11,20]. Most widely used open-source MIS EMR are WorldVistA (http://worldvista.org/), OpenEMR (http://www.open-emr.org/) and Open-MRS (http://openmrs.org/) [1,8]. Advantages of such MIS software are shown in [1,23]. Prospects for open-source and free MIS software usage in developing countries, or countries with financial problems has been considered by Aminpour, Fritz, Reynolds and others [1,8,20]. The approaches to implementing open-source MIS, especially OpenEMR, OpenMRS and OpenDental, in Ukraine healthcare system has been studied as well as methods of integrating these MIS EMR with other MIS software has been developed by authors of this work during last few years [22,23].

Clinical Decision Support Systems (CDSS) regular usage in physician's practice is strongly recommended for improving of the quality of care. This thesis was confirmed in [4,10,21]. Advantages of CDSS usage in healthcare systems of the developing countries was shown in [7]. The importance of integration of different types of MIS, and MIS EMR with CDSS especially, is provided in [9]. The CDSS theoretical approaches as well as software applications has been developed by TSMU Medical Informatics Department staff [3,14–16,25].

Approaches of the CDSS usage in obstetrics for early detection of pathologies of miscarriage of pregnancy are analyzed in [5,12,18]. A prototype of such CDSS has been developed by Semenets AV, Zhilyayev MM and Heryak SM in 2013. The effectiveness of proposed algorithm was confirmed by experimental exploitation of this CDSS prototype in the Ternopil state perinatal center "Mother and Child" during 2013–2015 that is proved in [6]. As a result, the fully functional CDSS application for miscarriage pathology diagnostic has been developed by authors in form of an information module (plugin) for free- and open-source MIS OpenEMR [13,24].

The *objective* of this work is to present an approach of code refactoring of the plugin, which implements dialog component of custom CDSS platform, for usage with free- and open-source MIS.

Results of practical implementation using MIS OpenMRS is presented in Sect. 2 including adaption of the information model of dialog component of the

CDSS module and development of user interface. Experimental research which is based on the decision tree induction algorithm applied for gestational age of birth is shown in Sect. 3.

2 Implementation of Code Refactoring for the CDSS Platform Dialog Component

The alternative method of the decision making process, based on the algorithm for induction of decision trees, was proposed by Martsenyuk as result of preceding investigations described in [3,14,15,25]. Finally, given decision-making diagnostic algorithm was implemented with Java programming language as a web-service for the Google App Engine platform. A web-service training database has been deployed to Google Datastore service, which is a form of no-SQL data warehouse [13,24]. This approach provide flexible way to integrate above Google App Engine (GAE) Decision Tree service with third-party MIS EMR by developing appropriate dialog components (modules, plugins) as well as administrative tools (Fig. 1). Therefore the feasibility of CDSS dialog component's plugin [13,24] code refactoring for usage with free- and open-source MIS OpenMRS is obvious.

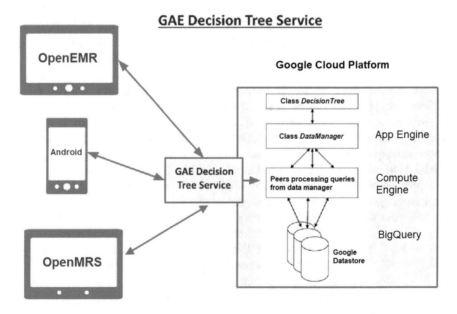

Fig. 1. Integration of the GAE Decision Tree CDSS web service with arbitrary EMR MIS

2.1   The OpenMRS Add-Ons (modules) Development Capabilities

OpenMRS is a free- and open source software platform dedicated to develop and integrate MIS EMR solutions (https://github.com/openmrs/). This MIS is focused on EMR automation of primary health care institutions like ambulances and small clinics. Several academics and non-governmental organizations, including the Institute Regenstrief (http://regenstrief.org/) and In Health Partners (http://pih.org/), are responsible to support and maintain OpenMRS core code. There are dozens of implementations [17] registered, mainly in Africa and Asia (https://atlas.openmrs.org/).

The OpenMRS core is written in Java programming language using Spring and Hibernate frameworks. An MySQL RDBMS is used as data storage. There are tree main way to perform OpenMRS customization and adoption process:

- The visual interactive editor for managing templates of patient registration forms and their components - Concepts, Form Metadata and Form Schema - Form Administrator (https://wiki.openmrs.org/display/docs/Administering+Forms).
- The tool for integration of forms, developed by InfoPath (http://www.infopathdev.com/) - InfoPath Administrator (https://wiki.openmrs.org/display/docs/InfoPath+Notes).
- Set of programming interfaces (API) for creating custom modules using Java programming language (https://wiki.openmrs.org/display/docs/API and https://wiki.openmrs.org/display/docs/Modules).

The first two tools are easy-to-use and do not require knowledge of programming languages. However, they do not have features which are required to implementation of given CDSS. Therefore, OpenMRS Modules API has been selected to develop a module that implements features of the dialog component of CDSS platform. Corresponded module architecture is shown on Fig. 2.

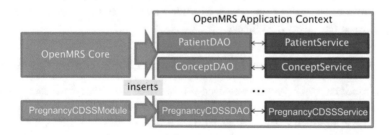

Fig. 2. Software architecture of Pregnancy CDSS module for OpenMRS that implements the dialog component of the CDSS platform

2.2 Adaption of the Information Model of Dialog Component of the CDSS Module

The external representations of the information model (IM) of CDSS dialog component, as well as the necessary data structures, are described in [13,24]. The internal representation of information model has been adapted according to OpenMRS database requirements for the custom modules (https://wiki. openmrs.org/display/docs/Data+Model):

- a mechanism of IM key concepts identification by the universal identifier (UUID) values assignment has been introduced (https://wiki.openmrs.org/ display/docs/UUIDs);
- some tables key field data types has been adopted according to OpenMRS coding guidelines (https://wiki.openmrs.org/display/docs/Conventions);
- module's database tables installation procedure according Liquibase technology (http://www.liquibase.org) description has been developed and set of special XML files has been formed.

Data structures for the recorded patient's data representation has been developed as the following Java-classes according to general (MVC, Model - View - Controller) approach adoption with the Spring framework usage.

- SymptCategoryModel.java - represent symptom's categories;
- SymptomModel.java - represent symptom's description;
- SymptomOptionModel.java - represent possible symptom's values;
- DiseasesSymptOptModel.java - represent information about probability of a certain diagnosis depending on the given symptom's value;
- PatientExamModel.java - represent general Patient questionnaire data model;
- PatientSymptomByExamModel.java - represent each patient's questionnaire submission.

The Java Hibernate framework should be used within OpenMRS to implement database management operations according coding guidelines (https:// wiki.openmrs.org/display/docs/For+Module+Developers). Therefore, necessary service classes has been developed.

2.3 Development of User Interface of the CDSS Dialog Component

Most of modern web- technologies could be used for user interface development of OpenMRS custom modules, including HTML 5, CSS 3, AJAX (JQuery usage is recommended). According to above, set of flexible forms and reports has been developed to effectively implement necessary Pregnancy CDSS module User Interface views according to IM external representations as it was shown in [24] and MVC paradigm. These views include:

- patientExamForm.jsp - the patient's survey main form;
- encounterPatientExamData.jsp - the portlet which represent pregnancy miscarriage pathology diagnostic data, provided by Pregnancy CDSS module, inside OpenMRS patient encounter form (Fig. 3);

- patientExamForm2Print.jsp - the survey report with patient's answers and diagnostic conclusion;
- series of forms under OpenMRS Administration section for the CDSS platform dialog component content management, settings adjustment and configuration customization.

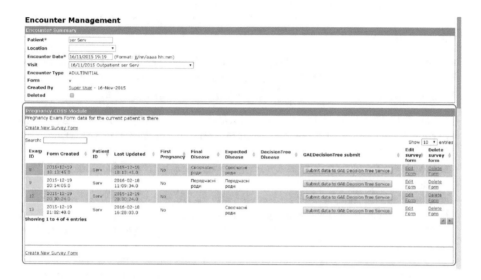

Fig. 3. Representation of pregnancy miscarriage pathology examination summary, provided by Pregnancy CDSS module, inside OpenMRS patient encounter form

Main decision-making algorithm are based on results of research obtained in [13,14,24]. This algorithm as well as common module's management activities has been implemented in form of Java servlets, according to general MVC approach.

- EncounterPatientExamDataPortletController.java - portlet controller to manage module data representation within OpenMRS patient's encounter form;
- PatientExamFormController.java - patient's survey form controller;
- GAEDecisionTreeController.java - provides interaction of the Pregnancy CDSS module with GAEDecisionTree diagnostic web-service;
- PregnancyCDSSManageController.java - provides Pregnancy CDSS module administrative features and customization capabilities.

The presented CDSS platform dialog's component and provided GAE Decision Tree web-service interaction procedure has been developed according to recommendations how to cross-site data request being performed (http://www.gwtproject.org/doc/latest/tutorial/Xsite.html#design). The following methods of the GAEDecisionTreeController.java controller are responsible for:

- getPatientDataJson2 - handles GET-type of HTTP request and returns data for the selected survey form as a JSON object;
- getAllPatientDataJson - handles GET-type of HTTP request and returns data for all survey forms, where final diagnosis is given, as a JSON object. It is used for the training dataset formation during GAE Decision Tree web-service education stage (http://decisiontree-1013.appspot.com);
- setGAEDecision - handles POST-type of HTTP request and store GAE Decision Tree diagnostic output in Pregnancy CDSS module database for appropriate patient's record.

Practically, Querying service GAE Decision Tree service has been queried directly from view (portlet encounterPatientExamData.jsp) with AJAX technology using jQuery library via the following code sniplet (listing 1):

- gaeDecisionTreeSubmitFunction - retrieves a survey form data by asynchronous calling of the getPatientDataJson2 method of the GAEDecisionTreeController.java servlet;
- submitData2GAE - submits a survey form data to the GAE Decision Tree service via asynchronous request;
- setDecisionTreeResponceFunction - receives a diagnostic conclusion provided by GAE Decision Tree service and redirect it to the GAEDecisionTreeController.java servlet by asynchronous calling of the setGAEDecision method.

A training dataset deployment to the GAE Decision Tree service has been implemented in the same way within the managepatientexams.jsp view in OpenMRS administrative panel of the Pregnancy CDSS module.

The Pregnancy CDSS module installation process has been performed according general OpenMRS administration guide (https://wiki.openmrs.org/display/docs/Administering+Modules):

- downloading the Pregnancy CDSS module compiled file (pregnancycdss-1.hh-SNAPSHOT.omod) from author's GitHub repository (https://github.com/semteacher/pregnacy_cdss).
- logging in to OpenMRS as administrator. Go to MIS module administration page (Administration - Manage Modules).
- pressing Add or Upgrade Module button. In "Add Module" dialog click Choose File in the Add Module section. Specify downloaded module file location and click OK than Upload.
- after installation will complete, new "Pregnancy CDSS Module" section will appears in OpenMRS patient Encounter form.

3   Experimental Research

In our experimental study we use data of 622 pregnant women which were investigated in work [19]. The data include 31 attributes concerning the following items

- *antibiotic* - taking antibiotics during pregnancy;
- *bpgest1 bpgest2 bpgest3 bpgest4* - gestational age at first-second-third-forth blood pressure reading (weeks);
- *map1 map2 map3 map4* - first-second-third-forth mean arterial blood pressure reading (mmHg);
- *sbp1 sbp2 sbp3 sbp4* - first-second-third-forth systolic blood pressure reading (mmHg);
- *dbp1 dbp2 dbp3 dbp4* - first-second-third-forth diastolic blood pressure reading (mmHg);
- *uti* - having a urinary tract infection in pregnancy;
- *uti_trim1 uti_trim2 uti_trim3* - having a urinary tract infection in the first-second-third trimester of pregnancy;
- *mumage* - mother's age;
- *parity* - parity;
- *gest_age_birth* - gestational age of the birth;
- *bweight* - birth weight of the baby;
- *sex* - sex of the baby;
- *maternalBMI* - pre-pregnancy BMI;
- *smoking* - mother smoked during pregnancy;
- *gdm* - mother had gestational diabetes during pregnancy;
- *ins0* - week 28 fasting insulin concentration (pmol/L);
- *gluc0* -week 28 fasting blood glucose concentration (mmol/L).

Some of the attributes are factors (taking antibiotics during pregnancy; parity; sex of the baby etc.). Others are numbers (mother's age; week 28 fasting insulin concentration (pmol/L) etc).

We have determined the gestational age of the birth as a class attribute for learning tuples. This class attribute was categorized using intervals for its values, namely ≤ 36, $[36, 37)$, $[37, 38)$, $[38, 39)$, $[39, 40)$, $[40, 41)$, ≥ 41 weeks.

As a result of application of decision tree induction algorithm (C5.0) we obtained the decision tree (see Listing 2)[1].

Thus, the size of the constructed tree is 29 levels. We have the following usage of attributes (in %): 100.00% - dbp4; 93.51% - parity; 56.28% - mumage; 38.10% - sbp1; 27.71% - sex; 18.61% - gdm; 17.75% - sbp3; 17.75% - ins0; 14.72% - dbp2; 11.26% - bweight; 8.23% - map3; 6.49% - sbp4; 3.03% - dbp3; 1.73% - map1; 1.73% - map2.

Further we investigated errors when using this decision tree for classification of pregnant due to class attribute values in the intervals mentioned above. If we accept the majority class in the leave as a predicted one, we get error in 45 cases (19.5%). This is a consequence of "rough" approach of such kind of prediction. If we analyze this error deeper, we can see that 33 of these 45 cases are in the intervals $[40, 41)$ and ≥ 41. In order to overcome this shortcoming and to decrease error size, we join these intervals. As a result we reduce the error to 12 cases (5.2%). Since the minimal value of testing error is not yet reached, the next ways

---

[1] Here we present decision tree in textual form. However, in general case decision tree can be displayed as an image.

of reducing classification errors should be dealt with the increasing of volume of training set and increasing of tree size.

4   Conclusions

Effectiveness of the Clinical Decision Support System (CDSS) application in the medical decision making process has been signed. An opportunities provided by CDDS in diagnostics of miscarriage pathologies with aim to prevent of preterm birth has been shown as a result of trial evaluation of the CDSS prototype in Ternopil regional perinatal center "Mother and Child". An approach to the decision making process which is based on the decision tree algorithm has been recommended. The implementation of the given above approach as separate web-service based on the GAE capabilities has been provided.

The results of code refactoring of the dialog subsystem of the CDSS platform which is made as module for the open-source MIS OpenMRS has been presented. The Model-View-Controller (MVC) based approach to the CDSS dialog subsystem architecture has been implemented with Java programming language using Spring and Hibernate frameworks. The OpenMRS Encounter portlet form for the CDSS dialog subsystem integration has been developed as a module. The data exchanging formats and methods to establish interaction between OpenMRS newly-developed Pregnancy CDSS module and GAE Decision Tree service are developed with AJAX technology via jQuery library.

Experimental research displayed opportunities of decision tree induction due to C5.0 algorithm for prediction of gestational age of the birth. In a similar way other data mining algorithms can be used (e.g., sequential covering for obtaining classification rules).

The prospects for the further research is to extend web-service core decision tree algorithm capabilities to support different types of diagnostic problems. Such achievements will allow to more comprehensive end more effective utilize of patient's health data which are collected within both supported MIS - OpenEMR and OpenMRS.

5   Appendix

Listing 1. Implementing of asynchronous interaction of the OpenMRS Pregnancy CDSS module with the GAE Decision Tree web-service

```
<script type="text/javascript">
  function submitData2GAE(formData){
    jQuery.ajax({
      type : 'GET',
      url : 'http://decisiontree-1013.appspot.com/patientdata',
        data : formData,
        dataType : 'json',
        success : function(response) {
          var mystr = JSON.stringify(response);
```

```
          setGAEDecision (response);
        },
        error : function(e) {
          alert('Error: ' + e);
        }
    });
  };
  function gaeDecisionTreeSubmitFunction(examId,encounterId,patientId){
    jQuery.ajax({
      type : 'GET',
      url : '${pageContext.request.contextPath}/module/
      pregnancycdss/gAEDecisionTree/single.json',
      data : 'examId=' + examId + '&encounterId=' + encounterId
      + '&patientId=' + patientId,
      dataType : 'json',
      success : function(response) {
         submitData2GAE(response);
      },
      error : function(e) {
        alert('Error: ' + e);
      }
    });
  };
  function setGAEDecision(GAEresponse){
    jQuery.ajax({
      type : 'POST',
      url : '${pageContext.request.contextPath}/module/
      pregnancycdss/gAEDecisionTree/setdisease.json',
      data : gAEresponse =' + GAEresponse,
      dataType : 'json',
      success : function(response) {
         alert('Sucessfully saved!');
      },
      error : function(e) {
        alert('Error: ' + e);
      }
    });
  };
</script>
```

Listing 2. Decision tree inducted for the experimental research in the Sect. 3

```
dbp4 > 86:
:...bweight <= 2.1: 36 (2/1)
:   bweight > 2.1:
:   :...dbp2 <= 66: 38 (2)
:       dbp2 > 66: 41 (11/3)
dbp4 <= 86:
:...parity = 4: 41 (4/1)
    parity = 5: 40 (2/1)
```

```
parity = 1:
:...mumage > 26.7: 41 (113/29)
:   mumage <= 26.7:
:   :...dbp3 <= 76: 40 (5)
:       dbp3 > 76: 39 (2/1)
parity = 3:
:...sbp1 > 110: 40 (4)
:   sbp1 <= 110:
:   :...ins0 > 59.8: 39 (3/1)
:       ins0 <= 59.8:
:       :...sbp4 > 105: 41 (9)
:           sbp4 <= 105:
:           :...dbp4 <= 58: 41 (2)
:               dbp4 > 58: 40 (4/1)
parity = 2:
:...dbp4 <= 54: 40 (2)
    dbp4 > 54:
    :...sbp1 > 130: 40 (2)
        sbp1 <= 130:
        :...sex = female:
            :...gdm = Yes:
            :   :...map1 <= 84: 41 (2)
            :   :   map1 > 84: 39 (2)
            :   gdm = No:
            :   :...sbp1 <= 95: 39 (3/1)
            :       sbp1 > 95:
            :       :...ins0 > 25.4: 41 (16/2)
            :           ins0 <= 25.4:
            :           :...sbp3 <= 100: 41 (3/1)
            :               sbp3 > 100: 40 (4)
            sex = male:
            :...sbp3 <= 115:
                :...gdm = Yes: 39 (2/1)
                :   gdm = No:
                :   :...bweight > 3.55: 41 (7)
                :       bweight <= 3.55:
                :       :...map2 <= 79.33334: 39 (2)
                :           map2 > 79.33334: 40 (2/1)
                sbp3 > 115:
                :...dbp2 > 75: 40 (2)
                    dbp2 <= 75:
                    :...map3 > 93: 41 (9)
                        map3 <= 93:
                        :...mumage <= 32: 41 (4)
                            mumage > 32: 40 (6/1)
```

References

1. Aminpour, F., Sadoughi, F., Ahamdi, M.: Utilization of open source electronic health record around the world: a systematic review. J. Res. Med. Sci. **19**(1), 57 (2014). the official journal of Isfahan University of Medical Sciences
2. Avramenko, V.I., Kachmar, V.O.: Creation of new ways for development of information technologies in medicine for Ukrainian health care grounded at worldwide approaches. Ukrainian Journal of Telemedicine and Medical Telematics, vol. 9(2), 124–133 (2011)
3. Borys, R.M., Martsenyuk, V.P.: Classification algorithm of polytrauma by induction of decision trees. Med. Inf. Eng. **2**, 12–17 (2013). https://doi.org/10.11603/mie.1996-1960.2013.2
4. Bright, T.J., et al.: Effect of clinical decision-support systems: a systematic review. Annal. Intern. Med. **157**(1), 29–43 (2012)
5. Edelman, E.A., et al.: Evaluation of a novel electronic genetic screening and clinical decision support tool in prenatal clinical settings. Matern. Child Health J. **18**(5), 1233–1245 (2014)
6. Effectiveness of computer screening system for diagnosing and prediction of preterm delivery. https://cyberleninka.ru/article/n/effektivnost-primeneniya-kompyuternoy-skriningovoy-sistemy-dlya-diagnostiki-i-prognozirovaniya-prezhdevremennyh-rodov. Accessed 12 Nov 2017
7. Esmaeilzadeh, P., et al.: Adoption of clinical decision support systems in a developing country: antecedents and outcomes of physician's threat to perceived professional autonomy. Int. J. Med. inf. **84**(8), 548–560 (2015)
8. Fritz, F., Tilahun, B., Dugas, M.: Success criteria for electronic medical record implementations in low-resource settings: a systematic review. J. Am. Med. Inf. Assoc. **22**(2), 479–488 (2015)
9. Goldspiel, B.R., et al.: Integrating pharmacogenetic information and clinical decision support into the electronic health record. J. Am. Med. Inf. Assoc. **21**(3), 522–528 (2013)
10. Jaspers, M., et al.: Effects of clinical decision-support systems on practitioner performance and patient outcomes: a synthesis of high-quality systematic review findings. J. Am. Med. Inf. Assoc. **18**(3), 327–334 (2011)
11. List of open-source health software. https://en.wikipedia.org/wiki/List_of_open-source_health_software#Electronic_health_or_medical_record. Accessed 12 Nov 2017
12. Martirosyan, H., et al.: A decision-support system for expecting mothers and obstetricians. In: 6th European Conference of the International Federation for Medical and Biological Engineering, pp. 703–706. Springer (2015)
13. Martsenyuk, V., Semenets, A.: System elektronicznych zapisów medycznych dla wspomagania decyzji z wykorzystaniem Google Application Engine (GAE). Studia Ekonomiczne **308**, 157–172 (2016)
14. Martsenyuk, V.P., Andrushchak, I.Y., Gvozdetska, I.S.: Qualitative analysis of the antineoplastic immunity system on the basis of a decision tree. Cybern. Syst. Anal. **51**(3), 461–470 (2015). https://doi.org/10.1007/s10559-015-9737-6
15. Martsenyuk, V.P., Stakhanska, O.O.: About clinical expert system based on rules using data mining technology. Med. Inf. Eng. **1**, 24–27 (2014). https://doi.org/10.11603/mie.1996-1960.2014.1
16. Martsnyuk, V.P., Semenets, A.V.: Medical informatics. Developer and expert systems. Ukrmedknyha, Ternopil (2004)

17. Mohammed-Rajput, N.A., et al.: OpenMRS, a global medical records system collaborative: factors influencing successful implementation. In: AMIA Annual Symposium Proceedings, vol. 2011. American Medical Informatics Association, p. 960 (2011)
18. Pahl, C., et al.: Role of OpenEHR as an open source solution for the regional modelling of patient data in obstetrics. J. Biomed. Inf. **55**, 174–187 (2015)
19. Petry, C.J., et al.: Associations between bacterial infections and blood pressure in pregnancy. Pregnancy Hypertens. **10**, 202–206 (2017). https://doi.org/10.1016/j.preghy.2017.09.004
20. Reynolds, C.J., Wyatt, J.C.: Open source, open standards, and health care information systems. J. Med. Internet Res. **13**(1) (2011)
21. Roshanov, P.S., et al.: Features of effective computerised clinical decision support systems: meta-regression of 162 randomised trials. Bmj **346**, f657 (2013)
22. Semenets, A.V.: About experience of the patient data migration during the open source EMR-system implementation. Med. Inf. Eng. **1**, 28–37 (2014). https://doi.org/10.11603/mie.1996-1960.2014.1
23. Semenets, A.V.: On organizational and methodological approaches of the EMR-systems implementation in public health of Ukraine. Med. Inf. Eng. **3**, 35–42 (2013)
24. Semenets, A.V., Martsenyuk, V.P.: On the CDSS platform development for the open-source MIS OpenEMR. Med. Inf. Eng. **3**, 22–40 (2015). https://doi.org/10.11603/mie.1996-1960.2015.3
25. Stakhanska, O.O.: Development of clinical diagnostic system based on rules with help of method of sequential covering. Med. Inf. Eng. **2**, 51–55 (2014). https://doi.org/10.11603/mie.1996-1960.2014.2

Programmable RDS Radio Receiver on ATMEGA88 Microcontroller on the Basis of RDA5807M Chip as the Central Module in Internet of Things Networks

Jakub Peksinski[(⊠)], Pawel Kardas, and Grzegorz Mikolajczak

West Pomeranian University of Technology in Szczecin, Szczecin, Poland
{Jakub.Peksinski, kp36848,
Grzegorz.Mikolajczak}@zut.edu.pl

Abstract. The article presents innovative approach to RDS (Radio Data System). In the proposed solution, the microcontroller communicates with the system of the radio receiver via I2C bus, checking individual memory addresses, where the current data from the radio data system are stored. Selection of a radio station or adjusting of radio frequency is made through the user menu on OLED I2C display (I2C – two-way serial bus used for data transfer). The transmission form uses only two data lines: SDA (serial data) and SCL (serial clock). These two lines are used to transfer all information between devices connected to I2C bus). By means of a microswitch or infrared remote control a user may select a radio station, set an alarm and change a mobile radio set into a comfortable alarm clock. The device has a connection for additional external Wi-Fi and Bluetooth modules, and thanks to them it may become an universal, mobile central station in IOT networks. Thanks to this, a user may always have at hand a small radio set, which may be used additionally - for instance in a smart home – as the controller that switches the light on and off, check temperature in individual rooms of a factory or is used as a remote controller of a gate to the premises.

Keywords: Internet of Things · Microcontroller AVR · Radio Data System
I2C transmission · C programming language

1 Introduction

The purpose of this article is to present a prototypical device, which performs functions of an RDS radio receiver and has an additional functionality – user-modifiable mobile central base in IOT networks.

At the moment there are numerous devices on the market that facilitate user's access to information, news, music, etc. and a possibility of communication among such electronic subassemblies is a kind of standard embedded in almost all basis items of household equipment, such as TV sets, printers, electronic clocks, etc. For example, let's look at any smartphone and a smart TV set. A user may use a smartphone to play selected movies, programmes, music and listen to the news in comfort. There are companies specialising in creation of so called smart homes, where by means of a

© Springer Nature Switzerland AG 2019
J. Pejaś et al. (Eds.): ACS 2018, AISC 889, pp. 209–219, 2019.
https://doi.org/10.1007/978-3-030-03314-9_19

phone or a special controller/center a user can make roller binds go down, turn the light on and off, control the heating level at home by means of a network of sensors and the above mentioned controller. Another example of a certain form of IoT in every-day life is a car equipped with a hands-free set, which couples a phone with a vehicle via bluetooth and enables comfortable search through lists of contacts by a driver and allowed form of phone conversation when driving a car. However, the majority of such systems is limited to communication of two, three devices (save for possible networks of sensors that are treated as subassemblies or elements of a larger project or a device).

But what is IOT [1, 2], that makes it possible for us to benefit from the above mentioned examples. On the basis of the article written by Prajsnar [3] and published on website www.forbes.pl, the Internet of things is most of all an idea enabling transfer of data by means of a transmission mode selected by a designer. Of course, the most popular one is wireless Wi-Fi transmission, that is one of the relatively easiest forms to be implemented. Obviously, there is no rule that the IoT system may be built only on the basis of Wi-Fi communication. There are other forms of data transmission among devices, like the above example of a smartphone connected with a car via bluetooth, or a set with a radio receiver and a radio transmitter. Increasingly more popular term of the Internet of Things seems to be the future in designing of modern devices and according to this idea the device discussed in this article is created.

The presented radio receiver will not be a small module of a radio, clock, alarm clock, but also a mobile version of universal central station of user IOT network that a user himself may develop by adding new functionalities according to his own ideas. The smart home is only one of many examples of possibilities of the project development. This device may be compared at first sight with a smartphone with an access to the Internet, it may be used to benefit from different forms of transmission, such as bluetooth or infrared, thanks to which IOT network may be to a certain extent implemented on the basis of independent devices. But the device presented in the title will be characterised mainly by a significantly lower cost of production as compared to the cheapest versions of smartphones, miniaturised size and larger programmable possibilities for a user.

According to the topic, the main function of our device is to play sound from selected ratio stations and read information from transmitting stations, enabling user's convenient and simple access to information and entertainment – the RDS technology will be implemented in the project [4, 5]. Another essential function of the project is the universal nature of a device in the IOT network as a mobile centre. Other functions, such as a watch, alarm clock, etc. have been described in Subsect. 2.1 containing the project assumptions. We can distinguish 8 typical functions of the discussed RDS technology (which were described in detailed in an article written by Sagan [6]), which define transmitted information:

- Program Services (PS) – name of the station,
- Program Type (PTY) – type of program,
- Radio Text (RT) – any text,
- Clock Time and Date (CT) – current time and date,

- Traffic Program/Traffic Announcement (TP/TA) – information for drivers,
- Program Identification (PI) – station identification code,
- Alternative Frequencies (AF) – list of alternative frequencies,
- Enhanced Other Networks (EON) – information about other programs stations.

On the basis of Krzysztof Sagan's article [7], a definition of radio data system may be presented as a subcarrier with a possibility of its modification by means of digital information, which is added to the traditional broadcasting of very high frequency waves (VHF) with Frequency Modulation (FM); it enables a separation of the data stream by means of duly adjusted receivers and consequently, improved reception and possibility to deliver additional information (which has been listed as RDS functions in the preceding paragraph).

2 Project Description

Universal small radio that can always be at hand. Listening to the news, listening to the music, checking time or setting an alarm. These are the most basic functions, which have occurred at the project designing stage. During the initial works we found out that it may be improved by adding functions, which make it possible for a user to develop the project on its own and add on ideas.

First of all, a user has a screen with OLED display, three micro switches and – what is most important – a coupling enabling transmission of RX and TX signals of the microcontroller, which facilitate use of Wi-Fi and Bluetooth transmission, thanks to which, consequently, a devices may be implemented in any IoT network. A user will dispose with all main functions of our radio receiver, that are described, for instance, below in the assumptions of the project prototype (Subsect. 2.1), while development of the device as the central station in the IOT network is open for the user, meaning that a user has programmable output at his disposal, thanks to which it may upload own programme code responsible for specific needs of ideas.

Obviously, it requires certain programming experience of the user to use fully the potential of a device. The miniaturisation of the device is based mainly on use of a small Lithium-polymer battery, which ensures mobility of a device and use of electronic elements that are adjusted for surface mounting, thanks to which the size of PCB plate will be maximally reduced. Mounting holes enable docing of the project in a selected place, which is one of the additional options.

2.1 Assumptions of the Project Prototype

The following project assumptions have been prepared on the stage of prototyping:

- Receipt of signals sent by UKF FM and RDS stations,
- Displaying information from RDS on OLED display,
- Possibility to play sound by means of an embedded speaker based on mono audio-type amplifier,
- Possibility to play sound via external headset by means of Jack 3.5 mm connection and signal of FM radio receive used,

- Displaying current time, date and temperature on OLED display on the basis of RTC chip,
- A possibility to set an alarm clock, time and date on the basis of RTC chip,
- Output connection for Bluetooth/Wi-Fi communication (a possibility to develop the device to include communication with external devices in IOT network),
- Powering the devices with a small Lithium-Polymer battery, using ATB-LION converter,
- A possibility to charge a device by means of a micro USB able on the basis of ATB-LION charger,
- Intuitive user menu based on OLED display and 3 micro switches and an infrared recipient (receiving signals from a remote device with RC5 coding), which enables setting of time, date, alarms, changing ratio stations,
- Output SPI bus and a connection equipped with RX, TX, signals enables user's modification of the programme through introduction of own changes in respect to handling an OLED display, three micro switches and use of infrared receiver. Moreover, a possibility to introduce own set of AT commands adjusted to communication through Wi-Fi and Bluetooth transmission (receiving and sending information).

2.2   Used Electronic Components in the Prototype

For the prototyping stage, mainly electronic parts adapter for surface mount SMD were used to minimize the prototype as much as possible.

- Microcontroller AVR ATMEGA88-AU, technical documentation by microchip [8],
- RTC system DS3231 – real-time clock system with built-in temperature sensor, maxim integrated technical documentation [9],
- RDA5807 M system – FM radio system with RDS, technical documentation by RDA microelectronics [10],
- TDA7052A system – mono audio amplifier system, NXP technical documentation [11],
- MCP1700T system – voltage stabilizer 3,3 V,
- ATB-LION system – DC-DC converter module (5 V output voltage) and Li-poly battery chargers (maximum 500 mA charging current), atnel technical documentation [12],
- TSOP311236 system – infrared receiver,
- OLED I2C 128x64 display,
- 3.5 mm jack connector,
- Resistors: 0 Ω, 180 Ω, 510 Ω, 10 kΩ, 39 kΩ, 56 kΩ, 100 kΩ, 330 kΩ, 1 M Ω,
- Ceramic capacitors: 22 pF, 1 μF, 10 μF,
- Bipolar capacitors: 22 μF, 100 μF,
- Inductors: 10 nH, 10 μH,
- Transistors: N-MOSFET BSS138 – conversion 5 V do 3,3 V for I2C communication with the RDA5808 M system,
- Quartz: 18.432 MHz,

- 2.54 mm pin connectors – universal RX, TX, VCC, GND line output/SPI transmission,
- Lithium-polymer battery (Li-Poly) 2,8 V–4,2 V.

2.3 Electronic Schemes of the Prototype System

The scheme of the device has been divided into functional parts. Individual of them are presented below:

- Figure 1 presents main block of our device. The electronic components responsible for the power supply are located on the left. On the right side are placed micro-switches, battery outputs and universal output for WIFI or Bluetooth modules, thanks to which the device can operate in IOT networks,

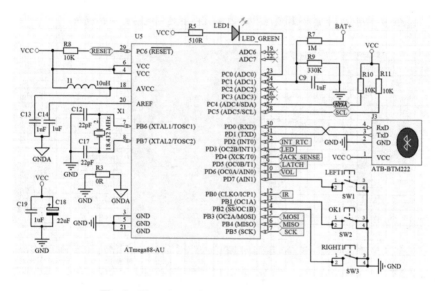

Fig. 1. The scheme for control unit – ATMEGA88

- Figure 2 presents RDS system with simple 5 V – 3.3 V converter based on BSS138 transistors on microcontroller lines – RDS system,

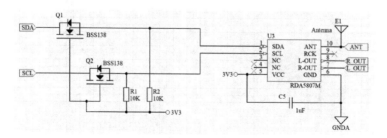

Fig. 2. The scheme for FM radio system with RDS and 5 V to 3.3 V voltage conversion

- Figure 3 presents connection of the system supplying the whole device with 5 V voltage and on the right side a voltage stabilizer 3.3 V,

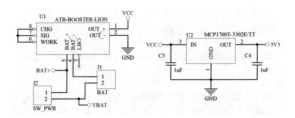

Fig. 3. The scheme for system power supply – 5 V DC/DC converter/charger and 3.3 V voltage stabilizer (for RDA5807M requirements)

- Figure 4 presents connection of the amplifier and headphone system,

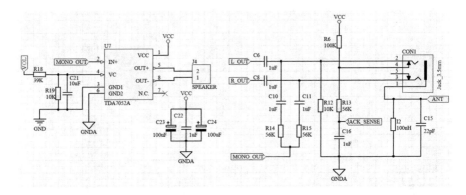

Fig. 4. The scheme for the mono audio amplifier circuit and the headphone input circuit

- Figure 5 presents connection of the RTC system on right side and on the left side SPI program bus output,

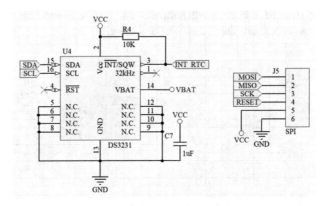

Fig. 5. The scheme for RTC clock chip and SPI program bus output

- Figure 6 presents connection of infrared receiver system and OLED I2C,

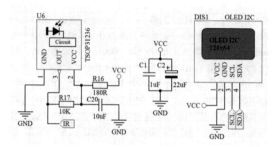

Fig. 6. The scheme for infrared receiver system and OLED I2C display connection

2.4   Operating Principle of the Device

AVR ATMEGA88 microcontroller software has been written in C programming language in Eclipse environment on the basis of books by Kardaś [13–15]. The code is uploaded and the system tested by means of SPI bus and USP-ASP programmer. The programme was written on the basis of the assumptions presented in such-chapter containing the list of functionalities. Communication between microcontroller and the system of RDA5807 M radio receiver is performed through I2C bus. As it can be noticed on scheme 2, ATMEGA88 is powered by 5 V current, while RDA5807 M – by 3.3 V current, scheme 2, 3. To communicate safety between systems powered with different voltage of 5 V and 3.3 V, a simple voltage conversion system should be installed that uses N-MOSFET BSS138-type transistors together with PULL-UP transistors for 3.3 V. The programme operates on the basis of principle of collection of addresses from the memory of radio receiver system, that stores, for instance, current RDS readouts. RDA5807 M system polling takes place in a non-blocking manner, that is on the basis of the programme timer created in the basis of the time of the equipment (TIMERX/COUNTERX). Read information is converted into string-type variables through their uploading to the buffered of characters (CHAR-type variables), and afterwards they are also transmitted by means of I2C bus to the OLED display. There, they are displayed also on the basis of programme timers, while universal menu operates according to the principle of multi-layer display, thanks to which, when creating own programme, a user operates on separate layers of OLED display and therefore when the device operates in the mode of user programme, layers with RDS information, or date, time and alarms may be displayed in intervals set by the user. Moreover, the real time clock (RTC) system is queried every second, which enables downloading of current date, time and temperature. Information is transferred to the display. The user menu may be opened by means of microswitches or infrared remote controllers (RC5 coding mode), and on this level the user may set current time, date, alarm and raise or reduce the volume level (depending on whether headset is connected or not) by means of TDA7052A or RDA5807 M amplifier. Despite entering into and leaving the menu, a user may change frequency – change of the radio station which is received by the device or switch into own programme mode that made operate according to a code written by the user. Communication of the microcontroller with DS3231 system take place according to the same principle as in case of the radio receiver. According to default setting, when it is connected to power, the system emits sounds from a built-in speaker linked with the system of mono audio TDA7052A amplifier. The programme checks on ongoing basis whether headphones are connected to the Jack slot. If it happens, it will entail an event that switches the emission of sound from the built-in speaker linked to an amplifier directly to the RDA5807 M system.

2.5 Fragment of the Source Code

Part of the main loop of the program, responsible for downloading time from the RTC system and displaying of individual RDS contents on the OLED display (Figs. 7 and 8).

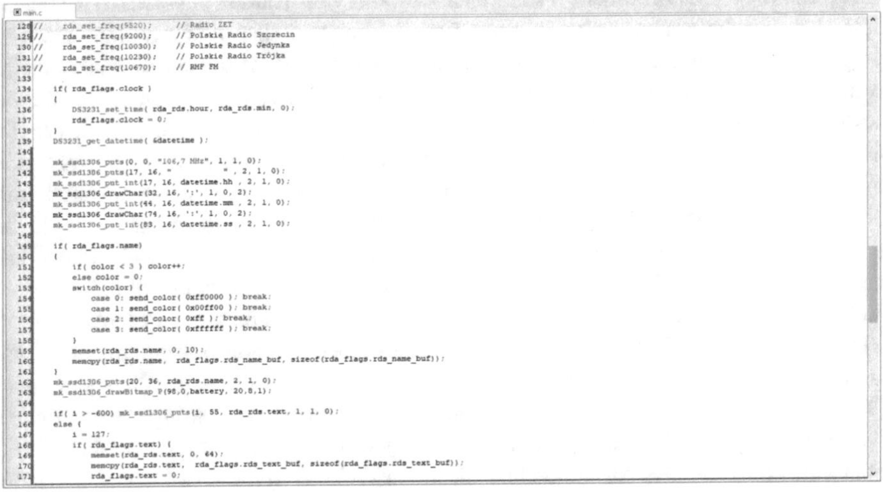


```
//    rda_set_freq(9520);     // Radio ZET
//    rda_set_freq(9200);     // Polskie Radio Szczecin
//    rda_set_freq(10030);    // Polskie Radio Jedynka
//    rda_set_freq(10230);    // Polskie Radio Trójka
//    rda_set_freq(10670);    // RMF FM

    if( rda_flags.clock )
    {
        DS3231_set_time( rda_rds.hour, rda_rds.min, 0);
        rda_flags.clock = 0;
    }
    DS3231_get_datetime( &datetime );

    mk_ssd1306_puts(0, 0, "106,7 MHz", 1, 1, 0);
    mk_ssd1306_puts(17, 16, "            " , 2, 1, 0);
    mk_ssd1306_put_int(17, 16, datetime.hh , 2, 1, 0);
    mk_ssd1306_drawChar(32, 16, ':', 1, 0, 2);
    mk_ssd1306_put_int(44, 16, datetime.mm , 2, 1, 0);
    mk_ssd1306_drawChar(74, 16, ':', 1, 0, 2);
    mk_ssd1306_put_int(83, 16, datetime.ss , 2, 1, 0);

    if( rda_flags.name)
    {
        if( color < 3 ) color++;
        else color = 0;
        switch(color) {
            case 0: send_color( 0xff0000 ); break;
            case 1: send_color( 0x00ff00 ); break;
            case 2: send_color( 0xff ); break;
            case 3: send_color( 0xffffff ); break;
        }
        memset(rda_rds.name, 0, 10);
        memcpy(rda_rds.name,  rda_flags.rds_name_buf, sizeof(rda_flags.rds_name_buf));
    }
    mk_ssd1306_puts(20, 36, rda_rds.name, 2, 1, 0);
    mk_ssd1306_drawBitmap_P(98,0,battery, 20,8,1);

    if( i > -600) mk_ssd1306_puts(i, 55, rda_rds.text, 1, 1, 0);
    else {
        i = 127;
        if( rda_flags.text) {
            memset(rda_rds.text, 0, 64);
            memcpy(rda_rds.text,  rda_flags.rds_text_buf, sizeof(rda_flags.rds_text_buf));
            rda_flags.text = 0;
```

Fig. 7. A fragment for the source in C – main.c

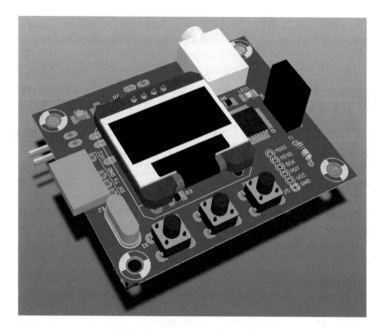

Fig. 8. Prototype – 3D model

2.6 Device Preview in 3D View

The prototype of our device in the design stage.

3 Summary

The device presented in the title may be in a way competitive to smartphones, stationary radio receivers, companies specialising in creation of IOT-related projects, and to launch and educational sets. The table below (Table 1) presents a comparison of our device to the average smartphone.

As compared to the companies on the market that deal with the above mentioned topics of the smart home and IOT in this respect (Table 1 - Usability in IOT networks), a new undertaking would be likely to occur due to smaller costs of production (Table 1 – Production cost) of a smart radio receiver and a base with tutorials that could present certain paths to be used in order to develop the project on one's own, while learning new programming languages.

The programme was written in a non-blocking form (using device and programme timers). Therefore, the user is free to introduce changes to the programme on its own (Table 1 – Possibility to modify the source code by user). It is an advantage of selected functional systems, to be used for the project, such as: DS3231, OLED display and RDA5807 M that they communicate with the microcontroller by means of I2C bus. Such solution makes it possible to save memory in respect to the volume of the code (handling one communication protocol for three systems).

This fact significantly affects the entire project, because a user, who wants to benefit from the entire potential of the device, can also use as much of free memory as possible.

Table 1. Comparison of the title device to the average smartphone

Functionalities	Our device	Average smartphone
Production cost	Low (~30–40 USD)	High (30–1000 USD)
Dimensions	Small (5 × 5 × 2 cm)	Medium (10 × 6 × 0.7 cm)
Possibility to modify the source code by user	Yes	Impossible
RDS radio function	Available immediately	User need to download and install dedicated application
Difficulty level of modifying code by user	Easy (knowledge of the basics of the C language)	Medium (knowledge of java, css, html, application development)
Usability in IOT networks	High (user is able to modify source code for his needs)	Medium (user can only create application)

Of course, every technological solution mentioned above has its strengths and weaknesses, each of them offers certain possibilities to the user, and limits certain solutions, devices have different dimensions and are based on different technology. But the device presented in the title relies on innovation, which means the user has a possibility to develop the project on his own.

Authors think about the further development of the projects as if the device enters into the market, an independent educational platform may be added to the project in form of an Internet website with descriptions, articles, instructions and video tutorials about creation and programming of devices for instance for the smart home that is so popular now. In such case, the entire project would gain an additional feature like educational for beginners in learning programming, which can highly influence on popularity of our project.

References

1. Petrenko, A.S., Petrenko, S.A., Makoveichuk, K.A., Chetyrbok, P.V.: The IIoT/IoT device control model based on narrow-band IoT (NB-IoT). In: 2018 IEEE Conference of Russian Young Researchers in Electrical and Electronic Engineering (EIConRus), 29 January–1 February 2018
2. Ruiz, M.C., Olivares, T., Lopez, J.: Evaluation of cloud platforms for managing IoT devices. In: 2017 8th International Conference on Information, Intelligence, Systems & Applications (IISA), 27–30 August 2017
3. Prajsnar, P.: https://www.forbes.pl/technologie/czym-jest-internet-rzeczy/egcvmr0
4. Barca, C., Neamtu, C., Popescu, H., Dumitrescu, S., Sandu, A.-S.: Implementation of RDS platform solutions for an emergency system. In: Proceedings of the International Conference on ELECTRONICS, COMPUTERS and ARTIFICIAL INTELLIGENCE - ECAI-2013, 27–29 June 2013

5. Heymann, C.H., Ferreira, H.C., Weber, J.H.: A Knuth-based RDS-minimizing multi-mode code. In: 2011 IEEE Information Theory Workshop (ITW), 16–20 October 2011
6. Sagan, K.: http://old.radiopolska.pl/portal/staticpages/index.php?page=rds-funkcje
7. Sagan, K.: http://old.radiopolska.pl/portal/staticpages/index.php?page=rds
8. Technical documentation of AVR ATMEGA88 microchip microcontroller (2016). https://www.microchip.com/wwwproducts/en/ATmega88
9. Documentation of the company maxim integrated, RTC DS3231 (2015). https://datasheets.maximintegrated.com/en/ds/DS3231.pdf
10. Documentation of the RDA microelectronics company, the RDA5801 M radio system (2011). http://cxem.net/tuner/files/tuner84_RDA5807M_datasheet_v1.pdf
11. NXP documentation, mono audio amplifier circuit (1991). https://www.nxp.com/docs/en/data-sheet/TDA7052A_AT.pdf
12. Documentation of the ATNEL company, DC/DC converter (2016). http://atnel.pl/digi-lion-3.html
13. Kardaś, M.: AVR Microcontrollers C – Programming Basics (ATNEL), Szczecin (2011). ISBN 978-83-931797-3-2
14. Kardaś, M.: Język C Pasja programowania mikrokontrelrów 8-bitowych (ATNEL), Szczecin (2014). ISBN 978-83-931797-4-9
15. Kowalski, J., Peksinski, J., Mikolajczak, G.: Detection of noise in digital images by using the averaging filter name COV. In: 5th Asian Conference on Intelligent Information and Database Systems, ACIIDS 2013, vol. 7803, pp. 1–8. Springer, Berlin (2013)

Business Process Modelling with "Cognitive" EPC Diagram

Olga Pilipczuk[1]([⊠]) and Galina Cariowa[2]

[1] University of Szczecin, Mickiewicza 64, 71-101 Szczecin, Poland
olga.pilipczuk@wneiz.pl
[2] West Pomeranian University of Szczecin, Żołnierska 49,
71-210 Szczecin, Poland
gcariowa@wi.zut.edu.pl

Abstract. This paper presents the conception of Cognitive Event Driven Chain Diagram (cEPC) based on integration of traditional EPC diagram and fuzzy cognitive maps. At the beginning the stages of evolution of business process modelling (BPM) tools are presented. The pyramid of business process classification in terms of cognitive BPM is discussed. The paper also describes the business process cognitive intensity evaluation method. An example of cEPC diagram of skin cancer diagnosis process is provided as well.

Keywords: Business process · Event-Driven Chain Diagram
Fuzzy cognitive map · ARIS methodology

1 Introduction

The market of business process management tools has gone through peaks and falls in recent decades. In the early nineties, the most popular concept of redesigning processes was reengineering. The popularity of this approach, however, gradually waned. Until recently, the BPM tools on the global market have been kept calm [1]. The stagnation period was interrupted by the introduction of new generation BPM systems [1]. At present, traditional BPM is weakening in strength, and its place is replaced by systems connected with Business Intelligence and Cognitive Computing creating the concept of iBPM (intelligent BPM).

Figure 1 presents the evolution of business process management systems based on changes in level of functionality advancement. It shows the transformation from simple executive models to advanced analytics and process intelligence. The goal of modern iBPM systems is to manage dynamics and improvisation processes. These processes have a lot of novelty and risk. They require quick recognition of the situation, making decisions and applying creativity during their implementation.

The change in the technology of creating BPM systems was caused by the changes in the perception of the nature of business processes. The era of cognitive business process management was initiated. At the moment there is a small number of works related to this issue. Mostly, these are conceptual works published by IBM managers. In the work of Hull and Nezhad, the components of the Cognitive BPM (cBPM) concept are described [4]. Nezhad and Akkiraju from the IBM Research Center in the

© Springer Nature Switzerland AG 2019
J. Pejaś et al. (Eds.): ACS 2018, AISC 889, pp. 220–228, 2019.
https://doi.org/10.1007/978-3-030-03314-9_20

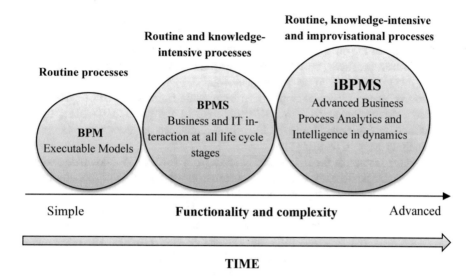

Fig. 1. The evolution of BPM systems (Source: [2, 3])

U.S.A are also talking about the pursuit towards cognitive process management [5, pp. 161–162].

Cognitive BPM means cognitive decision support, cognitive process learning, cognitive interaction with processes and cognitive process enablement.

Cognitive process enablement and learning require using the cognitive knowledge representation models. In this paper authors make attempts to create the cognitive EPC diagram (cEPC) based on combination of traditional EPC diagram and fuzzy cognitive maps. An example of cEPC implementation in ARIS methodology is provided as well.

2 The Cognitive BPM Concept

In 2016 Hulls and Nezhad have proposed the pyramid of business process classification [4]. The processes were divided into three categories: transaction-intensive processes, judgement-intensive processes and strategic processes with a high degree of cognitive intensity.

The first group are transactional processes. Transaction processes, i.e. well-structured and executed many times processes [4]. Typical examples of such operations are: payroll processing, supply chain management, receivables, as well as typical on-line shopping and self-service in the retail and service industry [4].

The second group consists of opinion-making intensive processes requiring many judgments regarding information, organization and systems, for example processes such as: customer relationship management, management of extensive projects in the IT industry, etc. [4]. In practice, many of these processes are performed on an ad-hoc basis, which is why they are managed using simple solutions, such as spreadsheets [4]. Planning and strategy processes that involve open creative decisions based on many fields of knowledge and types of analysis [4]. Examples include merger and acquisition planning, exploration of new markets, decision to build and purchase, as well as transformation of the business model of exploration [4]. Although these processes may come from best practices, they are often unorganized or temporary, due to the many possible decision options to be investigated [4].

In recent years, several attempts have been made to integrate BPM systems with Knowledge Management Systems (KMS) [5, 6]. In the years 2010–2011, the first specific methodologies and tools for Knowledge-Intensive business processes analysis and modelling have to appear [7].

One of the leaders of the iBPM market software is Software AG company, which is owner of ARIS platform. According to Gartner's "iBPM software quadrant" Software AG is the leader among visionaries [2].

In ARIS methodology an EPC diagram may be integrated with the Knowledge structure diagram. The knowledge symbols are used to describe the required knowledge to perform the function [8]. The knowledge map provides Coverage quality connection attribute, which can have the following values: Low, Average, High, and Maximum [8, p. 179]. This information can be visualized by graphic symbols. There is no direct relation between the values of the Degree of coverage and Coverage quality attributes [8, p. 179]. If both attributes are used, it is advisable that the qualification "Low" be used for a degree of coverage of up to 25%, "Average" for 26–50%, "High" for 51–75%, and "Maximum" for 76–100% [8, p. 179].

2.1 The Business Process Cognitive Intensity

The cognitive intensity could be estimated in terms of human capital cognitive abilities. Decision making is one of the basic cognitive processes [9–11]. Although the cognitive abilities of decision-makers can vary considerably, cognitive processes occurring in the human brain exhibit similarities in characteristics and mechanisms [12–14].

In order to measure the cognitive abilities of employees, specialized tests are usually used. Cognitive abilities reproduce the speed of information processing. They determine how quickly a person is able to go from the perception of new information to knowledge and its practical application in different situations. Employees' abilities depend on analytical, logical and problem-solving abilities. The cognitive abilities are the part of overall persons' intelligence.

A different method of measuring cognitive effort is the use of eye tracking technology, which includes: avoidance preference testing [15–17], feelings of emotional excitement [18], pupil diameter [19], cramps facial muscles [20], neural activity [21–25] etc.

The cognitive intensity of the business process could be estimated as the average of coverage cognitive attributes associated with functions:

$$Cint = \frac{\sum_i^n Cp}{n} \Big/ \frac{\sum_i^n Cr}{n} \qquad (1)$$

where:

Cint – business process cognitive intensity;
Cp – possessed cognitive abilities for the function;
Cr – required cognitive abilities for the function;
n – number of functions

2.2 The Concept of Cognitive EPC Diagram

The idea of cEPC is based on dividing the overall employee intelligence on two parts: domain knowledge and cognitive abilities. This division is created on the basis of McGrew Intelligence model [26]. For each part the appropriate symbol is used. The functions are linked with final concepts of fuzzy cognitive maps using the similar technology like the links between traditional eEPC models (Fig. 2). Each intelligence symbol is supported by the Coverage quality attribute symbol and percentage number [8].

The process cycle efficiency (E_t) can be calculated using following formulae:

$$E_t = \frac{(F_t + C_t)}{T_t} \tag{2}$$

where:

F_t - function processing time – is the time necessary for function execution;
C_t - concept processing time – is the working time on concept development;
T_t - total process time

3 An Example of Cognitive Map Analysis

At this part of the paper we provide the example of the single patient skin cancer diagnosis process. The main process functions are mentioned in Table 1. We selected the skin cancer risk estimation function to show the advances of connection of fuzzy cognitive map and an EPC diagram.

The cancer risk estimation function is connected with skin cancer risk concept. The following factors usually have influence on skin cancer risk: previous cancer cases; family history of cancer cases; skin hypersensitivity; immunosuppressants; dermatitis; skin infections; more than 50 skin marks; non-typical skin marks; self-examination, Bowen's disease, radiotherapy, exposure to chemical substances, UV radiation. After the dermatological analysis it was set that only four of above mentioned factors have influence on the patient.

Figure 3 shows a fuzzy cognitive map for skin cancer risk description. Verbal values are given in a result of consultations with specialists in the dermatological industry and may change depending on the experience, region and other factors. The stated verbal values constitute an aggregation of opinions of four experts. Figure 4

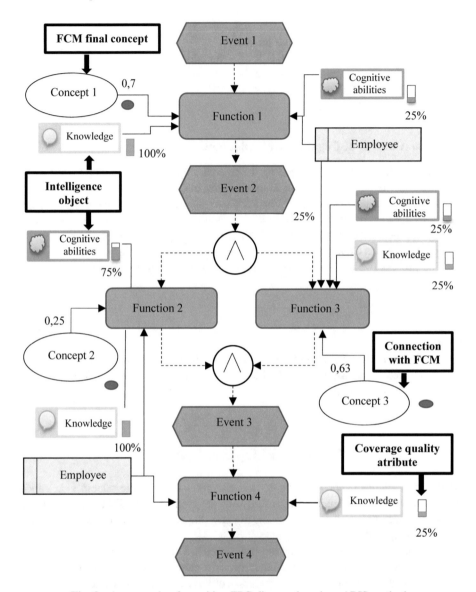

Fig. 2. An example of cognitive EPC diagram based on ARIS method

presents the probability distribution for the concept of "skin cancer risk" created using a triangular fuzzy model.

The Table 1 presents the verbal opinions acquired from the dermatology specialists together with the numerical interpretations.

Table 1. Defuzification of experts opinion on skin cancer risk

Relation number	Relation type	Expert 1	Expert 2	Expert 3	Expert 4	Numeric value
1	Positive	Often	Often	Often	Often	0,67
2	Positive	Often	Often	Often	Often	0,67
3	Positive	Often	Often	Often	Often	0,67
4	Positive	Sometimes	Average	Sometimes	Sometimes	0,31

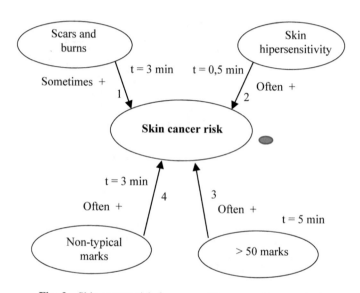

Fig. 3. Skin cancer risk fuzzy cognitive map of one patient

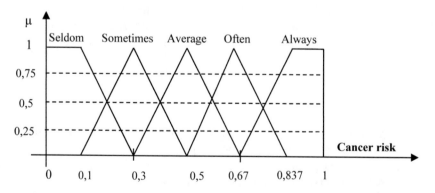

Fig. 4. The probability distribution for the concept of "skin cancer risk"

The probability of occurrence of the concept based on the sigmoidal function is calculated.

$$f(2,32) = \frac{e^{2,32}}{1+e^{2,32}} = 0,91 \tag{3}$$

The obtained results shows the patient high cancer risk.
We estimated the process cycle efficiency on the basis of the data from Table 2.

Table 2. Process of "skin cancer diagnosis" cycle time

Function name	Waiting time	Processing/decision time
Interview		5 min
Dermatoscopy		10 min
Cancer risk estimation		11,5 min
Symptoms assessment		3,6 min
Initial biopsy	10 min	7 min
Biopsy results interpretation		10 min
Seams making		10 min
Seams removing	7 days	3 min
Sending the skin slice		5 min
Diagnosis	3–5 weeks	2 min
Sending to additional tests		5 min
Determination of cancer stage	5–10 days	5 min
Issuing hospital referral		2 min

The process cycle efficiency = 0,89
The process of making a diagnosis runs with large intervals needed to obtain dermatological test results, which are not affected by the dermatological clinic. It has an adverse effect on its efficiency. Therefore, only the waiting time from the total process cycle time, which relates directly to the facility, was taken into account during the calculations.

4 Discussion

The results obtained from the example presented above should be used during the simulation process and presented on cEPC diagram.

Using cognitive map models, it is possible to determine the probability of decision concepts such as the risk of disease and the size of cancer symptoms affecting the diagnosis. The cognitive maps create the basis for further cognitive analytics.

Additionally, an cEPC diagram can be colored using color coded scales to show the current status of the coverage attributes [27].

References

1. Harmon, P.: BP Trends report. The State of Business Process Management 2016 (2016). www.bptrends.com
2. Gartner Business Transformation & Process Management Summit, 16–17 March 2016, London, UK. https://www.gartner.com/binaries/content/assets/events/keywords/business-process-management/bpme11/btpm_magicquadrantforintelligentbusinessprocess.pdf
3. Dunie, R.: Magic Quadrant for Intelligent Business Process Management Suites, Gartner (2015)
4. Hull, R., Nezhad, H.: Preprint from Proceedings of International Conference on Business Process Management, Rethinking BPM in a Cognitive World: Transforming How We Learn and Perform Business Processes, Business Process Management 14th International Conference, BPM 2016 Proceedings, Rio de Janeiro, Brazil, 18–22 September, pp. 3–19 (2016)
5. Marjanovic, O., Freeze, R.: Knowledge intensive business processes: theoretical foundations and research challenges. In: 44th Hawaii International Conference on System Sciences (HICSS) (2011). https://doi.org/10.1109/hicss.2011.271
6. Sarnikar, S., Deokar, A.: Knowledge management systems for knowledge-intensive processes: design approach and an illustrative example. In: Proceedings of the 43rd Hawaii International Conference on System Sciences (2010)
7. Rychkova, I., Nurcan, S.: Towards adaptability and control for knowledge-intensive business processes: declarative configurable process specifications. In: Proceedings of the 44th Hawaii International Conference on System Sciences (2011)
8. ARIS Method (2016). https://industryprintserver-aris9.deloitte.com/abs/help/en/documents/ARIS%20Method.pdf
9. Wang, Y., Wang, Y.: Cognitive informatics models of the brain. IEEE Trans. Syst. Man Cybern. Part C Appl. Rev. 36(2), 203–207 (2006)
10. Wang, Y.: Software Engineering Foundations: A Software Science Perspective. Auerbach Publications, Boston (2007a)
11. Wang, Y.: The theoretical framework of cognitive informatics. Int. J. Cogn. Inform. Nat. Intell. (IJCINI), 1(1), 1–27 (2007b)
12. Wang, Y., Gafurov, D.: The cognitive process of comprehension. In: Proceedings of the 2nd IEEE International Conference on Cognitive Informatics (ICCI 2003), London, UK, pp. 93–97 (2003a)
13. Wang, Y., Wang, Y., Patel, S., Patel, D.: A layered reference model of the brain (LRMB). IEEE Trans. Syst. Man Cybern. 36(2), 124–133 (2004)
14. Wang, Y.: On cognitive informatics. Brain Mind Transdisc. J. Neurosci. Neurophilos. 42, 151–167 (2003)
15. Kool, W., McGuire, J., Rosen, Z., Botvinick, M.: Decision making and the avoidance of cognitive demand. J. Exp. Psychol. Gen. 139, 665–682 (2010)
16. McGuire, J., Botvinick, M.: Prefrontal cortex, cognitive control, and the registration of decision costs. Proc. Natl. Acad. Sci. 107, 7922 (2010)
17. Westbrook, A., Kester, D., Braver, T.: What is the subjective cost of cognitive effort? load, trait, and aging effects revealed by economic preference. PLoS ONE 8(7), e68210 (2013)
18. Dreisbach, G., Fischer, R.: Conflicts as aversive signals: motivation for control adaptation in the service of affect regulation. In: Braver, T.S. (ed.) Motivation and Cognitive Control. Psychology Press, New York (2012)
19. Kahneman, D.: Maps of bounded rationality: A perspective on intuitive judgment and choice, Les Prix Nobel 2002, Almquist & Wiksell International, Sztokholm, Sweden (2003)

20. Elkins-Brown, N., Saunders, B., Inzlicht, M.: Error-related electromyographic activity over the corrugator supercilii is associated with neural performance monitoring. Psychophysiology **53**, 159–170 (2015)
21. Cavanagh, J., Masters, S., Bath, K., Frank, M.: Conflict acts as an implicit cost in reinforcement learning. Nat. Commun. **5**, 5394 (2014)
22. Cavanagh, J., Frank, M.: Frontal theta as a mechanism for cognitive control. Trends Cogn. Sci. **18**, 414–421 (2014)
23. Spunt, R., Lieberman, M., Cohen, J., Eisenberger, N.: The phenomenology of error processing: the dorsal anterior cingulate response to stop-signal errors tracks reports of negative affect. J. Cogn. Neurosci. **24**, 1753–1765 (2012)
24. Blain, B., Hollard, G., Pessiglione, M.: Neural mechanisms underlying the impact of daylong cognitive work on economic decisions. PNAS **113**, 6967–6972 (2016)
25. Westbrook, A., Kester, D., Braver, T.: What is the subjective cost of cognitive effort? load, trait, and aging effects revealed by economic preference. PLoS ONE **8**, e68210 (2013)
26. Schneider, W., McGrew, K.: The Cattell-Horn-Carroll model of intelligence. In: Flanagan, D., Harrison, P. (eds.) Contemporary Intellectual Assessment: Theories, Tests, and Issues (3rd ed.), pp. 99–144. Guilford, New York (2012)
27. Pilipczuk, O., Cariowa, G.: Opinion acquisition an experiment on numeric, linguistic and color coded rating scale comparison. In: Kobayashi, S., Piegat, A., Pejaś, J., El Fray, I., Kacprzyk, J. (eds.) Hard and Soft Computing for Artificial Intelligence, Multimedia and Security, Advances in Intelligent Systems and Computing, vol. 534, pp. 27–36. Springer, Cham (2016)

Algorithmic Decomposition of Tasks with a Large Amount of Data

Walery Rogoza[1](✉) and Ann Ishchenko[2](✉)

[1] Faculty of Computer Science and Information Technology,
West Pomeranian University of Technology, Zolnierska Str. 52,
71-210 Szczecin, Poland
wrogoza@wi.zut.edu.pl

[2] Educational and Scientific Complex "Institute of Applied Systems Analysis" -
ESC "IASA", The National Technical University of Ukraine
"Igor Sikorsky Kyiv Polytechnic Institute", Building 35,
Peremogy Av. 37-A, Kiev 03056, Ukraine
annvalish@gmail.com

Abstract. The transformation of models and data to the form that allows their decomposition is called algorithmic decomposition. It is a necessary preparatory stage in many applications, allowing us to present data and object models in a form convenient for dividing the processes of solving problems into parallel or sequential stages with significantly less volumes of data. The paper deals with three problems of modeling objects of different nature, in which algorithmic decomposition is an effective tool for reducing the amount of the data being processed and for flexible adjustment of object models performed to improve the accuracy and reliability of the results of computer simulation. The discussion is accompanied by simple examples that allow the reader to offers a clearer view of the essence of the methods presented.

Keywords: Algorithmic decomposition · Model reduction · Time series
Complex objects · Computer simulation

1 Introduction

The decomposition of mathematical models of objects into a number of simpler (in a certain sense) models, which can be investigated by conventional computational methods, is a traditional approach to overcoming difficulties of studying complex objects. In recent years, methods of model decomposition have acquired a new interpretation in connection with the development of computer platforms and software allowing the division of simulation processes into a number of concurrent computational flows. As an example, we can mention the MapReduce programming model and the Apache Hadoop open programming platform [1], which were designed for concurrent processing sets of big data using computer clusters.

In some cases, running parallel computing processes does not require preliminary preparation of input data (for example, when processing texts, data can usually be divided into parts in an arbitrary manner). In other cases, the task must be prepared

© Springer Nature Switzerland AG 2019
J. Pejaś et al. (Eds.): ACS 2018, AISC 889, pp. 229–243, 2019.
https://doi.org/10.1007/978-3-030-03314-9_21

beforehand so that it can be solved by dividing it into several subtasks (for example, when solving a big set of equations, it should be divided into several loosely coupled subsystems). The processes of transformation of models to a form that allows their decomposition we call the algorithmic decomposition. The advantages of this decomposition are that, as a rule, subtasks are less complex and can be solved in less time. Thanks to the parallelization of computational processes, the overall decision time is also reduced, therefore the above decomposition can be considered as a way of reduction of a complex problem.

In this paper, we discuss several tasks and methods for their solution, which clearly demonstrate the close relationship between decomposition and reduction and the advantages that these methods bring for us. The discussion is accompanied by examples of problems, whose solutions were proposed by the authors.

2 Model Decomposition Based on the Reduction of Singularly Perturbed Models

The mentioned math model can be represented in the matrix form as follows:

$$\begin{cases} a)\ \mu\dot{x} = f(x,y), & x(0) = x^0, & x \in \mathbf{R}^n, \\ b)\ \dot{y} = g(x,y), & y(0) = y^0, & y \in \mathbf{R}^m, \end{cases} \tag{1}$$

where $x(t)$ and $y(t)$ are the n- and m-dimensional sub-vectors of time-dependent state variables determined in real spaces, and μ is the n-dimensional diagonal matrix of small in magnitude parameters. It is assumed that model (1) represents the physical states of the considered object within a certain time interval $t \in [0, T]$ and the initial conditions for state variables $x(t)$ and $y(t)$ are given by the x^0 and y^0 vectors. The state equations of type (1) are characteristic, for example, in describing the behavior of large integrated circuits with allowance for second-order effects on the substrate of the semiconductor structure [2].

The theory of singularly perturbed ordinary differential equations (ODEs) [3] establishes that matrix equation (1,a) describes fast processes which take place within the relatively narrow boundary layer $t \in [0, \tau]$, $\tau \ll T$, and matrix equation (1,b) describes relatively slow processes beyond the boundary layer. In practical cases, the dimension of (1,a) is much greater than the dimension of (1,b). If the dimensionality of (1) is very large, the computer resources may not be sufficient for its numerical solution at an acceptable time. The simplest way of reduction (1) lies in setting small parameters to zero $\mu = 0$ to give:

$$\begin{cases} a)\ 0 = f(x,y), & x \in \mathbf{R}^n, \\ b)\ \dot{y} = g(x,y), & y(0) = y^0, & y \in \mathbf{R}^m, \end{cases} \tag{2}$$

where fast processes are ignored, and variables $x(t)$ can be expressed as functions of variables $y(t)$ within the entire interval of observation $[0, T]$. Since in most practical cases the dimension of the first subset is much greater than the dimension of the second subset, replacing (1) by (2) leads to a significant reduction in the total dimension of the set of differential equations. Unfortunately, despite the simplicity, this approach has significant drawbacks, the essence of which can be explained in the following simple consideration.

Usually (2,a) is a subset of non-linear algebraic equations, which has many roots. Some of them provide a good approximation for the exact solution of the original problem (1), and they are called the stables root of the reduced set of equations [3], but the remaining roots can be arbitrarily far from the actual exact solution of (1). According to Tikhonov's theorem [3], if the initial point (x^0, y^0) belongs to the attraction domain D of the stable roots, then the following asymptotic relationships are valid:

$$x(t, \mu) \rightarrow \bar{x}(t), \quad \tau \leq t \leq T,$$

$$y(t, \mu) \rightarrow \bar{y}(t), \quad 0 \leq t \leq T, \tag{3}$$

when $\mu \rightarrow +0$, where $(\bar{x}, \bar{y})$ is the solution of the reduced problem (2) and $[0, \tau]$ is a boundary layer. In order to find out which of the roots of subsystem (2,a) are stable, it is necessary to perform additional studies that may take a long time and require considerable computer memory.

The second disadvantage of this method of reduction is that we lose information about the behavior of the object within the boundary layer. At the same time, this information can be of great importance for the researcher (for example, if second-order effects in the integrated circuit are investigated).

The theory of singularly perturbed ODEs offers another approach to model reduction, which makes it possible to take into account small parameters in the reduced models [3]. Indeed, we can build the asymptotic solution of problem (1) using series:

$$z(t, \mu) \sim \sum_{k=0}^{\infty} \mu^k \bar{z}_k(t) + \sum_{k=0}^{\infty} \mu^k \pi_k \left(\frac{t}{\mu} \right), \tag{4}$$

where $\bar{z}_k(t)$ means variables $\bar{x}_k(t)$ or $\bar{y}_k(t)$, which correspond to the regular components of the solution of (1), as they are called. Those components vary mainly beyond the boundary layer. Terms $\pi_k(t/\mu)$ correspond to the singular components of the solution, as they are called, that is, those which vary mainly within the relatively narrow boundary layer $[0, \tau]$, where τ is determined by the values of small parameter μ. The first group of variables vary with time t and the second group vary with time t/μ, that is, in another time scale. In other words, representing the desired solution in the form (4), we deal both with "fast" and "slow" variables, which vary in different time scales.

It is noteworthy that the first and second groups of components in (4) can be computed in a certain sense independently of each other. This fact can be of great importance if we take into account that in actual singularly perturbed models, as a rule, there are a large number of small parameters with different values, that is, the diagonal matrix μ in (1) can take the form of a diagonal block matrix $\text{diag}(\mu_1, \mu_2, ..., \mu_i, ..., \mu_L)$, where each block μ_i includes small parameters with approximately the same values.

And respectively, the boundary layer can be split into a number of sub-layers $[t_0,$ $t_0 + \tau]$, $[t_0 + \tau, t_0 + 2\tau]$,..., $[t_0 + (L - 1)\tau, t_0 + L\tau]$, each is characterized by its own time scale.

The separation of the boundary layer into a number of boundary sublayers makes it possible to represent a singularly perturbed system of Eq. (1) in the form:

$$
\begin{cases}
a)\, diag(\mu_1, \mu_2, \ldots, \mu_L)\dot{x} = f(x, y), & x(0) = x^0, & x \in \mathbf{R}^n, \\
b)\, \dot{y} = g(x, y), & y(0) = y^0, & y \in \mathbf{R}^m,
\end{cases}
\tag{5}
$$

where vector x is split into L subvectors: $x = (x_1, \ldots, x_i, \ldots, x_L)$.

A model of the form (5) has two remarkable features. First, each subset in (5,a) can be solved independently of the rest subsets. Indeed, solving, say, a subsystem with respect to variables x_i, we can assume that all variables from subvectors $x_1, \ldots, x_{i-1}$ are algebraic variables dependent on x_i (just as variables x in (2) can be computed as functions of variables y), and the rest of variables $x_{i+1}, \ldots, x_L$, y are considered quasi-constant. Thus, the solution procedure of (5) can be divided into $L + 1$ successive procedures, and at each stage we deal with a set of equations having an essentially smaller dimension in comparison with the original system.

And second, solving (5), we do not lose information about the behavior of the "fast" variables inside the boundary layers, in contrast to the solution of (2). Some details of the computer implementation of the above method are discussed in [4], where the solution algorithm is offered based on the use of adaptive neuron networks.

3   Hierarchical Decomposition of Models Based on the Concepts of Deep Learning

There are a number of examples, where object models are formed by multilevel imbedding of simple models into more complex ones. At each level there are a number of alternative variants of particular models that are selected depending on the external conditions in which the object operates.

Such a multi-stage embedding of simpler models into more complex ones can be interpreted as a hierarchical decomposition of models. This approach forms the basis for the so-called deep learning techniques, which are actively used, for example, in the class of methods of pattern recognition [5].

Another example can be mentioned from the area of design of very-large integrated circuits (VLSI) [6]. The choice of alternative variants of particular models, formed at each level, can be entrusted to software components (we call them adapters), which, according to predetermined criteria, select those components of models that correspond to the actual physical conditions of operation of the VLSI regions. Since such a calculation scheme provides for a gradual refinement of the physical states of the object under study and, correspondingly, a gradual correction of the model being formed, then it should be constructed as an iterative procedure (Fig. 1).

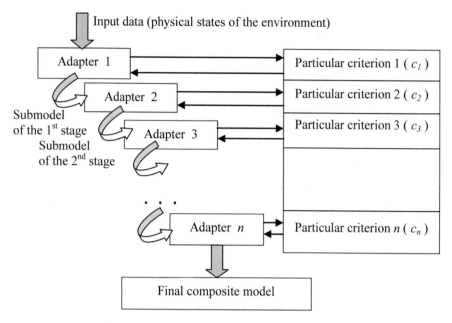

Input data (physical states of the environment)

Adapter 1 — Particular criterion 1 (c_1)

Adapter 2 — Particular criterion 2 (c_2)

Submodel of the 1st stage

Adapter 3 — Particular criterion 3 (c_3)

Submodel of the 2nd stage

. . .

Adapter n — Particular criterion n (c_n)

Final composite model

Fig. 1. Hierarchical structure of the adaptive system

Adapters are software modules designed to evaluate the external conditions of the environment and to select one of the alternative versions of model components that adequately meet these conditions.

In our opinion, an interesting feature of the above implementation is the possibility of correcting the previously selected model components as the properties of the simulated object are refined. This technique can be considered as a deep learning process, which is realized by means of the adaptive simulation, and the system in which the described idea is implemented as an adaptive system.

The effectiveness of the above system depends to a large extent on the consistency of the criteria for the chosen decomposition scheme. In adaptive systems, three categories of criteria can be distinguished: (1) theoretical (created on the basis of theoretical analysis of existing prototypes of the current tasks), (2) empirical (based on the classification of experimental data), and (3) heuristic (based on expert assessments).

Analyzing methods of representation of complex object models, we can distinguish the following types of decomposition: (1) structural, (2) discrete-event, and (3) algorithmic. The structural decomposition is based on the idea that a lot of objects are composed with the use of simpler structural elements, whose behavior can be described by rather simple functions. As a rule, the states of such objects can be described by the sets of differential and algebraic equations (DAEs).

The discrete-event model of decomposition represents the object behavior in the form of discrete sequence of events in time. Each event occurs at a particular instant in time and marks changes of object states.

And algorithmic decomposition is traditionally based on the programming techniques, which provide the decomposition of the software into modules that represent steps of a simulation process. Those modules can be implemented using programming language constructs, say, such as classes and objects in C#.

As can be seen, the multi-stage building of complex models using adaptors resembles closely the process of deep learning in methods of pattern recognition. This conclusion can be supported by the following example.

Example 1. The above technique of multi-stage building of complex models using adaptors can be illustrated by the example of the architecture of the computer-aided design system LINE dedicated to simulation of analogue very-large integrated circuits (VLSI) [6]. One of the modules of the mentioned system solves the problems of simulation of physical behavior of transistors, resistors, and capacitors on the semiconductor substrate. In the structure of a semiconductor substrate, each circuit component is divided into sections for which the systems of physical equations are formed and solved [7]:

1. Equations for the electron (n) and hole (p) current density

$$J_{n,p} = qn\mu_{n,p}E \pm qD_{n,p}\nabla_r n, p$$

2. Continuity equations for electrons (n) and holes (p)

$$divJ_{n,p}\mu q\partial n, p/\partial t = \pm qR \tag{6}$$

3. Poisson's equation

$$\nabla^2 \varphi = -\left(\frac{1}{\varepsilon_S}\right)(p - n + N),$$

where $q = 1.6 \times 10^{-19}C$; $n(p)$ is the density of electrons (holes); μ_n (μ_p) is the effective electron (hole) mobility; D_n (D_p) is the diffusion coefficient of electrons (holes); E is the electric field intensity; R is the rate of recombination; N is the resulting impurity concentration; ε_S is the dielectric permeability of semiconductor; φ is electric potential, $J_{n,p}$, is the current density of electrons or holes, and ∇ is the Laplace differential operator.

In view of the huge number of physical regions into which the entire substrate is broken, the solution of physical equations for all semiconductor regions is almost impossible on computers with limited memory and performance. These difficulties are aggravated by the fact that the values of the parameters in (6) (say, $n(p)$, μ_n (μ_p), and others) are not the same for different regions of the semiconductor structure, and moreover, they are determined by different equations, depending on the physical states of the local regions of the semiconductor. The choice of the corresponding equations is carried out on the basis of predetermined criteria and gradual iterative refinement of the physical states of the local regions of the semiconductor. Each iteration is divided into a number of stages, and each stage possesses a number of alternative versions of particular models, which are adjusted as the iterations are performed. Such an idea was realized in the experimental LINE system [6].

As an example, let us consider a scheme for computing the electron and hole densities $n(p)$ in the semiconductor substrate in the LINE system (these parameters are present in (6)). For better visualization of this idea, the scheme of computation is given in the diagram (Fig. 2) in terms of mappings. In order not to clutter up the presentation with excessive details, we do not present these equations in explicit form. Instead, relationships between physical parameters are represented in the form of mappings. The mapping of the view $(x_1, x_2, x_3, ...) \rightarrow y$ reflects the fact that there is an equation in the semiconductor physics [7], which expresses the value of parameter y as a function of parameters $x_1, x_2, x_3, ...$.

Stage 1.   *Mapping 1.1:* $(E_{Fn}, T) \rightarrow \Phi^{(n)}_{1/2}$, $(E_{Fp}, T) \rightarrow \Phi^{(p)}_{1/2}$;
Mapping 1.2: $(T) \rightarrow m^*_n$, $(T) \rightarrow m^*_p$;
Mapping 1.3: $(E, E_{Fn}, T) \rightarrow f_n$, $(E, E_{Fp}, T) \rightarrow f_p$;
Mapping 1.4: $(T, N_D, N_A) \rightarrow n_{ie}$, $(T, N_D, N_A, \varepsilon_S) \rightarrow n_{ie}$;
Mapping 1.5: $(T) \rightarrow a_1, a_2, a_3$, $(N) \rightarrow a$

Stage 2.   *Mapping 2.1:* $(\alpha, E_{g0}) \rightarrow \mu_E, \mu_C$;
Mapping 2.2: $(m^*_n, T) \rightarrow N_C$, $(m^*_p, T) \rightarrow N_V$;
Mapping 2.3: $(m^*_n, m^*_p, \varepsilon_S, N_D, N_A) \rightarrow \lambda$;
Mapping 2.4: $(\varepsilon_S, N_D, N_A, T_{ion}, \partial n/\partial E_{Fn}, \partial p/\partial E_{Fp}) \rightarrow \lambda$;
Mapping 2.5: $(a_1, a_2, a_3, N_D, N_A) \rightarrow n_{ie}$

Stage 3.   *Mapping 3.1:* $(E_{C0}, \mu_E, E_{g0}) \rightarrow E_C$;
Mapping 3.2: $(E_{V0}, \mu_V, E_{g0}) \rightarrow E_V$;
Mapping 3.3: $(N_D, N_A, \lambda, \varepsilon_S, \alpha) \rightarrow \sigma_G$;
Mapping 3.4: $(N_D, N_A, \lambda, \varepsilon_S) \rightarrow \sigma_{DA}$

Stage 4.   *Mapping 4.1:* $(m^*_n, E, E_C) \rightarrow g_C$;
Mapping 4.2: $(m^*_p, E, E_V) \rightarrow g_V$;
Mapping 4.3: $(E, E_V, \sigma_G) \rightarrow F(\eta)$;
Mapping 4.4: $(N_D, E_D, E, \sigma_{DA}) \rightarrow g_p$;
Mapping 4.5: $(N_A, E_A, E, \sigma_{DA}) \rightarrow g_A$

Stage 5.   *Mapping 5.1:* $(F(\eta), \sigma_G, m^*_n) \rightarrow g_C$;
Mapping 5.2: $(F(\eta), \sigma_G, m^*_p) \rightarrow g_V$

Stage 6.   *Mapping 6.1:* $(f_n, g_C) \rightarrow n$; $(f_p, g_V) \rightarrow p$;
Mapping 6.2: $(n_{i0}, E_{Fn}, T) \rightarrow n$; $(n_{i0}, E_{Fp}, T) \rightarrow p$;
Mapping 6.3: $(f_n, g_C, g_D) \rightarrow n$; $(f_p, g_V, g_A) \rightarrow p$;
Mapping 6.4: $(\Phi^{(n)}_{1/2}, N_C) \rightarrow n$; $(\Phi^{(p)}_{1/2}, N_V) \rightarrow p$;
Mapping 6.5: $(n_{ie}, \psi, \varphi_{Fn}, T) \rightarrow n$; $(n_{ie}, \psi, \varphi_{Fp}, T) \rightarrow p$
Output parameters = { n, p }

Fig. 2. Stages of computation of the n and p parameters

We can see that a large number of intermediate parameters of the physical structure of the semiconductor are included in the equations by means of which the required parameters of Eq. (6) are calculated, and they are: E_{g0} is the gap energy; E_{Fn} and E_{Fp} are the quasi-Fermi energies, respectively; E_{C0} and E_{V0} are the effective normalized energies of the bottom of the conductivity zone and ceiling of the valence zone, respectively; E_D and E_A are the activation energies of donors and acceptors, respectively; N is the summarized dopant density; N_D and N_A are the densities of donors and acceptors, respectively; T is temperature; T_{ion} is the temperature of ionized gas; N_{C0} and N_{V0} are the normalized densities of free levels in the conductivity and valence zones, respectively; $\phi_{1/2}^{(n)}$ and $\phi_{1/2}^{(p)}$ are the Fermi statistic functions for electrons and holes, respectively; m^*_n and m^*_p are the effective masses of electrons and holes, respectively; and n_{ie} is the intrinsic carrier density. Other function components are formal parameters used for model adjustment.

Possible input parameters: $\{E_{g0}, E_{Fn}, E_{Fp}, E_{C0}, E_{V0}, , E_D, E_A, N, \varepsilon_S, N_D, N_A, T, T_{ion}, N_{C0}, N_{V0}\}$

Thus, by combining different variants of intermediate equations over different stages of model formation, one can obtain a large number of different variants of the required parameter values.   ∎

It is quite obvious that the application of the above multi-level decomposition of the formation of models of regions of a semiconductor structure not only makes it possible to automate the simulation process, but also significantly improves the accuracy of simulation. The multi-stage nature of this process can be interpreted as a vertical (or in other words, hierarchical) decomposition of models. In our opinion, the development and generalization of this approach to the decomposition of complex models of objects of different nature deserves serious attention of researchers.

4   Decomposition Based on the Concurrent Use of Learning Samples

In technologies of computer data processing, the important role is played by the methods of model formation, which are based on the analysis of learning samples. The essence of the typical problem of this class is as follows.

Suppose that physical characteristics of the investigated object are described by n time-varying parameters $x_1, x_2, \ldots, x_n$, and we have in our disposal N experimental samples $S_0, S_1, \ldots, S_N$ with known values of the mentioned parameters obtained at a successive series of discrete time points $t_0, t_1, \ldots, t_N$, respectively (we can call the mentioned samples the learning samples). The challenge is to form a math model of data, which allow us to establish formalized functional relationships between the values of parameters. The very concept of a data model is not clearly defined, and depends on the specific type of problem being solved.

For greater certainty, let us consider one of the typical tasks in which learning sets of samples are actively used, namely, the problem of predicting time series. As shown in [8], the essence of the problem is to predict the values of the object parameters at a number of future instants $t_{N+1}, t_{N+2}, \ldots$ on the basis of the analysis of parameter values presenting in learning (experimental) samples $S_0, S_1,\ldots, S_N$. In this challenge, regression equations constructed to analyze the changes in parameter values can be considered as the desired data models. If changes in the parameter values observed in the learning samples are subject to certain regularities, then there is a theoretical possibility to construct models for predicting the values of parameters at future instants of time. Obviously, the longer the time interval at which experimental samples are collected, the more reliable and accurate models can be built. Based on practical considerations, it is generally assumed that the number of experimental samples should be at least an order greater than the number of time points for which it is required to calculate the predicted values of the parameters. One of the conventional forms of predictive models is the set of regression equations of the form of Kolmogorov-Gabor polynomials (KGP) [9]:

$$y = a_0 + \sum_{i=1}^{n} a_i x_i + \sum_{j=1}^{n} \sum_{i \leq j} a_{ij} x_i x_j + \sum_{i=1}^{n} \sum_{j \leq i} \sum_{k \leq j} a_{ijk} x_i x_j x_k + \cdots,$$

$$(7)$$

where $x_i, x_j, x_k, \ldots$ are the above object parameters (we call them the input variables of the model) that describe the behavior of the analyzed object at the current instant of time t_N, y is one of the mentioned parameters $(x_i, x_j, x_k,\ldots)$, whose value is to be predicted at the future instant t_{N+1} (we call it the output variable), and $a_0, a_i, a_{ij},\ldots$ are real coefficients.

An effective method for constructing and using models of the form (7) is realized in the Group Method of Data Handling (GMDH) [9]. The method is based on the inductive learning of models (7) with moving along the experimental samples. The effectiveness of the described approach is proved for a sufficiently large number of samples (say, a few dozen samples). Nevertheless, there are problems of complex object analysis, when it is necessary to predict the behavior of an object in a short-term perspective having a small set of experimental samples (for example, seven to ten samples). One of the reasons for this rigid restriction on the number of processed samples can be caused by the two circumstances: first, when it is necessary to predict the behavior of an object from the very beginning of its operation, and secondly, in the presence of a very large amount of data, which forces the researcher to limit himself (herself) to process only small amounts of the most fresh experimental results.

Under such conditions, we should restrict the regressive equations by low order polynomials of the form (7). Further we would like to show that the problem can be solved on the principles of the parallel processing of some subsets of experimental samples.

Following [10], let us consider the problem as a specific task of multi-dimensional function extrapolation with the use of the second order KGP polynomials:

$$y_i(x_i, x_j) = a_0 + a_1 x_i + a_2 x_j + a_3 x_i^2 + a_4 x_j^2 + a_5 x_i x_j, \qquad (8a)$$

where $y_i(x_i, x_j)$ means the i-th output variable $y_i(t_{N+1})$ evaluated at the instant t_{N+1} as a function of two input variables $x_i(t_N)$ and $x_j(t_N)$, whose values are known for the instant of time t_N, and where for simplicity a one-dimensional notation for polynomial coefficients is used, namely, $a_0,\ldots, a_5$ are real coefficients of the regressive equation. A peculiarity of

(8a) is that it expresses the dependence of the i-th output variable $y_i(t_{N+1}) \equiv x_i(t_{N+1})$ on the value of this variable $x_i(t_N)$ and other variable $x_j(t_N)$, where $i \neq j$. Hence, model (8a) is a minimal order polynomial that allows us to express mutual influences of any two object variables (variables x_i and x_j, $i, j = 1,\ldots, n$). For each pair of object variables, there are two polynomials of the above form with the same input variables (we call them the dual models). For example, with respect to (8a), the dual polynomial is as follows:

$$y_j(x_j, x_i) = b_0 + b_1 x_j + b_2 x_i + b_3 x_j^2 + b_4 x_i^2 + b_5 x_j x_i, \qquad (8b)$$

where $y_j(x_i, x_j)$ means the value of the j-th output variable $y_j(t_{N+1}) \equiv x_j(t_{N+1})$ evaluated at the instant of time t_{N+1}, and $b_0,\ldots, b_5$ are real coefficients. All our comments concerning the properties of model (8a) are applied equally to model (8b).

The conditional division of object parameters into the groups of input and output variables allows us to consider the formation of predictive models as a task of system identification. The coefficients of dual models (8a) and (8b) can be calculated on the basis of data on the parameter values given in the experimental samples.

The prediction equations can be represented in the form of equations in increments of input parameters [10]. For example, instead of (8a), we can form the equation in increments:

$$\begin{aligned} y_i(t_{N+1}) &\approx x_i(t_N) + S_{x_i}^{y_i} \cdot ddx_i(t_N) + S_{x_j}^{y_i} \cdot ddx_j(t_N) \\ &+ \frac{1}{2} \cdot S_{x_i^2}^{y_i} \cdot ddx_i^2(t_N) + \frac{1}{2} \cdot S_{x_j^2}^{y_i} \cdot ddx_j^2(t_N) + S_{x_i x_j}^{y_i} \cdot ddx_i(t_N) \cdot ddx_j(t_N), \end{aligned} \qquad (9)$$

where $y_i(t_N) \equiv x_i(t_N)$, $ddx_{i,j}(t_N) = \Delta x_{i,j}(t_N)/x_{i,j}(t_N)$, $\Delta x_{i,j} = x_{i,j}(t_{N+1}) - x_{i,j}(t_N)$, $\forall i,j$, are forward differences (increments) of input variables $x_i(t_N)$ and $x_j(t_N)$, and $\Delta x_{i,j}/x_{i,j}$ are relative forward differences (relative increments) of the mentioned variables.

It can be shown that coefficients of (9) are sensitivities of the output variable with respect to input variables, and can be computed using polynomial coefficients of (8a) and (8b) [10]. Say, using coefficients of (8a), we can compute the above sensitivities in (9):

$$a)\ S_{x_i}^{y_i} = \frac{\partial y_i}{\partial x_i} \cdot x_i = \left(a_1 + 2a_3 x_i + a_5 x_j\right) \cdot x_i,$$

$$b)\ S_{x_j}^{y_i} = \frac{\partial y_i}{\partial x_j} \cdot x_j = \left(a_2 + 2a_4 x_j + a_5 x_i\right) \cdot x_j,$$

$$c)\ S_{x_i^2}^{y_i} = \frac{\partial^2 y_i}{\partial x_i^2} \cdot x_i^2 = 2a_3 x_i^2, \qquad (10)$$

$$d)\ S_{x_j^2}^{y_i} = \frac{\partial^2 y_i}{\partial x_j^2} \cdot x_j^2 = 2a_4 x_j^2,$$

$$e)\ S_{x_ix_j}^{y_i} = \frac{\partial^2 y_i}{\partial x_ix_j} \cdot x_i \cdot x_j = a_5 \cdot x_i \cdot x_j,$$

where (10a) and (10b) represent the first-order sensitivities, and (10c) – (10e) the second-order sensitivities of output variable $y_i(t_{N+1})$ with respect to input variables $x_i(t_N)$ and $x_j(t_N)$.

If we have n parameters at our disposal, then there are $(n-1)$ equations of type (9) for the prediction of each parameter value, and the total number of such equations is n $(n-1)$, each we can call the particular (predicting) model. Since there are $(n-1)$ particular models (and consequently $(n-1)$ alternative versions) for the prediction of each variable x_i, $i = 1, \ldots, n$, we can find the most exact model. For this purpose, we can use the set of experimental samples available at our disposal.

Indeed, using the set of experimental samples $\Sigma = \{S_0, S_1, \ldots, S_{N-1}\}$, we can construct different combinations by two, three, and more experimental samples, each is called the learning subset. For example, the Ω_0 learning subset may include samples $\{S_0, S_1\}$, the Ω_1 subset may include $\{S_1, S_3, S_{N-1}\}$, and so on. In this way, we can form the set of learning subsets $\Omega = \{\Omega_1, \Omega_2, \ldots, \Omega_k, \ldots, \Omega_K\}$, and each subset is formed quite arbitrary, what mattered is only that the learning subsets should cover all the experimental samples presenting in Σ.

Each learning subset Ω_k is characterized by a certain set of particular increments for each particular prediction model. For example, assume that Ω_k includes three experimental samples from Σ, say, S_1, S_3, and S_5, that is, $\Omega_k = \{S_1, S_3, S_5\}$, where, according to our notation, the mentioned experimental samples are given for instants t_1, t_3, and t_5, respectively. Hence, we can compute three values of particular increments for each particular model associated with Ω_k. For example, for the particular model $y_i(x_i, x_j)$ they are $ddx_i(t_1)$, $ddx_i(t_3)$, $ddx_i(t_3)$, and $ddx_j(t_1)$, $ddx_j(t_3)$, $ddx_j(t_5)$.

Using the particular increments presenting in the above learning subsets and applying the Lagrange extrapolation method [11], we can extrapolate the particular forward increments $ddx_i(t_N)$, $\forall i$, for each predicted variable. And finally, we can compute the desired predicted values substituting the obtained extrapolated particular increments into particular models (9) for each output variable.

What gives us the use of learning subsets? By using different learning subsets from Ω as learning patterns, we get different extrapolated values of the same particular increments, and substituting them into particular models (9), we can calculate different predicted values of the same output variables.

In order to extract most plausible predicted values among those obtained by the above method, we introduce the requirement that the values of object variables should be normalized within the interval [0, 1]. If the predicted value of any variable is out of this range, we exclude it from our consideration. Following this strategy, we can obtain the truncated set of the plausible predicted values of object variables Ψ. If Ψ includes several predicted values for the same object variable, we can take the arithmetic mean of those values as the required predicted value for this variable. This option can be called a batch voting.

Example 2. Let us apply the above technique for the time series presented in [10]. Assume that the behavior of an object is reflected by the three object variables (x_1, x_2,

x_3), and we have the series of experimental samples $(S_0, S_1, S_2,..., S_8)$ with known values of the mentioned object variables obtained at instants of time $t_0, t_1,..., t_8$, respectively: S_0 (0.2366, 0.2068, 0.2496), S_1(0.2403, 0.2462, 0.2731), S_2(0.3033, 0.2406, 0.2856), S_3(0.2367, 0.2688, 0.2966), S_4(0.2679, 0.2607, 0.2951), S_5(0.3308, 0.2484, 0.2823), S_6(0.2410, 0.3136, 0.3370), S_7(0.3151, 0.2736, 0.3713), and S_8 (0.4963, 0.2788, 0.3282), where we use the notation S_i (x_1, x_2, x_3), $i = 0, 1, 2, ..., 8$. As an example, let us use those data to illustrate the prediction method under consideration.

For this purpose we can consider samples $\Sigma = \{S_0, S_1,..., S_6\}$ as learning samples to compute polynomial coefficients of the regression Eq. (8a), (8b) and particular prediction models (9) for each predicted variable. Samples S_7 and S_8 we can then use to compare the predicted values of object variables with the given actual values. For brevity, we confine ourselves to the application of the method to the prediction of one of the variables, say, x_1.

The regressive equations for x_1 obtained using samples $\Sigma = \{S_0, S_1,..., S_6\}$ are as follows:

$$y_1(x_1, x_2) = 2.0455 - 16.2210x_1 + 1.3324x_2 - 14.3273x_1^2 \\ - 49.3550x_2^2 + 95.4847x_1x_2, \tag{11a}$$

$$y_1(x_1, x_3) = 39.3362 - 250.489x_1 - 54.623x_3 + 96.1379x_1^2 \\ - 197.77x_3^2 + 689.9113x_1x_3, \tag{11b}$$

and dual equations:

$$y_2(x_2, x_1) = 1.5550 - 5.6708x_2 - 5.2520x_1 + 14.9988x_2^2 \\ + 12.8445x_1^2 - 5.1894x_2x_1, \tag{11c}$$

$$y_3(x_3, x_1) = 1.9487 + 2.0640x_3 - 15.5134x_1 \\ - 11.7447x_3^2 + 18.0084x_1^2 + 20.3766x_3x_1, \tag{11d}$$

where Eq. (11a) and (11b) represent the predicted value of $y_1 \equiv x_1(t_{N+1})$ as functions of variables $x_1 \equiv x_1(t_N)$, $x_2 \equiv x_2(t_N)$ and variables $x_1 \equiv x_1(t_N)$, $x_3 \equiv x_3(t_N)$, respectively, and Eq. (11c) and (11d) represent the predicted values of $y_2 \equiv x_2(t_{N+1})$ and $y_3 \equiv x_3(t_{N+1})$ as functions of variables $x_2(t_N)$, $x_1(t_N)$ and variables $x_3(t_N)$, $x_1(t_N)$, respectively.

Next, using Eq. (10), we can compute sensitivities of output variables with respect to input variables for each experimental sample, and substitute the obtained values into Eq. (9) to form particular prediction models in increments. It is evident that the values of sensitivities are different for different samples, since they depend on the current values of object variables. For example, the values of sensitivities of the $y_1(x_1, x_2)$ particular model for the samples S_5 and S_6 are as follows:

Sample S_5 : $S_{x_1}^{y_1(x_1,x_2)} = -0.6555$, $S_{x_2}^{y_1(x_1,x_2)} = 2.0839$, $S_{x_1^2}^{y_1(x_1,x_2)} = -3.1360$, $S_{x_2^2}^{y_1(x_1,x_2)} = -6.0900$, $S_{x_1,x_2}^{y_1(x_1,x_2)} = 7.8460$.

Sample S_6 : $S_{x_1}^{y_1(x_1,x_2)} = 1.6430$, $S_{x_2}^{y_1(x_1,x_2)} = -2.0760$, $S_{x_1^2}^{y_1(x_1,x_2)} = -1.6640$, $S_{x_2^2}^{y_1(x_1,x_2)} =$ -9.7080, $S_{x_1,x_2}^{y_1(x_1,x_2)} = 7.2165$.

The particular increments are the same for each dual pair of particular models, and for example, for models $y_1(x_1, x_2)$ and $y_2(x_2, x_1)$ they are: (t_0: $ddx_1 = -0.1028$, $ddx_2 = -0.1293$), (t_1: $ddx_1 = 0.1646$, $ddx_2 = 0.0805$), (t_2: $ddx_1 = 0.0602$, $ddx_2 = -0.0132$), (t_3: $ddx_1 = 0.0440$, $ddx_2 = -0.0125$), (t_4: $ddx_1 = 0.0195$, $ddx_2 = 0.0018$), (t_5: $ddx_1 = 0.0852$, $ddx_2 = -0.0084$), (t_6: $ddx_1 = -0.0372$, $ddx_2 = -0.0677$).

Next, we form learning subsets, for example: $\Omega_1 = \{S_0, S_1\}$, $\Omega_2 = \{S_0, S_1, S_2\}$, $\Omega_3 = \{S_0, S_1, S_2, S_3\}$, $\Omega_4 = \{S_0, S_1, S_2, S_3, S_4\}$, $\Omega_5 = \{S_1, S_2, S_3, S_4\}$, $\Omega_6 = \{S_2, S_3, S_4\}$, and $\Omega_7 = \{S_3, S_4\}$.

Now we can compute the predicted values of partial increments for each particular predicting model. For example, selecting the first learning subset Ω_1, we use the particular increments which correspond to samples S_0 and S_1, and applying the Lagrange extrapolation equation [11], we can obtain the following predicted increments for each particular predicting model and each learning subset.

For example, for dual particular models $y_1(x_1, x_2)$ and $y_2(x_2, x_1)$, the use of the learning subset Ω_1 gives the predicted values of particular increments $ddx_1(t_6) = 0.0596$ and $ddx_2(t_6) = 0.0265$. Substituting them into (9), we can obtain the predicted values of object variables at the instant of time t_7: $y_1(t_7) = 0.2889$, $y_2(t_7) = 0.3275$. We can write this fact in the following notation: Ω_1: ($ddx_1(t_6) = 0.0596$, $ddx_2(t_6) = 0.0265$) -> ($y_1(t_7) = 0.2889$, $y_2(t_7) = 0.3275$).

Using the same notation for the rest of learning samples, we can write the predicted values of object variables $y_1(t_7)$ and $y_2(t_7)$ obtained with the help of particular prediction models $y_1(x_1, x_2)$ and $y_2(x_2, x_1)$, :

Ω_1 : ($ddx_1(t_6) = 0.0596$, $ddx_2(t_6) = 0.0265$) -> ($y_1(t_7) = 0.2889$, $y_2(t_7) = 0.3275$),
Ω_2 : ($ddx_1(t_6) = 0.0652$, $ddx_2(t_6) = -0.0142$) -> ($y_1(t_7) = 0.3664$, $y_2(t_7) = 0.2956$),
Ω_3 : ($ddx_1(t_6) = 0.0705$, $ddx_2(t_6) = 0.0821$) -> ($y_1(t_7) = 0.1913$, $y_2(t_7) = 0.3773$),
Ω_4 : ($ddx_1(t_6) = 0.2130$, $ddx_2(t_6) = 0.0553$) -> ($y_1(t_7) = 0.5086$, $y_2(t_7) = 0.3551$),
Ω_5 : ($ddx_1(t_6) = 0.1880$, $ddx_2(t_6) = 0.0531$) -> ($y_1(t_7) = 0.4686$, $y_2(t_7) = 0.3504$),
Ω_6 : ($ddx_1(t_6) = -0.5538$, $ddx_2(t_6) = 0.0560$) -> ($y_1(t_7) = -1.2795$, $y_2(t_7) = 0.6948$),
Ω_7 : ($ddx_1(t_6) = 0.2190$, $ddx_2(t_6) = -0.0012$) -> ($y_1(t_7) = 0.5615$, $y_2(t_7) = 0.3123$).

As can be seen, all the learning subsets, except Ω_6 for $y_1(t_7)$, yield the predicted value of variable $y_1(t_7)$ within the admissible range of values, i.e. [0, 1]. In other words, the truncated set of learning subsets for the variable $y_1(t_7)$ is $\Psi_{y_1} = \{\Omega_1, \Omega_2, \Omega_3, \Omega_4, \Omega_5, \Omega_7\}$. In the same way, we can conclude that the truncated set of learning subsets for the variable $y_2(t_7)$ includes all the learning subsets formed above, that is, $\Psi_{y_2} = \{\Omega_1, \Omega_2, \Omega_3, \Omega_4, \Omega_5, \Omega_6, \Omega_7\}$.

In the closing stage we can compute the predicted values of object variables as average values of those which are obtained with the use of all the particular models for each output variable. The particular model $y_1(x_1, x_2)$ gives six possible values of the $y_1(t_7)$ variable presented above, and the average value is $y_1(t_7) = 0.3974$. Moreover, the $y_1(t_7)$ is computed by the particular prediction model $y_1(x_1, x_3)$, either, and the predicted

value obtained using the mentioned model is $x_1^P(t_7) = 0.3281$. Thus the desired predicted value of variable $x_1^P(t_7)$ with the use of the both particular models is the arithmetic mean of the above two values, that is, $x_1^P(t_7) = 0.3628$. Comparing the obtained
predicted value with the actual value $x_{1,act}(t_7) = 0.3151$ given by the sample S_7, we can
conclude that the relative error of prediction is $\delta_{x_1}(t_7) = 0.15$.

Using the same computation procedure for other variables, we can obtain the
following predicted values of the remaining variables: $y_2(t_7) = 0.3054$ (the actual value
is $x_{2,act}(t_7) = 0.2736$, and the relative error is $\delta_{x_2}(t_7) = 0.12$), and $y_3(t_7) = 0.3295$ (the
actual value is $x_{3,act}(t_7) = 0.3713$, and the relative error is $\delta_{x_3}(t_7) = 0.11$).

According to the above method, it is possible to determine the predicted values of
the object variables at the next time point, too. Omitting the details of computation, we
give the final results: $y_1(t_8) = 0.4968$ (the actual value given in sample S_8 is
$x_{1,act}(t_8) = 0.4963$, and the relative error is $\delta_{x_1}(t_8) = 0.001$), $y_1(t_8) = 0.2422$ (the actual
value is $x_{2,act}(t_8) = 0.2788$, and the relative error is $\delta_{x_2}(t_8) = 0.13$), and $y_3(t_8) = 0.3142$
(the actual value $x_{3,act}(t_8) = 0.3282$, and the relative error is $\delta_{x_3}(t_8) = 0.04$).

As can be seen, the actual accuracy of prediction of values of object variables (x_1,
x_2, x_3) is quite acceptable for the most practical applications. Thus, the described
approach assumes separate processing of various combinations of experimental samples with the subsequent summation of the results of the prediction, obtained independently in concurrent computational processes.   ■

Consequently, the above method shows that concurrent analysis of time series is a
winning alternative to methods, in which the forecast of time series is based on a
statistical analysis of large sets of experimental data.

5 Concluding Remarks on Building Computation

In the approaches considered in the paper, the decomposition of models is achieved by
applying two fundamentally different strategies. In the first two methods, it is assumed
that the object model is divided into a number of smaller models, and then those
models are formed and analyzed sequentially one after another. In the third approach,
the object is studied on the basis of the formation of particular models that can be
processed in parallel. Accordingly, we can talk about sequential and parallel decomposition of models. A common feature of these methods is that the model reduction is
based on the idea of the algorithmic decomposition, although the use of decomposition
algorithms does not exclude the possibility of applying for their implementation some
unified computing architectures that are invariant with respect to decomposition
algorithms. Moreover, a well-chosen architecture of the computer system can significantly improve the efficiency of the decomposition algorithm. As an example, we can
mention the architecture of the multi-agent system [12], which was designed to
implement algorithms for inductive building models using the GMDH method for
solving weather forecast problems. An important feature of such a system is the actual
independence of its architecture from the specifics of the problem being solved, and the
possibility of parallelizing computational processes, thanks to the specialization of
agents.

Thus, algorithmic decomposition of models can be considered as an effective tool for investigating complex objects, direct study of which by traditional methods can be fraught with difficulties in storing large volumes of information and the multivariate nature of possible models requires adaptive organization of computing processes.

References

1. Leskovec, J., Rajaraman, A., Ullman, J.D.: Mining of Massive Datasets. Cambridge University Press (2014)
2. Weste, N., Harris, D.: CMOS VLSI Design. Addison-Wesley (2004)
3. Tikhonov, A.N.: Systems of differential equations containing small parameters in the derivatives. Mat. sb. **73**(3), 575–586 (1952)
4. Rogoza, W.: Adaptive simulation of separable dynamical systems in the neural network basis. In: Pejas, J., Piegat, A. (eds.) Enhanced Methods in Computer Security, Biometrcic and Artificial Intelligence Systems, pp. 371–386. Springer, Heidelberg (2005)
5. Goodfellow, I., Bengio, Y., Courville, A.: Deep Learning. The MIT Press, Cambridge (2017)
6. Rogoza, W.: Some models of problem adaptive systems. Pol. J. Environ. Stud. **16**(#5B), 212–218 (2006)
7. Sze, S.M.: Physics of Semiconductor Devices, 2nd edn. Wiley (WIE), New York (1981)
8. Box, G., Jenkins, G.: Time Series Analysis: Forecasting and Control. Holden-Day, San Francisco (1970)
9. Madala, H.R., Ivakhnenko, A.G.: Inductive Learning Algorithms for Complex Systems Modeling. CRC Press, Boca Raton (1994)
10. Rogoza, W.: Deterministic method for the prediction of time series. In: Kobayashi, S., Piegat, A., Pejaś, J., El Fray, I., Kacprzyk, J (eds.) ACS 2016. AISC, vol. 534, pp. 68–80. Springer, Heidelberg (2017)
11. Miller, G.: Numerical Analysis for Engineers and Scientists. Cambridge University Press, Cambridge (2014)
12. Rogoza, W., Zabłocki, M.: A feather forecasting system using intelligent BDI multiagent-based group method of data handling. In: Kobayashi, S., Piegat, A., Pejaś, J., El Fray, I., Kacprzyk, J (eds.) Hard and Soft Computing for Artificial Intelligence, Multimedia and Security. AISC, vol. 534, pp. 37–48. Springer, Heidelberg (2017)

Managing the Process of Servicing Hybrid Telecommunications Services. Quality Control and Interaction Procedure of Service Subsystems

Mariia A. Skulysh$^{(\boxtimes)}$, Oleksandr I. Romanov$^{(\boxtimes)}$,
Larysa S. Globa$^{(\boxtimes)}$, and Iryna I. Husyeva$^{(\boxtimes)}$

National Technical University of Ukraine
«Igor Sikorsky Kyiv Polytechnic Institute», Kyiv, Ukraine
{mskulysh, a_i_romanov}@gmail.com, lgloba@its.kpi.ua,
iguseva@yahoo.com

Abstract. The principle of telecommunication system management is improved. Unlike the principle of software-defined networks, the functions of managing the subscriber service process, namely: subscriber search, the search for the physical elements involved in the transmission process, and the transfer of control to the corresponding physical elements, are transferred to the cloud. All subsystems of mobile communication will be managed from controllers located in the date center. The interaction between subsystems controllers for managing occurs only in the data center. It will reduce the number of service streams in telecommunications network. The procedure of interaction of the mobile communication control system and the virtualized environmental management system is proposed.

Keywords: NFV · VeCME · VBS · VeEPC · LTE · 5G · TC gibrid-service

1 Introduction

The work of the telecommunications network is inextricably linked with computer systems. According to [1], a hybrid telecommunications service is a service that includes components of cloud and telecommunications services. A mobile network consists of a local area network, a radio access network and a provider core network. The advent of cloud computing has expanded the possibilities for servicing telecommunications systems.

Specifications [2] represents the main architectural solutions in which complex hardware solutions are replaced by different ways of virtualizing network sections. This allows configuring the network computing resources in a flexible way. To do this, it is necessary to create new methods of managing the quality of service that will take into account the features of the process in the telecommunication system and in the computing environment for servicing hybrid services.

The purpose of this work is to improve the quality of service of hybrid telecommunication services. To this end, it is proposed to use methods to control the formation

© Springer Nature Switzerland AG 2019
J. Pejaś et al. (Eds.): ACS 2018, AISC 889, pp. 244–256, 2019.
https://doi.org/10.1007/978-3-030-03314-9_22

of service request flows and to manage the allocation of resources. Realization of the set goal is achieved by solving the following tasks:

1. Research of developments of the scientific community in the field of monitoring and ensuring the quality of service of hybrid services. Identification of processes regularities and features.
2. Development of a model of servicing hybrid services for a heterogeneous telecommunication environment.
3. Development of methods for ensuring the quality of service in access networks.
4. Development of methods for ensuring the quality of service in the local networks of the mobile operator and on their borders.
5. Development of a model and methods for the operation of the provider core network in a heterogeneous cloud infrastructure.
6. Development of a functioning model for provider charging system.

To realize these tasks, it is necessary to take into account such factors as the annual growth of traffic volumes in an exponential progression; the need for differentiated services for a multiservice flow and different quality of service requirements; the need for constant monitoring of quality indicators and timely response to their decline.

Thus, the operator's monitoring system collects and processes a large amount of information about the quality of each service. It also monitors the telecommunications operator subsystems, the number of failures, etc.

For an adaptive response to the decline in quality of service today, such mechanisms are used:

- Monitoring the workload of the network;
- Monitoring of queue service quality in communication nodes;
- Managing of subscriber data flows;
- Managing of queues for differentiated servicing of multi-service flows;
- Overload warning mechanisms;
- Methods of engineering the traffic for an equable distribution of resources.

2 Organization of a Heterogeneous Telecommunications Network

According to [3], all calculative functions that accompany transfer process are performed in data centers with cloud infrastructure. Virtualization of the base station will reduce the amount of energy consumed by the dynamic allocation of resources and load balancing. In addition to virtual base stations, radio access networks with cloud-based resources organization (Cloud-RAN) is required to create a resource base frequencies processing, which will combine different computing resources of centralized virtual environment. The specification offers virtualization of network functions for the router located on the border of provider local network. A router performs flow classification, routing management and providing firewall protection.

For the organization of virtual base stations and VeCPE it is necessary for data center to be close to base stations and to each output of the local network. So, the of the provider network represents a geographically distributed network of data centers with communication channels delivered primary information of mobile subscribers to each of them. The network requires conversion at the lowest level, so the signal requires recognition and decoding at higher levels of MAC, RLC, RRC, and PDCP. The specification is also propose provider core virtualization.

Based on this, it can be assumed that most of the network processes are performed in datacenters, and the network is only a means of delivering information messages [4]. In the conditions of program-controlled routers distribution, there is network structure shown in Fig. 1.

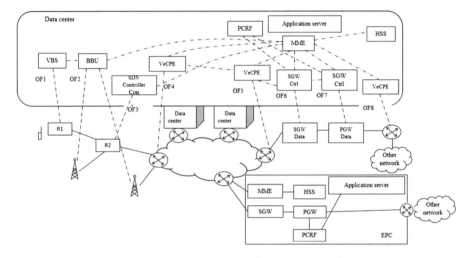

Fig. 1. Provider core network structure using software-controlled routers

Figure 1 shows how the mobile subscriber communicates with the R1 transponder, which converts the radio signal to optical, and then the signal reaches the R2 transponder managed by the SDN controller, which is also situated in the data center. After attaining the data center, the signal is processed by the virtual base station. Further, according to LTE technology, the flow is sent to the operator's core for further processing. The BBU subsystem is based on the technology of software-configurable networks/virtualization of network operation. This system supports either the work of virtual base stations or hybrid of 2G/3G/4G/Pre5G solutions.

Further direction of data channels is determined by servicing in the core. If the flow is directed to the provider internal network, it is immediately sent to the corresponding virtual base station in the data center for service, and then forwarded to the subscriber through the transponders R2 and R1. If the stream is to be sent outside the operator's local network, it is directed to the boundary virtual router, and then to external networks. This is the example of Next Generation Network

Thus, the data center combines a group of data centers that are connected to a single logical space for servicing virtualized network functions through a secure network.

The quality of end users service is influenced by the organization of processes in such a heterogeneous data center based on the cloud computing concept.

According to the ITU-T Y.3500 recommendation, cloud computing is the paradigm network access providing to a scalable and flexible set of shared physical and virtual resources with administration based on on-demand self-service.

The structure of described data center in which the group of functional blocks shown in Fig. 1 are servicing is shown in Fig. 2. There is a transport network and connected data centers, forming a single virtualized space.

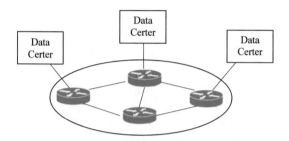

Fig. 2. The structure of the heterogeneous data center

Recommendation ITU-T Y.3511 defines this complex system of data center groups as multi-cloud computing. It is a paradigm for interaction between two or more providers of cloud services. Recommendation ITU-T Y.3520 presented the conceptual architecture of the multi-cloud and multi-platform cloud services management presented in Fig. 3 [5].

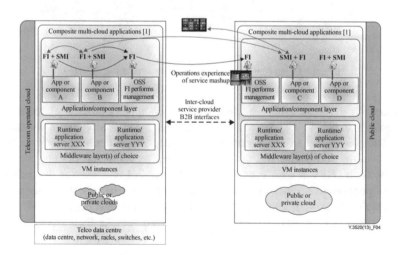

Fig. 3. Architectural vision for multi-cloud, multi-platform cloud management

During the work of provider data center, virtual BS system, the core subsystems and the virtual router are in a single logical space. In Fig. 3 we can see that at the middleware level XXX Server is presented in every data center that participates in the inter-cloud computing infrastructure. The corresponding programs that activate the provider functional blocks are performed at the application and component level.

To ensure the work of mobile network using virtualization technology, it is necessary to provide a distributed structure of data centers, organized in a single virtual space. The structure should include deployed logical elements of the mobile service network, process management and flow allocation carried out by the orchestrator (Fig. 4).

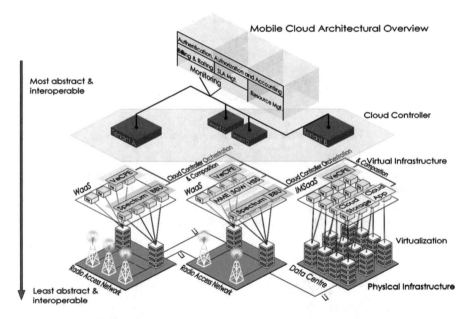

Fig. 4. Organization of service in new generation networks

According to the research, the effectiveness of computing processes organization in functional units affects the efficiency of end-users servicing of a mobile operator. The data processing center in this architectural solution is a complex organizational and technical set of computing and telecommunication systems that ensures the smooth operation of the NFV infrastructure. The effectiveness of its operation depends on the choice of physical data centers that will become part of the distributed center structure; the location of network functions in the infrastructure; the organization of flows between virtualized objects and the allocation of resources for their servicing.

3 The Principle of Flow Service with the Resource Virtualization in Public Telephone Network

Controllers located in the data center guide all subsystems of mobile communication. The interaction between controllers of subsystems for the purpose of control occurs only in the middle of the date center. The functions of managing the service process, namely: searching for the subscriber, searching for the physical elements involved in the transmission process, and passing the guidance on the corresponding physical elements, are transferred to the cloud.

The subscriber device for connection organization interacts with the base station controller located at the data center. According to the protocols, subsystem controllers interact at the level of the data center, sending the final hardware solutions to the physical equipment to start the data transmission process (Fig. 5).

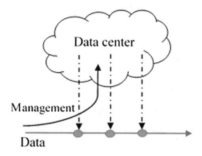

Fig. 5. The principle of flow service with the resource virtualization in public telephone network

There are two principles of virtualization of network resources. The first principle redirects through the cloud resources only control flows. The second principle is to use cloud-based data centers to process both network and information flows. In this paper, the first principle is considered. According to it, virtualization of network functions allows separating the control system of the mobile network nodes from the data transmission system. The main functions of the core subsystem were analyzed, and thefunctions associated with the control and data transfer were selected. Data transferfunctions are distributed into a virtualized environment deployed on the basis of datacenters group [6]. A number of research are devoted to the interaction processes ofcommunication networks and their cloud components [7, 8, 9, 10]. Proposed in thisresearch distribution of network core functions between physical and virtual devices ispresented on Fig. 6.

F1 – packet filtering by users and legitimate interception of traffic;
F2 – IP pool distribution functions for UE and PCEF Functionality;
F3 – basic routing and interception of packet traffic;
F4 – the function of an anchor point (traffic aggregation point) for a handover between the NodeBs within one access network in the base station service area according to a set of rules and instructions;
F5 – processing of BBERF functionality;

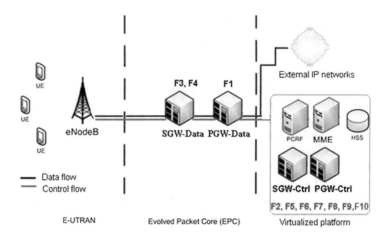

Fig. 6. Distribution of network core functions between physical and virtual devices

F6 – Traffic Detection Function;
F7 – User Data Repository (UDR);
F8 – Application Function (AF);
F9 – Online Charging System (OCS).

Figure 7 shows the processes of network subsystems interaction with the separation of control functions and data transmission with virtualization in the provision of data transfer functions. In fact, each arrow on this scheme is a service request in this virtual (or physical) node. The number of requests per time unit is the load intensity on given service node.

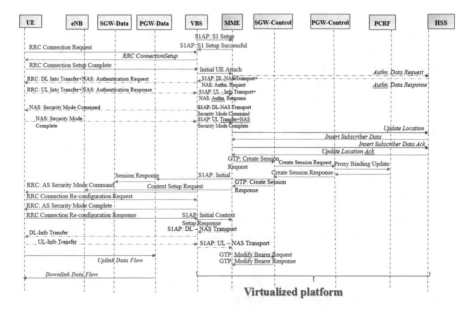

Fig. 7. Procedure of subsystems interaction during subscriber's service

Network structure and user service quality control take place in the nodes. Traditionally, the subsystems of LTE network perform a set of functions, in accordance with standards and specifications. The paper proposes to divide subsystem management functions and functions that are associated with the data transfer process directly to the LTE network. The feature is the expansion of subsystems functionality, compared with the networks of previous generations. More than half of the subsystem functions are connected not with the service process, but with the management of the communication system. Service quality control occurs in the subsystems eNodeB, SGSN, PCRF (Fig. 8). Delay control in virtualized network nodes, where service intensities depend on computing resources requires PCRF modification.

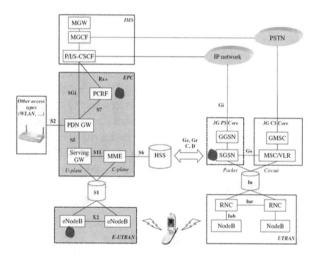

Fig. 8. General architecture of the standard LTE network

The efficiency of hybrid telecommunication services is estimated by quantitative indicators of service quality:

- t_d – time delay in the maintenance of the hybrid telecommunication management service $(t_{data} - t_{start})$, where t_{start} is the moment of request by the subscriber for permission to transmit data information flows, t_{data} is the moment when the subscriber begins to transmit information streams;
- $P-$ the probability of refusal in service.

$$P = \prod_{i=1}^{N} P_i$$

where P_i is failure probability in virtualized service node for one of the requests types to the subsystem of the heterogeneous telecommunication environment.

4 Procedure of Guaranteeing the Adjusted Quality of Service

The principle of dynamic quality control is as follows: the delay value in maintaining the application for connection (disconnection, recovery) is compared with the service quality policy of the subscriber. If the metric does not match, then the quality metrics in virtual nodes and VLANs are consistently compared with the thresholds of the corresponding policies stored in the PCRF subsystem. This principle analyzes the following quantitative indicators of the effective system operation, such as: the time of service flow request delay in the virtual node and the probability of queries loss in the service node. Service node is a virtual machine that performs the functions the network node managing.

After discovering the reason of service efficiency indicators problem, then appropriate measures are taken. If there is a problem in the time of transmission between service nodes, then it is recommended to reconfigure the system, namely to change the location of virtual nodes in physical nodes of the heterogeneous data center structure. If the problem is identified in one service nodes, then it is recommended to increase the number of service resources. If there is a decrease in service quality rates in a group of linked interface nodes, for example, which form a single core of the EPC network, then it is recommended to limit the flow of applications sent to service the corresponding core. For this purpose it is recommended to calculate the intensity of the load on the group of nodes. The algorithm of the procedure is shown in Fig. 9.

To implement the principle of dynamic quality control, a modification of the PCRF system subsystems is required. The "Single Policy Storage" subsystem is expanding, and the following policies regarding quality management service flow rates are added:

1. The allowable delay time for an application service flow in a virtual host.
2. Permissible loss of requests in the virtual node.
3. Permissible time for serving requests in groups of virtual nodes that provide a given service.
4. Permissible delays in transmission between service nodes.
5. The value of the admissible delivery delays of the guiding influence on network nodes.

An expanded subsystem is shown in Fig. 10.

- The "Policy Management" subsystem creates a set of requirements for implementing a set of policies in relation to different flows of management.
- The "Policy Server" subsystem detects a problem of inconsistency of the current quality metrics with the declared subscription service policies.
- In the "Application Server" subsystem, program modules in which calculations are performed according to the proposed methods are implemented. The source data for the methods is the statistics obtained from the monitoring system and policy data that is provided to respective subscribers.

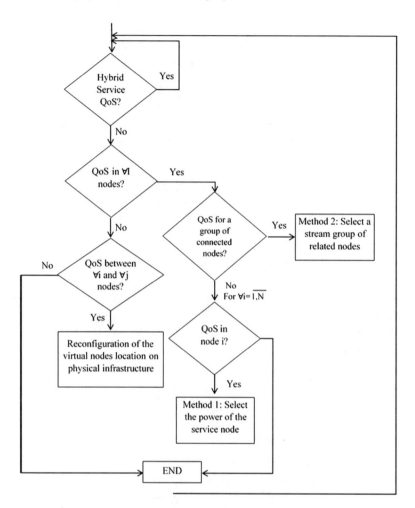

Fig. 9. Procedure for guaranteeing the preset quality of service

- The "Subscriber Data Store" subsystem is supplemented by information about virtual nodes, or a separate virtual network maintenance statistics database is created. This database collects information about service requests flows; the statistics of the relative dependence of the service intensity on service resources for each type of request.

The principle of dynamic quality control requires new procedures: it is necessary to arrange the interaction of mobile communication management system with virtualized resources management system (Fig. 11).

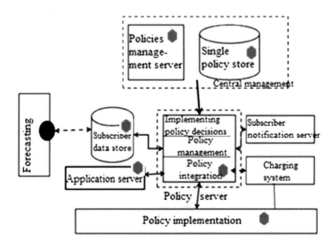

Fig. 10. PCRF subsystem modification

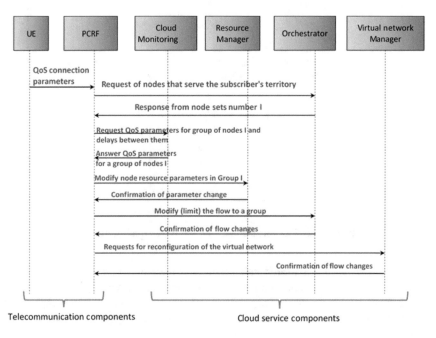

Fig. 11. Interaction of the mobile communication control system and the virtualized environment management system

The quality control of management procedures implementation is evaluated at the level of User Equipment:

The User Equipment records the time delay in execution of service procedures, namely the time from the moment of connection initialization to the moment of data transmission beginning, and transfers to the subsystem of PCRF.

The PCRF receives this information from the subscriber and analyzes the policy server; in the policy implementation sub-system it compares the received data to the correspondence of chosen subscriber policy that is stored in the "Subscriber data store".

If the delay values are not in accordance with the policy, PCRF requests the "Orchestrator" subsystem to identify the group of nodes i that serve the subscriber.

Orchestrator sends the numbers of nodes serving the subscriber, located in a given area. PCRF sends request to "Cloud Monitoring" for information on the delay and loss parameters in the nodes i, and information on the delay between nodes services. The Cloud Monitoring collects information regarding the latency and loss performance of hybrid services that are served on the nodes of the virtual network. The data about the service node group is transferred to the PCRF, where the principle of dynamic quality control of the service of hybrid services is realized. According to the management decisions, the PCRF subsystem sends inquiries:

- for reconfiguration of the virtual network, to the "Virtual Network Manager";
- to reconfigure resources to "Resource Manager";
- to change flows of service to "Orchestrator" streams over a virtual network.

When implementing the principle of dynamic quality control, most subsystems of the PCRF system are involved.

5 Conclusions

An approach to managing a heterogeneous telecommunication environment for increasing the efficiency of the service process of hybrid telecommunication services in new generation systems is proposed. A unified solution for telecommunication systems, where the maintenance of hybrid telecommunication services is carried out with the use of software is proposed. This approach allows to avoid reducing the quality of service during dash of overload and to maintain quality of service indicators at a given level, subject to compliance the resource utilization rate within the specified limits.

The modification of PCRF subsystems and new procedures for organizing the interaction of the mobile telecommunication network subsystems and the virtualized environmental management subsystems is proposed. It provides a process for monitoring the quality of service of hybrid telecommunication streams in the telecommunication environment, which allow providing the quality of service control and planning the amount of service resources for the efficient operation of heterogeneous telecommunication environment.

References

1. ITU-T Recommendation M.3371 of October 2016
2. ETSI GS NFV 001 v.1.1.1 (10/2013)
3. ETSI GS NFV 001 v.1.1.1 (10/2013)

4. Skulysh, M., Romonov, O.: The structure of a mobile provider network with network functions virtualization. In: 14th International Conference on Advanced Trends in Radioelectronics, Telecommunications and Computer Engineering, Conference Proceedings, TCSET 2018, 20–24 February 2018, Lviv, Slavske, pp. 1032–1034 (2018)
5. J. ITU-T Y.3520 Telecommunication standardization sector of ITU (06/2013). Series Y: Global information infrastructure, internet protocol aspects and next-generation networks (2013)
6. Skulysh, M., Klimovych, O.: Approach to virtualization of evolved packet core network functions. In: 2015 13th International Conference on Experience of Designing and Application of CAD Systems in Microelectronics (CADSM), pp. 193–195. IEEE (2015)
7. Globa L., et al.: Managing of incoming stream applications in online charging system. In: 2014 X International Symposium on Telecommunications (BIHTEL), pp. 1–6. IEEE (2014)
8. Skulysh, M.: The method of resources involvement scheduling based on the long-term statistics ensuring quality and performance parameters. In: 2017 International Conference on Radio Electronics & Info Communications (UkrMiCo) (2017)
9. Globa, L.: Method for resource allocation of virtualized network functions in hybrid environment. In: Globa, L., Skulysh, M., Sulima, S. (eds.) 2016 IEEE International Black Sea Conference on Communications and Networking, pp. 1–5 (2016). https://doi.org/10.1109/blackseacom.2016.7901546
10. Semenova, O., Semenov, A., Voznyak, O., Mostoviy, D., Dudatyev, I.: The fuzzy-controller for WiMAX networks. In: Proceedings of the International Siberian Conference on Control and Communications (SIBCON), 21–23 May 2015, Omsk, Russia, pp. 1–4 (2015). https://doi.org/10.1109/sibcon.2015.7147214

Information Technology Security

Validation of Safety-Like Properties for Entity-Based Access Control Policies

Sergey Afonin$^{(\boxtimes)}$ and Antonina Bonushkina

Moscow State University, Moscow, Russian Federation
serg@msu.ru

Abstract. In this paper safety problems for a simplified version of entity-based access control model are considered. By safety we mean the impossibility for a user to acquire access a given object by performing a sequence of legitimate operations over the database. Our model considers the database as a labelled graph. Object modification operations are guarded by FO-definable pre- and post-conditions. We show undecidability of the safety problem in general and describe an algorithm for deciding safety for a restricted class of access control policies.

Keywords: Access control · ABAC · EBAC · Safety · Decidability

1 Introduction

Access control management is an important part of most information systems. The ultimate goal of an access control *policy* is to define a collection of rules that allow *subjects* (users or software agents) to access *objects* of an information system, and to restrict any non-legitimate accesses. For example, a physician may only access medical records of his own patients, or patients he gave a prescription last week. Policies are usually specified in a natural language. In order to implement a policy in a software system, or to prove its correctness, the policy should be described in terms of some formal model. First models for access control go back to 70s and quite large number of models have been proposed since then [6].

There is a trade-off between models simplicity and usefulness in real life applications. Popular models, such as role-based access control (RBAC), are well studied. On the other hand, many natural access rules are hardly expressible in terms of such models. For example, "pure" RBAC can not express rules like "a user can modify *his own* files". In order to overcome such limitations, a number of extensions have been proposed in the literature, including the actively developing research area frequently referenced to as *attribute-based access control* (ABAC) [7]. In this approach, the security policy is specified by means of *rules* that depend on values of objects and subjects *attributes*, or properties, as well as the requests context, such as time of a day or physical location of the subject.

The reported study was supported by RFBR, research project No. 18-07-01055.

J. Pejaś et al. (Eds.): ACS 2018, AISC 889, pp. 259–271, 2019.
https://doi.org/10.1007/978-3-030-03314-9_23

For example, access to files may be defined by a rule like REQUEST.USER = FILE.OWNER.

When a policy is represented in terms of a formal model it is possible to check that the policy satisfies some desired properties. Examples of such properties, studied for RBAC [5], include *safety* (untrusted user can not gain membership in a given role), *availability* (a permission is always available to the user), or *liveness* (a permission is always available to at least one user). In case of RBAC data processing operations are not important, while granting and revoking of access rights or setting up security labels on objects are. In particular, the seminal paper [2] showing that there exists no algorithm for verifying impossibility of right "leakage" in an access control systems using object/subject matrices explicitly eliminate from consideration all data operations. In contrast, attributes values play central role for ABAC models, so it seems natural to model data operations in order to analyze an ABAC policy.

Formal analysis of ABAC policies, e.g. [3,4,8], are mainly focused on analysis of such properties as policies subsuming, separation of duties, etc. Many research on ABAC policies assume that attribute values are computable functions of object. This approach is attractive from practical point of view as one can implement procedures of arbitrary complexity. In recently proposed entity-based access control model (EBAC) [1] attributes are *selected* from database using a query language, rather then computed by a program in a Turing complete language. Such a restriction of expressive power of attribute evaluation procedure gives a hope for possibility of automated analysis of access control policies.

The contribution of this paper is the following. We introduce formal model for simplified version of EBAC and define safety-like policy validation problem. We show that this problem is undecidable in general and define a class of access control policies leading to decidable validation problem.

Our model consist of three parts: the model for database, data modification operations, and access control rules. Database is represented as a finite labeled directed graph. Vertices of this graph correspond to objects, the label of a vertex represents object's value, which is a rational number in our model, and edges define named relations between objects — if u and v are connected by an edge labeled by a, when object u has an attribute a, and the value of this attribute is the value of v. Users actions on a database are modeled by modification of labels of vertices. Access control policy is represented as a collection of predicates that specify when a vertex can be modified, deleted, or assigned a new attribute. The access decision on a vertex v depends on values of vertices in a finite neighborhood of v. Our policy validation problem consists in checking that some unsafe state of the database is not reachable from the current state by means of a sequence of allowed actions. In other words we are trying to verify that if a malefactor can not perform an operation on a object in the current state of the database then he can not transform the database, by a sequence of allowable actions, into a state such that the object become accessible.

Consider the following example. Let the database consist of three objects, say a, b, and c, and the only possible user action is the modification of object value. Let object a may be modified if $b > 0 \wedge c \leqslant 1$ (the value of b is positive, and value of c does not exceed 1), and b and c may be modified if $c \geqslant 1$ and $b < 0$, respectively. Assume that initial values of (a, b, c) are $(0, 1, 2)$ and our safety condition states that the user should not modify value of a. This particular initial state is unsafe, because a sequence $(0, 1, 2) \rightarrow_b (0, -1, 2) \rightarrow_c (0, -1, 1) \rightarrow_b (0, 1, 1)$ leads to a configuration when modification of a is allowed (here subscripts denote modified objects name). On the other hand, if the initial configuration is $(0, 1, 0)$, then there exists no possibility for a user to change value of a because none of the objects can be modified. Note that a successive sequence requires repetitive modification of some vertices.

Checking impossibility of getting access to a specific object may be considered as a reachability problem in a state transition system. A variety of reachability problems arise in connection with policy validation. This paper is devoted to the case when only one vertex can be modified at a time.

The remainder of this paper is organized as follows. In the next Section we give a formal definition of the problem. The algorithm for deciding reachability for access policies restricted by object values modification only is described in Sect. 3. In Sect. 4 we show that the safety problem is undecidable in the general case and consider graphs of bounded diameter. We conclude the paper with a discussion on a list of questions for future research.

2   Definitions and Notation

A *data graph* is a labeled directed graph $D = \langle O, A, R, l \rangle$, where $O = \{o_1, \ldots, o_N\}$ is a finite set of *objects*, $R \subseteq O \times O$, A is a finite set of attribute names, $l : R \rightarrow A$ is the edge labeling function. A *valuation* of objects is a mapping $\mu : O \rightarrow \mathbb{Q}$, where $\mathbb{Q}$ is the set of rational numbers. We will use both functional and vector notation, i.e. $\mu_i = \mu(o_i)$ for valuation of o_i, and $\mu = (\mu_1, \ldots, \mu_N)$ for a tuple of all valuations. A pair (D, μ) is called a *configuration* of the system. By $s(r)$ and $t(r)$ denote origin and target vertices of an edge $r \in R$. A vertex o' is *accessible by a path* $w \in A^*$ from vertex o, $w(o, o')$ in notation, if there exists a sequence of edges $r_1, \ldots, r_k \in R$ such that $s(r_1) = u$, $t(r_k) = v$, $s(r_{i+1}) = t(r_i)$ for all $1 \leqslant i < k$, and $w = l(r_1)l(r_2)\cdots l(r_k)$. We call data graph *deterministic*, if $|\{o' : w(o, o')\}| \leqslant 1$ for all $o \in O$ and $w \in A^*$.

We consider following *graph operations*: object editing, object or edge creation, and object or edge deletion. Object editing, $update(o, q)$, is the assignment of a new value $q \in \mathbb{Q}$ to object o. Object creation $create(o, a, q)$ creates new object with valuation q and connected to o by an a-labeled edge. Edge creation $createEdge(o_1, a, o_2)$ creates an a-labeled edge between objects o_1 and o_2. Object and edge deletion are $delete(o)$ and $deleteEdge(o_1, o_2)$, respectively. By $(D, \mu) \vdash (D', \mu')$, or by $\mu \vdash \mu'$ if the data graph is fixed, we denote that configuration (D', μ') may be obtained from (D, μ) using one graph operation. Transitive and reflexive closure of this relation is $\vdash^*$.

Access rules are defined using first order formulae. The signature consists of countable set of binary predicates w for all $w \in A^+$, distinguished binary predicates $\equiv$, and $<$, unary predicate obj, unary function symbol μ, and logical connectors $\wedge$, $\vee$, $\neg$, $\rightarrow$. The domain of interpretation I is the disjoint union of two sets: O, the set of objects of a data graph in consideration, and the set of rational numbers $\mathbb{Q}$. In order to simplify formulae we assume that quantifiers range over O, binary predicates $\{w, w \in A^+\}$ define relations on $O \times O$, binary predicate $<$ is a usual relation on $\mathbb{Q}$, function symbol μ maps elements of O to $\mathbb{Q}$, and obj is a characteristic function of O. We will use symbol $\equiv$ to compare variables as elements of O and $=$ to compare values, i.e. $x = y \leftrightarrow \exists v_1 \exists v_2 \ v_1 \equiv \mu(x) \wedge v_2 \equiv \mu(y) \wedge v_1 \equiv v_2$. For example, $\exists x \exists y \ abc(x, y) \rightarrow (\forall z \ cba(y, z) \wedge z = x \rightarrow z \equiv x)$ states that there exists a pair of vertices x and y, connected by a path abc, such that every vertex z reachable from y by a path cba coincides with x if $\mu(z) = \mu(x)$. The set of all formulae in this signature with k free variables is denoted by $\mathcal{F}^k$.

Access control *policy* defines conditions that should be fulfilled in order to execute data modification operation. Operations are guarded by pre- and post-conditions. Without loss of generality a policy may be considered as a mapping from objects to formulae. For *update* operation it is a mapping $P_{\text{update}} : O \rightarrow \mathcal{F}^1 \times \mathcal{F}^1$. Policy P_{update} permits assigning value q to object $o \in O$ in configuration (D, μ) if and only if $(D, \mu) \models pre$ and $(D', \mu') \models post$, where $(pre, post) = P_{\text{update}}(o)$, and $(D, \mu) \vdash_{\text{update}(o,q)} (D', \mu')$. We shall write $pre(o)$ and $post(o)$ to denote pre- and post-conditions associated by the policy to object o. Other operations, such as *createEdge*, depends on two objects. This is modeled by mappings like $P_{\text{createEdge}} : O \times O \rightarrow \mathcal{F}^2 \times \mathcal{F}^2$. Note that requirement to create an edge with a specified label may be verified in post-conditions. An object o is called accessible for a given operation in configuration μ, or just accessible, if the pre-condition defined by the policy holds in this configuration, $(D, \mu) \models pre(o)$. A policy P is a collection of mappings P_{op}, where op is the name of data modification operation. We shall call a policy *conjunctive* if every formula is an existentially quantified conjuncts of the form $\exists x_1 \ldots \exists x_k \ w_1(o, x_1) \wedge p_1(x_1) \wedge \ldots \wedge w_k(o, x_k) \wedge p_k(x_k)$.

In our model we intentionally omit users and their properties. For sake of simplicity we consider situation when only one user works with a system.

Policy Safety Problem. Given a configuration (D, μ), an access control policy P, and the *target object* t of the data graph one should decide whether there exists configuration (D', μ') such that $(D, \mu) \vdash^* (D', \mu')$ and object t is accessible in (D', μ').

3   Fixed-Structure Data Graphs

In this section we restrict the set of graph editing operations to a singleton set containing vertex editing only. In this case a data graph $D = \langle O, A, R, l \rangle$ does not change and the system state is completely described by a tuple $\mu =$

$(\mu(o_1), \ldots, \mu(o_N))$ of objects values. For a configuration μ by $\mu/o_i \to x$ denote a configuration μ' such that $\mu'_k = \mu_k$ for all k not equals i, and $\mu'_i = x$. The problem is to find a sequence of operations that transforms a given initial configuration μ^0 into a one that admits modification of target object $t \in O$.

3.1 Factorization of System Configurations for Conjunctive Policies

Let us start with consideration of one of the simplest validation problem. Assume that: (1) data graph is *deterministic*, (2) access control policy does not contain rules with post-conditions, (3) pre-conditions are existentially quantified conjuncts of the form $F(o) = \exists x_1 \ldots \exists x_k \; w_1(o, x_1) \wedge p_1(x_1) \wedge \ldots \wedge w_k(o, x_k) \wedge p_k(x_k)$, where for all $i \in \{1, \ldots, k\}$ predicate w_i defines reachability of x_i from the object vertex o, and $p_i(x_i)$ is a quantifier free formula restricting possible values of x_i, and (4) the policy consists of edit operations only. Under such assumptions for every object o one can associate a set $Dep(o)$ of vertices which values affect access decision for object o. Note that $Dep(o)$ is determined by the policy, by the formula $F(o)$, and does not depend on objects values μ.

Given a data graph $D = \langle O, A, R, l \rangle$ we construct the *dependency graph* $G = \langle O, E, c \rangle$, which is a directed graph with the same set of vertices as the data graph. A pair of objects (o, o') constitute a directed edge in the dependency graph if $o' \in Dep(o)$. Each edge is labeled by $c(o, o') = p_{oo'}$ — an unary predicate corresponding to this pair of objects. If policy P assigns access rule $\exists x_1 \ldots \exists x_k \; w_1(o, x_1) \wedge p_1(x_1) \wedge \ldots \wedge w_k(o, x_k) \wedge p_k(x_k)$ to o, and o' is the interpretation of x_i, then $c(o, o') = p_i$. An example of a data graph and the corresponding dependency graph is presented in Fig. 1.

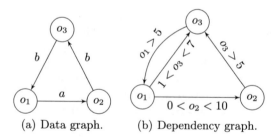

(a) Data graph.   (b) Dependency graph.

Fig. 1. Example of a data graph (a) and corresponding dependency graph (b) for a policy of two rules $P_1(o) = \exists x \exists y \; ab(o, x) \wedge x > 1 \wedge x < 7 \wedge a(o, y) \wedge y > 0 \wedge y < 10$ (applicable to vertices with an outgoing a-edge), and $P_2(o) = \exists x \; b(o, x) \wedge x > 5$ (applicable to vertices with b-edge).

Let $p_1, \ldots, p_k$ be a tuple of predicates appearing as labels of incoming edges to object $o \in O$ in a dependency graph. Call two values $v_1, v_2 \in \mathbb{Q}$ *dep-equivalent for* o if $p_i(v_1)$ holds if and only if $p_i(v_2)$ holds for all $i \in \{1, \ldots, k\}$. Dep-equivalent values does not affect accessibility of objects: if current configuration assigns value v to object o, i.e. $v = \mu(o)$, then this value may be replaced by any value from the set $[v]_o = \{v' \in \mathbb{Q} \mid v' \text{ and } v \text{ are dep-equivalent for } o\}$ without changing accessibility of other object. Two configurations μ^1 and μ^2 are *dep-equivalent*, $\mu^1 \sim_{dep} \mu^2$, if for all $i \in \{1, \dots, N\}$ values $\mu^1(o_i)$ and $\mu^2(o_i)$ are dep-equivalent for o_i. Let $[\mu]$ denotes the set of all configurations that are dep-equivalent to μ. If a vertex of the dependency graph has k incoming edges, then there exist up to 2^k dep-equivalence classes for this object. Clearly, the set of configurations of a fixed-structure data graph policy splits into finitely many dep-equivalence classes.

Now consider the directed *states graph* $G_s = \langle S, E_s \rangle$ with the set of dep-equivalence classes of (D, P) as a set of vertices. Two vertices s_1 and s_2 are connected by an edge if there are exist a configuration $\mu \in s_1$, an index $i \in \{1, \dots, N\}$, and a rational number x such that (1) object o_i is accessible in μ, and (2) $[(\mu_1, \mu_2, \dots, \mu_{i-1}, x, \mu_{i+1}, \dots, \mu_N)] = s_1$. That means that it is possible to transform a configuration in s_1 into a configuration in s_2 by a single edit operation. It is clear that if there exists a sequence of configurations $\mu^0, \mu^1, \dots, \mu^m$ such that target object t is accessible in μ^m, then vertices $[\mu^0]$ and $[\mu^m]$ of G_s are connected. The converse statement holds as well.

Proposition 1. *Let $G_s = \langle S, E_s \rangle$ be the states graph for a conjunctive policy P over deterministic data graph D. Then the following statements hold:*

(a) $\mu \vdash^* \mu'$ if and only if vertices $s = [\mu]$ and $s' = [\mu']$ are connected in G_s;
(b) if two vertices $s_1, s_2 \in S$ are connected in G_s, then for every configuration $\mu \in s_1$ there exists a configuration $\mu' \in s_2$ such that $\mu \vdash^* \mu'$.

Theorem 1. *Safety problem is decidable for conjunctive policies.*

It is worth noting that if pre-conditions are arbitrary functions $pre : \mathbb{Q}^N \to \{0, 1\}$ then simple factorization argument does not suffice. For example, we can define *access-deny equivalence* of configurations $\mu \sim_{AD} \mu'$ as coincidence of sets of accessible objects for both configurations (note that $[\mu] = [\mu'] \to \mu \sim_{AD} \mu'$). Nevertheless, it is possible that for some two pairs of adjusting configurations $\mu^1 \to \mu^2$, and $\mu^3 \to \mu^4$ the equivalence $\mu^2 \sim_{AD} \mu^3$ holds but μ^1 can not be transformed into μ^4 by any sequence of edit operations.

3.2 Heuristic Algorithm

Decidability result is based on an upper bound for the number of equivalence classes. If a given initial configuration is unsafe, then there exists a sequence of operations bounded in length by the number of vertices of G_s, which is $O(2^{K^N})$, where N is the number of objects and K is the maximum in-degree of dependency graph. While K may be assumed a constant (it is a property of the policy), exponential growth with respect to number of objects is not feasible for any reasonable application. Nevertheless, one can expect that in real-life situations safety property may be established in a reasonable time. In this section we

---

Algorithm 1. Construction of dependency subgraph at t.

Input: Data graph $D = \langle O, A, R, l \rangle$, policy P, target $t \in O$, initial state μ.
Output: Subgraph (V, B, W) rooted at t.

1 $V \leftarrow \{t\}$, $B \leftarrow \varnothing$, $W \leftarrow \varnothing$, $F \leftarrow \{t\}$ // F is a front
2 **while** $F \neq \varnothing$ **do**
3       $F' \leftarrow \varnothing$
4   **foreach** $u \in F$ **do**
5   **foreach** $v \in dep(u)$ **do**
6   **if** $(D, \mu) \models \neg p_{uv}(v)$ **then**
7   **if** $(v, u) \in B^*$ *(black loop)* **then return** $\varnothing$
8                   $F' \leftarrow F' \cup \{v\}$
9                   $B \leftarrow B \cup \{(u, p_{uv}, v)\}$
10   **else**
11                  $W \leftarrow W \cup \{(u, p_{uv}, v)\}$
12              $V \leftarrow V \cup \{v\}$

13      $F \leftarrow F'$
14 **return** $\langle V, B, W \rangle$

---

describe a quite natural algorithm of "ordered search" for a proof of non-safety of an object t.

The first stage consists of construction of a subgraph of the dependency graph starting from object t (Algorithm 1). It is a breadth first search algorithm that verifies accessibility of objects in the current configuration. The dependency graph is explored at vertex o only if there exists an *unsatisfied* incoming edge. If an unsatisfied edge discovered by BFS algorithm completes a loop of unsatisfied edges, then a proof of safety is found. Recall, that we are dealing with conjunctive policies and a cycle of unsatisfied edges in dependency graph indicates impossibility of changes to any object in the chain. The output of Algorithm 1 is a graph with colored edges. Black edges are edges of the dependency graph with unsatisfied, in the initial configuration μ^0, predicates. Edges with satisfied predicates marked by white color. Note that subgraph induced by black edges is a connected directed acyclic graph.

The second stage, Algorithm 2, takes the constructed colored dependency graph as an input, and yields a sequence of operations leading to modification of the target object t, if such a sequence exists. This is a backtracking algorithm that keeps all visited classes of configurations. The main idea is to process the dependency graph in a bottom-to-up manner, considering its black edges. Leaf vertices, that have no outgoing black edges, are accessible objects. By choosing correct values for these objects one can process one level up, and so on, until the target object become accessible, or a proof for safety will be found. The problem is, that once an unsatisfied edge is "fixed" by a change of configuration from μ to μ', the some other edges, that were satisfied in μ, might become unsatisfied in μ'.

Algorithm 2. Checking accessibility of a target object for a conjunctive policy.

Input: Spanning tree $G = \langle V, B, W \rangle$, target $t \in O$, initial configuration μ^0.
Output: Sequence of operations leading to modification of t.

```
1   μ ← μ⁰, S ← {[μ]}, M ← ∅, T ← ∅
2   Loop
3   |   A ← {o | ∃o'(o', o) ∈ B ∧ ∀o'(o, o') ∈ B ∪ W → (o, o') ∈ W}    // accessible
4   |   if t ∈ A then return T
5   |   if A ≠ ∅ then
6   |   |   choose o from A and x ∈ ℚ such that [μ/o → x] ∉ S
7   |   |   if (o, x) was selected then
8   |   |   |   T.push((o, x)), S.push([μ])
9   |   |   |   M.push(⟨μ, B, W⟩)
10  |   |   |   μ ← μ/o → x
11  |   |   |   update colors of edges incident to o
12  |   if A = ∅ or (o, x) was not selected then
13  |   |   if T is empty then return ∅
14  |   |   T.pop(), S.pop()
15  |   |   ⟨μ, B, W⟩ ← M.pop()
16  return ∅
```

Heuristics may be used for next object and its value selection (line 5). For example, choose object with largest black-edges depth (the longest black-path from the root t), and choose value that satisfies edge leading from an object of largest black depth. The choice of value x may be performed by a reduction to formula satisfiability. If $p_1, \ldots, p_k$ are predicates of incoming edges, then one can check satisfiability of a formula $p_1 \wedge \neg p_2 \wedge \neg p_3 \wedge p_4 \wedge \ldots \wedge p_k$ to find a value that makes all edges white, with the exception of edges 2 and 3.

4 Policies with Objects Creation

In this section we consider policies that admints new objects to be created. It is not surprising that the safety property is much harder to verify for such policies. We show, by a simple reduction to halting problem of a Turing machine, that the safety problem is undecidability in general and describe a class of data graphs with decidable safety problem.

4.1 Undecidability in the Presence of Objects Creation

If multiple objects may be updated as a result of single operation, then a arbitrary Turing machine can be simulated by the system easily, see Fig. 2 for an example. Objects of a data graph encode both cells of the machine tape and internal states. Values of cell encoding objects are letters of the machine tape alphabet. Value of the state-encoding object, which is distinguished from the other objects by the presence of "is a state" relation, is the number of the current machine internal state. A policy allows edit operations for "the current state and cell" only, which is enforced by pre- and post-conditions. Undecidability of the safety problem, in terms of our definition, follows from the fact that one can construct a policy in such a way that it is unsafe if and only if a given Turing machine reaches its terminal state.

More formally, let $M = \langle Q, \Sigma, q_0, \delta \rangle$ be a Turing machine, where Q is finite set of machine *states*, Σ is finite *tape alphabet*, $q_0 \in Q$ is an *initial state*, and $\delta : Q \times \Sigma \to Q \times \Sigma \times \{L, R\}$ is a *transition function*. Consider a data graph $D = \langle O, A, R, l \rangle$, where $\{t, s, d, f\} \subseteq O$, $A = \{\text{TM, iaastate, head, next}\}$, $R \supseteq \{(s, d), (s, f), (t, s)\}$, and edge labeling l contain mappings $(s, d) \mapsto$ isaState, $(s, f) \mapsto$ head, and $(t, s) \mapsto$ TM. Define some encoding of both machine states and tape alphabet symbols as an injective function $d : Q \cup \Sigma \to \mathbb{Q}$. For example, d may be a enumeration of elements, i.e. a bijection between the finite set $Q \cup \Sigma$ and the set of first $|\Sigma| + |Q|$ natural numbers. We are ready to describe an access control policy that simulates Turing machine M showing that target object t is accessible if and only if M halts.

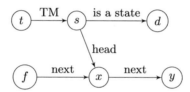

Fig. 2. Encoding of a Turing machine configuration by a data graph. The machine is in state $q = \mu(s)$ with the head at a cell x holding tape alphabet symbol $\mu(x)$. The pre-condition $pre(t)$ for the target object t is $\exists s\, \text{TM}(t, q) \land s = q_{\text{halt}}$.

Every transition $(q, a) \mapsto (q', a', R)$ may be encoded by the following composed rules (rules performing several data modification operations). The first rule operates if there exists a cell to the right from the current position (here $isaState(s) := \exists z\, isaState(s, z)$).

> $pre(s)$ $\exists x \exists y\, isaState(s) \land \text{head}(s, x) \land \text{next}(x, y) \land s = q \land x = a$
> $post(s)$ $s = q'$
> $body$
>> $update(x, a')$
>> $createEdge(s, \text{head}, y)$
>> $deleteEdge(s, x)$
>> $update(s, q')$

If M is at the rightmost position on the tape, i.e. cells to the right from the head position were never visited by the machine so far, a new object representing a blank cell may be created if the policy contains a rule.

$pre(x)$ $\exists s \forall y$ $isaState(s) \wedge \text{head}(s, x) \wedge \neg \text{next}(x, y) \wedge s = q \wedge x = a$
$body$
$$create(x, \text{next}, q_{\text{blank}})$$

Similar rules are used to simulate transitions moving machine head to the left, except it is not required to create new cell as the tape is semi-infinite. Policy P contains up to $2|Q| * |\Sigma|$ rules, instantiated from the above "templates" by replacing occurrences of a, q, a', and q' by corresponding constants.

Now, define target object $t \in O$ to be accessible if $\exists s$ $\text{TM}(t, q) \wedge s = q_{\text{halt}}$, where $q_{\text{halt}} \in \mathbb{Q}$ is the encoding of a halting state $halt \in Q$ of M, $q_{halt} = d(halt)$. If the initial configuration μ assigns $d(q_0), d(blank)$ to s and f, respectively, then the safety property for object t is equivalent to checking halting of M on the empty input word. At every moment only one rule may be performed by the system, and the sequence of configurations $\mu^0, \mu^1, \ldots$ in a one-to-one correspondence with configurations of M.

Composed rules might be considered too powerful and such rules do not satisfy our definition of the policy. We show now, that a mono-operational policy can simulate Turing machine behavior as well.

Theorem 2. *Safety is an undecidable property of unrestricted policies over non-deterministic data graphs.*

Proof. Let we have a transition $(q, a) \mapsto (q', a', R)$. Our goal is to split the composed rule presented early into several atomic operations. To this purpose encode next state q' and symbol a' in neighbors of state object s. Composed rule evaluation will be simulated by a sequence of atomic operations grouped into four *stages*: fill (recording of q', a', R), perf (performing updates), clear (clearing data recorded during the fill stage), done (processing completed). In order to track stages we introduce two more special objects connected to s, r and e, holding the information on the current rule and stage, respectively.

Let $P(s) := \exists x \exists y$ $isaState(s) \wedge \text{head}(s, x) \wedge \text{next}(x, y) \wedge s = q \wedge x = a$ be a predicate verifying that transition rule under consideration matches current state encoded in the data graph, i be an unique identifier of the transition rule $(q, a) \mapsto (q', a', R)$, and $inStage(s, x) := \exists r \exists z$ $\text{rule}(s, r) \wedge \text{stage}(s, z) \wedge z = x \wedge r = i$.

The following rules implement filling neighbors of s by values q', a' and move direction. The purpose for storing this known parameters (we are translating specific transition rule, so q, q', a, a' are known constants) in the data graph is to track update procedure described later. Column obj below stores free variable of corresponding pre-condition which interpretation is object referenced by the data modification operation listed in the rightmost column. All operations require assignment of constant values, which can be enforced by post-conditions. When we write that an operation is $update(r,i)$ we mean that we allow modification of object r with post-condition $post(r) := r = i$.

obj	pre-condition	operation
r	$\exists s \exists e\, P(s) \wedge \text{stage}(s,e) \wedge e = \text{done}$	$update(r,i)$
e	$\exists s\, P(s) \wedge inStage(s,\text{done})$	$update(e,\text{fill})$
s	$inStage(s,\text{fill}) \wedge \forall z\, \neg\text{state}(s,z)$	$create(s,\text{state},q')$
s	$inStage(s,\text{fill}) \wedge \exists z\, \text{state}(s,z) \wedge \forall z\, \neg\text{sym}(s,z)$	$create(s,\text{sym},a')$
s	$inStage(s,\text{fill}) \wedge \exists z\, \text{sym}(s,z) \wedge \forall z\, \neg\text{move}(s,z)$	$create(s,\text{move},1)$
e	$\exists s \exists z\, \text{stage}(s,e) \wedge e = \text{fill} \wedge \text{move}(s,z)$	$update(e,\text{perf})$

Once all data describing the next state of the Turing machine are recorded in the data graph, one can perform update operations as follows (we consider head position movement only).

obj	pre-condition	operation
s	$\exists z \exists x \exists y\, inStage(s,\text{perf}) \wedge \text{move}(s,z) \wedge \text{head}(s,x)$ $\wedge\, \text{next}(x,y) \wedge \forall x'(\text{head}(s,x') \rightarrow x' \equiv x)$	$createEdge(s,\text{head},y)$
m	$\exists x \exists y \exists s\, inStage(s,\text{perf}) \wedge \text{move}(s,m)$ $\wedge\, \text{head}(s,x) \wedge \text{head}(s,y) \wedge x \not\equiv y$	$delete(m)$
s	$\exists x \exists y\, inStage(s,\text{perf}) \wedge \forall m \neg\text{move}(s,m)$ $\wedge\, \text{head}(s,x) \wedge \text{head}(s,y) \wedge x \not\equiv y$	$deleteEdge(s,x)$
e	$\exists s \exists z\, \text{stage}(s,e) \wedge e = \text{fill} \wedge \text{sym}(s,z)$	$update(e,\text{clear})$

On the clearing stage we simply removes all technical objects.   □

This construction shows that atomic operations are quite flexible. The only non-atomic action we used in this construction is object creation, which introduces new object *and* creates an edge to it. Edge creation allows us to identify newly created object. Alternatively, if new object will be created without connection to any other objects, but with a special value the simulation of Turing machine is possible as well.

It is worth noticing that pre-conditions appeared in the proof rely on checking for edge existence, absence or uniqueness only, a quite restricted subset of FO language.

4.2   Graphs of Bounded Diameter

The main component of the proof of safety undecidability is a chain of vertices representing Turing machine tape cells. In this section we consider graphs with bounded diameter. If data graph is strongly connected, then bounded diameter means that there exists an upper bound on number of vertices for this graph. In this case we can reduce the problem to fixed structure data graph. When graph is not strongly connected then the number of vertices may be arbitrary large. If graph diameter is bounded by N, then a graph containing arbitrary many components of diameter $N - 1$ and connected to a single root has bounded diameter. Such graphs could be of practical interset as they model objects with arbitrary many unordered dependent objects.

If data graph diameter is bounded by N and there exists a successive sequence of operations, i.e. a sequence leading to an unsafe configuration, then there exists a successive sequence of operations that modifies no more then $f(N)$ objects. Thus, safety problem could be decidable for such graphs.

5   Conclusion

In this paper we have considered a specific form of an attribute-based access control policy validation problem, when impossibility of getting access to a specific object should be verified for a give initial configuration of the system. The problem, which is motivated by a recently proposed Entity-based access control model, was shown decidable for a restricted case of access control policies, and undecidable if a policy admits objects creations. It is worth noticing that safety problem is undecidable if graph operations are restricted to creation or modification of one object or edge at a time, provided that newly created object is connected to another one. Both decidability and undecidability results are not surprising by themselves. When objects creation is not allowed then the system is finite in some sense, regardless of cardinality of the object values domain. The resulting algorithm for checking safety with respect to a given initial configuration and the target object enumerates equivalence classes of system configurations. That procedure could be difficult in general. Nevertheless, one can expect that for many reasonable policies safety property, as we have defined it, could be established fast.

Possible directions of future work include the following. Different heuristics proposed for safety checking algorithm should be analyzed in more details and compared on real life policied. Necessity conditions that fixed-structure policy should satisfy for a decidable safety problem (not only FO-definable pre- and post conditions) should be established. As it is unlikely that real-life information systems admit arbitrary relations between objects, conditions, similar to bounded diameter, on data graph leading to decidable problems should be established. Finally, a more general versions of safety should be considered. Instance-based checking, like the one addressed in this paper, is not of very large practical interest because only one object may be verified at a time.

References

1. Bogaerts, J., Decat, M., Lagaisse, B., Joosen, W.: Entity-based access control: supporting more expressive access control policies. In: Proceedings of the 31st Annual Computer Security Applications Conference, pp. 291–300. ACM (2015)
2. Harrison, M.A., Ruzzo, W.L., Ullman, J.D.: Protection in operating systems. Commun. ACM **19**(8), 461–471 (1976)
3. Hughes, G., Bultan, T.: Automated verification of access control policies using a SAT solver. Int. J. Softw. Tools Technol. Transf. **10**(6), 503–520 (2008)

4. Kolovski, V., Hendler, J., Parsia, B.: Analyzing web access control policies. In: Proceedings of the 16th International Conference on World Wide Web, pp. 677–686. ACM (2007)
5. Li, N., Tripunitara, M.V.: Security analysis in role-based access control. ACM Trans. Inf. Syst. Secur. (TISSEC) **9**(4), 391–420 (2006)
6. Samarati, P., de Vimercati, S.C.: Access control: policies, models, and mechanisms. In: International School on Foundations of Security Analysis and Design, pp. 137–196. Springer (2000)
7. Servos, D., Osborn, S.L.: Current research and open problems in attribute-based access control. ACM Comput. Surv. **49**(4), 65:1–65:45 (2017)
8. Turkmen, F., den Hartog, J., Ranise, S., Zannone, N.: Formal analysis of XACML policies using SMT. Comput. Secur. **66**, 185–203 (2017)

Randomness Evaluation of PP-1 and PP-2 Block Ciphers Round Keys Generators

Michał Apolinarski[(⊠)] [iD]

Institute of Control, Robotics and Information Engineering,
Poznan University of Technology, ul. Piotrowo 3a, 60-965 Poznań, Poland
michal.apolinarski@put.poznan.pl

Abstract. Round keys in block ciphers are generated from a relatively short (64-, 128-, 256-, and more bits) master key and are used in encryption and decryption process. The statistical quality of round keys impact difficulty of block cipher cryptanalysis. If round keys are independent (not-related) then cryptanalysis need more resources. To evaluate key schedule's statistical quality we can use NIST 800-22 battery test. PP-1 key schedule with 64 bits block size and 128-bit master key generates 22 64-bits round keys that gives cryptographic material length of 1408 bits. PP-2 with 64-bits block size generates in single run from 128-bits master key only 13 round keys, which give 832-bits sample from single master key. Having such short single samples we can perform only couple of NIST 800-22 tests. To perform all NIST 800-22 tests at least 106 bits length samples are required. In this paper we present results of randomness evaluation including all NIST 800-22 tests for expanded PP-1 and PP-2 round key generators.

Keywords: Key schedule · Round keys · Block cipher · NIST 800-22
Statistical tests · PP-1 block cipher · PP-2 block cipher · Round keys generator

1 Introduction

Key schedule algorithm in block ciphers can be treated as a pseudorandom generator used to generate a set of round keys from a relatively short master key (main key/user key). Round keys are used in the encryption and decryption process in the ciphers rounds. Key schedule algorithm is a collection of simple linear and/or non-linear operations – depending on the operations used, both the generation time and the quality of the generated keys may be different. Generating round keys usually take place once before encrypting or decrypting and is a time consuming process. The important property is that the generated round keys should be independent (not-related). Independence of round keys affects the process of cryptanalysis [5]. If the round keys in the cipher are independent, cryptanalysis of the ciphertext is more difficult and requires more resources [3, 4, 6–9]. Designing a key schedule, we need to find a compromise between the speed of key generation and the quality (the independence of the round keys generated by the key schedule) [12, 13].

© Springer Nature Switzerland AG 2019
J. Pejaś et al. (Eds.): ACS 2018, AISC 889, pp. 272–281, 2019.
https://doi.org/10.1007/978-3-030-03314-9_24

2 Statistical Tests

Statistical tests package NIST 800-22 [14] allows to evaluate the quality of the PRNG, by examining how the generated bit sequence is different from the random sample.

Among other things, the NIST 800-22 statistical test package was used to evaluate the finalists for the AES block cipher [15, 16]. In the articles [1, 2] was presented the possibility of using selected NIST tests to evaluate key schedule algorithms for generating block ciphers round keys. Selected tests because to carry out all NIST 800-22 tests the single sample sequence must be of length $n > 10^6$. If single sample sequence is bigger than 10^6 then all 15 tests can be performed:

- Frequency Test– determines whether the number of 1s and 0s in a sequence is approximately the same as would be expected for a truly random sequence.
- Cumulative Sum Test – determines whether the sum of the partial sequences occurring in the tested sequence is too large or too small.
- Spectral DFT Test – checks whether the test sequence does not appear periodic patterns.
- Binary Matrix Rank Test – checks for linear dependence among fixed length substrings of the original sequence.
- Longest Run of One's Test – determines whether the length of the longest run of ones within the tested sequence is consistent with that would be expected in a random sequence.
- Random Excursions Test – determines if the number of visits to a particular state within a cycle deviates from what one would expect in a random sequence.
- Random Excursions Variant Test – detects deviations from the expected number of visits to various states in the random walk.
- Runs Test – counts strings of ones and zeros of different lengths in the sequence and checks if these numbers correspond to the random sequence.
- Block Frequency Test – determines whether the number of 1 s and 0 s in each of m non-overlapping blocks created from a sequence appear to have a random distribution.
- Overlapping Template Matching Test – rejects sequences that show deviations from the expected number of runs of ones of a given length.
- Non-overlapping Template Matching Test – rejects sequences that exhibit too many occurrences of a given non-periodic (aperiodic) pattern.
- Parameterized Serial Test – checks whether the number of m-bit overlapping blocks is suitable.
- Approximate Entropy Test – compares the frequency of overlapping blocks of length m and $m + 1$, checks if any of the blocks does not occur too often.
- Linear Complexity Test – determines whether or not the sequence is complex enough to be considered random.
- Universal Test – detects whether or not the sequence can be significantly compressed without loss of information.

The result of each test must be greater than the acceptance threshold to be considered as sequence with good statistical properties and obtained results can be interpret-ed as the proportion of sequences passing a test:

$$\hat{p} \pm \sqrt[3]{\frac{\hat{p}(1-\hat{p})}{z}}, \tag{1}$$

where $\hat{p} = 1 - \alpha$, and z denote number of samples (tested sequences).

In our research the level of significance was set for $\alpha = 0.01$ so in this case acceptance threshold was: 0.980561.

As the input to NIST battery in our research we take 1000 bits long sequences generated by expanded PP-1 and PP2 key schedule [see Sect. 4]. A single bit sequence was obtained by concatenating set of round keys received from a single master key. Each successive bit sequence was generated from a master key incremented by 1 bit in relation to the previous one.

$$MK_i = \begin{cases} \text{random}(0, 2^b - 1), & \text{for } i = 1 \\ (MK_{i-1} + 1) \bmod 2^b, & \text{for } i = 2, \ldots, z \end{cases} \tag{2}$$

where $b = |MK_i|$ and b denotes length of master key MK_i.

3 Standard PP-1 and PP-2 Key Schedules

Scalable PP-1 [10] block cipher operates on n-bit blocks. The key schedule from the master key (n or $2n$ bits length) generate $2r$ n-bits round keys. The round keys are generated in $2r + 1$ iterations, where n block size r is the number of rounds (in the first iteration of the key schedule round key is not produced, so $k_1, k_2, \ldots, k_{2r}$ are round keys). Figure 1 shows one iteration of PP-1 key schedule.

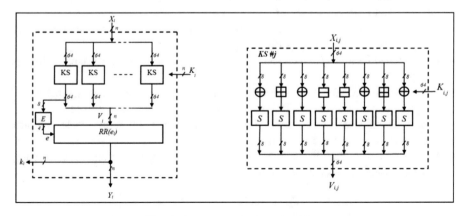

Fig. 1. PP-1 key schedule algorithm.

For iteration #0 an input X_i is n-bit constant: $B = B_1 \| B_2 \| \ldots \| B_t$, where $B_1 = 0x91B4769B2496E7C$, $B_j = Prm(B_j - 1)$ for $j = 2, 3, \ldots, t$, where Prm is auxiliary permutation describe in [10].

Input K_i for iteration #0 and #1 is computed depending on master key length:

- if the length of the master key is equal to n, then $K_0 = k$ and $K_1 = 0^n$ (concatenation of zeros),
- if the length of the master key is equal to $2n$, then the key is divided into 2 parts k_H and k_L, giving $K_0 = k_H$ and $K_1 = k_L$.

Value K_i for iterations #2 is: $K_2 = RL(B \oplus (A \wedge (K_0 \oplus K_1)))$, where $\wedge$ means Boolean AND operation, and RL is left rotation by 1 bit, value A depends on master key length:

- if the master key is equal to n then $A = 0^n$,
- if the master key is equal to $2n$, then $A = 1^n$. Value K_i for iteration #3…#2r is computed as $K_i = RL(K_i - 1)$.

Rest of key schedule components are:

- KS – main element consisting of S-block, XOR, add, sum mod 256 performed on 8-bit values, derived from 64-bit input from n-bit block X_i;
- $RR(e_i)$ – right rotation by e_i bits of n-bit V_i block;

E – component that computes 4-bit value $e_i = E(b_1, b_2, \ldots b_n) = (b_1 \oplus b_8)$ $(b_2 \oplus b_{10})(b_3 \oplus b_{11})(b_4 \oplus b_{12})$, based on 8-bit input, which is concatenation of 4 most significant bits outputs of 2 left most S-boxes in KS element.

If we consider a PP-1 key schedule with block size 64-bit and 128-bit master key that generate in single run 22 round keys with a length of 64 bits. That gives us cryptographic material length of 1408 bits (concatenated 22 round keys are treated as a single sequence sample for NIST battery). Like was said in the previous chapter for samples of this length can be carried out only 7 of 15 NIST 800-22 test [1]:

- Frequency Test,
- Block Frequency Test,
- Cumulative Sums Test,
- Runs Test,
- Spectral DFT Test,
- Approximate Entropy Test,
- Parameterized Serial Test.

The similar situation is with block cipher PP-2 [11] where the key schedule generates in a single run from 128 master key only 13 round keys, which gives the sample length of 832 bits and such sample is too short to perform all NIST 800-22 tests.

The PP-2 cipher is a scalable cipher and the number of rounds of the PP-2 cipher depends on the size of the block n being processed and the size of the master key. The master key k has the size $|k| = d \cdot n$ bits, where $d = 1; 1.5; 2; \ldots$ If the key size $|k|$, such that $(d - 1) \cdot n < |k| < d \cdot n$, this key is padded with zeros to the size $d \cdot n$. The key k is divided into d subkeys, each of size n, such that $k = \kappa_1 \| \kappa_2 \| \ldots \| \kappa_{\lceil d \rceil}$, where $\lceil d \rceil$ is the lowest integer not lower than d. If the size of the subkey $\kappa_{2\lceil d\rceil} = n/2$, this key is supplemented with zeros to the size of n.

The Fig. 2 shows a one iteration of the PP-2 key schedule. The components of this algorithm are:

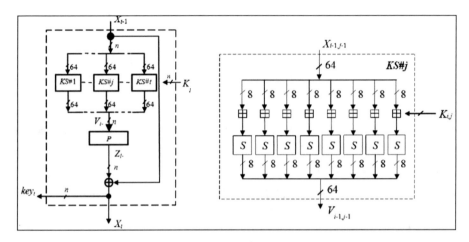

Fig. 2. One iteration of PP-2 key schedule algorithm.

- KS – operations on 8-bit data blocks,
- adding modulo 256,
- S-blocks,
- $P(V)$ - multiple rotations,
- XOR operation.

The PP-2 key schedule has run-in rounds and not every iteration produces a round key. The constants c_0 and c_1 used in the cipher are scalable as an entire PP-2 cipher:

- $c_0 = RR(0, (E3729424EDBC5389)) \parallel RR(1, (E3729424EDBC5389)) \parallel \ldots \ldots \parallel RR(t-1, (E3729424EDBC5389))$,
- $c_1 = RR(0, (59F0E217D8AC6B43)) \parallel RR(1, (59F0E217D8AC6B43)) \parallel \ldots \ldots \parallel RR(t-1, (59F0E217D8AC6B43))$,

where $RR(b, x)$ means rotation of the binary word x to the right by b bits. For assumed constants c_0 and c_1 and for $i = 1, 2, \ldots, \lceil d\rceil \cdot t + r$ is calculated:

$$K_i = K_i^* \oplus RR(i - 1, \; c_0), \tag{3}$$

$$\left(K_i^*\right)_{i=1}^{\lceil d\rceil \cdot t + r} = \left(\kappa_1, \underbrace{0,0,\ldots,0}_{t}, \kappa_2, \underbrace{0,0,\ldots,0}_{t}, \ldots \kappa_{\lceil d\rceil}, \underbrace{0,0,\ldots,0}_{t}, \underbrace{0,0,\ldots,0}_{r-\lceil d\rceil}\right) \tag{4}$$

Furthermore, it is assumed that:

$$k_i = \left\{ \begin{array}{ll} key_{i(t+1)}, & \text{for } i = 1, 2, \ldots, \lceil d \rceil \\ key_{\lceil d \rceil \cdot (t+1)+i}, & \text{for } i = \lceil d \rceil + 1, \lceil d \rceil + 2, \ldots, r \end{array} \right\} \tag{5}$$

The Fig. 3 presents generation algorithm of round keys for PP-2 with 64-bit block and 128-bit master key, thus $r = 13$, $d = 2$, $t = 1$;

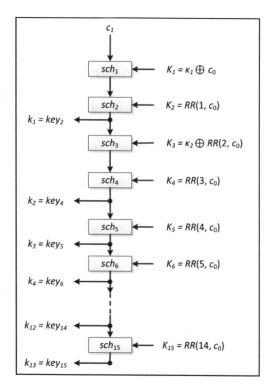

Fig. 3. The schema of the generation of round keys for PP-2 with 64-bit block and 128-bit master key.

4   Expanded PP-1 and PP-2 Key Schedules

An idea for presented research is to expand (by increasing) number of iterations of PP-1 and PP-2 key schedules to generate "unlimited" number of round keys based on single input data (single master key).

Instead of evaluating bit samples constructed (concatenated) from standard PP-1 or PP-2 key schedule, we evaluate bit samples constructed from 15642 round keys (64-bit length) generated by an expanded PP-1 or PP-2 key schedule. Such expanded version of key schedule can provide samples longer than 10^6 bits (precisely 1 001 088 bits) and we can evaluate key generators using all tests from NIST 800-22 package.

Also extended evaluation can show if there are any statistical defects or periods in the algorithm when we try to generate more round keys – defects that could be not identified in standard operation mode.

The Fig. 4 shows an example of an expanded PP-2 key schedule that generates 15642 round keys from 128-bit master key.

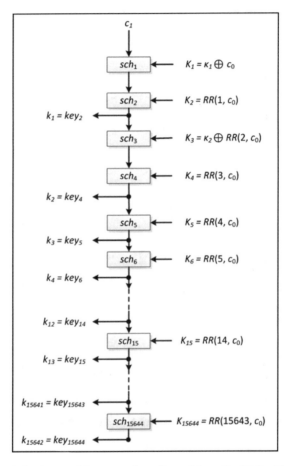

Fig. 4. The expanded schema of the generation of round keys for PP-2 with 64-bit block and 128-bit master key

In Fig. 5 we can see example round keys and bitstream of output generated from two master keys (MK) differing in 1 bit.

Results of all performed tests are presented in Fig. 6. As the input for NIST 800-22 tests was taken 1000 bit streams of 1 001 088 lengths generated from 1000 different master keys like described in Sect. 2.

MK	0000000000000000 0000000000000000	0000000000000000 0000000000000001
k_1	0x553DC29DD68BC880	0x553DC29DD68BC880
k_2	0xAFCF9688025D3E89	0xA74CA28992597E8D
...	...	...
k_{12}	0xF9A4A91B320E86C0	0x0278379D5C697624
k_{13}	0xAA279AD8EA6116D2	0x5D615BC3447D3DB9
...	...	...
k_{15641}	0x2B640E6B9B68411C	0xD2FD0FB00B0C16AA
k_{15642}	0x58D5FF55F364D989	0xF7F99FCAF84FCA57
the number of 0's	501344	500812
the numer of 1's	499744	500276

Fig. 5. Round keys example from expanded PP-2 key schedule.

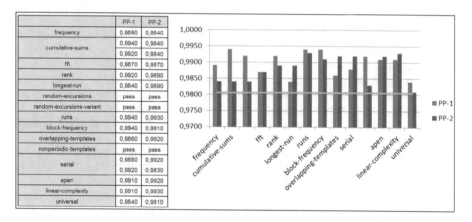

Fig. 6. Results of all NIST 800-22 tests for expanded PP-1 and PP-2 key schedules

We can see that for both PP-1 and PP-2 all tests met the acceptance threshold 0.980561 and gave positive2 results. We can also notice that proportion of passing tests were slightly better for PP-1 key schedule than PP-2.

Tests like: random-excursions, random-excursions-variants and nonperiodic-templates consist of many subtests, so detailed (average) value of pass-rate was omitted.

5 Conclusions

Presented methodology and research results show that all performed NIST 800-22 tests for expanded PP-1 and PP-2 versions were positive and met acceptance thresh-old 0.980561 for 1000 samples generated from different 1000 master keys. So both PP-1 and PP-2 key schedule algorithms generates statistically good round keys (with no statistical defects) for block ciphers and also can be used as classical PRGN, for example as session key generators.

Based on our researches we also propose to consider statistical evaluation of existing and designed in the future key schedules for block ciphers (for original and for extended version if such modification is possible).

Acknowledgements. This research has been supported by Polish Ministry of Science and Higher Education under grant 04/45/DSPB/0163.

References

1. Apolinarski, M.: Statistical properties analysis of key schedule modification in block cipher PP-1. In: Wiliński, A., et al. (ed.) Soft Computing in Computer and Information Science. Advances in Intelligent Systems and Computing, vol. 342, pp. 257–268. Springer, Cham (2015)
2. Apolinarski, M.: Quality evaluation of key schedule algorithms for block ciphers. Studia z Automatyki i Informatyki – tom 37, Poznań (2012)
3. Biham, E., Dunkelman, O., Keller, N.: Related-key boomerang and rectangle attacks. In: Proceedings of the 24th Annual International Conference on Theory and Applications of Cryptographic Techniques, 22–26 May 2005, Aarhus, Denmark (2005)
4. Biham, E., Dunkelman, O., Keller, N.: A unified approach to related-key attacks. In: Fast Software Encryption: 15th International Workshop, FSE 2008, Lausanne, Switzerland, 10–13 February 2008, Revised Selected Papers. Springer, Heidelberg (2008)
5. Biham, E., Shamir, A.: Differential Cryptanalysis of the Data Encryption Standard. Springer, New York (1993)
6. Biryukov, A., Nikolić, I.: Automatic search for related-key differential characteristics in byte-oriented block ciphers: application to AES, Camellia, Khazad and Others. In: Gilbert, H. (ed.) EUROCRYPT 2010. LNCS, vol. 6110, pp. 322–344. Springer, Heidelberg, (2010)
7. Biryukov, A., Khovratovich, D., Nikolic, I.: Distinguisher and related-key attack on the full AES-256. In: Halevi, S. (ed.) Advances in Cryptology – CRYPTO 2009. LNCS, vol. 5677. Springer (2009)
8. Biryukov, A., Khovratovich, D.: Related-key cryptanalysis of the full AES-192 and AES-256. In: Asiacrypt 2009. LNCS, vol. 5912, pp. 1–18. Springer (2009)
9. Bogdanov, A., Tischhauser, E.: On the wrong key randomisation and key equivalence hypotheses in Matsui's algorithm 2. In: Moriai, S. (ed.) FSE 2013. LNCS, vol. 8424, pp. 19–38. Springer, Heidelberg (2014)
10. Bucholc, K., Chmiel, K., Grocholewska-Czuryło, A., Idzikowska, E., Janicka-Lipska, I., Stokłosa, J.: Scalable PP-1 block cipher. Int. J. Appl. Math. Comput. Sci. **20**(2), 401–411 (2010)
11. Bucholc, K., Chmiel, K., Grocholewska-Czurylo, A., Stoklosa, J.: PP-2 block cipher. In: 7th International Conference on Emerging Security Information Systems and Technologies (SECURWARE 2013), pp. 162–168. XPS Press, Wilmington (2013)
12. Huang, J., Lai, X.: Revisiting key schedule's diffusion in relation with round function's diffusion. Des. Codes Cryptogr. **73**, 1–19 (2013)
13. Kim, J., Hong, S., Preneel, B., Biham, E., Dunkelman, O., Keller, N.: Related-Key Boomerang and Rectangle Attacks. IACR eprint server, 2010/019 January (2010)

14. Rukhin, A., et al.: A Statistical Test Suite for Random and Pseudorandom Number Generators for Cryptographic Applications. NIST Special Publication 800-22, revision 2 (2008)
15. Soto, J.: Randomness Testing of the Advanced Encryption Standard Candidate Algorithms. NIST IR 6390 (1999)
16. Soto, J., Bassham, L.: Randomness Testing of the Advanced Encryption Standard Finalist Candidates. NIST IR 6483 (2000)

New Results in Direct SAT-Based Cryptanalysis of DES-Like Ciphers

Michał Chowaniec[✉], Mirosław Kurkowski, and Michał Mazur

Institute of Computer Sciences, Cardinal Wyszynski University, Warsaw, Poland
chomich1995@gmail.com, m.kurkowski@uksw.edu.pl, mazurmichal1993@gmail.com

Abstract. SAT based cryptanalysis is one of efficient ways to investigate about desire properties of symmetric ciphers. In this paper we show our research and new experimental results in the case of SAT based, direct cryptanalysis of DES-like ciphers. For this, having a given cipher, we built firstly propositional logical formula that encode the cipher's algorithm. Next, having a randomly generated plaintext, and a key we compute the proper ciphertext. Finally, using SAT solvers, we explore cipher properties in the case of plaintext and ciphertext cryptanalysis. In our work we compare several SAT solvers: new ones and some rather old but so far efficient. We present our results in the case of original version of DES cipher and its some modifications.

Keywords: Symmetric ciphers · Satisfiability
SAT based cryptanalysis

1 Introduction

Boolean satisfiability (SAT) is a well-known NP-complete problem. In the whole case solving satisfiability of big formulas is hard. Although, satisfiability of many boolean formulas with hundreds or thousands variables can be solved surprisingly efficiently. Most of implemented algorithms for this purpose used for computing satisfying valuation are optimized versions of the DPLL procedure [7,8]. Usually SAT solvers, special programs that answer the question about boolean satisfiability, takes input formulas in the conjunctive normal form (CNF). It is a conjunction of clauses, where a clause is a disjunction of literals, and a literal is a propositional variable or its negation.

SAT is used for solve many decision, computing problems [2]. In these approaches investigated problem is encoded as boolean, propositional formula. If this formula is satisfiable then answer the question about the problem is positive. SAT is used among others for cryptanalysis of some cryptographic algorithms, especially symmetric ciphers [10,12–15,18].

In this work we develop concepts introduced in [9], where the efficiency of SAT based cryptanalysis of the Feistel Network and the DES cipher was shown. We try to increase investigations in this area trying to check how SAT solvers

© Springer Nature Switzerland AG 2019
J. Pejaś et al. (Eds.): ACS 2018, AISC 889, pp. 282–294, 2019.
https://doi.org/10.1007/978-3-030-03314-9_25

work with some modifications of DES cipher. We also checked how several new SAT solvers work in this case.

The rest of this paper is organized as follows. In Sect. 2, we introduce all basic information on both ciphers mentioned, to the extent necessary for explaining our boolean encoding method. Section 3 gives a process of a direct, boolean encoding of the ciphers we consider. In Sect. 4, we introduce several optimization and parallelization ideas used in our method. In Sect. 5, we present some experimental results we have obtained. Finally, some conclusion and future directions are indicated in the last section.

2   Feistel Network and DES Cipher

In this section, we present basic information on the Feistel and the DES ciphers needed for understanding our methodology of SAT based cryptanalysis of symmetric cryptographic algorithms.

The Feistel Network (FN) is a block symmetric cipher introduced in 1974 by Horst Feistel. Firstly FN was used in IBM's cipher named Lucifer, designed by Feistel and Coppersmith. Thanks to iterative character of FN, implementing the cipher in hardware is easy. It is important to note that with respect to simple structure provide to using Feistel-like networks to design various cipher, such as DES, MISTY1, Skijack, early mentioned Lucifer or Blowfish [17]. An idea of this algorithm is the following.

Let F denote the round function and $K_1, \ldots, K_n$ denote a sequence of keys obtained in some way from the main key K for the rounds $1, \ldots, n$, respectively. We use symbol $\oplus$ for denoting the exclusive-OR (XOR) operation.

The basic operations of FN are specified as follows:

1. break the plaintext block into two equal length parts denoted by (L_0, R_0),
2. for each round $i = 0, \ldots, n$, compute $L_{i+1} = R_i$ and $R_{i+1} = L_i \oplus F(R_i, K_i)$.

Then the ciphertext sequence is (R_{n+1}, L_{n+1}).

The structure of FN allows easy method of decryption. Lets recall basic properties of operation $\oplus$ for all $x, y, z \in \{0, 1\}$:

- $x \oplus x = 0$,
- $x \oplus 0 = x$,
- $x \oplus (y \oplus z) = (x \oplus y) \oplus z$.

A given ciphertext (R_{n+1}, L_{n+1}) is decrypted by computing $R_i = L_{i+1}$ and $L_i = R_{i+1} \oplus F(L_{i+1}, K_i)$, for $i = n, \ldots, 0$. It is easy to observe that (L_0, R_0) is the plaintext again. Observe additionally that we have the following equations:

$$
\begin{aligned}
R_{i+1} \oplus F(L_{i+1}, K_i) &= (L_i \oplus F(R_i, K_i)) \oplus F(L_i, K_i) \\
&= L_i \oplus (F(R_i, K_i) \oplus F(L_i, K_i)) = L_i \oplus 0 = L_i.
\end{aligned}
$$

Data Encryption Standard (DES) is a symmetric block cipher that uses a 56-bit key. In 1970s US National Bureau of Standards chose DES as an official Federal Information Processing Standard. For over 20 years it had been considered secure. In 1999 the *distributed.net* and the *Electronic Frontier Foundation* collaborated to break a DES key in 22 h and 15 min. This lead to assumption that the 56-bit key size has been too small. Now we know few attacks that can break the full 16 rounds of DES, which are less complex than a brute-force search. Due to huge progress in designing hardware some of those can be verified experimentally. For example, linear cryptanalysis discovered by Mitsuru Matsui requires 2^{43} known plaintexts and differential cryptanalysis [16], discovered by Eli Biham and Adi Shamir needs 2^{47} chosen plaintexts to break the full 16 rounds [5]. Now with some modifications DES is believed to be strongly secure. One of these modified form is called Triple DES.

The algorithm consists of 16 rounds. Before all rounds block is split into halves (each for 32 bits), which are processed separately with respect to some alterations FN. Using FN assure us that coding process have much alike computational time cost. Although, there is difference between decryption and encryption- subkeys are provided in reverse order. Due to those similarities (between decryption and encryption), implementation is easier. We do not have to have different units for decryption and encryption.

F-function takes halves of the main block and mixes them with one of the subkeys. The output from the F-function is then combined with the second portion of the main block, and both portions are swapped before the next round. After the last round, the portions are not swapped.

F-function takes one half of the block (32 bits) and consists as follow:

Expansion. The 32-bit half-block is enlarged into 48 bits using some special function by duplicating half of the bits. The output consists of eight 6-bit ($8 \cdot 6 = 48$ bits) pieces, each containing a copy of 4 corresponding input bits, plus a copy of the immediately adjacent bit from each of the input pieces to either side.

Key Mixing. The result is combined with a sub-key using operation $\oplus$. Sub-keys are obtained from the main initial encryption key using a special key schedule - one for each round. The schedule used consists of some rotations of bits. For each round a different subset of key bits is chosen.

Substitution. After mixing with the subkey, the block is divided into eight 6-bit portions, before processing using the S-boxes. Each of the eight S-boxes is a matrix with four rows and six columns. It can be treated as a non-linear function from $\{0,1\}^6$ into $\{0,1\}^4$. Each S-box replaces a six-tuple input bits with some four output bits. The S-boxes provide a high level of security - without them, the cipher would be linear, and easily susceptible to be broken.

Permutation. Finally, the 32 output bits ($8 \cdot 4$) from the S-boxes are mixed with a next fixed permutation, called P-box. This is designed in such a way that after expansion, each S-box's output bits go across 6 different S-boxes in the next round of the algorithm.

The key schedule for decryption procedure is similar. The subkeys are in the reversed order than in the encryption procedure.

As we can see from boolean encoding point of view, all of the basic operations in DES can be represented by some equivalences (i.e. permutations, rotations, expansions). On the other hand, S-box can be described by proper implication. In the next section will be described the full encoding process.

3   Boolean Encoding for Cryptanalysis

After presenting what FN and DES ciphers are, now we can show, proposed in [9], method of direct, boolean encoding of the two benchmark ciphers. Firstly we show encoding FN. Then, we present the encoding of the main steps of DES, particularly permutations and S-box computations.

In this paper we consider the Feistel Network with a 64-bit block of a plaintext and a 32-bit key. Let the propositional variables representing a plaintext, a key, and the ciphertext be $q_1, \ldots, q_{64}, l_1, \ldots, l_{32}$ and $a_1, \ldots, a_{64}$ respectively. Observe that following the Feistel algorithm for the first half of ciphertext we have:

$$\bigwedge_{i=1}^{32} (a_i \Leftrightarrow q_{i+32}).$$

As a simple instantiation of function F (occurred in FN) we use function XOR, denoted by $\oplus$. (Clearly this is a simplest possible example of function F, but at this point we only show our encoding method for the FN structure.) It is easy to observe that for the second half of ciphertext we have:

$$\bigwedge_{i=33}^{64} (a_i \Leftrightarrow (q_i \oplus l_{i-32} \oplus q_{i+32}).$$

Hence, the encoding formula for one round of FN is this:

$$\Psi_{FN}^1 : \bigwedge_{i=1}^{32} (a_i \Leftrightarrow q_{i+32}) \ \wedge \ \bigwedge_{i=33}^{64} (a_i \Leftrightarrow (q_i \oplus l_{i-32} \oplus q_{i+32}).$$

Let us now consider the case of j rounds of FN. Let $(q_1^1, \ldots, q_{64}^1)$, $(l_1, \ldots, l_{32})$ are a plaintext and a key vectors of variables, respectively. By $(q_1^k, \ldots, q_{64}^k)$ and $(a_1^i, \ldots, a_{64}^i)$ we describe vectors of variables representing input of k-th round for $k = 2, \ldots, j$ and output of i-th round for $i = 1, \ldots, t-1$. We denote by $(a_1^j, \ldots, a_{64}^j)$ the variables of a cipher vector after j-th round, too.

The formula which encodes the whole j-th round of a Feistel Network is as follows:

$$\Psi_{FN}^j : \bigwedge_{i=1}^{32} \bigwedge_{s=1}^{j} (a_i^s \Leftrightarrow q_{i+32}^s) \ \wedge \ \bigwedge_{i=1}^{32} \bigwedge_{s=1}^{j} [a_{i+32}^s \Leftrightarrow (q_i^s \oplus q_{i+32}^s \oplus l_i)]$$

$$\wedge \bigwedge_{i=1}^{64} \bigwedge_{s=1}^{j-1} (q_i^{s+1} \Leftrightarrow a_i^s).$$

Observe that the last part of the formula states that the outputs from s-th rounds are the inputs of the $(s+1)$-th.

As we can see, the formula obtained is a conjunction of ordinary, or rather simple, equivalences. It is important from the translating into CNF point of view. The second advantage of this description is that we can automatically generate the formula for many investigated rounds.

In the case of DES, we show an encoding procedure in some detail of the most important parts only for the cipher. An advantage of our method is a direct encoding of each bit in the process of a DES execution, with no redundancy from the size of the encoding formula point of view. For describing each bit in this procedure we use one propositional variable. We encode directly all parts of DES.

The whole structure of the encoding formula is similar to FN. We can consider DES as a sequence of permutations, expansions, reductions, XORs, S-box computations and key bits rotations. Each of these operations can be encoded as a conjunction of propositional equivalences or implications.

For example, consider σ - the initial permutation function of DES. Let $(q_1, \ldots, q_{64})$ be a sequence of variables representing the plaintext bits. Denote by $(p_1, \ldots, p_{64})$ a sequence of variables representing the block bits after permutation σ. Easy to observe that we can encode P as the following formula:

$$\bigwedge_{i=1}^{64} (q_i \Leftrightarrow p_{\sigma(i)}).$$

In a similar way, we can encode all the permutations, expansions, reductions, and rotations of DES.

In the case of S-box encoding, observe that S-box is the matrix with four rows and sixteen columns where in each row we have one different permutation of numbers belonging to Z_{16}. These numbers are denoted in binary form as four-tuples of bits. Following the DES algorithm we can consider each S-box as a function of type $S_{box} : \{0,1\}^6 \rightarrow \{0,1\}^4$.

For simplicity let us denote a vector $(x_1, \ldots, x_6)$ by $\overline{x}$ and by $S_{box}^k(\overline{x})$ the k-th coordinate of value $S_{box}(\overline{x})$, for $k = 1, 2, 3, 4$.

We can encode each S-box as the following boolean formula:

$$\bigwedge_{\overline{x} \in \{0,1\}^6} (\bigwedge_{i=1}^{6} (\neg)^{1-x_i} q_i \Rightarrow \bigwedge_{j=1}^{4} (\neg)^{1-S_{box}^j(\overline{x})} p_j),$$

where $(q_1, \ldots, q_6)$ is the input vector of S-box and $(p_1, \ldots, p_4)$ the output one. Additionally, by $(\neg)^0 q$ and $(\neg)^1 q$ we mean q and $\neg q$, respectively. Using this we can encode each of the S-boxes used in all considered rounds of DES as 256 simple implication. This number is equal to the size of S-box matrix. Due to the strongly irregular and random character of S-boxes, we are sure that this is the simplest method of boolean encoding of the S-boxes.

Having these procedures, we can encode any given number of rounds of DES algorithm as a boolean formula. Our encoding gave formulas shorter than those of Massacci [14]. We got 3 times less variables and twice less clauses. Observe that from the computational point of view, it is important to decrease as far as possible the number of variables and connectives used in the formula. In the next section we briefly describe a method of decreasing the parameters of the formula obtained, preserving its equivalences.

The cryptanalysis procedure we propose in this paper is the following. Firstly we encode a single round of the cipher considered as a boolean propositional formula. Then the formula encoding a desired number of iteration rounds (or the whole cipher) is automatically generated. Next we convert the formula obtained into CNF. Here we randomly choose a plaintext and the key vector as a 0, 1-valuation of the variables representing them in the formula. Next the chosen valuation into the formula is inserted. Now we calculate the corresponding ciphertext using an appropriate key and insert it into the formula. Finally we run SAT-solver with the plaintext and its ciphertext bits inserted, to find a satisfying valuation of the key variables.

4 Experimental Results

To our investigations we use formulas that encode a specific number of rounds of the DES algorithm in a three versions. In the first approach for each stage of the algorithm new variables are created. This encoding method causes significant overlapping of unnecessary variables and clauses. Such encoding will be referred later as **Base Form**.

The second version of the encoding formula will be labeled as **Optim 1**. In this case, the specified number of rounds of the algorithm is encoded exactly the same as in Base Form, but before converting it to the form of CNF and DIMACS, the redundant variables and clauses are reduced. All unnecessary subforms in the formula of literal equivalence are removed from base formula, using the well known logical properties: $(\alpha \Leftrightarrow \beta \wedge \beta \Leftrightarrow \gamma) \rightarrow (\alpha \Leftrightarrow \gamma)$.

Then the number of variables is reduced. Removal of equivalence results in the fact that the some variables do not appear in the formula. In this case, the indexes of the remaining variables should be changed in such a way that they are successive natural numbers.

The third version is called **Optim 2**. Here, the reduction takes place after adding in conjunction the variables valuation represents bits of the plain text and ciphertext. If the variable has a positive value and in the clause is not negated, then the whole clause is a tautology, and therefore, regardless of the valuation, it will not cause conflicts so can be removed from the encoding formula. The same applies to variables with a negative value and being negated in clauses.

In cases when the variable is true and is negated in the clause and when it has false value and appears in the clause, but it is not negated, this variable is removed from the clause.

The Table 1 shows the number of variables and clauses depending on the round for each of the three forms of the encoding formula.

All our experiments were carried out in the environment Kali Linux, version 2018.2. The physical machine was equipped with 4 core (8 logical CPU) processor

Table 1. Variables and clauses in encode formulas.

Rounds	Base form		Optim 1		Optim 2	
	Var	Cl	Var	Cl	Var	Cl
2	968	6112	408	4992	408	2496
4	1688	11840	632	9728	632	9216
6	2408	17568	856	14464	856	13952
8	3128	23296	1080	19200	1080	18685
10	3848	29024	1304	23936	1304	23421
12	4568	34752	1528	28672	1528	28157
14	5288	40480	1752	33408	1752	32893
16	6008	46208	1976	38144	1976	37632

from the Intel Haswell family - Intel Core i7-4770K frequency 3.4–3.9 GHz with 8MB SmartCache.

For our work we decided to check several SAT solvers. We used recognized and popular solutions (like MiniSAT), SAT solvers used by us in earlier works (Clasp) as well as the best programs taking part in SAT Competitions. The solutions have been tested using a problem which complexity was similar to Base Form of the 4 round of DES. The results are presented in Table 2.

Table 2. Reference problem results for sequential SAT solvers.

SAT solver	Time [s.]	SAT solver	Time [s.]	SAT solver	Time [s.]
SAT4J 2.3.4	630	SPLATZ-078	396	RSAT 2.02	366
MiniSAT 2.2	221	Glucose 4.0	70.8	Glucose-Syrup	48.8
CaDiCal-06w	28.9	PicoSAT 965	24.8	LingeLing	23.1
CryptoMiniSAT	12.8	pLingeLing	9.10	glu_vc	6.72

The obtained results show that the best from sequential solvers for our problem were: glu_vc, CryptoMiniSAT, LingeLing, PicoSAT oraz CaDiCal and this solvers will be used in further experiments.

The popular SAT solver Glucose, obtained a comparable result with the bests, but for a given problem it turned out to be slightly inferior and will not be used in the experiments. The remaining solvers were significantly worse. It is worth noting that the MiniSAT and RSAT solutions that achieved high positions in the SAT Competition 10 years ago [SAT 2007 Competition], for this problem obtained results many times worse in comparison to the bests programs.

The SAT Competition is a competitive event for solvers of the SAT problem. It is organized yearly at the International Conference on Theory and Applications of Satisfiability Testing. The goal of this is to motivate implementors to present their work to a broader audience and to compare it with that of others [6].

Here we present basic information about chosen SAT solvers. It is important to note that they were awarded past few years in mentioned competition.

glu_vc is a SAT solver submitted to the hack track of the SAT Competition 2017. It updates Glucose 3.0 in the following aspects: phase selection, learnt clause database reduction and decision variable selection [6].

CryptoMiniSAT was presented in 2009 [18]. Authors extended the solver's input language to support the XOR operation, which with few others modifications allows to optimize solver for cryptographic problems.

Lingeling is a SAT solver created on Johannes Kepler University (JKU) in Linz [3]. It use some techniques to save space by reduction of some literals [4]. First time it was presented on SAT Competition in 2010. Through years it has been developing and latest version was presented on SAT competition in 2013.

PicoSat was also created on JKU [1]. It has many similar solutions as MiniSAT 1.14, which is a well-known SAT solver. First time shown in 2007. Low-level optimization saves memory and efficiently increase this SAT solver.

CaDiCal, created on JKU. It's a solver originally developed to simplify the design and internal data structures [11]. First time it was presented in 2017 on SAT Competition and it's the latest created SAT solver from JKU considered in this paper.

Experiment 1. The first experiment rely on investigation the time of solving the SAT based cryptanalysis of a given number of DES algorithm rounds in three encoding variants: Base Form, Optim1, Optim2. For this we use methodology introduced above.

Table 3. Sequential SAT solvers results.

Rounds	Problem	glu_vc	CryptoMiniSAT	LingeLing	PicoSAT	CaDiCaL
3	Base Form	0.71	0.1	0	0.1	0.08
3	Optim 1	0.398	0.05	0.2	0.1	0.07
3	Optim 2	0.038	0.08	0.2	0.1	0.06
4	Base Form	29.4	81	50.4	163	36.7
4	Optim 1	20.6	42.2	23.8	24	29.9
4	Optim 2	36.4	59.4	23.9	131	40.6

It can be seen that for 3 rounds all SAT solvers returned a solution in negligible time. A significant increase in times occurred in case of rounds 4 and more (Tables 3 and 4).

It is worth analyzing the results for the 4th round. For all sequential solvers there was a reduction in the time of solution of the formula in the Optim 1, in relation to Base Form and increased problem solving time in the form of Optim 2,

Table 4. Parallel SAT solvers results.

Round	Problem	pLingeLing	Glucose-Syrup
3	Base Form	0.1	0.0454
3	Optim 1	0.1	0.0394
3	Optim 2	0.1	0.0351
4	Base Form	11.3	50.1
4	Optim 1	22.4	44.8
4	Optim 2	23.5	17.4

in relation to Optim 1. For glu_vc and CaDiCaL there was a slight deterioration of results for Optim 2, compared to Base Form. In the case of parallel solvers, the results are different. For Glucose-Syrup there has been a significant improvement for the Optim 1 and Optim 2 probes. For pLingeLing, the Optim 1 and Optim 2 scores were worse than Base Form. Attempts to break the fifth round for all solvers failed.

Experiment 2. In the previous experiment, attempts to solve the problem for the fifth round of the algorithm were unsuccessful. Therefore, variables representing the valuation of the key were added in conjunction to the encode formulas. The results of the experiment are presented in the Tables 5 and 6 below.

glu_vc dealt best with the given problem. We managed to solve the problem for the fifth round of the algorithm with the value of 4 key bits. The remaining sequential solvers found the matching valuation with the given 7 key bits for

Table 5. Results for sequential SAT-solvers.

Added key bits	glu-vc	CryptoMiniSAT	LingeLing	PicoSAT	CaDiCaL
15	0.462	7.91	5.4	2.0	0.89
14	3.14	10.46	6.7	122.3	1.10
13	7.03	18.8	17.0	0.4	2.39
12	10.1	9.69	24.8	19.8	8.24
11	14.6	49.3	38.4	29.7	8.72
10	47	82.4	107	92.3	72.3
9	118	146	47.1	202	34.1
8	359	153	332	485	347
7	213	90.7	97.7	568	445
6	620	860	-	-	5000
5	1450	2430	-	-	-
4	10700	-	-	-	-
3	-	-	-	-	-

Table 6. Results for parallel SAT-solvers.

Added key	pLingeLing	Glucose-Syrup	Added key	pLingeLing	Glucose-Syrup
15	1.4	1.40	9	12.6	272
14	1.4	0.206	8	24.6	107
13	1.8	5.66	7	44.7	1490
12	1.7	8.37	6	25.8	495
11	29.7	36.5	5	10000	-
10	20.0	8.56	4	-	-

LingeLing and PicoSAT. CaDiCaL found a solution with values of 6 key bits, and CryptoMiniSAT with five.

Experiment 3. Here we study of the SBOX influence on the complexity of the SAT problem. In this experiment, we investigate the time necessary to resolve SAT problem for 4th round of DES with several variants of SBOXes. In the first case we examined formula with the standard DES SBOXes (**Normal SBOX**). In the second one, standard SBOXes were replaced with identical ones (**Same SBOX**). The third variant algorithm is equipped with newly constructed linear SBOXes (**Linear SBOX**).

In our work to simplify analysis we replaced original S-boxes by permutations that can be represented by linear functions, such that $f : \{0, \ldots, 15\} \rightarrow \{0, \ldots, 15\}$, and $f(x) = (a_1 x + a_0) mod 16$, where $a_i = 0, \ldots, 15$ for $i = 0, 1$ and $mod 16$ means *modulo 16* (it takes the remainder after division by 16).

In the fourth case (**No SBOX**), SBOXes were removed. It caused a significant reduction in the complexity of all three forms of the coding formulas. In case of Base Form amount of variables did not changed due to redundant coding method, but there was a huge difference in clauses number, reduction from 11840 to 3904. Equally large decreases in the number of clauses took place in the case of Optim 1 and Optim 2. The number of clauses is 1536 and 1024, respectively. In both cases, the number of variables was 504.

The linear SBOXes resulted in a significant reduction in the time of solving the problem for all of tested SAT solvers. After removing SBOXes, the duration of solving is negligible. All our results in this case are presented in the Tables 7 and 8 below.

From SAT point of view we have expected, that solving times in the case of linear SBOXes should be rather similar to the original ones because sizes of formulas used are very close. Obtained results shows that some SAT-solvers can work faster with some linear dependencies with values of some literals. They must have some heuristics that work in this case faster. It is interesting for next research because some fragments of SBoxes can be described by linear functions.

Table 7. Results for sequential SAT-solvers.

SBOX Type	Problem	glu_vc	CryptoMiniSAT	LingeLing	PicoSAT	CaDiCaL
Normal SBOX	Base Form	29.4	82.9	50.4	163	36.7
Normal SBOX	Optim 1	20.6	42.2	23.8	24	29.9
Normal SBOX	Optim 2	36.4	59.3	23.7	131	40.6
Same SBOX	Base Form	53.7	86.2	109	509	26.3
Same SBOX	Optim 1	16.5	25.2	17.2	55.2	31.9
Same SBOX	Optim 2	16.7	41.1	17.1	312	24.8
Linear SBOX	Base Form	3.83	15.3	10.3	10.7	12.1
Linear SBOX	Optim 1	10.9	17	12.5	9.4	3.59
Linear SBOX	Optim 2	9.26	12.8	12.6	22.1	7.96
No SBOX	Base Form	0.00486	0.02	0	0	0.01
No SBOX	Optim 1	0.00209	0.03	0	0	0.01
No SBOX	Optim 2	0.0031	0.02	0	0	0.01

Table 8. Results for parallel SAT-solvers.

SBOX Type	Problem	pLingeLing	Glucose-Syrup
Normal SBOX	Base Form	52.7	22.5
Normal SBOX	Optim 1	23.2	24.5
Normal SBOX	Optim 2	44.1	52
Same SBOX	Base Form	37.9	24.4
Same SBOX	Optim 1	35.5	18.7
Same SBOX	Optim 2	38.4	49.5
Linear SBOX	Base Form	6.6	4.06
Linear SBOX	Optim 1	7	6.15
Linear SBOX	Optim 2	17.8	3.53
No SBOX	Base Form	0.1	0.0145
No SBOX	Optim 1	0	0.0111
No SBOX	Optim 2	0	0.00414

5   Conclusion and Future Directions

In this paper we have presented our investigations about SAT-based, direct cryptanalysis of symmetric ciphers. We compare results obtained from several well known and efficient SAT-solvers. Our main goal was not to create the fastest method of cryptanalysis in this case. Rather we have checked how new solvers work and how they solve some problems with modifications of DES cipher.

During our experiments we have showed that in this case the best solver is glu_vc. One of future research directions is trying to modify the solvers' code to solve SAT cryptanalysis problem for a given cipher.

Also interesting seems to be observation that DES with linearly constructed SBOXes is much easier to SAT cryptanalysis than original one. Probably in solvers' algorithms are special heuristics that can solve big formulas with linear dependencies between values of some variables.

In our next research we will try to apply our experience for SAT cryptanalysis of several others ciphers like Blowfish, Twofish, and AES. We will also try to apply this cryptanalysis technique for checking security properties of some hash functions.

References

1. Biere, A.: PicoSAT essentials. J. Satisf. Boolean Model. Comput. (JSAT) **4**, 75 – 97 (2008). Delft University
2. Biere, A., Heule, M., van Maaren, H., Walsh, T. (eds.) Handbook of Satisfiability. Frontiers in Artificial Intelligence and Applications, vol. 185. IOS Press, Amsterdam (2009)
3. Biere, A.: Lingeling, Plingeling, Picosat and Precosat at SAT Race 2010. Technical Report FMV Reports Series 10/1, Institute for Formal Models and Verification, Johannes Kepler University, Linz, Austria (2010)
4. Biere, A.: Lingeling, Plingeling and Treengeling entering the SAT competition 2013. In: Balint, A., Belov, A., Heule, M., Jarvisalo, M. (eds.) Proceedings of SAT Competition 2013, vol. B-2013-1, Department of Computer Science Series of Publications B, pp. 51–52, University of Helsinki (2013)
5. Biham, E., Shamir, A.: Differential cryptanalysis of DES-like cryptosystems. J. Cryptol. **4**(1), 3–72 (1991)
6. Chen, J.: Proceedings of SAT Competition 2017: Solver and Benchmark Descriptions, vol. B-2017-1, Department of Computer Science Series of Publications B, University of Helsinki (2017)
7. Davis, M., Putnam, H.: A computing procedure for quantification theory. J. ACM **7**(3), 201–215 (1960)
8. Davis, M., Logemann, G., Loveland, D.W.: A machine program for theorem-proving. Commun. ACM **5**(7), 394–397 (1962)
9. Dudek, P., Kurkowski, M., Srebrny, M.: Towards parallel direct SAT-based cryptanalysis. In: PPAM 2011 Proceedings. LNCS, vol. 7203, pp. 266-275. Springer (2012)
10. Dwivedi, A.D., et al.: SAT-based cryptanalysis of authenticated ciphers from the CAESAR Competition. In: Proceedings of the 14th International Joint Conference on e-Business and Telecommunications (ICETE 2017). SECRYPT, vol. 4, pp. 237–246 (2017)

11. https://github.com/arminbiere/cadical
12. Lafitte, F., Lerman, L., Markowitch, O., van Heule, D.: SAT-based cryptanalysis of ACORN, IACR Cryptology ePrint Archive, vol. 2016, p. 521 (2016)
13. Lafitte, F., Nakahara Jr., J., van Heule, D.: Applications of SAT solvers in cryptanalysis: finding weak keys and preimages. JSAT **9**, 1–25 (2014)
14. Massacci, F.: Using Walk-SAT and Rel-SAT for cryptographic key search. In: Dean, T. (ed.) IJCAI, pp. 290–295. Morgan Kaufmann (1999)
15. Massacci, F., Marraro, L.: Logical cryptanalysis as a SAT problem. J. Autom. Reason. **24**(165), 165–203 (2000)
16. Matsui, M.: The first experimental cryptanalysis of the data encryption standard. In: Desmedt, Y. (ed.) CRYPTO. LNCS, vol. 839, pp. 1–11. Springer (1994)
17. Menezes, A., van Oorschot, P.C., Vanstone, S.A.: Handbook of Applied Cryptography. CRC Press, Boca Raton (1996)
18. Soos, M., Nohl, K., Castelluccia, C.: Extending SAT solvers to cryptographic problems. In: Proceedings of 12th International Conference on Theory and Applications of Satisfiability Testing - SAT 2009, Swansea, UK, pp. 244 – 257 (2009)

Secure Generators of q-Valued Pseudo-random Sequences on Arithmetic Polynomials

Oleg Finko[1(✉)], Sergey Dichenko[1], and Dmitry Samoylenko[2]

[1] Institute of Computer Systems and Information Security of Kuban State
Technological University,
Krasnodar Moskovskaya St., 2, 350072, Russia
ofinko@yandex.ru
[2] Mozhaiskii Military Space Academy,
Zhdanovskaya St., 13, St. Petersburg 197198, Russia
19sam@mail.ru

Abstract. A technique for controlling errors in the functioning of nodes for the formation of q-valued pseudo-random sequences (PRS) operating under both random errors and errors generated through intentional attack by an attacker is provided, in which systems of characteristic equations are realized by arithmetic polynomials that allow the calculation process to be parallelized and, in turn, allow the use of redundant modular codes device.

Keywords: q-valued pseudo-random sequences
Secure generators of q-valued pseudo-random sequences
Primitive polynomials · Galois fields
Linear recurrent shift registers · Modular arithmetic
Parallel logical calculations by arithmetic polynomials
Error control of operation · Redundant modular codes

1 Introduction

In the theory and practice of cryptographic information protection, one of the key tasks is the formation of PRS which width, length and characteristics meet modern requirements [1]. Many existing solutions in this area aim to obtain a binary PRS of maximum memory length with acceptable statistical characteristics [2]. However, recently it is considered that one of the further directions in the development of means of information security (MIS) is the use of multi-valued functions of the algebra of logic (MFAL), in particular, using the PRS over the Galois field GF(q) ($q > 2$), which have a wider spectrum of unique properties comparing to binary PRS [3].

The nodes of the formation of the q-valued PRS, like the others, are prone to failures and malfunction, which leads to the occurrence of errors in their functioning. In addition to random errors occurrence in the generation of PRS related

J. Pejaś et al. (Eds.): ACS 2018, AISC 889, pp. 295–306, 2019.
https://doi.org/10.1007/978-3-030-03314-9_26

to "unintentional" failures and malfunctions caused by various causes: aging of the element base, environmental influences, severe operating conditions, etc. (reasons typical for reliability theory), there are deliberate actions of an attacker aimed to create massive failures of electronic components of the formation nodes of PRS due to the hardware errors generation (one of the types of information security threats) [4].

Many methods have been developed to provide the necessary level of reliability of the digital devices functioning; the most common are backup methods and methods of noise-immune coding. However, backup methods do not provide the necessary levels of operation reliability with limitations on hardware costs, and methods of noise-immune coding are not fully adapted to the specifics of the construction and operation of MIS, in particular, generators of q-valued PRS.

The work [5] offers a solution that overcomes the complexity of using code control for the nodes of the binary PRS generation, based on the "arithmetic" of logical count and the application of the redundant modular code device, which provides the necessary level of security for their functioning. However, the solution obtained is limited to exclusive applicability in the formation of binary PRS. At the same time, work [6], is known where by means of "arithmetic" of logical count the task of parallelizing the nodes of forming of binary PRS is solved, but without monitoring their functioning. As a result, it becomes necessary to generalize the solutions obtained to ensure the security of the functioning of the nodes of q-valued PRS formation.

2   General Principles of Building Generators of q-Valued PRS

The most common and tested methods for PRS are algorithms and devices of PRS generation — linear recurrent shift registers (q-LFSR) with feedback — based on the use of recurrent logical expressions [2].

The construction of the q-LFSR over the field $GF(q)$ is carried out from the given generating polynomial:

$$K(x) = \sum_{i=0}^{m} k_{m-i} x^{m-i}, \tag{1}$$

where m — is the polynomial degree $K(x)$, $m \in N$; $k_i \in GF(q)$, $k_m = 1$, $k_0 \neq 0$.

Thus, the q-LFSR element is formed in accordance with the following characteristic equation [7]:

$$a_{p+m} = -k_{m-1} a_{p+m-1} - k_{m-2} a_{p+m-2} - \ldots - k_1 a_{p+1} - k_0 a_p. \tag{2}$$

The Eq. (2) is a recursion which describes an infinite q-valued PRS with period $q^m - 1$ (with nonzero initial state, as well as under condition that the polynomial (1) is primitive over the field $GF(q)$), each nonzero state appears once per period.

A homogeneous recurrent Eq. (2) can be presented in the following form:

$$a_{p+m} = k_{m-1}a_{p+m-1} \oplus k_{m-2}a_{p+m-2} \oplus \ldots \oplus k_1 a_{p+1} \oplus k_0 a_p$$

or

$$a_{p+m} = \bigoplus_{i=1}^{m} k_{i-1} a_{p+i-1}, \tag{3}$$

where $\oplus$ — is the symbol of addition on module q.

The q-LFSR corresponding to the polynomial (3) is shown in Fig. 1, whose cells contain field GF(q) elements: $a_p, \ldots, a_{p+m-1}$.

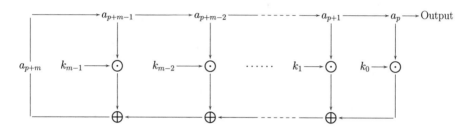

Fig. 1. Structural diagram of the operation of the sequential q-LFSR in accordance with formula (3) ($\oplus$ and $\odot$ — according to transaction of addition and multiplication of the mod q)

3   Analysis of Possible Modifications q-Valued PRS Caused by the Error Occurred

It is known that the consequences of accidental errors that occur during the PRS generation associated with "unintentional" failures, as well as the consequences of intentional actions by an attacker based on the use of thermal, high-frequency, ionizing or other external influences in order to obtain mass malfunctions of the equipment by initiation of calculation errors, lead to similar types of PRS modification.

Figure 2 shows main types of modification of PRS over the GF(q) field. The attacker's actions based on error generation are highly effective for most of the known and currently used algorithms for generating q-valued PRS [8–10]. It is known [11] that the probability of error generation is proportional to the irradiation time of the respective registers in a favorable state for the error occurrence and to the number of bits within which an error is expected. This type of impact has not been sufficiently studied and therefore represents a threat to the information security of modern and promising MIS functioning.

One of the ways to solve this problem is to develop a technique for improving the safety of the operation of the MIS nodes most susceptible to these effects, in particular, the nodes of q-valued PRS formation.

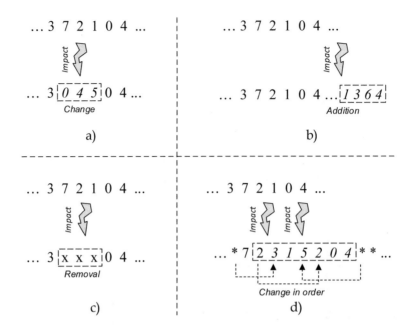

Fig. 2. The main types of PRS modification: (a) change in the elements of the PRS, (b) addition of new PRS elements, (c) removal of the CAP elements, (d) change in the order of the PRS elements

4   Analysis of Ways to Control the Generation of q-Valued PRS

Currently, the necessary level of security for the functioning of the nodes for the q-valued PRS formation is achieved both through the use of redundant equipment (structural backup) and temporary redundancy due to various calculations repetition.

In the field of digital circuit design solutions based on the use of block redundant coding methods are known. To apply these methods to q-valued PRS generators it is necessary to solve the problem of parallelizing the calculation process of the q-valued PRS.

The solution of the problem is based on the use of classical parallel recursion calculation algorithms [12], for which the characteristic Eq. (3) corresponding to the generating polynomial (2) can be represented as a system of characteristic equations:

$$
\begin{cases}
a_{t,\,m-1} = \displaystyle\bigoplus_{i=1}^{m} k_{i-1}^{(m-1)} a_{t-1,\,p+i-1}, \\[2mm]
a_{t,\,m-2} = \displaystyle\bigoplus_{i=1}^{m} k_{i-1}^{(m-2)} a_{t-1,\,p+i-1}, \\[2mm]
\cdots\cdots\cdots\cdots\cdots\cdots\cdots\cdots \\[2mm]
a_{t,\,1} = \displaystyle\bigoplus_{i=1}^{m} k_{i-1}^{(1)} a_{t-1,\,p+i-1}, \\[2mm]
a_{t,\,0} = \displaystyle\bigoplus_{i=1}^{m} k_{i-1}^{(0)} a_{t-1,\,p+i-1},
\end{cases}
\tag{4}
$$

where $k_{i-1}^{(j)} \in \mathrm{GF}(q)$; $j = 0, 1, \ldots, m-2, m-1$.

The system (4) forms an information matrix:

$$
\mathbf{G_{Inf}} =
\begin{Vmatrix}
k_0^{(m-1)} & k_1^{(m-1)} & \cdots & k_{m-2}^{(m-1)} & k_{m-1}^{(m-1)} \\
k_0^{(m-2)} & k_1^{(m-2)} & \cdots & k_{m-2}^{(m-2)} & k_{m-1}^{(m-2)} \\
\vdots & \vdots & \ddots & \vdots & \vdots \\
k_0^{(1)} & k_1^{(1)} & \cdots & k_{m-2}^{(1)} & k_{m-1}^{(1)} \\
k_0^{(0)} & k_1^{(0)} & \cdots & k_{m-2}^{(0)} & k_{m-1}^{(0)}
\end{Vmatrix}.
$$

Similar result can be obtained in another convenient way [1]:

$$
\mathbf{G_{Inf}} =
\begin{Vmatrix}
k_{m-1} & k_{m-2} & \cdots & k_1 & k_0 \\
1 & 0 & \cdots & 0 & 0 \\
0 & 1 & \ddots & 0 & 0 \\
0 & 0 & \cdots & 0 & 0 \\
0 & 0 & \cdots & 1 & 0
\end{Vmatrix}^{m},
$$

where the elements raised to the power m are of a matrix which is created according to the known rules of linear algebra for the calculation of the next q-valued element of the PRS a_{p+m}:

$$
\begin{Vmatrix}
a_{p+m} \\
a_{p+m-1} \\
\vdots \\
a_{p+2} \\
a_{p+1}
\end{Vmatrix}
=
\begin{Vmatrix}
a_{p+m-1} \\
a_{p+m-2} \\
\vdots \\
a_{p+1} \\
a_{p}
\end{Vmatrix}
\cdot
\begin{Vmatrix}
k_{m-1} & \cdots & k_0 \\
1 & \cdots & 0 \\
0 & \cdots & 0 \\
0 & \cdots & 0 \\
0 & \cdots & 0
\end{Vmatrix}_q,
$$

where $|\cdot|_q$ — is the smallest nonnegative deduction of the number "·" on module q.

The technique for raising a matrix to the power can be performed with help of symbolic calculations in any computer algebra system with the subsequent simplification (in accordance with the axioms of the algebra and logic) of the

elements of the resulting matrix of the form $Y k_j^b = k_j$ according to the rules:
1) $k_j^b = k_j$; 2) $Y = 0$, for even Y and $Y = 1$, for odd Y. Thus, we obtain the t-block of PRS:

$$\mathbf{A}_t = \left| \mathbf{G_{Inf}} \cdot \mathbf{A}_{t-1} \right|_q,$$

where

$$\mathbf{A}_t = \left[a_{t,\,p+m-1}\; a_{t,\,p+m-2}\; \cdots\; a_{t,\,1}\; a_{t,\,0} \right]^{\mathsf{T}},$$

$$\mathbf{A}_{t-1} = \left[a_{t-1,\,p+m-1}\; a_{t-1,\,p+m-2}\; \cdots\; a_{t-1,\,1}\; a_{t-1,\,0} \right]^{\mathsf{T}}.$$

To create conditions for the use of a separable linear redundant code, we obtain a generating matrix $\mathbf{G_{Gen}}$, consisting of the information and verification matrixes by adding in the (4) test expressions:

$$\begin{cases} a_{t,\,p+m-1} = \displaystyle\bigoplus_{i=1}^{m} k_{i-1}^{(m-1)} a_{t-1,\,p+i-1}, \\ \cdots\cdots\cdots\cdots\cdots\cdots\cdots \\ a_{t,\,0} = \displaystyle\bigoplus_{i=1}^{m} k_{i-1}^{(0)} a_{t-1,\,p+i-1}, \\ a_{t,\,p+r-1}^{*} = \displaystyle\bigoplus_{i=1}^{r} c_{i-1}^{(r-1)} a_{t-1,\,p+i-1}, \\ \cdots\cdots\cdots\cdots\cdots\cdots\cdots \\ a_{t,\,0}^{*} = \displaystyle\bigoplus_{i=1}^{r} c_{i-1}^{(0)} a_{t-1,\,p+i-1}, \end{cases}$$

where $k_{i-1}^{(j)}, c_{i-1}^{(z)} \in \mathrm{GF}(q)$; $z = 0, \ldots, r-1$; r — is the number of redundant symbols of the applied linear code; $j = 0, \ldots, m-1$.

The forming matrix takes the form:

$$\mathbf{G_{Gen}} = \left\| \begin{matrix} k_0^{(m-1)} & k_1^{(m-1)} & \cdots & k_{m-2}^{(m-1)} & k_{m-1}^{(m-1)} \\ \vdots & \vdots & \ddots & \vdots & \vdots \\ k_0^{(0)} & k_1^{(0)} & \cdots & k_{m-2}^{(0)} & k_{m-1}^{(0)} \\ c_0^{(r-1)} & c_1^{(r-1)} & \cdots & c_{r-2}^{(r-1)} & c_{r-1}^{(r-1)} \\ \vdots & \vdots & \ddots & \vdots & \vdots \\ c_0^{(0)} & c_1^{(0)} & \cdots & c_{r-2}^{(0)} & c_{r-1}^{(0)} \end{matrix} \right\|.$$

Then the t-block of the q-valued PRS with test digits (linear code block)

$$\mathbf{A}_t^{*} = \left[a_{t,\,p+m-1}\; \cdots\; a_{t,\,0}\; a_{t,\,p+r-1}^{*}\; \cdots\; a_{t,\,0}^{*} \right]^{\mathsf{T}}$$

is calculated as:

$$\mathbf{A}_t^{*} = \left| \mathbf{G_{Gen}} \cdot \mathbf{A}_{t-1} \right|_q.$$

The anti-jamming decoding procedure is performed using known rules [13].

The use of linear redundant codes and "hot" backup methods is not the only option for realizing functional diagnostics and increasing the fault tolerance of digital devices. Important advantages for these purposes are found in arithmetic redundant codes, in particular, the so-called AN-codes and codes of modular arithmetic (MA). However, arithmetic redundant codes are not applicable to logical data types. In logical calculations, their structure collapses, which leads to the impossibility of monitoring errors in logical calculations.

The use of arithmetic redundant codes to control logical data types must be ensured by the introduction of additional procedures related to the "arithmetic" of the logical count.

5 The Procedure for Parallelizing the Generation of q-Valued PRS by Means of Arithmetic Polynomials

Parallelizing the "calculation" processes of complex systems or minimizing the number of operations involving the use of all resources makes it possible to achieve any utmost characteristic or quality index, which in turn is necessary in most practically important cases. In turn, the new direction formed at the end of the last century – parallel-logical calculations through arithmetic (numerical) polynomials [14], also allowed to provide "useful" structural properties. It became possible to use arithmetic redundant codes to control logical data types and increase the fault tolerance of implementing devices by representing arithmetic expressions [14] as logical operations, in particular, by linear numerical polynomials (LNP) and their modular forms [15].

In [5] an algorithm for parallelizing the generation of binary PRS is presented based on the representation of systems of generating recurring logical formulas by means of LNP offered by V. D. Malyugin, which allowed using the redundant modular code device to control the errors of the functioning of the PRS generation nodes and, ensure the required safety of their functioning in the MIS.

To ensure the possibility of applying code control methods to generators of q-valued PRS, it is necessary to solve the problem of parallelizing the process of calculating them, while in [6] in general terms, approach for the synthesis of parallel generators of q-valued PRS on arithmetic polynomials is presented, the essence of which is the following.

Let a_0, a_1, a_2, ..., a_{m-1},... — be the elements of the q-valued PRS satisfying the recurrence Eq. (3). Knowing that random element a_p $(p \geq m)$ of the sequence a_0, a_1, a_2, ..., a_{m-1}, ... is determined by the preceding m elements, let us present the elements a_{p+m}, a_{p+m+1}, ..., a_{p+2m-1} of the section of the q-valued PRS by the length m in the form of a system of characteristic equations:

$$
\begin{cases}
a_{p+m} = \bigoplus_{i=1}^{m} k_{i-1} a_{p+i-1}, \\
a_{p+m+1} = \bigoplus_{i=1}^{m} k_{i-1} a_{p+i}, \\
\cdots\cdots\cdots\cdots\cdots\cdots\cdots\cdots \\
a_{p+2m-1} = \bigoplus_{i=1}^{m} k_{i-1} a_{p+i+m-2},
\end{cases}
\tag{5}
$$

where $[a_{p+m}\ a_{p+m+1}\ \cdots\ a_{p+2m-1}]$ — is the vector of the m-state of the q-valued PRS (or the internal state of the q-LFSR on m-cycle of work).

By analogy with [5] let us express the right-hand sides of the system (5) through the given initial conditions and let us write it as the m MFAL system of m variables:

$$
\begin{cases}
f_1\left(a_p, a_{p+1}, \ldots, a_{p+m-1}\right) = \bigoplus_{i=1}^{m} k_{i-1}^{(0)} a_{p+i-1}, \\
f_2\left(a_p, a_{p+1}, \ldots, a_{p+m-1}\right) = \bigoplus_{i=1}^{m} k_{i-1}^{(1)} a_{p+i-1}, \\
\cdots\cdots\cdots\cdots\cdots\cdots\cdots\cdots\cdots\cdots\cdots\cdots\cdots\cdots \\
f_m\left(a_p, a_{p+1}, \ldots, a_{p+m-1}\right) = \bigoplus_{i=1}^{m} k_{i-1}^{(m-1)} a_{p+i-1},
\end{cases}
\tag{6}
$$

where the coefficients $k_{i-1}^{(j)} \in \{0, 1, \ldots, q-1\}$ $(i = 1, \ldots, m;\ j = 0, \ldots, m-1)$ are formed after expressing the right-hand parts of the system (5) through given initial conditions.

It is known that random MFAL can be represented in the form of an arithmetic polynomial in simple way [16,17]:

$$
L\left(a_p, a_{p+1}, \ldots, a_{p+m-1}\right) = \sum_{i=0}^{q^{m-1}-1} l_i\, a_p^{i_0} a_{p+1}^{i_1}\ \cdots\ a_{p+m-1}^{i_{m-1}},
\tag{7}
$$

where $a_u \in \{0, 1, \ldots, q-1\}$; $u = 0, \ldots, m-1$; l_i — i-coefficient of an arithmetic polynomial; $(i_0\ i_1\ \cdots\ i_{m-1})_q$ — representation of the parameter i in the q-scale of notation:

$$
(i_0\ i_1\ \cdots\ i_{m-1})_q = \sum_{u=0}^{m-1} i_u q^{m-u-1} \quad (i_u \in 0, 1, \ldots, q-1);
$$

$$
a_u^{i_u} = \begin{cases} 1, & i_u = 0, \\ a_u, & i_u \neq 0. \end{cases}
$$

Similar to [16, 17] let us implement the MFAL system (6) by computing some arithmetic polynomial. In order to do this, we associate the MFAL system (6) with a system of arithmetic polynomials of the form (7), we obtain:

$$
\begin{cases}
L_1\left(a_p, a_{p+1}, \ldots, a_{p+m-1}\right) = \displaystyle\sum_{i=0}^{q^{m-1}-1} l_{1,i}\, a_p^{i_0} a_{p+1}^{i_1} \cdots a_{p+m-1}^{i_{m-1}}, \\[4mm]
L_2\left(a_p, a_{p+1}, \ldots, a_{p+m-1}\right) = \displaystyle\sum_{i=0}^{q^{m-1}-1} l_{2,i}\, a_p^{i_0} a_{p+1}^{i_1} \cdots a_{p+m-1}^{i_{m-1}}, \\[4mm]
\cdots\cdots\cdots\cdots\cdots\cdots\cdots\cdots\cdots\cdots\cdots\cdots\cdots\cdots \\[2mm]
L_m\left(a_p, a_{p+1}, \ldots, a_{p+m-1}\right) = \displaystyle\sum_{i=0}^{q^{m-1}-1} l_{m,i}\, a_p^{i_0} a_{p+1}^{i_1} \cdots a_{p+m-1}^{i_{m-1}}.
\end{cases} \tag{8}
$$

Let us multiply the polynomials of the system (8) by weights q^{e-1} $(e = 1, 2, \ldots, m)$:

$$
\begin{cases}
L_1^*\left(a_p, a_{p+1}, \ldots, a_{p+m-1}\right) = q^0 L_1\left(a_p, a_{p+1}, \ldots, a_{p+m-1}\right) \\[2mm]
\quad = \displaystyle\sum_{i=0}^{q^{m-1}-1} l_{1,i}^*\, a_p^{i_0} a_{p+1}^{i_1} \cdots a_{p+m-1}^{i_{m-1}}, \\[4mm]
L_2^*\left(a_p, a_{p+1}, \ldots, a_{p+m-1}\right) = q^1 L_2\left(a_p, a_{p+1}, \ldots, a_{p+m-1}\right) \\[2mm]
\quad = \displaystyle\sum_{i=0}^{q^{m-1}-1} l_{2,i}^*\, a_p^{i_0} a_{p+1}^{i_1} \cdots a_{p+m-1}^{i_{m-1}}, \\[4mm]
\cdots\cdots\cdots\cdots\cdots\cdots\cdots\cdots\cdots\cdots\cdots\cdots\cdots\cdots \\[2mm]
L_m^*\left(a_p, a_{p+1}, \ldots, a_{p+m-1}\right) = q^{m-1} L_m\left(a_p, a_{p+1}, \ldots, a_{p+m-1}\right) \\[2mm]
\quad = \displaystyle\sum_{i=0}^{q^{m-1}-1} l_{m,i}^*\, a_p^{i_0} a_{p+1}^{i_1} \cdots a_{p+m-1}^{i_{m-1}},
\end{cases}
$$

where $l_{e,i}^* = q^{e-1} l_{e,i}$ $(e = 1, 2, \ldots, m; \quad i = 0, \ldots, q^m - 1)$.

Then we get:

$$
L\left(a_p, a_{p+1}, \ldots, a_{p+m-1}\right) = \sum_{i=0}^{q^{m-1}-1} \sum_{e=1}^{d} l_{e,i}^*\, a_p^{i_0} a_{p+1}^{i_1} \cdots a_{p+m-1}^{i_{m-1}} \tag{9}
$$

or using the provisions of [18]:

$$
D\left(a_p, a_{p+1}, \ldots, a_{p+m-1}\right) = \left| \bigoplus_{i=0}^{q^{m-1}-1} v_i\, a_p^{i_0} a_{p+1}^{i_1} \cdots a_{p+m-1}^{i_{m-1}} \right|_{q^m}, \tag{10}
$$

where

$$
v_i = \bigoplus_{e=1}^{m} l_{e,i}^* \quad (i = 0, 1, \ldots, q^{m-1} - 1).
$$

Let us calculate the values of the desired MFAL. For this, the result of the calculation (10) is presented in the q-scale of notation and we apply the camouflage operator $\Xi^w\{D(a_p, a_{p+1}, \ldots, a_{p+m-1})\}$:

$$\Xi^w\{D(a_p, a_{p+1}, \ldots, a_{p+m-1})\} = \left\|\left[\frac{D(a_p, a_{p+1}, \ldots, a_{p+m-1})}{q^w}\right]\right\|_q,$$

where w — is the desired q-digit of the representation $D(a_p, a_{p+1}, \ldots, a_{p+m-1})$.

The presented method, based on the MFAL arithmetic representation, makes it possible to control the q-valued PRS generation errors by means of arithmetic redundant codes.

6   Control of Errors in the Operation of Generators of q-Valued PRS by Redundant MA Codes

In MA, the integral nonnegative coefficient $l^*_{e,i}$ of an arithmetic polynomial (9) is uniquely presented by a set of balances on the base of MA ($s_1, s_2, \ldots, s_\eta < s_{\eta+1} < \ldots < s_\psi$ — simple pairwise):

$$l^*_{e,i} = (\alpha_1, \alpha_2, \ldots, \alpha_\eta, \alpha_{\eta+1}, \ldots, \alpha_\psi)_{\text{MA}}, \qquad (11)$$

where $\alpha_\tau = \left|l^*_{e,i}\right|_{s_\tau}$; $\tau = 1, 2, \ldots, \eta, \ldots, \psi$. The working range $S_\eta = s_1 s_2 \ldots s_\eta$ must satisfy $S_\eta > 2^g$, where $g = \sum\limits_{1 \leq \varepsilon \leq \sigma} \theta_\varepsilon$ — is the number of bits required to represent the result of the calculation (9).

Balances $\alpha_1, \alpha_2, \ldots, \alpha_\eta$ are informational, and $\alpha_{\eta+1}, \ldots, \alpha_\psi$ — are control. In this case, MA is called extended and covers the complete set of states presented by all the ψ balances. This area is the full MA range $[0, S_\psi)$, where $S_\psi = s_1 s_2 \ldots s_\eta s_{\eta+1} \ldots s_\psi$, and consists of the operating range $[0, S_\eta)$, defined by the information bases of the MA, and the range defined by the redundant bases $[S_\eta, S_\psi)$, representing an invalid area for the results of the calculations. This means that operations on numbers $l^*_{e,i}$ are performed in the range $[0, S_\psi)$. Therefore, if the result of the MA operation goes beyond the limits S_η, then the conclusion about the calculation error follows.

Let us study the MA given by the $s_1, s_2, \ldots, s_\eta, \ldots, s_\psi$ bases. Each coefficient $l^*_{e,i}$ of a polynomial (9) is presented in the form (11) and we obtain an MA redundant code, represented by a system of polynomials:

$$
\begin{cases}
U^{(1)} = L^{(1)}\left(a_p,\, a_{p+1},\, \ldots,\, a_{p+m-1}\right) = \sum_{i=0}^{q^{m-1}-1} \sum_{e=1}^{d} l_{e,\,i}^{*(1)}\, a_p^{i_0} a_{p+1}^{i_1} \cdots a_{p+m-1}^{i_{m-1}}, \\[2mm]
U^{(2)} = L^{(2)}\left(a_p,\, a_{p+1},\, \ldots,\, a_{p+m-1}\right) = \sum_{i=0}^{q^{m-1}-1} \sum_{e=1}^{d} l_{e,\,i}^{*(2)}\, a_p^{i_0} a_{p+1}^{i_1} \cdots a_{p+m-1}^{i_{m-1}}, \\[2mm]
\cdots\cdots\cdots\cdots\cdots\cdots\cdots \\[2mm]
U^{(\eta)} = L^{(\eta)}\left(a_p,\, a_{p+1},\, \ldots,\, a_{p+m-1}\right) = \sum_{i=0}^{q^{m-1}-1} \sum_{e=1}^{d} l_{e,\,i}^{*(\eta)}\, a_p^{i_0} a_{p+1}^{i_1} \cdots a_{p+m-1}^{i_{m-1}}, \\[2mm]
\cdots\cdots\cdots\cdots\cdots\cdots\cdots \\[2mm]
U^{(\psi)} = L^{(\psi)}\left(a_p,\, a_{p+1},\, \ldots,\, a_{p+m-1}\right) = \sum_{i=0}^{q^{m-1}-1} \sum_{e=1}^{d} l_{e,\,i}^{*(\psi)}\, a_p^{i_0} a_{p+1}^{i_1} \cdots a_{p+m-1}^{i_{m-1}}.
\end{cases}
\tag{12}
$$

Substituting in (12) the values of the MA balances for the corresponding bases for each coefficient (9) and the values of the variables $a_p, a_{p+1}, \ldots, a_{p+m-1}$, we obtain the values of the polynomials of the system (12), where $U^{(1)}, U^{(2)}, \ldots, U^{(\eta)}, \ldots, U^{(\psi)}$ — are nonnegative integrals. In accordance with the Chinese balances theorem, we solve the system of equations:

$$
\begin{cases}
U^* = \left|U^{(1)}\right|_{s_1}, \\[1mm]
U^* = \left|U^{(2)}\right|_{s_2}, \\[1mm]
\cdots\cdots\cdots\cdots \\[1mm]
U^* = \left|U^{(\eta)}\right|_{s_\eta}, \\[1mm]
\cdots\cdots\cdots\cdots \\[1mm]
U^* = \left|U^{(\psi)}\right|_{s_\psi}.
\end{cases}
\tag{13}
$$

Since $s_1, s_2, \ldots, s_\eta, \ldots, s_\psi$ are simple pairwise, the only solution (13) gives the expression:

$$
U^* = \left| \sum_{d=1}^{\psi} S_{d,\,\psi} \mu_{d,\,\psi} U^{(d)} \right|_{S_\psi},
\tag{14}
$$

where $S_{d,\,\psi} = \dfrac{S_\psi}{s_d}$, $\mu_{d,\,\psi} = \left|S_{d,\,\psi}^{-1}\right|_{s_d}$, $S_\psi = \prod_{d=1}^{\psi} s_d$.

The occurrence of the calculation result (14) in the range (test expression)

$$
0 \le U^* < S_\eta,
$$

means no detectable calculation errors.

Otherwise, the procedure for restoring the reliable functioning of the q-valued PRS generator can be implemented according to known rules [19].

7   Conclusion

A secure parallel generator of q-valued PRS on arithmetic polynomials is presented. The implementation of generators of q-valued PRS using arithmetic polynomials and redundant MA codes makes it possible to obtain a new class of solutions aimed to safely implement logical cryptographic functions. At the same time, both functional monitoring of equipment (in real time, which is essential for MIS) and its fault tolerance is ensured due to the possible reconfiguration of the calculator structure in the process of its degradation. The classical q-LFSR, studied in this work, forms the basis of more complex q-valued PRS generators.

References

1. Klein, A.: Stream Ciphers. Springer (2013). http://www.springer.com
2. Schneier, B.: Applied Cryptography. Wiley, New York (1996)
3. Lidl, R., Niederreiter, H.: Introduction to Finite Fields and Their Applications. Cambridge University Press, Cambridge (1987)
4. Yang, B., Wu, K., Karri, R.: Scan based side channel attack on data encryption standard. Report **2004**(324), 114–116 (2004)
5. Finko, O.A., Dichenko, S.A.: Secure pseudo-random linear binary sequences generators based on arithmetic polynoms. In: Advances in Intelligent Systems and Computing, Soft Computing in Computer and Information Science, vol. 342, pp. 279–290. Springer, Cham (2015)
6. Finko, O.A., Samoylenko, D.V., Dichenko, S.A., Eliseev, N.I.: Parallel generator of q-valued pseudorandom sequences based on arithmetic polynomials. Przeglad Elektrotechniczny **3**, 24–27 (2015)
7. MacWilliams, F., Sloane, N.: Pseudo-random sequences and arrays. Proc. IEEE **64**, 1715–1729 (1976)
8. Canovas, C., Clediere, J.: What do DES S-boxes say in differential side channel attacks? Report **2005**(311), 191–200 (2005)
9. Carlier, V., Chabanne, H., Dottax, E.: Electromagnetic side channels of an FPGA implementation of AES. Report **2004**(145), 111–124 (2004)
10. Page, D.: Partitioned cache architecture as a side-channel defence mechanism. Report **2005**(280), 213–225 (2005)
11. Gutmann, P.: Software generation of random numbers for cryptographic purposes. In: Usenix Security Symposium, pp. 243–25. Usenix Association, Berkeley (1998)
12. Ortega, J.M.: Introduction to Parallel & Vector Solution of Linear Systems. Plenum Press, New York (1988)
13. Hamming, R.: Coding and Information Theory. Prentice-Hall, Upper Saddle River (1980)
14. Malyugin, V.D.: Representation of boolean functions as arithmetic polynomials. Autom. Remote Control **43**(4), 496–504 (1982)
15. Finko, O.A.: Large systems of Boolean functions: realization by modular arithmetic methods. Autom. Remote Control **65**(6), 871–892 (2004)
16. Finko, O.A.: Modular forms of systems of k-valued functions of the algebra of logic. Autom. Remote Control **66**(7), 1081–1100 (2005)
17. Kukharev, G.A., Shmerko, V.P., Zaitseva, E.N.: Algorithms and Systolic Processors of Multivalued Data. Science and Technology, Minsk (1990). (in Russian)
18. Aslanova, N.H., Faradzhev, R.G.: Arithmetic representation of functions of many-valued logic and parallel algorithm for finding such a representation. Autom. Remote Control **53**(2), 251–261 (1992)
19. Omondi, A., Premkumar, B.: Residue Number System: Theory and Implementation. Imperial Collegt Press, London (2007)

A Hybrid Approach to Fault Detection in One Round of PP-1 Cipher

Ewa Idzikowska[✉]

Poznań University of Technology, pl. M. Skłodowskiej-Curie 5,
60-965 Poznań, Poland
ewa.idzikowska@put.poznan.pl

Abstract. Deliberate injection of faults into cryptographic devices is an effective cryptanalysis technique against symmetric and asymmetric encryption algorithms. In this paper we describe concurrent error detection (CED) approach against such attacks in substitution-permutation network symmetric block ciphers on the example of PP-1 cipher. The specific objective of the design is to develop a method suitable for compact ASIC implementations targeted to embedded systems such as smart cards, cell phones, PDAs, and other mobile devices, such that the system is resistant to fault attacks. To provide the error detection it is proposed to adopt a hybrid approach consisting of multiple parity bits in combination with time redundancy. Taking such an approach gives a better ability to detect faults than simple parity codes. The proposed hybrid CED scheme is aimed at area-critical embedded applications, and achieves effective detection for single faults and most multiple faults. The system can detect the errors shortly after the faults are induced because the detection latency is only the output delay of each operation.

Keywords: Concurrent error detection · PP-1 block cipher · Parity bit code
Fault detection · Time redundancy

1 Introduction

Security is only as strong as its weakest link. To provide high security features, ciphers are implemented in an increasing number of consumer products with dedicated hardware; e.g., smart cards. Although the cipher used is usually difficult to break mathematically, its hardware implementation, unless carefully designed, may result in security vulnerabilities.

Hardware implementations of crypto-algorithms leak information via side-channels such as time consumed by the operations, power dissipated by the operators, electromagnetic radiation emitted by the device and faulty computations resulting from deliberate injection of faults into the system. Traditional cryptanalysis techniques can be combined with such side-channel attacks to break the secret key of the cipher. Even a small amount of side-channel information is sufficient to break ciphers.

Intentional intrusions and attacks based on the malicious injection of faults into the device are very efficient in order to extract the secret key [3, 5]. Such attacks are based

© Springer Nature Switzerland AG 2019
J. Pejaś et al. (Eds.): ACS 2018, AISC 889, pp. 307–316, 2019.
https://doi.org/10.1007/978-3-030-03314-9_27

on the observation that faults deliberately introduced into a crypto-device leak information about the implemented algorithms. First fault injection attack is presented in [4]

There are different types of faults and methods of fault injection in encryption algorithms. The faults can be transient or permanent. The methods of inducing faults using white light, laser and X-rays methods are discussed in detail in [1]. Even a single fault like change a flip-flop state or corruption of data values transferred from one digest operation to another can result in multiple errors in the end of a digest round.

It is well understood that one approach to guarding against fault attacks on ciphers is to implement concurrent error detection (CED) circuitry along with the cipher functional circuit so that suitable action may be taken if an attacker attempts to acquire secret information about the circuit by inducing faults.

The objective of the research in this paper is to investigate a compact implementation of PP-1 cipher with concurrent error detection. The PP-1 was designed for platforms with very limited resources. It can be implemented for example in simple smart cards. We try to create a bridge between the area requirements of embedded systems and effective fault attack countermeasure. The design goal is to achieve 100% error detection with minimal area overhead.

This paper is organized as follows. Sections 2 and 3 present the idea of concurrent error detection and PP-1 symmetric block cipher, respectively. Possible faults and faults models are described in Sect. 4. In Sect. 5 there are presented CED schemes for linear and non-linear functions of PP-1 and for one round of PP-1 cipher. Simulation results are shown in Sect. 6 and in Sect. 7 this paper is concluded.

2   Concurrent Error Detection

Concurrent error detection (CED) checks the system during the computation whether the system output is correct. If an erroneous output is produced, CED will detect the presence of the faulty computation and the system can discard the erroneous output before transmission. Thus, the encryption system can achieve resistance to malicious fault-based attacks. Any CED technique will introduce some overhead into the system and can be classified into four types of redundancy: information, hardware, time, and hybrid [2, 11–13].

CED with information redundancy are based on error detecting codes. In these techniques, the input message is encoded to generate a few check bits, and these bits are propagated along with the input message. The information is validated when the output message is generated. A simple error detecting code is parity checking. The fault detection coverage and detection latency depend on how many parity bits the system uses and the locations of the checking points.

In case of hardware redundancy the original circuit is duplicated, and both original and duplicated circuits are fed with the same inputs and the outputs are compared with each other. It requires more than 100% hardware overhead, it means that this method is not suitable for embedded systems.

The time redundancy technique involves the same data a second time using the same data-path and comparing the two results. This method has more than 100% time overhead and is only applicable to transient faults.

Hybrid redundancy techniques combine the characteristics of the previous CED categories, and they often explore certain properties in the underlying algorithm and/or implementation.

3 The PP-1 Cipher

The scalable PP-1 cipher is a symmetric block cipher designed at the Institute of Control Robotics and Information Engineering, Poznań University of Technology. It was designed for platforms with limited resources, and it can be implemented for example in simple smart cards.

The PP-1 algorithm is an SP-network. It processes in r rounds data blocks of n bits, using cipher keys with lengths of n or $2n$ bits, where $n = t*64$, and $t = 1, 2, 3, \ldots$ One round of the algorithm is presented in Fig. 1. It consists of $t = n/64$ parallel processing paths. In each path the 64-bit nonlinear operation NL is performed (Fig. 2). The 64-bit block is processed as eight 8-bit sub-blocks by four types of transformations:

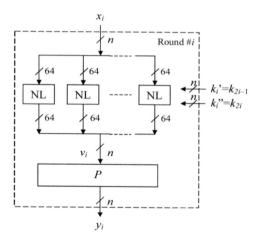

Fig. 1. One round of PP-1 ($i = 1, 2, \ldots, r - 1$) [6]

8×8 S-box S, XOR, addition and subtraction. These are modulo 256 transformations of integers represented by respective bytes. Additionally the n-bit permutation P is used. In the last round, the permutation P is not performed. These algorithm is presented in [6].

The same algorithm is used for encryption and decryption because two components, S-box S and permutation P are involutions, i.e. $S^{-1} = S$, and $P^{-1} = P$. However, if in the encryption process round keys $k_1, k_2, \ldots, k_{2r}$ are used then in the decryption process they must be used in the reverse order, i.e. $k_{2r}, k_{2r-1}, \ldots, k_1$. The round key scheduling is also performed in [6].

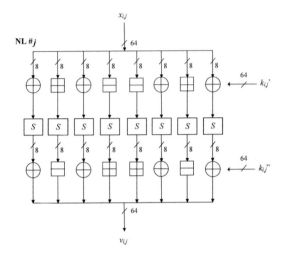

Fig. 2. Nonlinear element NL (j = 1, 2, …,t) [6]

4   Fault Models

Fault attack tries to modify the functioning of the computing device in order to retrieve the secret key. The attacker induces a fault during cryptographic computations. The efficiency of a fault attack depends on the exact capabilities of the attacker and the type of faults he can induce.

In our considerations we use a realistic fault model wherein either transient or permanent faults are induced randomly into the device. We consider single and multiple faults. Fault simulations were performed for two kind of fault models. In one model the fault flips the bit, and the other model introduces bit stuck-at faults (stuck-at-1 and stuck-at-0) [7–9].

5   CED Architecture for PP-1

Concurrent error detection followed by suppression of the corresponding faulty output can thwart fault injection attacks.

In this paper, we examine the application of a hybrid concurrent error detection scheme in the context of an actual compact design of PP-1. The proposed CED design approach uses parity codes and time redundancy. A simple parity check, with the advantage of low hardware overhead, has been proposed as a CED method for linear elements, and time redundancy method for non-linear elements. The detection latency and fault detection coverage depend on how many parity bits the system uses and the locations of the checking points.

5.1   CED for Linear Operations

For linear operations the parity checking schemes are effective with small cost, so parity checking is adopted for these operations. The proposed scheme is implemented to the whole PP-1 system including the encryption/decryption data path and key expander. A multiple-bit parity code is adopted instead of the 1-bit parity code even though the 1-bit parity code has smaller hardware overhead. As it shown in [10], errors spread quickly throughout the encryption/decryption block and, on the average, about half of the state bits become corrupt. Hence, the fault coverage of the parity bits would be at best around 50%, which is unacceptable in practice. The multiple-bit parity code achieves better fault detection coverage for multiple faults. We propose to associate one parity bit with each input/output data byte of exclusive-or (Fig. 3), addition and subtraction elements. If the input data are correctly processed by the fault-free hardware into the output Y, the parity $P(Y)$ is equal $P(A) \oplus P(K)$, where:

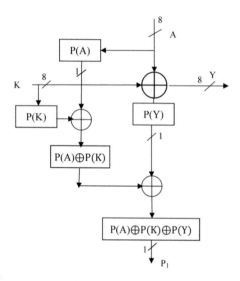

Fig. 3. Parity based CED for exclusive-or operation

A – input data byte, K – key byte, Y – output data byte.

If $P1 = P(A) \oplus P(K) \oplus P(Y)$ is not equal 0 there is an fault in this operation (Fig. 3). In the same way there is generated output parity bit for addition and subtraction elements.

The permutation P of the PP-1 block cipher is an n-bit involution. Its main role is to scatter 8-bit output subblocks of S-boxes S in the n-bit output block of a round. For permutation P only 1 parity bit for a n-bit data block is used.

Since the key scheduling uses similar functions as the data-path, a similar CED approach has been applied to the key expander. The additional operation is the rotation of the n-bit data block, but it is a linear operation and preserves parity.

5.2   CED for Non-linear Operation

The simple parity checking is not sufficient for the s-boxes, therefore the CED scheme is based on the duplication of S-box computation. The CED technique proposed in [8] exploits involution property of S-box designed for PP-1, to detect permanent as well as transient faults. This CED scheme is shown in Fig. 5. Function S is an involution, it means that $S(S(x)) = x$. It means also, that S-box input parity $P(X)$, if the input data is correctly processed by the fault-free hardware, after duplication of S-box computation (Fig. 4) is equal output parity after second computation, The S function is fault free if $P(X) = P(S(S(X)))$, it means that $P(X) \oplus P(S(S(X)))$ is equal 0.

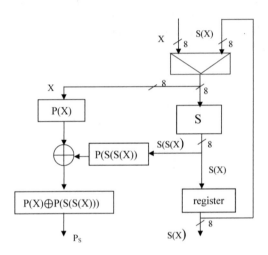

Fig. 4. CED for function S of PP-1 cipher

5.3   CED for One Round of PP-1 Cipher

The architecture of a symmetric block cipher contains an encryption/decryption module and key expansion module. Using the round keys, the device encrypts/decrypts the plain/cipher text to generate the cipher/plain text. PP-1 is an symmetric block cipher it means that has an iterative looping structure. All the rounds of encryption and decryption are identical in general, with each round using several operations and round key(s) to process the input data. Protection of PP-1 cipher entails protecting the encryption/decryption data paths as well as the key expansion module. The proposed CED design concept uses the parity code, but also the time redundancy, because not all operations are linear.

There are following operations in the PP-1 round: linear transformations exclusive-or, addition and subtraction with the round key, bit-permutation and nonlinear transformations - substitution boxes. S- box is a basic component of block ciphers and is used to obscure the relationship between the plaintext and the ciphertext. It should possess some properties, which make linear and differential cryptanalysis as difficult as possible [6]. These s-boxes do not maintain the parity from their inputs to their outputs.

The bit parity protection scheme for linear transformations is shown in Fig. 3. If there is not fault in the operation, the generated parity bit P_1 is equal zero. Non-linear substitution boxes are protected as it shown in Fig. 4. The S function is calculated twice (time redundancy). If there is no error in this operation, the input data is equal to the output data after the second calculation, and generated parity bit P_S jest equal 0.

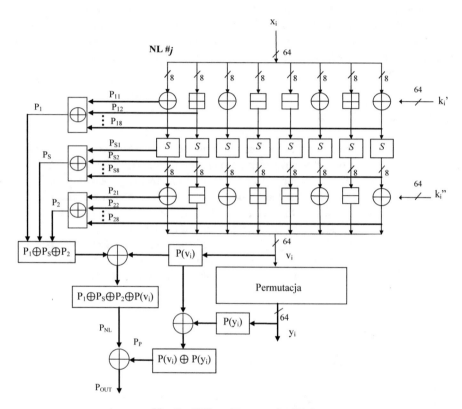

Fig. 5. CED architecture for PP-1

The complete CED architecture for PP-1 is shown in Fig. 5. During the operation of the cipher a parity vector is determined, the elements of which are:

- P_{11}, P_{12}... P_{18} — parity bits for linear operations (8 bite exclusive-or, addition, subtraction) preceding s-boxes,
- P_{S1}, P_{S2}... P_{S8} — parity bits for non-linear S-boxes,
- P_{21}, P_{22}... P_{28} — parity bits for linear operations following s-boxes,
- P_1, P_2 — parity bits for 64 bits of linear operations,
- P_S — parity bit for 64 bits of non-linear operations,
- P_{NL} — parity bit for non-linear element NL,
- P_P — parity bit for permutation,
- P_{OUT} — output parity bit.

If all parity bits have the value 0, no error was detected. If some of the parity bits are equal 1 it indicates that an error has been detected and also it is possible a partial localization of the error.

In this CED scheme multiple-bit parity code is adopted instead of the 1-bit parity code even though the 1-bit parity code has smaller hardware overhead, because the multiple-bit parity code achieves better fault detection coverage for multiple faults.

Check points are placed within each round to achieve good detection latency and higher fault detection coverage. The objective of the design is to yield fault detection coverage of 100% for the single faulty bit model and high coverage for multiple faults assuming a fault model of a bit-flip, stuck-at-0 or stuck-at-1 fault as a transient or permanent fault.

6   Simulation Results

We used VHDL to model the CED scheme shown in Fig. 5. Simulation was realized using Active-HDL simulation and verification environment. The faults were introduced on inputs, outputs of all operations and into internal memory of S-boxes. In our considerations we used a realistic fault model wherein faults are induced randomly into the device at the beginning of the rounds. In this experiment we focused on transient and permanent, single and multiple stuck-at faults and bit flips faults.

As it shown in Fig. 6 all single, and most of multiple faults ware detected. Percentage of undetected permanent errors is less as 0.15% for stuck-at and 0.1% for bit-flip errors. For transient errors percentage of undetected errors is greater, but not greater as 1%.

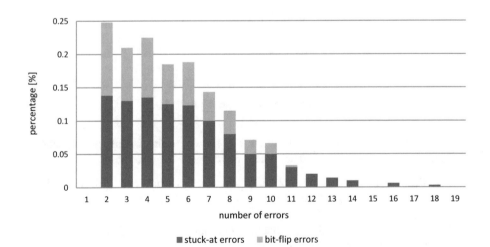

Fig. 6. Permanent faults undetected at the end of round

7 Conclusion

In this section, we now consider the application of an effective error detection scheme to the compact PP-1 cipher described in the Sect. 3. The implementation is aimed at area-critical embedded applications, such as smart cards, PDAs, cell phones, and other mobile devices. The proposed hybrid CED scheme achieves effective detection for single faults and most multiple faults. The system can detect the errors shortly after the faults are induced because the detection latency is only the output delay of each operation. Once an error is detected, the data currently being processed is discarded.

Since the key scheduling uses similar functions as the data-path, a similar CED approach has been applied to the key expander.

Acknowledgements. This research has been supported by Polish Ministry of Science and Higher Education under grant 04/45/DSPB/0163.

References

1. Bar-El, H., Choukri, H., Naccache, D., Tunstall, M., Whelan, C.: The sorcerer's apprentice guide to fault attacks. Proc. IEEE **94**, 370–382 (2006)
2. Bertoni, G., Breveglieri, L., Koren, I., Maistri, P., Piuri, V.: On the propagation of faults and their detection in a hardware implementation of the advanced encryption standard. In: Proceedings of Conference on Application-Specific Systems, Architectures, and Processors, pp. 303–312 (2002)
3. Biham, E., Shamir, A.: Differential fault analysis of secret key cryptosystems. In: Proceedings of Cryptology (1997)
4. Boneh, D., DeMillo, R., Lipton, R.: On the importance of checking cryptographic protocols for faults. In: Proceedings of Eurocrypt. LNCS, vol. 1233, pp. 37–51. Springer (1997
5. Boneh, D., DeMillo, R., Lipton, R.: On the importance of eliminating errors in cryptographic computations. J. Cryptol. **14**, 101–119 (2001)
6. Bucholc, K., Chmiel, K., Grocholewska-Czuryło, A., Stokłosa, J.: PP-1 block cipher. Pol. J. Environ. Stud. **16**(5B), 315–320 (2007)
7. Idzikowska, E., Bucholc, K.: Error detection schemes for CED in block ciphers. In: Proceedings of the 5th IEEE/IFIP International Conference on Embedded and Ubiquitous Computing EUC, Shanghai, pp. 22–27 (2008)
8. Idzikowska, E.: CED for involutional functions of PP-1 cipher. In: Proceedings of the 5th International Conference on Future Information Technology. Busan (2010)
9. Idzikowska, E.: CED for S-boxes of symmetric block ciphers. Electr. Rev. **56**(10), 1179–1183 (2010)
10. Idzikowska, E.: An operation-centered approach to fault detection in key scheduling module of cipher. Electr. Rev. **93**(1), 96–99 (2017)
11. Joshi, N., Wu, K., Karri, R.: Concurrent error detection schemes for involution ciphers. In: Proceedings of the 6th International Workshop CHES 2004. LNCS, vol. 3156, pp, 153–160. Springer (2004)

12. Wu, K., Karri, R., Kouznetzov, G., Goessel, M.: Low cost concurrent error detection for the advanced encryption standard. In: International Test Conference 2004, pp. 1242–1248 (2004)
13. Yen, C.-H., Wu, B.-F.: Simple error detection methods for hardware implementation of advanced encryption standard. IEEE Trans. Comput. **55**(6), 720–731 (2006)

Protection of Information from Imitation on the Basis of Crypt-Code Structures

Dmitry Samoylenko[1], Mikhail Eremeev[2], Oleg Finko[3(✉)],
and Sergey Dichenko[3]

[1] Mozhaiskii Military Space Academy, St. Petersburg 197198, Russia
`19sam@mail.ru`
[2] Institute a Comprehensive Safety and Special Instrumentation of Moscow
Technological University, Moscow 119454, Russia
`mae1@rambler.ru`
[3] Institute of Computer Systems and Information Security of Kuban State
Technological University, Krasnodar 350072, Russia
`ofinko@yandex.ru`

Abstract. A system is offered for imitation resistant transmitting of encrypted information in wireless communication networks on the basis of redundant residue polynomial codes. The particular feature of this solution is complexing of methods for cryptographic protection of information and multi-character codes that correct errors, and the resulting structures (crypt-code structures) ensure stable functioning of the information protection system in the conditions simulating the activity of the adversary. Such approach also makes it possible to create multi-dimensional "crypt-code structures" to conduct multi-level monitoring and veracious restoration of distorted encrypted information. The use of authentication codes as a means of one of the levels to detect erroneous blocks in the ciphertext in combination with the redundant residue polynomial codes of deductions makes it possible to decrease the introduced redundancy and find distorted blocks of the ciphertext to restore them.

Keywords: Cryptographic protection of information
Message authentication code · Redundant residue polynomial codes
Residue number systems

1 Introduction

The drawback of many modern ciphers used in wireless communication networks is the unresolved problem of complex balanced support of traditional requirements: cryptographic security, imitation resistance and noise stability. It is paradoxical that the existing ciphers have to be resistant to random interference, including the effect of errors multiplication [1–3]. However, such regimes of encrypting as cipher feedback mode are not only the exception, but, on the contrary, initiate the process of error multiplication. The existing means to withstand imitated actions of the intruder, which are based on forming authentication

© Springer Nature Switzerland AG 2019
J. Pejaś et al. (Eds.): ACS 2018, AISC 889, pp. 317–331, 2019.
https://doi.org/10.1007/978-3-030-03314-9_28

codes and the hash-code – only perform the indicator function to determine conformity between the transmitted and the received information [1,2,4], and does not allow restoring the distorted data.

In some works [5–8] an attempt was made to create the so-called "noise stability ciphers". However, these works only propose partial solutions to the problem (solving only particular types of errors "insertion", "falling out" or "erasing" symbols of the ciphertext etc.), or insufficient knowledge of these ciphers, which does not allow their practical use.

2 Imitation Resistant Transmitting of Encrypted Information on the Basis of Crypt-Code Structures

The current strict functional distinction only expects the ciphers to solve the tasks to ensure the required cryptographic security and imitation resistance, while methods of interference resistant coding is expected to ensure noise stability. Such distinction between the essentially inter-related methods to process information to solve inter-related tasks will decrease the usability of the system to function in the conditions of destructive actions of the adversary, the purpose of which is to try to impose on the receiver any (different from the transmitted) message (imposition at random). At the same time, if these methods are combined, we can obtain both new information "structures" – crypt-code structures, and a new capability of the system for protected processing of information – *imitation resistance* [9], which we consider to be the ability of the system for *restoration* of veracious encrypted data in the conditions of simulated actions of the intruder, as well as unintentional interference.

The synthesis of crypt-code structures is based on the procedure of complexing of block cypher systems and multi-character correcting codes [10–12]. In one of the variants to implement crypt-code structures as a multi-character correcting code, *redundant residue polynomial codes* (RRPC) can be used, whose mathematical means is based on fundamental provisions of the Chinese remainder theorem for polynomials (CRT) [13–15].

2.1 Chinese Remainder Theorem for Polynomials and Redundant Residue Polynomial Codes

Let $F[z]$ be ring of polynomials over some finite field $\mathbb{F}_q, q = p^s$. For some integer $k > 1$, let $m_1(z), m_2(z), \ldots, m_k(z) \in F[z]$ be relatively prime polynomials sorted by the increasing degrees, i.e. $\deg m_1(z) \leq \deg m_2(z) \leq \ldots \leq \deg m_k(z)$, where $\deg m_i(z)$ is the degree of the polynomial. Let us assume that $P(z) = \prod_{i=1}^{k} m_i(z)$. Then the presentation of φ will establish mutually univocal conformity between polynomials $a(z)$, that do not have a higher degree than $P(z)$ $\left(\deg a(z) < \deg P(z)\right)$, and the sets of residues according to the above-described system of bases of polynomials (modules):

$$\varphi : {}^{F[z]}\!/_{(P(z))} \to {}^{F[z]}\!/_{(m_1(z))} \times \ldots \times {}^{F[z]}\!/_{(m_k(z))} :$$
$$: a(z) \mapsto \varphi\big(a(z)\big) := \big(\varphi_1\big(a(z)\big), \varphi_2\big(a(z)\big), \ldots, \varphi_k\big(a(z)\big)\big),$$

where $\varphi_i\big(a(z)\big) := a(z) \mod m_i(z)$ $(i = 1, 2, \ldots, k)$.

In accordance with the CRT, there is a reverse transformation φ^{-1}, that makes it possible to transfer the set of residues by the system of bases of polynomials to the positional representation:

$$\varphi^{-1} : {}^{F[z]}\!/_{(m_1(z))} \times \ldots \times {}^{F[z]}\!/_{(m_k(z))} \rightarrow {}^{F[z]}\!/_{(P(z))} :$$

$$: \big(c_1(z), \ldots, c_k(z)\big) \mapsto a(z) = \sum_{i=1}^{k} c_i(z) B_i(z) \mod \big(p, \; P(z)\big), \quad (1)$$

where $B_i(z) = k_i(z) P_i(z)$ are polynomial orthogonal bases, $k_i(z) = P_i^{-1}(z)$ $\mod m_i(z)$, $P_i(z) = m_1(z) m_2(z) \ldots m_{i-1}(z) m_{i+1}(z) \ldots m_k(z)$ $(i = 1, 2, \ldots, k)$.

Let us also introduce, in addition to the existing number k, the number r of redundant bases of polynomials while observing the condition of sortednes:

$$\deg m_1(z) \le \ldots \le \deg m_k(z) \le \deg m_{k+1}(z) \le \ldots \le \deg m_{k+r}(z), \quad (2)$$

and

$$\gcd\big(m_i(z), m_j(z)\big) = 1, \quad (3)$$

for $i \ne j$; $i, j = 1, 2, \ldots, k + r$, then we obtain the expanded RRPC—an array of the kind:

$$C := \big(c_1(z), \ldots, c_k(z), c_{k+1}(z), \ldots, c_n(z)\big) : c_i(z) \equiv a(z) \mod m_i(z), \quad (4)$$

where $n = k + r$, $c_i(z) \equiv a(z) \mod m_i(z)$ $(i = 1, 2, \ldots, n)$, $a(z) \in {}^{F[z]}\!/_{(P(z))}$.

Elements of the code $c_i(z)$ will be called symbols, each of which is the essence of polynomials from the quotient ring of polynomials over the module $m_i(z) \in {}^{F[z]}\!/_{(m_i(z))}$. At the same time, if $a(z) \notin {}^{F[z]}\!/_{(P(z))}$, then it is considered that this combination contains an error. Therefore, the location of the polynomial $a(z)$ makes it possible to establish if the code combination $a(z) = \big(c_1(z), \ldots, c_k(z), c_{k+1}(z), \ldots, c_n(z)\big)$ is allowed or it contains erroneous symbols.

2.2   Crypt-Code Structures on Based RRPC

Now, the sender-generated message M shall be encrypted and split into blocks of the fixed length $M = \{M_1 \| M_2 \| \ldots \| M_k\}$, where "$\|$" is the operation of concatenation. Introducing a formal variable z number i block of the open text M_i, we will represent in the polynomial form:

$$M_i(z) = \sum_{j=0}^{s-1} m_j^{(i)} z^j = m_{s-1}^{(i)} z^{s-1} + \ldots + m_1^{(i)} z + m_0^{(i)},$$

where $m_j^{(i)} \in \{0, 1\}$ $(i = 1, 2, \ldots, k; \quad j = s - 1, s - 2, \ldots, 0)$.

In order to obtain the sequence of blocks of the ciphertext $\Omega_1(z)$, $\Omega_2(z)$, ..., $\Omega_k(z)$ we need to execute k number of encrypting operations, and to obtain blocks of the open text $M_1(z)$, $M_2(z)$, ..., $M_k(z)$, we need to execute k number of decrypting operations. The procedures of encrypting and decrypting correspond to the following presentations:

$$\begin{cases} \Omega_1(z) \to E_{\kappa_{e,1}} \; : \; M_1(z), \\ \Omega_2(z) \to E_{\kappa_{e,2}} \; : \; M_2(z), \\ \dots \dots \dots \dots \dots \dots \\ \Omega_k(z) \to E_{\kappa_{e,k}} \; : \; M_k(z); \end{cases} \qquad \begin{cases} M_1(z) \to D_{\kappa_{d,1}} \; : \; M_1(z), \\ M_2(z) \to D_{\kappa_{d,2}} \; : \; M_2(z), \\ \dots \dots \dots \dots \dots \dots \\ M_k(z) \to D_{\kappa_{d,k}} \; : \; M_k(z), \end{cases}$$

where $\kappa_{e,i}, \kappa_{d,i}$ are keys (general case) for encrypting and decrypting ($i = 1, 2, \ldots, k$); if $\kappa_{e,i} = \kappa_{d,i}$—the cryptosystem is symmetric, if $\kappa_{e,i} \neq \kappa_{d,i}$—it is asymmetric.

We will express the adopted blocks of the ciphertext and blocks of the open text correspondingly as $\Omega_i^*(z)$ and $M_i^*(z)$ ($i = 1, 2, \ldots, k$), as they can contain distortions. The formed blocks of the ciphertext $\Omega_i(z)$ will be represented as the minimum residues (deductions) on the pairwise relatively prime polynomials (bases) $m_i(z)$. Here, $\deg \Omega_i(z) < \deg m_i(z)$. The set of blocks of the ciphertext $\Omega_1(z)$, $\Omega_2(z)$, ..., $\Omega_k(z)$ will be represented as a single super-block of elements of the RRPC by the system of bases-polynomials $m_1(z), m_2(z), \ldots, m_k(z)$. In accordance with CRT for the set array of polynomials $m_1(z)$, $m_2(z)$, ..., $m_k(z)$, that meet the condition that $\gcd\big(m_i(z), m_j(z)\big) = 1$, and polynomials $\Omega_1(z)$, $\Omega_2(z)$, ..., $\Omega_k(z)$, such that $\deg \Omega_i(z) < \deg m_i(z)$, the system of congruences

$$\begin{cases} \Omega(z) \equiv \Omega_1(z) \mod m_1(z), \\ \Omega(z) \equiv \Omega_2(z) \mod m_2(z), \\ \dots \dots \dots \dots \dots \dots \dots \\ \Omega(z) \equiv \Omega_k(z) \mod m_k(z) \end{cases} \qquad (5)$$

has the only one solution $\Omega(z)$.

Then, we execute the operation of expansion (Base Expansion) of the RRPC by introducing r of redundant bases-polynomials $m_{k+1}(z)$, $m_{k+2}(z)$, ..., $m_{k+r}(z)$ that meet the condition (2), (3) and obtaining in accordance with Eq. (4) redundant blocks of data (residues), which we will express as $\omega_{k+1}(z)$, $\omega_{k+2}(z)$, ..., $\omega_n(z)$ ($n = k + r$). The combination of "informational" blocks of the ciphertext and redundant blocks of data form crypt-code structures identified as a code word of the expanded RRPC: $\{\Omega_1(z), \ldots, \Omega_k(z), \omega_{k+1}(z), \ldots, \omega_n(z)\}_{\text{RRPC}}$.

Here, we define a single error of the code word of RRPC as a random distortion of one of the blocks of the ciphertext; correspondingly the b-fold error is defined as a random distortion of b blocks. At the same time, it is known that RRPC detects b errors, if $r \geq b$, and will correct b or less errors, if $2b \leq r$ [10,13,14].

The adversary, who affects communication channels, intercepts the information or simulates false information. At the same time, in order to impose false, as applied to the system under consideration, the adversary has to intercept a set of information blocks of the ciphertext to detect the redundant blocks of data.

In order to eliminate the potential possibility that the adversary may impose false information, we need to ensure the "mathematical" gap of the procedure (uninterrupted function) of forming redundant elements of code words of the RRPC. Moreover, code words of RRPC have to be distributed randomly, i.e. uniform distribution of code words in the set array of the code has to be ensured. In order to achieve that, the formed sequence of redundant blocks of data $\omega_j(z)$ $(j = k + 1,\, k + 2,\, \ldots,\, n)$ undergoes the procedure of encrypting:

$$
\begin{cases}
\vartheta_{k+1}(z) \rightarrow E_{\kappa_{e,k+1}} &: \ \omega_{k+1}(z), \\
\vartheta_{k+2}(z) \rightarrow E_{\kappa_{e,k+2}} &: \ \omega_{k+2}(z), \\
\cdots \cdots \cdots \cdots \cdots \cdots \cdots \\
\vartheta_n(z) \rightarrow E_{\kappa_{e,n}} &: \ \omega_n(z),
\end{cases}
$$

where $\kappa_{e,\,j}$ $(j = k + 1,\, k + 2,\, \ldots,\, n)$ are the keys for encrypting.

The process of encrypting of redundant symbols of the code word of the RRPC executes transposition of elements of the vector $\{\omega_{k+1}(z),$ $\omega_{k+2}(z),\, \ldots \ldots,\, \omega_n(z)\} \in \mathcal{A}$ onto the formed elements of the vector of redundant encrypted symbols $\{\vartheta_{k+1}(z),\, \vartheta_{k+2}(z),\, \ldots,\, \vartheta_n(z)\} \in \mathcal{B}$, where $\mathcal{A}$ is the array of blocks of the ciphertext, $\mathcal{B}$ is a finite array.

The operation of transposition excludes the mutually univocal transformation and prevents the adversary from interfering on the basis of the intercepted informational super-block of the RRPC (the "informational" constituent) $\Omega_i(z)$ $(i = 1,\, 2,\, \ldots,\, k)$ by forming a verification sequence $\omega_j(z)$ $(j = k+1,\, k+2,\, \ldots,\, n)$ for overdriving the protection mechanisms and inserting false information. At the same time, it is obvious that, for the adversary, the set of keys $\kappa_{e,\,j}$ and functions of encrypting $E_i(\bullet)$ of the vector of redundant blocks of data forms a certain array $\mathcal{X}$ of the transformation rules, out of whose many variants, the sender and the addressee will only use a certain one [4, 16, 17].

We should also note the exclusive character of the operation of encrypting the sequence of redundant blocks of data, due to this, its implementation requires a special class of ciphers that do not alter the lengths of blocks of the ciphertext (endomorphic ones) and not creating distortions (like omissions, replacements or insertions) of symbols, for example, ciphers of permutation.

3   Imitation Resistant Transmitting of Encrypted Information on the Basis of Multidimensional Crypt-Code Structures

A particular feature of the above-described system is the necessity to introduce redundant encrypted information in accordance with the RRPC characteristics and specified requirements to the repetition factor of the detected or corrected distortions in the sent data. The theory of coding tells us of solutions to obtain quite long interference-resistant codes with good correct ability on the basis of composing shorter codes that allow simpler implementation and are called composite codes [18]. Such solutions can be the basis for the procedure to create multidimensional crypt-code structures.

Similarly to the previous solution, the open text M undergoes the procedure of encrypting. The formed sequence of blocks of the ciphertext $\Omega_1(z)$, $\Omega_2(z)$, ..., $\Omega_k(z)$ is split into k_2 number of sub-blocks, contain k_1 number of blocks of the ciphertext $\Omega_i(z)$ in each one and it is expressed in the form of a matrix $\mathbf{W}$ sized $k_1 \times k_2$:

$$\mathbf{W} = \begin{bmatrix} \Omega_{1,1}(z) & \Omega_{1,2}(z) & \dots & \Omega_{1,k_2}(z) \\ \Omega_{2,1}(z) & \Omega_{2,2}(z) & \dots & \Omega_{2,k_2}(z) \\ \vdots & \vdots & \ddots & \vdots \\ \Omega_{k_1,1}(z) & \Omega_{k_1,2}(z) & \dots & \Omega_{k_1,k_2}(z) \end{bmatrix},$$

where the columns of the matrix $\mathbf{W}$ are sub-blocks made of k_1 number of blocks of the ciphertext $\Omega_i(z)$.

For each line of the matrix $\mathbf{W}$, redundant blocks of data are formed, for example, using non-binary codes of Reed-Solomon (code RS [particular case]) over $\mathbb{F}_q$, that allow the 2-nd level of monitoring.

The mathematical means of the RS codes is explained in detail in [19], where one of the ways to form it is based on the deriving polynomial $g(z)$. In $\mathbb{F}_q$ the minimal polynomial for any element α^i is equal to $M^{(i)} = z - \alpha^i$, then, the polynomial $g(z)$ of the RS code corresponds to the equation:

$$g(z) = (z - \alpha^t)(z - \alpha^t) \dots (z - \alpha^{t+2b-1}), \tag{6}$$

where $2b = n - k$; usually $t = 0$ or $t = 1$.

At the same time, the RS code is cyclic and the procedure of forming the systematic RS code is described by the equation:

$$C(z) = U(z)z^{n-k} + R(z), \tag{7}$$

where $U(z) = u_{k-1}z^{k-1} + \dots + u_1z + u_0$ informational polynomial, and $\{u_{k-1}, \dots, u_1, u_0\}$ informational code blocks; $R(z) = h_{r-1}z^{r-1} + \dots + h_1z + h_0$ the residue from dividing the polynomial $U(z)z^{n-k}$ by $g(z)$, a $\{h_{r-1}, \dots, h_1, h_0\}$ the coefficients of the residue. Then the polynomial $C(z) = c_{n-1}z^{n-1} + \dots + c_1z + c_0$ and, therefore $\{c_{n-1}, \dots, c_1, c_0\} = \{u_{k-1}, \dots, u_1, u_0, h_{r-1}, \dots, h_1, h_0\}$ a code word.

Basing on the primitive irreducible polynomial, setting the characteristic of the field $\mathbb{F}_q$ in accordance with the Eq. (6) a deriving polynomial $g(z)$ of the RS code is formed.

Blocks of the ciphertext $\Omega_{i,1}(z)$, $\Omega_{i,2}(z)$,..., $\Omega_{i,k_2}(z)$ are elements $\mathbf{W}$ expressed as elements of the sorted array, at the same time a formal variable

x is introduced and a set of "informational" polynomials is formed:

$$\mho_i(x) = \sum_{j=1}^{k_2}\big(\Omega_{i,j}(z)\big)x^{j-1} = \big(\Omega_{i,k_2}(z)\big)x^{k_2-1} + \ldots + \big(\Omega_{i,2}(z)\big)x + \Omega_{i,1}(z),$$

where $i = 1, 2, \ldots, k_1$.

For $\mho_i(x)$ $(i = 1, 2, \ldots, k_1)$ in accordance with the Eq. (7) a sequence of residues is formed

$$R_i(x) = \sum_{j=1}^{r_2}\big(\omega_{i,j}(z)\big)x^{j-1} = \big(\omega_{i,r_2}(z)\big)x^{r_2-1} + \ldots + \big(\omega_{i,2}(z)\big)x + \omega_{i,1}(z),$$

where $\omega_{i,j}(z)$ are coefficients of the polynomial $R_i(x)$ $(i = 1, 2, \ldots, k_1)$ assumed as redundant blocks of data of the 2-nd level of monitoring; n_2 is the length of the RS code, k_2 is the number of "informational" symbols (blocks) of the RS code, r_2 is the number of redundant symbols (blocks) of the RS code; $n_2 = k_2 + r_2$.

Matrix $\mathbf{W}$ with generated redundant blocks of data of the 2-nd level of monitoring will take the form:

$$\Psi = \big[\mathbf{W}_{k_1 \times k_2} | \mathbf{\Upsilon}_{k_1 \times r_2}\big] = \left.\begin{bmatrix} \overbrace{\Omega_{1,1}(z) \; \ldots \; \Omega_{1,k_2}(z)}^{k_2} & \overbrace{\omega_{1,k_2+1}(z) \; \ldots \; \omega_{1,n_2}(z)}^{r_2} \\ \Omega_{2,1}(z) \; \ldots \; \Omega_{2,k_2}(z) & \omega_{2,k_2+1}(z) \; \ldots \; \omega_{2,n_2}(z) \\ \ldots \quad \ldots \quad \ldots & \ldots \quad \ldots \quad \ldots \\ \Omega_{k_1,1}(z) \ldots \Omega_{k_1,k_2}(z) & \omega_{k_1,k_2+1}(z) \ldots \omega_{k_1,n_2}(z) \end{bmatrix}\right\} k_1.$$

The lines of the matrix $\mathbf{\Upsilon}$ are redundant blocks of data of the 2-nd level of monitoring that undergo the procedure of encrypting:

$$\begin{cases} \vartheta_{1,\gamma}(z) \rightarrow E_{\kappa_{e_{1,\gamma}}} & : \; \omega_{1,\gamma}(z), \\ \vartheta_{2,\gamma}(z) \rightarrow E_{\kappa_{e_{2,\gamma}}} & : \; \omega_{2,\gamma}(z), \\ \ldots \; \ldots \; \ldots \; \ldots \; \ldots \; \ldots \; \ldots \\ \vartheta_{k_1,\gamma}(z) \rightarrow E_{\kappa_{e_{k_1,\gamma}}} & : \; \omega_{k_1,\gamma}(z), \end{cases}$$

where $\kappa_{e_{i,\gamma}}$ $(i = 1, 2, \ldots, k_1;$ $\gamma = k_2 + 1, k_2 + 2, \ldots, n_2)$ are the keys for encrypting.

The generated sequence of blocks of the redundant ciphertext of the 2-nd level of monitoring $\vartheta_{i,k_2+1}(z), \vartheta_{i,k_2+2}(z), \ldots, \vartheta_{i,n_2}(z)$ $(i = 1, 2, \ldots, k_1)$ form a matrix $\mathbf{V}$ sized $k_1 \times r_2$ redundant blocks of the ciphertext of the 2-nd level of monitoring:

$$\mathbf{V} = \begin{bmatrix} \vartheta_{1,k_2+1}(z) & \vartheta_{1,k_2+2}(z) & \ldots & \vartheta_{1,n_2}(z) \\ \vartheta_{2,k_2+1}(z) & \vartheta_{2,k_2+2}(z) & \ldots & \vartheta_{2,n_2}(z) \\ \ldots & \ldots & \ldots & \ldots \\ \vartheta_{k_1,k_2+1}(z) & \vartheta_{k_1,k_2+2}(z) & \ldots & \vartheta_{k_1,n_2}(z) \end{bmatrix}.$$

Now, each column of the matrix $\mathbf{W}$ and $\mathbf{V}$ as a sequence of blocks of the ciphertext $\Omega_{1,j}(z), \Omega_{2,j}(z), \ldots, \Omega_{k_1,j}(z)$ $(j = 1, 2, \ldots, k_2)$ and

$\vartheta_{1,\gamma}(z)$, $\vartheta_{2,\gamma}(z)$, ..., $\vartheta_{k_1,\gamma}(z)$ ($\gamma = k_2 + 1$, $k_2 + 2$, ..., n_2) are expressed in the form of minimal residues on the bases-polynomials $m_i(z)$, such that $\gcd(m_i(z), m_j(z)) = 1$ ($i \neq j$; $i, j = 1, 2, ..., k_1$). At the same time $\deg \Omega_{i,j}(z) < \deg m_i(z)$, and $\deg \vartheta_{i,\gamma}(z) < \deg m_i(z)$. Then, as we have noted above, the arrays of blocks of the ciphertext $\Omega_{1,j}(z), \Omega_{2,j}(z), \ldots, \Omega_{k_1,j}(z)$ ($j = 1, 2, ..., k_2$) and $\vartheta_{1,\gamma}(z), \vartheta_{2,\gamma}(z), \ldots, \vartheta_{k_1,\gamma}(z)$ ($\gamma = k_2 + 1$, $k_2 + 2, ..., n_2$) are expressed as united informational super-blocks of RRPC on the system of bases $m_1(z), m_2(z), \ldots, m_{k_1}(z)$. In accordance with CRT for the specified array of polynomials $m_1(z)$, $m_2(z)$, ..., $m_{k_1}(z)$ that meet the condition $\gcd(m_i(z), m_j(z)) = 1$, polynomials $\Omega_{1,j}(z), \Omega_{2,j}(z), \ldots, \Omega_{k_1,j}(z)$ ($j = 1, 2, \ldots, k_2$) and $\vartheta_{1,\gamma}(z), \vartheta_{2,\gamma}(z), \ldots, \vartheta_{k_1,\gamma}(z)$ ($\gamma = k_2 + 1, k_2 + 2, \ldots, n_2$) such that $\deg \Omega_{i,j}(z) < \deg m_i(z)$, $\deg \vartheta_{i,\gamma}(z) < \deg m_i(z)$, the system of congruences (5) will take the form:

$$
\begin{cases}
\begin{cases}
\Omega_1(z) \equiv \Omega_{1,1}(z) \mod m_1(z), \\
\Omega_1(z) \equiv \Omega_{2,1}(z) \mod m_2(z), \\
\cdots \cdots \cdots \cdots \cdots \cdots \cdots \cdots \\
\Omega_1(z) \equiv \Omega_{k_1,1}(z) \mod m_{k_1}(z);
\end{cases} \\
\cdots \cdots \cdots \cdots \cdots \cdots \cdots \cdots \\
\begin{cases}
\Omega_{k_2}(z) \equiv \Omega_{1,k_2}(z) \mod m_1(z), \\
\Omega_{k_2}(z) \equiv \Omega_{2,k_2}(z) \mod m_2(z), \\
\cdots \cdots \cdots \cdots \cdots \cdots \cdots \cdots \\
\Omega_{k_2}(z) \equiv \Omega_{k_1,k_2}(z) \mod m_{k_1}(z);
\end{cases}
\end{cases}
\tag{8}
$$

$$
\begin{cases}
\begin{cases}
\vartheta_{k_2+1}(z) \equiv \vartheta_{1,k_2+1}(z) \mod m_1(z), \\
\vartheta_{k_2+1}(z) \equiv \vartheta_{2,k_2+1}(z) \mod m_2(z), \\
\cdots \cdots \cdots \cdots \cdots \cdots \cdots \cdots \\
\vartheta_{k_2+1}(z) \equiv \vartheta_{k_1,k_2+1}(z) \mod m_{k_1}(z);
\end{cases} \\
\cdots \cdots \cdots \cdots \cdots \cdots \cdots \cdots \\
\begin{cases}
\vartheta_{n_2}(z) \equiv \vartheta_{1,n_2}(z) \mod m_1(z), \\
\vartheta_{n_2}(z) \equiv \vartheta_{2,n_2}(z) \mod m_2(z), \\
\cdots \cdots \cdots \cdots \cdots \cdots \cdots \cdots \\
\vartheta_{n_2}(z) \equiv \vartheta_{k_1,n_2}(z) \mod m_{k_1}(z),
\end{cases}
\end{cases}
\tag{9}
$$

where $\Omega_j(z)$, $\vartheta_\gamma(z)$ are the only solutions for $j = 1, 2, \ldots, k_2; \gamma = k_2 + 1, \ldots, n_2$.

Now, according to the additionally formed r_1 redundant bases of polynomials $m_{k_1+1}(z), m_{k_1+2}(z), \ldots, m_{n_1}(z)$ ($n_1 = k_1 + r_1$), meeting the condition (2), (3) and in accordance with the Eq. (4) redundant blocks of data are formed, that belong to the *1-st* level of monitoring, expressed as $\omega_{k_1+1,j}(z)$, $\omega_{k_1+2,j}(z), \ldots, \omega_{n_1,j}(z)$ ($j = 1, 2, \ldots, k_2$), as well as reference blocks of data $\omega_{k_1+1,\gamma}(z), \omega_{k_1+2,\gamma}(z), \ldots, \omega_{n_1,\gamma}(z)$ ($\gamma = k_2 + 1, k_2 + 2 \ldots, n_2$).

The formed redundant blocks of data o the *1-st* level of monitoring $\omega_{k_1+1,j}(z), \omega_{k_1+2,j}(z), \ldots, \omega_{n_1,j}(z)$ $(j = 1, 2, \ldots, k_2)$ are encrypted:

$$
\begin{cases}
\vartheta_{k_1+1,\gamma}(z) \rightarrow E_{\kappa_{e_{k_1+1,\gamma}}} : \omega_{k_1+1,\gamma}(z), \\
\vartheta_{k_1+2,\gamma}(z) \rightarrow E_{\kappa_{e_{k_1+2,\gamma}}} : \omega_{k_1+2,\gamma}(z), \\
\cdots \cdots \cdots \cdots \cdots \cdots \cdots \cdots \cdots \\
\vartheta_{n_1,\gamma}(z) \rightarrow E_{\kappa_{e_{n_1,\gamma}}} : \omega_{n_1,\gamma}(z),
\end{cases}
$$

where $\kappa_{e_{\iota,\gamma}}$ $(\iota = k_1 + 1, k_1 + 2, \ldots, n_1; \gamma = k_2 + 1, k_2 + 2, \ldots, n_2)$ are the keys for encrypting.

Now, the arrays of informational blocks of the ciphertext $\Omega_1(z)$, $\Omega_2(z), \ldots\ldots, \Omega_k(z)$, blocks of the redundant encrypted text of the *1-st* and *2-nd* levels of monitoring $\vartheta_{k_1+1,j}(z), \vartheta_{k_1+2,j}(z), \ldots, \vartheta_{n_1,j}(z)$ $(j = 1, 2, \ldots, k_2)$ and $\vartheta_{i,k_2+1}(z), \vartheta_{i,k_2+2}(z), \ldots, \vartheta_{i,n_2}(z)$ $(i = 1, 2, \ldots, k_1)$, as well as reference blocks of data $\omega_{k_1+1,\gamma}(z), \omega_{k_1+2,\gamma}(z), \ldots, \omega_{n_1,\gamma}(z)$ $(\gamma = k_2 + 1, k_2 + 2 \ldots, n_2)$ form multidimensional crypt-code structures, whose matrix representation correspond to the expression:

$$
\mathbf{\Phi} =
\left[
\begin{array}{ccccccc}
\overbrace{\Omega_{1,1}(z) \ldots \Omega_{1,k_2}(z)}^{k_2} & \overbrace{\vartheta_{1,k_2+1}(z) \quad \ldots \quad \vartheta_{1,n_2}(z)}^{r_2} \\
\cdots \quad \cdots \quad \cdots & \cdots \quad \cdots \quad \cdots \\
\Omega_{k_1,1}(z) \quad \ldots \quad \Omega_{k_1,k_2}(z) & \vartheta_{k_1,k_2+1}(z) \quad \ldots \quad \vartheta_{k_1,n_2}(z) \\
\vartheta_{k_1+1,1}(z) \ldots \vartheta_{k_1+1,k_2}(z) & \omega_{k_1+1,k_2+1}(z) \ldots \omega_{k_1+1,n_2}(z) \\
\cdots \quad \cdots \quad \cdots & \cdots \quad \cdots \quad \cdots \\
\vartheta_{n_1,1}(z) \quad \ldots \quad \vartheta_{n_1,k_2}(z) & \omega_{n_1,k_2+1}(z) \quad \ldots \quad \omega_{n_1,n_2}(z)
\end{array}
\right]
\left.\begin{array}{c} \\ \\ \\ \end{array}\right\} k_1
\left.\begin{array}{c} \\ \\ \end{array}\right\} r_1
.
$$

The formed multidimensional crypt-code structures correspond to the following parameters (a particular case for 2 levels of monitoring):

$$
\begin{cases}
n = n_1 n_2, \\
k = k_1 k_2, \\
r = r_1 n_2 + r_2 n_1 - r_1 r_2, \\
d_{\min} = d_{\min_1} d_{\min_2},
\end{cases}
$$

where n, k, r, $d_{\min}$ are generalized monitoring parameters; n_i, k_i, r_i, $d_{\min_i}$ are parameters of the level of monitoring number i $(i = 1, 2)$ [18].

On the receiving side, multidimensional crypt-code structures undergo the procedure of reverse transformation. In order to achieve that, the received sequence of blocks of the ciphertext $\Omega_i(z)$ $(i = 1, 2, \ldots, k)$ is split into k_2 number of sub-blocks containing k_1 blocks of the ciphertext and expressed in the form of the matrix $\mathbf{W}^*$ with the parameters identical to the parameters of the sending side:

$$
\mathbf{W}^* =
\begin{bmatrix}
\Omega_{1,1}^*(z) & \Omega_{1,2}^*(z) & \ldots & \Omega_{1,k_2}^*(z) \\
\Omega_{2,1}^*(z) & \Omega_{2,2}^*(z) & \ldots & \Omega_{2,k_2}^*(z) \\
\vdots & \vdots & \ddots & \vdots \\
\Omega_{k_1,1}^*(z) & \Omega_{k_1,2}^*(z) & \ldots & \Omega_{k_1,k_2}^*(z)
\end{bmatrix},
$$

where the columns of the matrix $\mathbf{W}^*$ are sub-blocks of k_1 blocks of the ciphertext $\Omega_i^*(z)$. The arrays of blocks of the redundant ciphertext of the *1-st* and *2-nd* levels of monitoring $\vartheta_{k_1+1,\,j}^*(z)$, $\vartheta_{k_1+2,\,j}^*(z)$, ..., $\vartheta_{n_1,\,j}^*(z)$ $(j = 1, 2, \ldots, k_2)$, $\vartheta_{i,\,k_2+1}^*(z)$, $\vartheta_{i,\,k_2+2}^*(z)$, ..., $\vartheta_{i,\,n_2}^*(z)$ $(i = 1, 2, \ldots, k_1)$ that were obtained in the parallel process undergo procedure of decrypting:

$$
\begin{cases}
\omega_{k_1+1,\,j}^*(z) \to D_{\kappa_{d_{k_1+1,\,j}}} & : \ \vartheta_{k_1+1,\,j}^*(z), \\
\omega_{k_1+2,\,j}^*(z) \to D_{\kappa_{d_{k_1+2,\,j}}} & : \ \vartheta_{k_1+2,\,j}^*(z), \\
\ldots \ \ldots \ \ldots \ \ldots \ \ldots \ \ldots \\
\omega_{n_1,\,j}^*(z) \to D_{\kappa_{d_{n_1,\,j}}} & : \ \vartheta_{n_1,\,j}^*(z);
\end{cases}
\qquad
\begin{cases}
\omega_{1,\,\gamma}^*(z) \to D_{\kappa_{d_{1,\,\gamma}}} & : \ \vartheta_{1,\,\gamma}^*(z), \\
\omega_{2,\,\gamma}^*(z) \to D_{\kappa_{d_{2,\,\gamma}}} & : \ \vartheta_{2,\,\gamma}^*(z), \\
\ldots \ \ldots \ \ldots \ \ldots \ \ldots \ \ldots \\
\omega_{k_1,\,\gamma}^*(z) \to D_{\kappa_{d_{k_1,\,\gamma}}} & : \ \vartheta_{k_1,\,\gamma}^*(z),
\end{cases}
$$

where $\kappa_{d_{\iota,\,j}}$ and $\kappa_{d_{i,\,\gamma}}$ $(\iota = k_1 + 1, k_1 + 2, \ldots, n_1; \ j = 1, 2, \ldots, k_2)$, $(i = 1, 2, \ldots, k_1; \ \gamma = k_2 + 1, k_2 + 2, \ldots, n_2)$ are the keys for decrypting.

Now, every column $\Omega_{1,\,j}^*(z)$, $\Omega_{2,\,j}^*(z)$, ... $\Omega_{k_1,\,j}^*(z)$ of the matrix $\mathbf{W}^*$ that is interpreted as an informational super-block of the RRPC is put into the conformity to the sequence of redundant blocks of data of the *1-st* level of monitoring $\omega_{k_1+1,\,j}^*(z)$, $\omega_{k_1+2,\,j}^*(z)$, ..., $\omega_{n_1,\,j}^*(z)$ $(j = 1, 2, \ldots, k_2)$ on the bases-polynomials $m_i(z)$ $(i = 1, 2, \ldots, n_1)$ resulting in forming the code vector of the expanded RRPC $\{\Omega_{1,\,j}^*(z), \ldots, \Omega_{k_1,\,j}^*(z), \omega_{k_1+1,\,j}^*(z), \ldots, \omega_{n_1,\,j}^*(z)\}_{\text{RRPC}}$.

Besides that, the columns of the *2-nd* level of monitoring $\vartheta_{1,\,\gamma}^*(z), \ldots, \vartheta_{k_1,\,\gamma}^*(z)$ are put into the conformity to the reference blocks of data $\omega_{k_1+1,\,\gamma}^*(z), \ldots, \omega_{n_1,\,\gamma}^*(z)$ $(\gamma = k_2 + 1, \ldots, n_2)$ on the bases-polynomials $m_i(z)(i = 1, 2, \ldots, n_1)$ and a code vector of the expanded RRPC $\{\vartheta_{1,\,\gamma}^*(z), \ldots, \vartheta_{k_1,\,\gamma}^*(z), \omega_{k_1+1,\,\gamma}^*(z), \ldots, \omega_{n_1,\,\gamma}^*(z)\}_{\text{RRPC}}$ is formed. Then, the procedure is started to detect the RRPC elements distorted (simulated) by the adversary, basing on the detection capability conditioned by the equation $d_{\min_1} - 1$. At the same time, if $\Omega_j^*(z), \vartheta_\gamma^*(z) \in {}^{F[z]}\!/_{(P(z))}$, then we assume that there are no distorted blocks of the ciphertext, where $\Omega_j^*(z), \vartheta_\gamma^*(z)$ solution of the comparison system (8), (9) in accordance with the Eq. (4), for $j = 1, 2, \ldots, k_2; \gamma = k_2 + 1, \ldots, n_2$. Considering the condition $\lfloor (d_{\min_1} - 1)2^{-1} \rfloor$, the procedure of restoring the distorted elements of RRPC can be executed with the help of calculating the minimal residues or with any other known method of RRPC decoding.

The corrected (restored) elements number j of the sequence of the ciphertext blocks $\Omega_{1,\,j}^{**}(z), \Omega_{2,\,j}^{**}(z), \ldots, \Omega_{k_1,\,j}^{**}(z)$ "replace" the distorted number i (of the ciphertext blocks) of the lines $\Omega_{i,\,1}^*(z), \Omega_{i,\,2}^*(z), \ldots, \Omega_{i,\,k_2}^*(z)$ $(i = 1, 2, \ldots, k_1)$ of the matrix $\mathbf{W}^*$. The symbols "**" indicate the stochastic character of restoration.

Now, each line $\Omega_{i,\,1}^*(z), \Omega_{i,\,2}^*(z), \ldots, \Omega_{i,\,k_2}^*(z)$ is put into conformity of the blocks of the redundant ciphertext of the *2-nd* level of monitoring $\omega_{i,\,k_2+1}^*(z)$, $\omega_{i,\,k_2+2}^*(z), \ldots, \omega_{i,\,n_2}^*(z)$ $(i = 1, 2, \ldots, k_1)$ and code vectors are formed for the RS code $\{\Omega_{i,\,1}^*(z), \ldots, \Omega_{i,\,k_2}^*(z), \omega_{i,\,k_2+1}^*(z), \ldots, \omega_{i,\,n_2}^*(z)\}_{\text{RS}}$.

According to the code vectors, polynomials are formed

$$C_i^*(x) = \mho_i^*(x) + R_i^*(x) = \sum_{j=1}^{k_2} (\Omega_{i,j}^*(z)) x^{j-1} + \sum_{\gamma=k_2+1}^{n_2} (\omega_{i,\gamma}^*(z)) x^{\gamma-1}$$

and their values are calculated for the degrees of the primitive element of the field α^ℓ :

$$\mathcal{S}_{i,\ell} = C_i^*(\alpha^\ell) = \sum_{j=1}^{k_2} \left(\Omega_{i,j}^*(z)\right)\left(\alpha^{(j-1)}\right)^\ell + \sum_{\gamma=k_2+1}^{n_2} \left(\omega_{i,\gamma}^*(z)\right)\left(\alpha^{(\gamma-1)}\right)^\ell,$$

where $i = 1, 2, \ldots, k_1$; $\ell = 0, 1, \ldots, r_2 - 1$, $r_2 = n_2 - k_2$.

At the same time, if the values of checksums $\mathcal{S}_{i,\ell}$ with α^ℓ for each vector of the line are equal to zero, then we assume that there are no distortions. Otherwise, the values $\mathcal{S}_{i,0}, \mathcal{S}_{i,1}, \ldots, \mathcal{S}_{i,r_2-1}$ for $i = 1, 2, \ldots, k_1$ are used for further restoration of the blocks of the ciphertext $\Omega_{i,1}^*(z), \Omega_{i,2}^*(z), \ldots, \Omega_{i,k_2}^*(z)$ with the help of well-known algorithms for decoding RS codes (of Berlekamp-Massey, Euclid, Forney and etc.).

The corrected (restored) sequences of redundant blocks of the ciphertext of the 2-nd level of monitoring $\vartheta_{1,\gamma}^{**}(z), \ldots, \vartheta_{k_1,\gamma}^{**}(z)$ are subject of the second transformation (decryption) of redundant blocks of the ciphertext of the 2-nd level of monitoring into redundant blocks of data of the 2-nd level of monitoring $\omega_{1,\gamma}^{**}(z), \ldots, \omega_{k_1,\gamma}^{**}(z)$. The redundant blocks of data of the 2-nd level of monitoring $\omega_{1,\gamma}^{**}(z), \ldots, \omega_{k_1,\gamma}^{**}(z)$ $(\gamma = k_2+1, k_2+2, \ldots, n_2)$ that have been formed again are used for forming code combinations of the RS code and their decoding.

4 Imitation Resistant Transmitting of Encrypted Information on the Basis of Crypt-Code Structures and Authentication Codes

Currently, to detect simulation by the adversary in the communication channel, an additional encryption regime is used to simulate imitated insertion (forming an authentication code [Message Authentication Code]) [1,2,4]. A drawback of this method to prevent imitation by the adversary is the lack of possibility to restore veracious information in the systems for transmitting information. Complexing the method to protect from imitating of data on the basis of message authentication codes (MAC) and the above-described solution based on expanding the RRPC with encrypting the redundant information, it shall make it possible to overcome the drawback of the known solution. Let us assume that MAC are formed as usual from the sequence consisting of k_2 number of sub-blocks containing k_1 blocks each of the ciphertext $\Omega_i(z)$ in each one. Then the procedure of generation of MAC $H_i(z)$ $(i = 1, \ldots, k_1)$ can be expressed:

$$\begin{cases} H_1(z) \to I_{h_1} : \Omega_1, \\ H_2(z) \to I_{h_2} : \Omega_2, \\ \ldots \ldots \ldots \ldots \ldots \ldots \\ H_{k_1}(z) \to I_{h_k} : \Omega_{k_1}, \end{cases}$$

where I_{h_i} is the operator of generation of an MAC on the key h_i $(i = 1, \ldots, k_1)$, $\mathbf{\Omega}_i = \{\Omega_{i,1}(z), \ldots, \Omega_{i,k_2}(z)\}$ is a vector equation of the super-block of the ciphertext, k_2 is the length of the super-block. Purposeful interfering of the adversary into the process of transmitting super-blocks of the ciphertext with the MAC calculated from them can cause their distorting. Correspondingly, on the receiving side, the super-blocks $\mathbf{\Omega}_i^* = \{\Omega_{i,1}^*(z), \ldots, \Omega_{i,k_2}^*(z)\}$ of the ciphertext are the source for calculating MAC:

$$\begin{cases} \widetilde{H}_1(z) \to I_{h_1} : \mathbf{\Omega}_1^*, \\ \widetilde{H}_2(z) \to I_{h_2} : \mathbf{\Omega}_2^*, \\ \cdots \cdots \cdots \cdots \cdots \\ \widetilde{H}_{k_1}(z) \to I_{h_{k_1}} : \mathbf{\Omega}_{k_1}^*, \end{cases}$$

where $\mathbf{\Omega}_i^* = \{\Omega_{i,1}^*(z), \ldots, \Omega_{i,k_2}^*(z)\}$ is the received super-block of the ciphertext; $\widetilde{H}_i(z)$ are MAC from the received blocks of the ciphertext, for $i = 1, 2, \ldots, k_1$.

Similarly to the previous solution for restoring the messages simulated by the adversary from the transmitted sequence of blocks of the ciphertext with MAC $\left\{ \{\mathbf{\Omega}_1, H_1(z)\}; \ldots; \{\mathbf{\Omega}_{k_1}, H_{k_1}(z)\}; \{\boldsymbol{\vartheta}_{k_1+1}, H_{k_1+1}(z)\}; \ldots; \{\boldsymbol{\vartheta}_{n_1}, H_{n_1}(z)\} \right\}_{\text{RRPC}}$, an extended RRPC is formed.

The sub-system of imitation-resistant reception of encrypted information on the basis of the RRPC and using MAC implements the following algorithm.

Input: the received sequence of vectors of encrypted message blocks with MAC: $\left\{ \{\mathbf{\Omega}_1^*, H_1^*(z)\}; \ldots; \{\mathbf{\Omega}_{k_1}^*, H_{k_1}^*(z)\}; \{\boldsymbol{\vartheta}_{k_1+1}^*, H_{k_1+1}^*(z)\}; \ldots; \{\boldsymbol{\vartheta}_{n_1}^*, H_{n_1}^*(z)\} \right\}_{\text{RRPC}}$.

Output: a corrected (restored) array of super-blocks of the ciphertext $\mathbf{\Omega}_1^{**}, \mathbf{\Omega}_2^{**}, \ldots, \mathbf{\Omega}_{k_1}^{**}$.

Step 1. Detection of the possible simulation by the adversary in the received sequence of blocks of the ciphertext with localization of the number i row vector with the detected false blocks of the ciphertext, is executed by comparing the MAC received from the communication channel $H_1^*(z), \ldots, H_{k_1}^*(z)$, $H_{k_1+1}^*(z), \ldots, H_{n_1}^*(z)$ and MAC $\widetilde{H}_1^*(z), \ldots, \widetilde{H}_{k_1}^*(z), \widetilde{H}_{k_1+1}^*(z), \ldots, \widetilde{H}_{n_1}^*(z)$ calculated in the sub-system of data reception. Next, a comparison procedure is performed for all row vectors $(i = 1, \ldots, k_1, k_1 + 1, \ldots, n_1)$:

$$\begin{cases} 1, & if \quad H_i^*(z) = \widetilde{H}_i(z); \\ 0, & if \quad H_i^*(z) \neq \widetilde{H}_i(z). \end{cases}$$

Step 2. Restoring veracious data by solving the congruences systems:

$$
\begin{cases}
\begin{cases}
\Omega_1^{**}(z) \equiv \Omega_{J_1,1}^{*}(z) \mod m_{J_1}(z), \\
\dots \dots \dots \dots \dots \dots \dots \dots \dots \dots \\
\Omega_1^{**}(z) \equiv \Omega_{J_{k_1},1}^{*}(z) \mod m_{J_{k_1}}(z), \\
\Omega_1^{**}(z) \equiv \omega_{J_{k_1+1},1}^{*}(z) \mod m_{J_{k_1+1}}(z), \\
\dots \dots \dots \dots \dots \dots \dots \dots \dots \dots \\
\Omega_1^{**}(z) \equiv \omega_{J_{n_1},1}^{*}(z) \mod m_{J_{n_1}}(z); \\
\end{cases} \\
\dots \dots \dots \dots \dots \dots \dots \dots \dots \dots \\
\begin{cases}
\Omega_{k_2}^{**}(z) \equiv \Omega_{J_1,k_2}^{*}(z) \mod m_{J_1}(z), \\
\dots \dots \dots \dots \dots \dots \dots \dots \dots \dots \\
\Omega_{k_2}^{**}(z) \equiv \Omega_{J_{k_1},k_2}^{*}(z) \mod m_{J_{k_1}}(z), \\
\Omega_{k_2}^{**}(z) \equiv \omega_{J_{k_1+1},k_2}^{*}(z) \mod m_{J_{k_1+1}}(z), \\
\dots \dots \dots \dots \dots \dots \dots \dots \dots \dots \\
\Omega_{k_2}^{**}(z) \equiv \omega_{J_{n_1},k_2}^{*}(z) \mod m_{J_{n_1}}(z), \\
\end{cases}
\end{cases}
\tag{10}
$$

where $J_1, J_2, \ldots, J_{n_1}$ are row vector numbers, if the comparison result for these MAC showed absence of distortions in sequence of blocks of the ciphertext $\Omega_j^{*}(z) = \{\Omega_{j,1}^{*}(z), \Omega_{j,2}^{*}(z), \ldots, \Omega_{j,k_2}^{*}(z)\}$. In accordance with the CRT solutions of systems (10) is the following:

$$
\Omega_j^{**} = \Omega_{J_1,j}^{*}(z)B_{J_1}(z) + \ldots + \Omega_{J_{k_1},j}^{*}(z)B_{J_{k_1}}(z) + \ldots
$$
$$
\ldots + \omega_{J_{k_1+1},k}^{*}(z)B_{J_{k_1+1}}(z) + \ldots + \omega_{J_{n_1},k}^{*}(z)B_{J_{n_1}}(z) \mod d\,(p, P_{k_v}(z)),
$$

where $B_{J_i}(z) = k_{J_i}(z)P_i(z)$ are polynomial orthogonal bases; $P_{k_v}(z) = \prod_{i=1,\ldots,k;i\neq v} m_i(z)$; v is the number of the detected "distorted" row vector; $P_{J_i}(z) = P_{k_v}(z)m_i^{-1}(z)$; $k_{J_i}(z) = P_{J_i}^{-1}(z) \mod m_{J_i}(z)$ $(j = 1, \ldots, k_2; i = 1, \ldots, n_1)$.

The values of polynomial orthogonal bases are calculated beforehand and are stored in the memory of the RRPC decoder. Restoring veracious blocks can be done by calculating the minimal deductions or by any other known method.

In a comparative evaluation of the effectiveness of the methods under consideration for providing imitation resistant transmission of encrypted information, we will assume that the adversary distorts the ciphertext blocks in the generated crypt-code structures with probability $p_{adv} = 2 \cdot 10^{-2}$. Probability p_{adv} distortion of each ciphertext block is constant and does not depend on the results of receiving the preceding elements of crypt-code structures. The probability $P(b)$ of reception crypt-code structures with b and more errors are presented in the Table 1, in accordance with which a higher recovery power is provided multi-dimensional crypt-code structures (RRP codes and RS codes). At what at the given values k_1, k_2, the closer the matrix being formed $\Phi_{n_1 \times n_2}$ to the square shape, the less the level of redundancy introduced.

Table 1. Effectiveness crypt-code structures

Method of construction	Structures	n	k	$d_{\min}$	$\frac{k}{n}$	$P(b)$
Crypt-code structures	(6, 3, 4)	6	3	4	0.5	0.1141
(RRPC)	(8, 4, 5)	8	4	5	0.5	0.01033
Multidimensional crypt-code	(6, 3, 4); (11, 5, 7)	66	15	28	0.227	0.000133
Structures: (RRPC); (RS)	(8, 4, 5); (8, 4, 5)	64	16	25	0.25	0.000106
Multidimensional crypt-code	(4, 3, 2); (6, 3, 4)	24	9	8	0.375	0.008862
Structures: (RRPC); (MAC)	(4, 3, 2); (8, 4, 5)	32	12	10	0.375	0.000802

5   Conclusion

The methods of information protection examined in this article (against simulation by the adversary) are based on the composition of block ciphering system and multi-character codes that correct errors by forming crypt-code structures with some redundancy. This redundancy is usually small and it makes it possible to express all the possible states of the protected information. Forming multidimensional crypt-code structures with several levels of monitoring makes it possible to not only detect simulating actions of the intruder but also, if necessary, to restore the distorted encrypted data with the set probability and their preliminary localization.

References

1. Ferguson, N., Schneier, B.: Applied Cryptography. Wiley, New York (2003)
2. Menezes, A.J., van Oorschot, P.C., Vanstone, S.A.: Handbook of Applied Cryptography. CRC Press, London (1997)
3. Knudsen, L.R.: Block chaining modes of operation. Reports in Informatics No. 207, Dept. of Informatics, University of Bergen, Norway (2000). October
4. Paar, C., Pelzl, J.: Understanding Cryptography. Springer, Heidelberg (2010)
5. McEliece, R.J.: A public-key cryptosystem based on algebraic coding theory. DSN Progress Report 42-44, pp. 114–116, JPL, Caltech (1978)
6. Niederreiter, H.: Knapsack-type cryptosystems and algebraic coding theory. Probl. Control Inf. Theory **15**(2), 159–166 (1986)
7. Samokhina, M.A.: Modifications of Niederreiter cryptosystems, its cryptographically strong and practical applications. In: Papers of the Proceedings of Moscow Institute of Physics and Technology, vol. 1(2), 121–128 (2009)
8. van Tilborg, H.: Error-correcting codes and Cryptography. Code-based Cryptography Workshop, Eindhoven (2011). May
9. Petlevannyj, A.A., Finko, O.A., Samoylenko, D.V., Dichenko, S.A.: Device for spoofing resistant coding and decoding information with excessive systematic codes. RU Patent No. 2634201 (2017)
10. Finko, O.A.: Group control of asymmetric cryptosystems using modular arithmetic methods. In: Papers of the XIV Inter. school-seminar "Synthesis and complexity of control systems", pp. 85–87 (2003)

11. Finko, O.A. Samoylenko, D.V.: Designs that monitor errors based on existing cryptographic standards. In: Papers of the VIII Intern. conf. "Discrete models in the theory of control systems", pp. 318–320 (2009)
12. Finko, O.A., Dichenko, S.A., Samoylenko, D.V.: Method of secured transmission of encrypted information over communication channels. RU Patent No. 2620730 (2017)
13. Bossen, D.C., Yau, S.S.: Redundant residue polynomial codes. Inf. Control **13**(6), 597–618 (1968)
14. Mandelbaum, D.: On efficient burst correcting residue polynomial codes. Inf. Control **16**(4), 319–330 (1970)
15. Yu, J-H., Loeliger, H-A.: Redundant Residue Polynomial Codes. In: Papers of the IEEE International Symposium of Inform. Theory Proceed, pp. 1115–1119 (2011)
16. Simmons, G.J.: Authentication theory/coding theory. In: Blakley, G.R., Chaum, D. (eds.) Advances in Cryptology. CRYPTO 1984. Lecture Notes in Computer Science. Springer, Heidelberg (1985)
17. Zubov, A.Y.: Authentication codes. Gelios-ARV, Moscow (2017)
18. Bloch, E.L., Zyablov, B.B.: Generalized Concatenated Codes. Sviaz, Moscow (1976)
19. MacWilliams, F.J., Sloane, N.J.A.: The Theory of Error-Correcting Codes. North-Holland Mathematical Library (1977)

On a New Intangible Reward
for Card-Linked Loyalty Programs

Albert Sitek[(✉)] and Zbigniew Kotulski

Institute of Telecommunications of WUT, Nowowiejska 15/19,
00-665 Warsaw, Poland
{a.sitek,z.kotulski}@tele.pw.edu.pl

Abstract. Card-Linked Loyalty is an emerging trend observed in the
market to use payment card as a unique identifier for Loyalty Programs.
This approach allows to redeem goods and collect bonus points directly
during a payment transaction. In this paper, we proposed additional,
intangible reward, that can be used in such solutions: shorter transaction
processing time. We presented a complete solution for it: Contextual
Risk Management System, that can make a dynamic decision whether
Cardholder Verification is necessary for the current transaction, or not.
It is also able to maintain an acceptable level of risk approved by the
Merchant. Additionally, we simulated the proposed solution with real-
life transaction traces from payment terminals and showed what kind of
information can be determined from it.

Keywords: Card-Linked Loyalty · Context · Risk Management
Transaction security · Payment card

1 Introduction

A loyalty program (LP) is an integrated system of marketing actions that aims to
reward and encourage customers' loyal behavior through incentives [1,2]. LPs,
in a variety of their forms, are widely spread across the world. According to
the recent report [3], an average customer in the U.S. belongs to 14 Loyalty
Programs. Moreover, 73% of the U.S. customers are more likely to recommend
brands with good LP [3]. The ubiquity of loyalty programs has made them a
seeming "must-have" strategy for organizations. Hence, it is no surprise that
most retailers have introduced LPs to remain competitive [4].

There are a lot of research papers, that analyzed Loyalty Programs from
different angles. For example, authors of [4] discussed what do customers get
and give in return for being a member of the loyalty program. Additionally,
they examined the effect of program and brand loyalty on behavioral responses,
including a share of wallet, the share of purchase, word of mouth, and willingness

© Springer Nature Switzerland AG 2019
J. Pejaś et al. (Eds.): ACS 2018, AISC 889, pp. 332–345, 2019.
https://doi.org/10.1007/978-3-030-03314-9_29

to pay more. On the other hand, authors of [5] analyzed effects of loyalty program rewards on store loyalty and divided them into following groups:

- Tangible (hard benefits): monetary incentives like discounts, vouchers, free hotel stays, tickets,
- Intangible (soft rewards): e.g., preferential treatment, an elevated sense of status, services, special events, entertainment, priority check-in, and so on.

Their research shows that the underlying effects of reward types on preferences and intended store loyalty differ depending on the level of consumers' personal involvement. In case of high personal involvement, compatibility with the store's image and intangible rewards increase LP preference and loyalty. Also, the time required to obtain the reward (delayed/immediate) has no impact. In the event of low personal involvement, immediate and tangible rewards increase LP preference and loyalty. Compatibility with the store image is not important. Finally, authors of [6] sketches the loyalty trends for the twenty-first century. They emphasized the role of new technologies by claiming that *"without sophisticated technology, the loyalty program operator is confined to a punch card or a stamp program - anonymous versions of reward and recognition that our grandparents may have liked, but which simply will not work in the wired world"*. That's completely true; one can observe constant migration from legacy dedicated loyalty cards to the cards stored digitally on the application installed on the smartphone [7]. According to the new statistics [8], 57% of consumers want to engage with their loyalty programs via mobile devices.

There is also an emerging trend observed on the market to resign from dedicated loyalty cards and switch directly to the payment cards. This technique is called Card-Linked Loyalty [9] and works in the following way: during the payment transaction, Point-of-Sale (POS) terminal reads the card number and verifies on the dedicated server if there are some discounts/promotions to be proposed to the customer. If yes, the customer can decide whether he wants to redeem an offered reward or not. Also, some bonus points can be added automatically after the transaction to the customer's account. Such an approach has plenty of advantages:

- Payment cards are widely spread across the world,
- No need to carry another plastic card,
- No need to print off rewards or coupons: just redeem your rewards during standard payment process,
- No need to enroll manually or online: sign up to the Loyalty Program during your payment,
- No need to download dedicated applications,
- Does not interrupt payment process; *Loyalty should be part of the payment process and not interrupt it* [9].

In this paper we present the Contextual Risk Management System for Payment Transactions that can be used together with Card-Linked Loyalty Program. It is capable to make a dynamic decision whether the PIN verification is necessary or not during the present payment transaction. The decision is made based on Cardholder's reputation calculation based on historical transactions, and other contextual factors like length of the queue, local promotions etc. Thanks to that, loyal and trustworthy customers can be awarded new intangible reward: shorter processing time during a card-based payment transaction. Moreover, our approach assures that the acceptable level of risk will be maintained during its operation. To build our solution, we used the dedicated Reputation System previously presented in [10]. Additionally, we simulated and verified the whole System using productive transaction traces from Polish market, described in [11].

The rest of this paper is organized as follows: Sect. 2 provides technical background to fully understand consecutive sections, Sect. 3 presents System's architecture, Sect. 4 describes performed tests and validations of the systems in details and Sect. 5 contains tests results, while Sect. 6 concludes the paper and maps out future work.

2 Card-Present Transaction Overview

Transactions performed with a payment card can be divided into two groups:

- Card-not-present (CNP): transactions perform without the physical presence of the card, for instance, via Internet (so-called eCommerce),
- Card-present (CP): transactions performed with a physical card by entering in (or tapping) to the payment terminal.

On the other hand, during the card-present transaction, card's data can be read directly from the magstripe card (deprecated), or from a smartcard. For this article we are focusing only on card-present transaction made with a smartcard.

Such a transaction is compliant with EMV specification [12]. This standard has been firstly proposed by Europay, MasterCard, and Visa in 1993. Currently it is promoted by EMVCo which associates all major Payment Card Schemes: Mastercard, Visa, Discover, Japan Credit Bureau (JCB), China UnionPay (CUP) and American Express (AmEx), and covers both contact and contactless payment cards. According to some statistics [12], the transaction made with the contactless card is 63% faster than using cash and 53% faster than a traditional magnetic stripe credit card transaction. There is also emerging trend observed on the market to emulate Contactless Payment Card with a smartphone [13]. It is thanks to services like Samsung Pay [14], ApplePay [15] or Google Pay [16] that uses Near Field Communication (NFC) interface [17] and Host Card Emulation technique (HCE) [18]. Thankfully, a smartphone emulating payment card is treated and read by the payment terminal as a physical card, so, no changes are required in the payment infrastructure to handle those devices correctly.

Payment transaction compliant with the EMV specification consists of several steps [19]. In [10] one can find a figure that depicts in details all possible transaction flows that can happen for both contact and contactless cards. The most remarkable steps that have a significant impact on the transaction processing time are Cardholder Verification (CV) and Transaction Authorization. In this article we are focusing only on the first one. The Cardholder can be verified by following Cardholder Verification Methods (CVMs): No CVM (no verification at all), Online PIN (verified by the Issuer), Offline PIN (verified by the card, only for contact EMV), Consumer Device CVM (CDCVM, verified by the device, only for HCE transactions), Signature (verified by the Merchant). The decision which Cardholder Verification Method should be used is being made based on terminal's configuration and data retrieved from the card (encoded on the card by the Issuer during its personalization phase). In case of Cardholder Verification, those parameters are: Terminal Capabilities (indicates which Cardholder Verification Method is supported by the terminal), and Cardholder Verification Limit (CVL, only for contactless transactions, the amount above which the Cardholder must be verified: currently 50 PLN in Poland). One can easily spot, that transaction processing rules are constant for every transaction: it means that each Cardholder is treated equally, no matter what's his history and the context of a current transaction. There are also clear rules regarding risk related to the transaction. If a disputed transaction has been authorized:

- With PIN verification, then it would be charged to the customer,
- With signature verification, then it would be charged to the merchant,
- By the card (Offline Authorization), it would be charged to issuing bank.

Such an approach is effortless, but it causes that a lot of transactions are processed "time and user experience-ineffectively" [10]. One can imagine that the transaction flow could be tailored to the Cardholder and to the particular transaction, based on various contextual factors. It may give a lot of profits, e.g. greater Cardholder's loyalty, better user experience, shorter transaction processing time, etc. It should also assure an acceptable level of transaction security. This is the main motivation why the context-aware solution for payment transactions started to appear [10,20,21]. They enable merchants to take some risk by allowing some payment transactions being authorized, for example, without any verification in exchange for above-mentioned profits. Such systems could be very useful in the markets, where the level of fraudulent transactions is low. For instance, such an information can be found in the European Central Bank's report [22], which says that the level of deceptions is very low in certain countries.

3   Contextual Risk Management System

The usage of contextual information during payment transaction processing has been firstly discussed in [20], where a new Cardholder Verification Method: One-time PIN verification was proposed. This method assumed, that each transaction was authorized online and the decision if PIN verification should be performed by the Issuer based on various contextual factors (like: place and time of the transaction, Cardholder's reputation, etc.). In the case of positive decision, encrypted PIN (or One-time PIN) was sent to the terminal and a payment application verified, if the encrypted PIN entered by the Cardholder was the same as the one received from the Issuer.

Another approach has been proposed in [21]. This Contextual Risk Management System allows performing dynamic decision whether the transaction should be authorized'offline' or'online'. To make the decision, simple algorithm and reputation system was proposed. Unfortunately, this reputation system is not capable to consider all possible transaction flows.

To extend and improve the previous solution, in the paper [10] a new Cardholder's Reputation System was proposed. It covers all possible transaction flows, and assumes, that each transaction flow has a constant rating assigned to it. After the transaction with a certain flow, Cardholder receives a proper rating. To determine Cardholder's reputation, a weighted average of ratings from last N transactions is calculated before the forthcoming transaction.

All mentioned papers presented various enhancements for current card payment ecosystem, however, all of them were tested using synthetic sets of data (prepared based on experts' knowledge), because of the lack of realistic production data. That is why a new approach to gather and analyze transaction traces collected directly from a payment terminal was proposed in [11]. Moreover, it describes an experiment performed on productive transaction traces gathered from 68 payment terminals through 6 months.

The proposed Contextual Risk Management System (CRMS) has been designed based on best ideas presented in above-mentioned papers. Its main features are as follows:

1. It is dedicated for huge merchants,
2. It allows performing dynamic decision whether the Cardholder should be verified with a PIN, or not,
3. During the decision-making process, it uses Cardholder's reputation calculated according to the algorithm presented in [10],
4. It was simulated and verified with productive transaction traces gathered within the experiment described in [11].

One must be aware, that utilization of the CRMS must be performed in compliance with General Data Protection Regulation (GDPR), because it can be classified as profiling tool that utilizes pseudonymized personal data.

In the rest part of this section, we present a high-level architecture of the CRMS, describe Risk Calculation and Decision-Making algorithms used in it, and try to estimate the Fraud Probability associated with the usage of this system.

3.1 High-Level Architecture

Figure 1 presents a high-level architecture of the CRMS. Whole transaction process should look as follows:

1. During the transaction, payment terminal reads card's data, tokenizes it and sends transaction data (amount, tokenized card) to the CRMS,
2. CRMS calculates the decision how the current transaction should be processed: with or without Cardholder Verification,

3. CRMS sends back the final decision to the terminal, and the transaction is completed according to it.

Such an approach has a few important features: the CRMS is located inside internal network together with payment terminals, so the delay caused by telecommunication overhead is negligibly small, and it handles only tokenized card's data, so, it is not obliged to be compliant with Payment Card Industry Data Security Standard (PCI DSS). The decision whether the current transaction should be processed with Cardholder Verification, or not, is being made based on Risk Calculation described in the next sections.

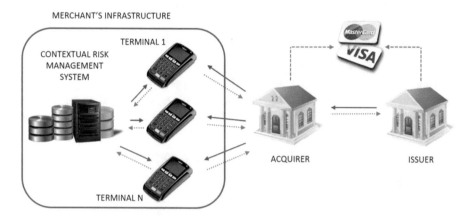

Fig. 1. High-level contextual risk management system architecture.

3.2 Risk Calculation

In general, the risk associated with a usage of the proposed system can be calculated as follows:

$$Risk = a * p, \tag{1}$$

where a is the amount of current transaction, and p denotes the probability that the current transaction will become fraudulent. One can easily spot, that the calculated risk denotes the maximal theoretical loss per each transaction.

To get Cardholder's Reputation (R) into account, above equation can be extended to the following form:

$$Risk = a * p * f(R), \tag{2}$$

where $f(R)$ indicates an impact of Cardholder's Reputation on theoretical risk. $f(R)$ function should fulfill following requirements: it should approach infinity for $R \to R_{min}$, and should have its minimum value for $R = R_{max}$. It is worth noticing that the shape of $f(R)$ function has an impact on a few important facts:

– For which R, $f(R) = 1$: it means for what reputation, the calculated risk is equal to theoretical one,

- What is $f(R_{max})$: e.g., if $f(R_{max}) = 1/2$, it means that maximal reputation causes that calculated risk is half of the theoretical one.

Assuming that the reputation $R \in \langle 0, R_{max} \rangle$, a good example of function f, that fulfills above-mentioned requirements, can be:

$$f(R) = \begin{cases} \infty & \text{if } R = 0 \\ \frac{a}{b*R} & \text{if } R \in (0, R_{MAX}) \end{cases}, \tag{3}$$

where a and b are the parameters which determine the shape of function f and which can be chosen dynamically, based on some contextual factors. A similar function will be used for further simulations presented in this paper.

3.3   Decision-Making

During the Decision-Making process, CRMS will set maximal risk ($Risk_{max}$) accepted to be taken by the Merchant during current transaction. This can be done based on some contextual factors, e.g. current length of the queue, content of the basket etc. Next, CRMS will calculate the risk associated with the current transaction ($Risk_{curr}$):

$$Risk_{curr} = a_{curr} * p * f(R). \tag{4}$$

Then, the final decision is made as follows:

$$\begin{cases} Risk_{curr} \leq Risk_{max} \Rightarrow & \text{without Cardholder Verification} \\ Risk_{curr} > Risk_{max} \Rightarrow & \text{with Cardholder Verification} \end{cases}. \tag{5}$$

3.4   Fraud Probability

There are a few types of Card Frauds: usage of lost or stolen card, cloned card (skimming), or stolen card data to perform eCommerce transaction. In practice, presented CRMS is only vulnerable to the transactions made with the lost or stolen card, because it operates only with EMV compliant smartcards (these cards are not prone to cloning), and because it works only for CP transactions.

Next, we will try to estimate the fraud probability by the example of the Polish market. It will also be used for further simulations. Analysis created by National Bank of Poland [23] presents the level of fraudulent transactions based on data gathered from Issuers (Banks) and Acquirers. It shows that:

- The number of transactions made with lost or stolen cards accounts for approximately 13% of all fraudulent transactions recorded by Issuers,
- According to Acquirers, the number of fraudulent transaction accounts for 0.001% of all processed transactions,
- An average amount of fraudulent transaction is 830.40 PLN.

It is worth to mention that we predict the presented system to operate (on Polish market) for transactions with the maximum amount of 200 PLN. Based on that, we estimate fraud probability at the level of 0.0001%.

4 The Experiment

As described in Sect. 1, we verified the proposed CRMS with productive data described in [11]. This dataset contains of 1048382 transactions' traces made using 189898 unique payment cards, collected within 6 months, in 18 shops belonging to one of the retail chain. All those shops are located in Northwest region of Poland, near the border with Germany.

4.1 Experiment's Details

The aim of our experiment was to simulate "what will happen if the proposed CRMS was deployed in given retail chain". Specifically, what could be the benefits from the usage of such system productively, and what would be an impact of acceptable level of the risk on those benefits. To do so, we implemented all algorithms described before, took the transaction history of each card token, and simulated which transaction from the history would be processed without CV. Then, we calculated the gain of time that could be achieved from the usage of

	time	amount	event_sequence	pt[s]	rate	sel.	rate_sel
0	2017-06-01 10:55:08	15.89	[crs, cr, pofs, pofv, onr]	4	7	0	7
1	2017-06-12 10:01:36	50.44	[crs, cr, pofs, pofv, onr]	5	7	0	7
2	2017-06-30 10:44:59	18.92	[crs, cr, pofs, pofv, onr]	5	7	0	7
3	2017-07-03 13:05:22	38.02	[crs, cr, pofs, pofv, onr]	4	7	0	7
4	2017-07-07 10:55:53	5.18	[crs, cr, pofs, pofv, onr]	7	7	0	7
5	2017-07-10 10:55:56	13.98	[crs, cr, pofs, pofv, onr]	4	7	1	0
6	2017-07-27 11:00:30	5.98	[crs, cr, pofs, pofv, onr]	7	7	1	0
7	2017-07-29 13:19:43	2.99	[crs, cr, pofs, pofv, onr]	4	7	1	0
8	2017-08-09 06:53:39	40.41	[crs, cr, pofs, pofv, onr]	4	7	0	7
9	2017-08-19 13:17:04	24.23	[crs, cr, pofs, pofv, onr]	4	7	1	0
10	2017-09-04 08:52:21	41.96	[crs, cr, pofs, pofv, onr]	4	7	0	7
11	2017-09-08 11:26:12	7.42	[crs, cr, pofs, pofv, onr]	4	7	1	0
12	2017-09-09 09:35:01	18.67	[crs, cr, pofs, pofv, onr]	6	7	1	0
13	2017-09-15 13:12:13	30.88	[crs, cr, pofs, pofv, onr]	4	7	0	7
14	2017-09-18 13:59:18	18.93	[crs, cr, pofs, pofv, onr]	8	7	1	0
15	2017-09-25 14:27:39	22.54	[crs, cr, pofs, pofv, onr]	4	7	0	7
16	2017-10-27 11:22:14	3.99	[crs, cr, pofs, pofv, onr]	5	7	1	0

Selected: 8
Total amount: 96.19
Saved time: 42s

Fig. 2. Transaction history of exemplary card token.

the system. Figure 2 presents an example transaction history of an exemplary card token, together with simulation details. The description of each column is as follows:

- time: transaction time; amount: transaction amount, in PLN,
- event_sequence: the detailed trace of given transaction. For example, [crs, cr, pofs, pofv, onr] denotes that during the transaction there were following events: Card Read Started, Card Read, PIN Offline Started, PIN Offline Verifies, Online Result received,
- pt: PIN input time, in seconds. It indicates what would be the gain of time if the certain transaction was processed without PIN,
- rate: indicates the score that will be given to the Cardholder for performing the transaction with given sequence of events. It is the parameter of Reputation System. All scores taken for the simulation can be found in [10],
- sel.: indicates, if given transaction will be selected to be processed without Cardholder Verification, if the CRMS was enabled,
- rate_sel: shows the score that will be given for the Cardholder considering, that CRMS was enabled, and some transaction could be processed without Cardholder Verification.

As we can see from the example illustrated in Fig. 2, there were 8 transactions selected from the transaction history to be processed without Cardholder Verification, what gave 42 s of time gain. Summary value of those transactions was 96.19 PLN. Additionally, Fig. 3 shows the Cardholder's Reputation in time in case of the CRMS is disabled (derived from column "rate"), while Fig. 4 presents the simulated situation when it is enabled (see column "rate_sel").

To perform above-mentioned simulation, we implemented a set of dedicated Python's scripts. We used following libraries: NumPy (fundamental package for scientific computing [24]), pandas (the library providing high-performance, easy-to-use data structures, and data analysis tools [25]), and Matplotlib (plotting library [26]). We wrote our scripts in IPython [27] (the system for interactive scientific computing). As an IDE (Integrated Dev. Env.) we used Jupyter [28].

During our simulations we used following algorithms and parameters:

- Reputation Calculation: the one mentioned in Sect. 3, with all parameters proposed in [10]. For instance, $R_{min} = 0$, $R_{max} = 10$,
- Fraud Probability: to mitigate some risk connected to probability estimation, we took additional security factor and multiplied it by 100. So, Fraud Probability taken to our simulation was equal to 0.0001,
- $f(R)$ Function: the one proposed in Sect. 3.2, with parameters $a = 10$ and $b = 1$. Such an approach caused, that for great reputation (equal to 10), the risk calculated will be equal to the theoretical one. For poorer reputation, the risk will approach infinity.

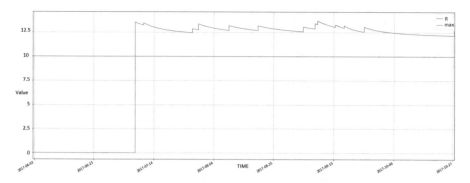

Fig. 3. Example Cardholder's reputation when CRMS is disabled.

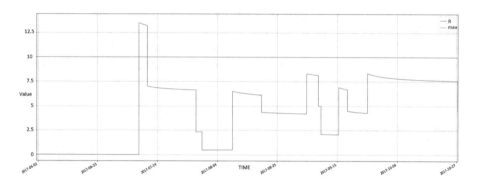

Fig. 4. Example Cardholder's reputation when CRMS is enabled.

So, selected parameters gave us a clear view on the relationship between risk taken by the Merchant and max. amount of transaction that will be allowed to be processed without PIN verification in case of excellent Cardholder's Reputation. For example, when the Merchant accepted the risk at the level of 0.008 PLN per transaction, it denotes that the transaction for max. 80 PLN will be processed without PIN verification, for a Cardholder with Reputation equal to 10.

During our Experiment, we simulated what would be an impact on benefits from the usage of the proposed system, depends on the risk accepted by the Merchant. We verified the range of risks from 0.005 PLN up to 0.02 PLN, what corresponds to the range of amounts between 50 PLN and 200 PLN.

5 Experiment's Results

Figure 7 presents the number of Customers with at least 1 transaction selected by the CRMS for processing without PIN verification, during simulated period of time. This number varies from 11700 to 23724, what gives from 6.19% to 12.49% of all recorded card tokens. On the other hand, the number of all selected transactions can be found in

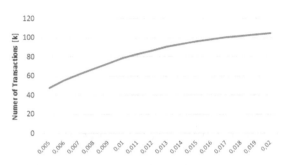

Fig. 5. Number of selected transactions.

Fig. 5. It shows that this number varies from 48055 up to 104905, what fives from 4,58% to 10% of all transactions. In our opinion such a situation could happen because the Experiment has been conducted in Poland, near the Polish-German border, where there are a lot of tourists visiting this area and buying things occasionally. Moreover, nowadays majority of Cardholders are using more than one payment card. A great improvement for the proposed CRMS would be a dedicated web service where Customers can log-in and link several payment cards to one account.

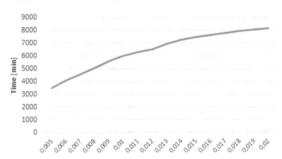

Fig. 6. Time gained because of the usage of the system.

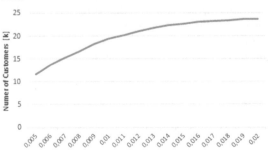

Fig. 7. Customers with at least 1 transaction selected.

After that, the CRMS could operate on the level of a client rather than on pure token. Figure 6 shows the time gained because of usage of proposed CRMS. This time varies from 3511 up to 8129 min, what stands for 58.5 to 135.5 h. We must admit that this time is quite impressive, considering that analyzed the transaction traces from 18 stored collected within 6 months. Next, in Fig. 8 one can see the collation between theoretical maximal loss caused by the usage of the CRMS and maximal loss calculated from the results of our simulation. In other world, it shows an impact of Cardholder Reputation and $f(R)$ function on maximal losses.

Such perspective is valuable during setting the CRMS's parameters. Finally, Fig. 9 shows maximal cost that must be paid for rewarding single Cardholder, selection of one transaction or for gaining one minute of processing time. Such an information is crucial for the Merchant during selection on accepted risk for the proposed CRMS.

6   Conclusion

Loyalty Programs are an immanent part of modern marketing strategy. They are using more and more sophisticated techniques to increase satisfaction of the customer (Quality of Experience). An emerging trend in this field is usage of payment card as a unique identifier that identifies the customer in Loyalty Program. In this paper we proposed a New Intangible Reward for Card-Linked Loyalty Program: shorted transaction processing time for frequent and trusted buyers. It uses dedicated CRMS that decides whether Cardholder Verification step should be perform during certain transaction, or not. This decision is made based on Cardholder Reputation calculated with from the transaction history and other contextual factors like length of the queue, content of the basket etc.

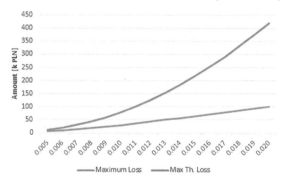

Fig. 8. Maximal losses.

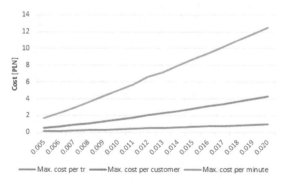

Fig. 9. Max cost per one promoted Cardholder, selected transaction and gained minute

We created special simulation environment and simulated it with the productive data collected in 18 shops from single retail chain located in Northwest part of Poland. The results show what type of information can be gathered from such simulations: what are potential profits from usage of such a system, and what are the risks connected to it. They also help to set up adequate CRMS's parameters according to preferences of the Merchant.

We believe, that analogous simulation should be performed on real-life data collected from the Merchant and locations where the similar system will be planned to deploy. In our future work, we would like to perform similar simulation changing the parameters of

Reputation System and finding it's optimal settings. Moreover, it would be valuable to collect an analogous simulation dataset in a different region of the country, which is not impacted by many occasional consumers and tourists.

References

1. Kang, J., Brashear, T., Groza, M.: Customer-company identification and the effectiveness of loyalty programs. J. Bus. Res. **68**, 464–471 (2015)
2. Leenheer, J., van Heerde, H.J., Bijmolt, T.H., Smidts, A.: Do loyalty programs really enhance behavioral loyalty? An empirical analysis accounting for self-selecting members. Int. J. Res. Mark. **24**(1), 31–47 (2007)
3. Bond. Brand Loyalty, Visa: The Loyalty Report 2017 (2017)
4. Theng So, J., Danaher, T., Gupta, S.: What do customers get and give in return for loyalty program membership? Aust. Mark. J. (AMJ) **23**, 196–206 (2015)
5. Meyer-Waarden, L.: Effects of loyalty program rewards on store loyalty. J. Retail. Consum. Serv. **24**, 22–32 (2015)
6. Capizzi, M.T., Ferguson, R.: Loyalty trends for the twenty-first century. J. Consum. Market. **22**(2), 72–80 (2005)
7. Marquardt, P., Dagon, D., Traynor, P.: Impeding individual user profiling in shopper loyalty programs. In: Danezis, G. (ed.) Financial Cryptography and Data Security, pp. 93–101. Springer, Heidelberg (2012)
8. Everything you need to know about customer loyalty [statistics], January 2018. https://revelsystems.com/blog/2018/01/27/customer-loyalty-statistics/. Accessed 12 Mar 2018
9. How does card-linking loyalty work? vPromos, May 2016. https://cardlinx. org/wordpress_8-2015/wp-content/uploads/2016/05/4-vPromos-Pres.pptx..pdf. Accessed 12 Mar 2018
10. Sitek, A., Kotulski, Z.: Cardholder's reputation system for contextual risk management in payment transactions. In: Rak, J., Bay, J., Kotenko, I., Popyack, L., Skormin, V., Szczypiorski, K. (eds.) Computer Network Security: 7th International Conference on Mathematical Methods, Models, and Architectures for Computer Network Security, MMM-ACNS 2017, Warsaw, 28–30 August 2017, Proceedings, pp. 158–170. Springer (2017)
11. Sitek, A., Kotulski, Z.: POS-originated transactions traces as a source of contextual information for Risk Management Systems in EFT transactions. EURASIP J. Inf. Secur. **1**, 5 (2018). https://doi.org/10.1186/s13635-018-0076-9
12. EMVCo: EMV Specifications. http://www.emvco.com/specifications.aspx. Accessed 24 Mar 2018
13. Press Information About HCE Development on the Market. http://www. bankier.pl/wiadomosc/Eksperci-Platnosci-HCE-to-rynkowy-przelom-3323308. html. Accessed 24 Mar 2018
14. Samsung Pay Homepage. http://www.samsung.com/us/samsung-pay/. Accessed 24 Mar 2018
15. Apple Pay Homepage. https://www.apple.com/apple-pay/. Accessed 25 May 2018
16. Google Pay Homepage. https://pay.google.com/intl/pl_pl/about/. Accessed 24 Mar 2018
17. Near Field Communication. http://nfc-forum.org/what-is-nfc/. Accessed 24 Mar 2018

18. Host   Card   Emulation.   https://en.wikipedia.org/wiki/Host_card_emulation. Accessed 24 Mar 2018
19. EMV Transaction Steps. https://www.level2kernel.com/flow-chart.html. Accessed 24 Mar 2018
20. Sitek, A.: One-time code cardholder verification method in electronic funds transfer transactions. In: Annales UMCS ser. Informatica, vol. 14, no. 2, pp. 46–59. Universitatis Mariae Curie-Skłodowska, Lublin (2014)
21. Sitek, A., Kotulski, Z.: Contextual management of off-line authorisation in contact EMV transactions. Telecommun. Rev. Telecommun. News **88**(84), 8–9, 953–959 (2015). (in polish)
22. European Central Bank, Germany: Fourth report on card fraud (2015)
23. Department of Payment System, National Bank of Poland, Warsaw, Poland: An assessment of the functioning of Polish payment system in 1st quarter 2017 (2017). (in Polish)
24. Numpy Homepage. http://www.numpy.org/. Accessed 24 Mar 2018
25. Pandas Homepage. http://pandas.pydata.org/. Accessed 24 Mar 2018
26. Matplotlib Homepage. https://matplotlib.org/. Accessed 24 Mar 2018
27. Pérez, F., Granger, B.E.: IPython: a system for interactive scientific computing. Comput. Sci. Eng. **9**(3), 21–29 (2007). http://ipython.org
28. Jupyter IDE Homepage. http://jupyter.org/. Accessed 24 Mar 2018

KaoChow Protocol Timed Analysis

Sabina Szymoniak[(⊠)]

Institute of Computer and Information Sciences, Czestochowa University
of Technology, Dabrowskiego 69, 42-200 Czestochowa, Poland
sabina.szymoniak@icis.pcz.pl

Abstract. This paper discusses the problem of timed security protocols'
analysis. Delay in the network and encryption and decryption times are
very important from a security point of view. This operations' times
may have a significant influence on users' security. The timed analysis
is based on a special formal model and computational structure. For
this theoretical assumptions, a special tool has been implemented. This
tool allows to calculate the correct protocol's execution time and carry
out simulations. Thanks to this, it was possible to check the possibility of Intruder's attack including various time parameters. Experimental
results are presented on KaoChow protocol example. These results show
how significant for security is time.

Keywords: KaoChow protocol · Timed analysis · Security protocols
Simulations

1 Introduction

Security protocols (SP) are an integral element of Internet communication.
Thanks to them, an appropriate level of security is assured. The SP's operation involves the execution of a sequence of steps. These steps can be aimed at
passing on confidential information or mutual authentication of users. Appropriately selected elements and security of communication can make the identity of
users and their data remain secret.

Security protocols are vulnerable to wicked persons called Intruders. The
Intruder aims to launch an attack to steal information sent between honest users
and then use it. One of the typical attacks carried out in computer networks is
the man in the middle attack. In this attack, the Intruder mediates between two
honest users. The messages sent do not reach their recipients immediately. The
messages reach the Intruder first. Intruder acquires knowledge about messages
and tries to decrypt the messages as much as he can. Then he sends the messages
to the correct recipient, impersonating the sender of the message.

Due to the appearance of wicked users on the network, it is necessary to study
security protocols and check their vulnerability to attacks by Intruders [16]. So
far, many methods for verify security protocols have been developed. Among
them are inductive methods [2], deductive methods [3], model checking [4] and

© Springer Nature Switzerland AG 2019
J. Pejaś et al. (Eds.): ACS 2018, AISC 889, pp. 346–357, 2019.
https://doi.org/10.1007/978-3-030-03314-9_30

other methods [5,6,13,14,18,19]. There were also many tools used to verify SP. Among them are ProVerif [8], Scyther [9] and AVISPA [7].

SP' security also depends on time. Sometimes fractions of seconds can decide on the security of communication participants. If the Intruder has more time to process the message properly, it may find that he can decipher the message and get confidential information. This action is another argument pointing to need to conduct the verification process of protocols.

The analysis of time impact on SP security appeared only in the works of Jakubowska and Penczek [10,11]. These works were related to the calculation of correct communication session duration and its impact on Intruder's activity. Unfortunately, these studies were not continued. In the paper [12] a formal model was proposed. This model allowed to define a security protocol as an algorithm, and then to determine a set of specific in time executions of this protocol.

The combination of the methods described in [10,11] and formal model from [12] has become the basis for a new method of verifying security protocols including time parameters. In our approach, we try to calculate the duration of the communication session and check the impact of various time parameters' values on the security of honest users and the Intruder's capabilities. The time parameters examined here are times of encryption and decryption as well as delays in the network. We analyze the fixed and random values of these parameters to enable a real image of Internet communication.

The rest of this paper is organized as follows. In the second section, we present the KaoChow protocol, which we used to show the results of our research. Next section shows our research methodology. The fourth section consists of experimental results for KaoChow protocol. The last section includes our conclusions and plans for the future.

2   KaoChow v.1 Protocol

One of SP is KaoChow (v.1) protocol. It was described by Long Kao and Randy Chow in [1]. This protocols' task is to establish a new symmetric (session) key and mutual authentication of two users, using a trusted server. The new session key is generated by a trusted server. A protocol should guarantee the secrecy of the new session key, which means that only users A and B and trusted server should know it. In addition, the authenticity of the session key should be guaranteed, which means that key will be generated and sent by server S for encryption and decryption in the current communication session. KaoChow protocol must also ensure the mutual authentication of users A and B. The scheme of the KaoChow protocol in Common Language is as follows [21]:

$$\alpha_1 \ A \to S : I_A, I_B, N_A$$
$$\alpha_2 \ S \to B : \{I_A, I_B, N_A, K_{AB}\}_{K_{AS}}, \{I_A, I_B, N_A, K_{AB}\}_{K_{BS}}$$
$$\alpha_3 \ B \to A : \{I_A, I_B, N_A, K_{AB}\}_{K_{AS}}, \{N_A\}_{K_{AB}}, N_B$$
$$\alpha_4 \ A \to B : \{N_B\}_{K_{AB}}$$

In the first communications' step, the user A sends to the trusted server S the identifiers I_A and I_B and his random number N_A. Server composes two ciphertexts and sends them in one message to the user B. Both ciphertexts contain the same cryptographic objects, i.e. identifiers of both users, a random number of the user A and symmetric key, generated by the server, which will be shared by the users A and B. However, the first ciphertext will be encrypted with a symmetric key shared between the server and user A and the second with key shared between user B and server. The user B creates his message, which contains the ciphertext of the previous step, addressed to A, and also random number N_A, encrypted key K_{AB} and its random number. In the last step of this protocol, A returns B to the random number N_B encrypted with the key K_{AB}.

KaoChow protocol exposed to an attack in which the old symmetric key will be reused. The execution scheme for this attack in Common Language is as follows [1]:

$$\alpha_1 \ A \to S : \qquad\qquad I_A, I_B, N_A$$
$$\alpha_2 \ S \to B : \qquad\qquad \{I_A, I_B, N_A, K_{AB}\}_{K_{AS}}, \{I_A, I_B, N_A, K_{AB}\}_{K_{BS}}$$
$$\beta_2 \quad I(S) \to B \ : \{I_A, I_B, N_A, K_{AB}\}_{K_{AS}}, \{I_A, I_B, N_A, K_{AB}\}_{K_{BS}}$$
$$\beta_3 \quad B \to I(A) \ : \{I_A, I_B, N_A, K_{AB}\}_{K_{AS}}, \{N_A\}_{K_{AB}}, N_B$$
$$\beta_4 \quad I(A) \to B \ : \{N_B\}_{K_{AB}}$$

In this attack messages from α_2 step are reused in second session (β).

In the rest of this paper timed version of KaoChow protocol will be used. A timed version is formed by exchange random numbers by timestamps.

3   Research Methodology

Our research was based on the formal model and a computational structure presented in [12]. We expanded definitions included in [12] by the time parameters. Thanks to this, it is possible to make a full specification of step and protocol in both versions, timed and untimed.

The new formal model allows to prepare the following definitions:

- time conditions' set, which includes delays in the network,
- step, which includes the protocol's external and internal actions,
- set of steps (protocol).

The new computational structure allows to define:

- real protocol's executions (including the Intruder),
- protocol's interpretations, which ensure generation of executions different in the time,
- timed step,
- user's knowledge,
- protocol's calculation,
- time dependencies.

In structure timestamps, message sending times and delays in the network were mapped into non-negative real numbers.

According to the definition of timed protocol's step described in [20] formal definition of timed KaoChow protocol is presented:

- $\alpha_1 = (\alpha_1^1, \alpha_1^2)$:

  - $\alpha_1^1 = (A; S; I_A, I_B, \tau_A)$,
  - $\alpha_1^2 = (\tau_1; D_1; \{I_A, I_B, \tau_A\}; \{\tau_A\}; \tau_1 + D_1 - \tau_A \leq \mathcal{L}_\mathcal{F})$.

- $\alpha_2 = (\alpha_2^1, \alpha_2^2)$:

  - $\alpha_2^1 = (S; B; \langle I_A, I_B, \tau_A, K_{AB} \rangle_{K_{AS}}, \langle I_A, I_B, \tau_A, K_{AB} \rangle_{K_{BS}})$,
  - $\alpha_2^2 = (\tau_2; D_2; \{\tau_A, K_{AB}, I_A, I_B, K_{AS}, K_{BS}\}; \{K_{AB}\}; \tau_2 + D_2 - \tau_A \leq \mathcal{L}_\mathcal{F})$.

- $\alpha_3 = (\alpha_3^1, \alpha_3^2)$:

  - $\alpha_3^1 = (B; A; \langle I_A, I_B, \tau_A, K_{AB} \rangle_{K_{AS}}, \langle \tau_A \rangle_{K_{AB}}, \tau_B)$,
  - $\alpha_3^2 = (\tau_3; D_3; \{\langle I_A, I_B, \tau_A, K_{AB} \rangle_{K_{AS}}, \tau_A, \tau_B, K_{AB}\}; \{\tau_B\}; \tau_3 + D_3 - \tau_A \leq \mathcal{L}_\mathcal{F} \wedge \tau_3 + D_3 - \tau_B \leq \mathcal{L}_\mathcal{F})$.

- $\alpha_4 = (\alpha_4^1, \alpha_4^2)$:

  - $\alpha_4^1 = (A; B; \langle \tau_B \rangle_{K_{AB}})$,
  - $\alpha_4^2 = (\tau_4; D_4; \{\tau_B, K_{AB}\}; \{\emptyset\}; \tau_4 + D_4 - \tau_A \leq \mathcal{L}_\mathcal{F} \wedge \tau_4 + D_4 - \tau_B \leq \mathcal{L}_\mathcal{F})$.

In the first step of KaoChow protocol, α_1^1 includes information similar to the protocol's specification in Common Language. There are designations of a sender (A), receiver (B) and also message sent between users (I_A, I_B, τ_A). α_1^2 includes information about cryptographic objects which are necessary to execute protocol's step:

- τ_1 signifies time of sending first message,
- D_1 signifies delay in the network in first step,
- $\{I_A, I_B, \tau_A\}$ signifies set of elements which step's message are constructed (first message consist of I_A, I_B, τ_A),
- $\{\tau_A\}$ signifies set of elements which must be generate by sender (A must generate his timestamp τ_A),
- $\tau_1 + D_1 - \tau_A \leq \mathcal{L}_\mathcal{F}$ signifies set of time conditions which must be met (time of sending first message increased by delay in the network in first step and reduced by A's timestamp, this value must be lower or equal then lifetime).

Next steps should be considered in this same way. Please note that the notation $\langle I_A, I_B, \tau_A, K_{AB} \rangle_{K_{AS}}$ (in third step) means that I_A, I_B, τ_A and K_{AB} were encrypted by symmetric key K_{AS}, which is shared between users A and S (server).

During the protocol's execution, users can acquire knowledge. Each of the users has initial knowledge which consists of publicly available elements and elements shared between them. Special operators define knowledge changes during the protocol's execution.

In computational structure time, dependencies were defined. We used dependencies about:

- message composing,
- step times,
- session times,
- lifetime.

In [15] symbols, which describe dependencies, have been defined. We consider three delays in the network values: minimal, current and maximal. Minimal and maximal values are related to the range of delays in the network's values. Current value means a delay in the network's value in executed step. Minimal, current and maximal values of step time are also associated with this assumption. A similar situation occurs in the case of session times. Minimal, current and maximal session times depends on used delay's value. These dependencies make it possible to check time influence on security protocols' correctness. Properly selected time parameters and time constraints may allow Intruder to interrupt attack and also prevent it.

Lifetime's value will be calculated according to following formula:

$$T_k^{out} = \sum_{i=k}^{n} T_i^{max} \tag{1}$$

In this notation k signify step number, i signify step counter (for $i = k...n$), n signify number of the step in the protocol, T_k^{out} signify lifetime in the k-th step and T_i^{max} signify maximum step time. Maximum step time is sum of encryption time, generation time, maximal delay in the network and decryption time.

Some aspects of the formal model and computational structure were described in details in [17].

4   Experimental Results

For the needs of the research, a proprietary modeling and verification protocols verification tool was implemented. This tool has been described in [15]. The research was carried out in several stages. In the first of them, a set of all executions of examined security protocols using the proprietary tool was determined. Next, a set of real executions using the SAT-solver was determined. In the next stage, analysis of the impact of particular times on the possibility of an attack by the Intruder was carried out. At this stage, fixed values of time parameters were included. In the last stage of the research, simulations of real protocols executions were carried out. During this stage, delays in the network were drawn according to selected probability distributions. The probability distributions have been selected to reflect the different load on the computer network. The tests were carried out using a computer unit with the Linux Ubuntu operating system, Intel Core i7 processor, and 16 GB RAM. During the research, an abstract time unit ([tu]) to determine the time was used. The experimental results will be presented on the example of KaoChow protocol.

At the beginning of this protocol's study, it was assumed that the Intruder could impersonate only honest users. This assumption had a huge impact on the course of the attackers' executions. Due to the structure of the protocol, these executions were a combination of a regular attack and a *man in the middle* attack.

Trying to acquire knowledge about the timestamps of honest users, Intruder could use his identity. However, while it was necessary for Intruder to use honest users' cryptographic keys, also it was necessary to send entire messages. Also in the situation when the Intruder (in the second protocol's step) was not able to decrypt received message from the server, he could not send it further due to the restriction of privileges. These executions ended with an error.

Table 1. Summary of KaoChow protocol's executions

Parts	Parameters	Execution	Parts	Parameters	Execution
A→S→B		1	B→S→A		10
I→S→B	T_I, K_{IS}	2	I→S→A	T_I, K_{IS}	11
I→S→B	T_A, K_{IS}	3	I→S→A	T_B, K_{IS}	12
I(A)→S→B	T_I, K_{AS}	4	I(B)→S→A	T_I, K_{BS}	13
I(A)→S→B	T_A, K_{AS}	5	I(B)→S→A	T_B, K_{BS}	14
A→S→I	T_I, K_{IS}	6	B→S→I	T_I, K_{IS}	15
A→S→I	T_B, K_{IS}	7	B→S→I	T_A, K_{IS}	16
A→S→I(B)	T_I, K_{BS}	8	B→S→I(A)	T_I, K_{AS}	17
A→S→I(B)	T_B, K_{BS}	9	B→S→I(A)	T_A, K_{AS}	18

For the KaoChow protocol eighteen executions have been generated. A list of these executions can be found in Table 1. Column *Parts* means protocol's participants (A, B - honest users, S - server, I, I(A), I(B) - Intruder. Column *Parameters* includes cryptographics object, which are used by Intruder during execution. Column *Execution* includes ordinal number assigned to execution in order to simplifying the reference to it.

4.1   Timed Analysis

The timed analysis was related to checking the impact of particular times on the possibility of Intruder's attack. Firstly, the impact of the encryption time value on attacker's executions correctness was checked, then the impact of delay in the networks' values on attacker's executions correctness was examined. Executions no. 5, 7, 9, 14, 16, and 18 have been designated as the attacking executions. However, due to the structure of protocol and restrictions imposed on Intruder, it was impossible to carry out executions no. 9, 16 and 18, which was confirmed by the SAT-solver.

During testing the impact of the encryption time value on the Intruder's executions correctness delay in the network range from 1 to 3 [tu] was assumed, and the lower limit of this range was used to calculate the session times. The encryption time increased by 1 [tu] starting from 2 [tu] to 10 [tu]. The obtained results showed that the encryption time made it impossible to carry out attacks by the Intruder in all executions. The steps were also important when conducting executions.

Table 2. List of execution results depending on delay in the network's value for the KaoChow protocol

Delay's range [tu]	Execution no. 5	Execution no. 14	Execution no. 7
1–3	!4	!4	!3
1–4	!4	!4	!3
1–5	!max	!max	!3
1–6	!max	!max	!3
1–7	!max	!max	!3
1–8	!max	!max	+
1–9	+	+	+
1–10	+	+	+

During testing the delay in the network's influence on Intruder's attack possibility, the encryption time was 2 [tu], while a delay in the network changed in each test series by 1 [tu], starting from the range 1–3 [tu] to the range 1–10 [tu]. The results obtained for the real executions of the attackers were collected in the Table 2. The first column includes a set of examined delay in the network's ranges. Other columns include results for tested executions. Designations !3 and !4 means that in such steps timed conditions were not met. Designation !max means that execution ended with session time upper then T_{ses}^{min} and + means execution ended in correct session time.

For the attacking executions no. 5 and no. 14 and delay in the network range 1–8 [tu] KaoChow protocol proved to be safe. In situations where the upper limit of delay in the network exceeded to 8 [tu], the Intruder was able to successfully perform the attack. When the upper limit was equal to 3 or 4 [tu], the execution ended with an error in the fourth step, because the Intruder did not have enough knowledge to make it. When the upper limit of delay in the network ranged between 5 and 8 [tu], these executions kept the imposed time conditions, but the session times exceeded T_{ses}^{max}.

For execution no. 7 it turned out that protocol's security can be provided only until an upper limit of the delay in the network value 7 [tu]. Below this value, Intruder will not be able to gather relevant knowledge to perform third protocol's step. When an upper limit of delay in the network was at least 8 [tu], Intruder could easily execute an attack on protocol.

For obtained results implemented tool proposed changes in lifetimes' values in selected steps. These changes prevent against attack. Changes are presented in the Table 3.

Table 3. List of changes in lifetime's values for KaoChow protocol

Delay's range [tu]	Step number	Lifetime	New lifetime
1–5	1	35	20
1–6	1	39	21
1–7	1	43	21
1–8	3	23	21
1–9	3	27	23
1–10	3	29	23

The proposed changes start from the interval 1–5 [tu] because for the smaller intervals there was no possibility of maintaining time conditions. The experimental results obtained with new lifetimes' values excluded attack's possibility.

4.2 Simulations

KaoChow protocol's simulations were carried out with the following assumptions:

- $T_e = T_d = 2$ [tu],
- $T_g = 1$ [tu],
- delay in the networks' range 1–10 [tu].

Minimal session time was set to 19 [tu], maximal session time was 55 [tu]. Executions no. 4, 5, 7, 8, 9, 13, 14, 16, 17 and 18 were marked as impossible to carry out. Those executions were not included in simulations. Delay in the network's values was generated according to uniform, normal, Cauchy's and exponential probability distributions. Simulations experimental results will be presented on a uniform probability distribution example.

First KaoChow protocol simulations' phase was made using a delay in the network's values generated according to a uniform probability distribution. The obtained results are as follows.

Each execution was tested in a thousand test series. For each of them, a status informing about the end of execution's result has been designated. The *correct* status indicated those executions that ended in correct session time. The *!min* status has been selected for executions that ended below set T_{ses}^{min},

Table 4. Experimental results for KaoChow protocol and uniform probability distribution

Execution no.	Correct	!min	!max	Error
1	1000	0	0	0
2	975	0	25	0
3	0	0	675	325
6	1000	0	0	0
10	1000	0	0	0
11	985	0	15	
12	0	0	691	309
15	1000	0	0	0

and status *!max* - for executions over T_{ses}^{max}. These three statuses meant that time conditions imposed on individual protocol steps were met. The last status (*Error*) referred to the situation in which one of the imposed time conditions was not met and the execution ended with an error. This distinction is necessary to verify various aspects of Intruder's activities.

A summary of the test series' number for real executions and statuses is presented in Table 4.

Table 5. Timed values for KaoChow protocol (series completed in correct time)

Execution no.	Session time [tu]			Average delay in the network [tu]
	Minimal	Average	Maximal	
1	26.3	41.28	55	5.57
2	29.4	44.01	54.8	5.51
6	19.8	36.11	51.5	5.55
10	26.1	41.5	54.6	5.63
11	28.4	44.35	54.5	5.56
15	20.5	34.09	50.9	5.56

The summary of timed values for KaoChow protocol and series completed in correct time was presented in Table 5. Summary consist of minimal, average and maximal session time and average delay in the network for all test series of each real execution. For example, for execution no. 1 minimal session time was equal 26.3 [tu], average session time was equal 41.28 [tu], maximal session time was equal 55 [tu] and average delay in the network was equal 5,57 [tu].

The summary of timed values for KaoChow protocol and series completed above the T_{ses}^{max} was presented in Table 6. Summary consist of minimal, average and maximal session time and average delay in the network for all test series of

Table 6. Timed values for KaoChow protocol (series completed above the T_{ses}^{max})

Execution no.	Session time [tu]			Average delay in the network [tu]
	Minimal	Average	Maximal	
2	55.1	56.9	60.4	8.73
3	60	80.48	103.5	5.53
11	55.1	56.54	58.8	5.76
12	62.3	79.93	97.8	5.52

each real execution. For example, for execution no. 2 minimal session time was equal 55.1 [tu], average session time was equal 56.9 [tu], maximal session time was equal 60.4 [tu] and average delay in the network was equal 8.73 [tu].

In the case of KaoChow protocol and delays in networks' values generated according to a uniform probability distribution, there were no sessions below the set T_{ses}^{min}. The remaining errors were caused by failure to meet time conditions in individual steps. All test series for honest executions ended correctly.

5   Conclusion

In this paper was presented analysis and verification of the KaoChow protocol's timed version. Analysis of time parameters' influence on protocol's security was related. Encryption and decryption times and delays in the network were taken into account. The research was based on a formal model and computational structure proposed in [12]. This model and structure were extended by time parameters.

The research was carried out using the implemented tool and SAT-solver MiniSAT. Tests took place in two phases. In the first phase, the possibility of KaoChow protocol's attack was checked. In this phase, constant delay in the network's values was used. In the second phase simulations of real KaChow protocol's executions were carried out. Current delays in the network's values were generated according to uniform, normal, Cauchy's and exponential probability distributions.

Delays in the network are crucial for Internet communication. Any delay in the network can be used by the Intruder. During this time, Intruder may try to decrypt the previously received ciphertexts. Thanks to this, Intruder may have the opportunity to use the information acquired to carry out an attack on authenticity or authentication.

Carried out research showed time parameters' influence on users' security and protocol's correctness. Badly selected time parameters and time constraints may allow Intruder to interrupt attack on protocol. On the other hand, properly selected time parameters and time constraints may prevent it. Badly selected time parameters and time constraints may also make that honest user will not execute protocol without errors. Also, Intruder can have enough time to increase your knowledge and prepare an attack in the future. Delay in the network limits should be adjusted so that the honest user can execute the protocol and the Intruder was unable to acquire additional knowledge.

According to this problems, it is necessary to regularly verify computer network's work and set appropriately adopted lifetime restrictions. If the imposed restrictions have been exceeded, communication should be terminated immediately, as the protocol is not secure. These actions make protocols safer. It should also be borne in mind that the acceptable limits may depend on the current network overload.

In further research, we will take into account random encryption and decryption times values. These values will be generated with a selected probability distribution.

References

1. Kao, I.L., Chow, R.: An efficient and secure authentication protocol using uncertified keys. Oper. Syst. Rev. **29**(3), 14–21 (1995)
2. Paulson, L.: Inductive analysis of the internet protocol TLS. ACM Trans. Inf. Syst. Secur. (TISSEC) **2**(3), 332–351 (1999)
3. Burrows, M., Abadi, M., Needham, R.: A logic of authentication. Proc. R. Soc. Lond. A **426**, 233–271 (1989)
4. Lowe, G.: Breaking and fixing the Needham-Schroeder public-key protocol using FDR. In: TACAS. LNCS, pp. 147–166. Springer (1996)
5. Steingartner, W., Novitzka, V.: Coalgebras for modelling observable behaviour of programs. J. Appl. Math. Comput. Mech. **16**(2), 145–157 (2017)
6. Dolev, D., Yao, A.: On the security of public key protocols. IEEE Trans. Inf. Theor. **29**(2), 198–207 (1983)
7. Armando, A., Basin, D., Boichut, Y., Chevalier, Y., Compagna, L., Cuellar, J., et. al.: The AVISPA tool for the automated validation of internet security protocols and applications. In: Proceedings of 17th International Conference on Computer Aided Verification (CAV 2005). LNCS, vol. 3576, pp. 281–285. Springer (2005)
8. Blanchet, B.: Modeling and verifying security protocols with the applied Pi Calculus and ProVerif. Found. Trends Priv. Secur. **1**(1–2), 1–135 (2016)
9. Cremers, C., Mauw, S.: Operational semantics and verification of security protocols. In: Information Security and Cryptography. Springer, Heidelberg (2012)
10. Jakubowska, G., Penczek, W.: Modeling and checking timed authentication security protocols. In: Proceedings of the International Workshop on Concurrency, Specification and Programming (CS&P 2006), Informatik-Berichte, vol. 206, no. 2, pp. 280–291. Humboldt University (2006)
11. Jakubowska, G., Penczek, W.: Is your security protocol on time? In: Proceedings of FSEN 2007. LNCS, vol. 4767, pp. 65–80. Springer (2007)
12. Kurkowski, M.: Formalne metody weryfikacji własności protokolow zabezpieczajacych w sieciach komputerowych, Exit, Warsaw (2013). (in Polish)
13. Kurkowski, M., Penczek, W.: Applying timed automata to model checking of security protocols. In: Wang, J. (ed.) Handbook of Finite State Based Models and Applications, pp. 223–254. CRC Press, Boca Raton (2012)

14. Siedlecka-Lamch, O., Kurkowski, M., Piatkowski, J.: Probabilistic model checking of security protocols without perfect cryptography assumption. In: Proceedings of 23rd International Conference on Computer Networks, Brunow, 14–17 June 2016. Communications in Computer and Information Science, vol. 608, pp. 107–117. Springer (2016)

15. Szymoniak, S., Siedlecka-Lamch, O., Kurkowski, M.: Timed analysis of security protocols. In: Proceedings of 37th International Conference ISAT 2016, Karpacz, 18–20 September 2017. Advances in Intelligent Systems and Computing, vol. 522, pp. 53–63. Springer (2017)

16. Klasa, T., Fray, I.E.: Data scheme conversion proposal for information security monitoring systems. In: Kobayashi, S., Piegat, A., Pejaś, J., El Fray, I., Kacprzyk, J. (eds.) Hard and Soft Computing for Artificial Intelligence, Multimedia and Security. ACS 2016. Advances in Intelligent Systems and Computing, vol. 534. Springer, Cham (2017)

17. Szymoniak, S., Kurkowski, M., Piatkowski, J.: Timed models of security protocols including delays in the network. J. Appl. Math. Comput. Mech. **14**(3), 127–139 (2015)

18. Chadha, R., Sistla, P., Viswanathan, M.: Verification of randomized security protocols. In: 32nd Annual ACM/IEEE Symposium on Logic in Computer Science (LICS), pp. 1–12 (2017)

19. Basin, D., Cremers, C., Meadows, C.: Model checking security protocols. In: Handbook of Model Checking, pp. 727–762. Springer (2018)

20. Szymoniak, S., Siedlecka-Lamch, O., Kurkowski, M.: SAT-based verification of NSPK protocol including delays in the network. In: Proceedings of the IEEE 14th International Scientific Conference on Informatics, Poprad, Slovakia, 14–16 November 2017. IEEE (2017)

21. Security Protocols Spen Repository. http://www.lsv.fr/Software/spore/table.html

Electronic Document Interoperability in Transactions Executions

Gerard Wawrzyniak[1(✉)] and Imed El Fray[1,2]

[1] Faculty of Computer Science and Information Technology,
West Pomeranian University of Technology, Szczecin, Szczecin, Poland
{gwawrzyniak,ielfray}@zut.edu.pl
[2] Faculty of Applied Informatics and Mathematics,
Warsaw University of Life Sciences, Warsaw, Warsaw, Poland
imed_el_fray@sggw.pl

Abstract. Transaction as a general human activity is always associated with the flow and processing of information. The electronic document is the form of legally binding information which is being exchanged between the transaction parties. Both humans and information systems take part in transaction executions especially in the area of information transfer and processing. Therefore the ease of implementation of services processing electronic forms using standard programming tools is extremely important for electronic support of transactions execution. Also, the meaning of data (information) stored in the electronic form must be unambiguously and uniformly understood by processing parties (humans and systems). Moreover, services supporting electronic documents transfer and processing must be standardised to make them accessible for a large number of transactions and participants. All considered problems are related to the concept of interoperability.

Keywords: Electronic document · Electronic form · Digital signature
Transaction · Interoperability

1 Introduction

Generally speaking, a transaction is each organized human activity. Execution of transaction is always associated with the flow of legally effective information. This information takes a form of a document or a form as a special type of document dedicated for interaction with humans. To ensure the effective collaboration several parties represented by humans and information systems it is necessary to ensure the proper level of interoperability.

Regardless of the origin of the word "transaction" presented in [1, 2], it is necessary to focus on the essence of this concept.

The term "transaction" [3] is referred to an agreement, contract, exchange, understanding, or transfer of cash or property that occurs between two or more parties and establishes a legal obligation. The term "transaction" is also called booking or reservation. In article [4] authors present a more precise definition by giving the features (properties) of the transaction: "a transformation of a state which has the properties of

© Springer Nature Switzerland AG 2019
J. Pejaś et al. (Eds.): ACS 2018, AISC 889, pp. 358–372, 2019.
https://doi.org/10.1007/978-3-030-03314-9_31

atomicity (all or nothing), durability (effects survive failures) and consistency (a correct transformation)". In article [7] the authors of the article define the transaction as:

1. The commercial operation associated with the purchase or sale of material assets, intangible assets or services and agreement associated with this operation,
2. Transfer of material goods, services or intangible goods between the parties resulting from various relations binding the parties, may be economic, commercial, financial, social or any other relation,
3. An agreement (contract) between the parties the subject of which are goods, services or other agreements and commitments.

Presented explanation of the *transaction* concept is compatible with the points of view presented in [3] and definitions [2, 4, 5] or [8].

Each legitimate transaction must be secure. To ensure a secure transaction it is required to use a flow of secure information. This information expresses, for example, intentions of parties to the transaction, obligations, notifications and confirmations which appear during the execution of the transaction. It also expresses all relevant information on the change of status of the transaction and information that enables the track the course of the transaction.

In addition, it should be noted that information not only supports the execution of the a transaction, but it also can be the subject of a transaction (such as intangible assets, obligations etc.).

To meet aforementioned requirements for secure transaction, information must follow specific features [7]: authenticity – reliability of information, non-repudiation of origin – indubitability of the origin of information, integrity - guarantee that the document has not been changed (tampered), durability – possibility to use information afterwards.

As presented in article [6] information that complies with these features can be named a *document*. Two forms of documents can be distinguished: a traditional paper document and an electronic document. Both have the same immanent features constituting a document, both can be used in transactions but an electronic document exists in the form of a file and hence can be transferred using electronic means. For an electronic document to be effectively used to support the transaction, it must have certain features:

- the ability to be used regardless of the maturity of IT used by users,
- document format must be independent of industry or activity sector of the economy,
- document format and software must be technology-neutral,
- ease of integration with various user's systems,
- autonomy – the ability to use a document on a device without access to the network,
- the ability to interpret the document both automatically and by human (the concept of such document is called "semantic document" and is presented in article [19]).

Because variety and multitude of both IT systems and people involved in the transaction, interoperability is an important problem.

The following parts are presented in the article: Motivation, to present the relevance of an electronic forms in transaction implementations and execution. The third chapter consists Interoperability concept and its influence on information systems in various aspects. The problem of interoperability of electronic forms in the light of general considerations (Sect. 3) is presented in fourth Sect. 4. Particularly design solutions implementing interoperability guidelines discussed in the Sect. 4 are presented in this chapter. The article ends with a short discussion and conclusions.

2 Motivation

There are (and will be) many different implementations of systems and services that support execution of transactions. This diversity is related to information technologies, communication protocols, processes, data formats and other fields. At the same time, it is required to ensure the safety and legal effectiveness of the tasks being performed. To ensure the possibility of practical transaction support, a high level of interoperability is required not only for services and software but also for electronic documents. In this article, an electronic document integrating various services, systems and in particular people, is a central object when considering transaction execution in the light of interoperability concepts.

Therefore, evaluation of an electronic document in the light of specific interoperability rules becomes important while building solutions supporting the implementation of transactions in the area of the documents application in the transaction execution.

The concept of the electronic form formulated by the authors assumes the use of standard solutions in the area of basic formats, internet communication protocols or electronic signature structures. A novelty is the introduction of the concept of a three-layer structure of an electronic form that is a single file and consisting of a data layer, a presentation layer and the logic layer. Therefore, the dogma of dividing a document into a presentation layer and a data layer has been abandoned. This approach gives the possibility of a new approach to transactions execution in the virtual world.

The main motivation of the article is to formulate guidelines (requirements) for interoperability for electronic forms and presentation of design solutions for the most important elements of electronic forms. The greatest emphasis was put on the interoperability while maintaining the legal effectiveness, recognizing the significance of these features in transactions.

3 Interoperability

As noticed in article [9] the problem of interoperability is older than the term itself and it occurred to be important when the problem of data exchange between programs appeared. It became relevant because of the necessity of exchanging and sharing data between organisations. In European directive [10] "interoperability" was officially defined as "the ability to exchange information and mutually use the information, which has been exchanged". Then, up to the digital agenda for Europe 2020 the growing role of interoperability can be observed [9] and interoperability is considered as a mean to allow trans-border exchange of data within a common market and between units of government in the different Member States. In the fourth chapter the role of interoperability

The term "interoperability" is not new and there are currently many definitions [11–15] for which the common denominator is the ability of a system, equipment or process to use information and/or exchange data assuming compliance with common standards. The interoperability architecture consists of a number of complementary technical specifications, standards, guidelines and principles.

The ETSI definition extends interoperability to three aspects [15]:

- Technical interoperability: covers technical issues of connecting computers, interfaces, data formats and protocols.
- Semantic interoperability: related to the precise meaning and understanding of exchanged information by other applications (not initially designed for this purpose).
- Organisational interoperability: concerned with modelling business, aligning information architectures with organisational goals and helping business to cooperate.

Presented taxonomy of interoperability is commonly known, but ETSI introduced a distinction between technical and syntactic interoperability [13]:

- "Technical Interoperability is usually associated with hardware/software components, systems, and platforms that enable machine-to-machine communication to take place."
- "Syntactic Interoperability is usually associated with data formats. Messages transferred via communication protocols need to have a well-defined syntax and encoding, even if only in the form of bit tables. This can be represented using high-level transfer syntaxes such as HTML, XML or ASN.1".

As result of considering the subject of interoperability is the fact, that data and services can be defined and applied regardless of a computer system, programming language, operating system or computing platform. Following examples are given in article [16]: EDI, OM like Microsoft's COM and DCOM, Java Beans, OMG Object and Component Models. Further, the authors mention Virtual Machines with Java Virtual Machines and at last Service Oriented Architectures with the use of XML to define data and message formats. The approach based on SOA is the preferred one.

The interoperability level can be measured. For example in article [17] the Maturity Model for Enterprise Maturity is presented. The authors present the framework for Enterprise Interoperability (referring to [18]) which defines three basic dimensions [17]:

- Interoperability concerns, defining the content of interoperation that may take place at various levels of the enterprise (data, service, process, business).
- Interoperability barriers, identifying various obstacles to interoperability in three categories (conceptual, technological, and organizational).
- Interoperability approaches, representing the different ways in which barriers can be removed (integrated, unified, and federated).

These three dimensions led to the development of a framework and then to determine the taxonomy of the organisational maturity of interoperability assessment in the form of five levels:

Level 0 (Unprepared) - resources are not prepared for sharing with others, cooperation is not possible, communication takes place as a manual data exchange, systems function independently.

Level 1 (Defined) - systems are still separated, some automatic interactions can be organised ad hoc, data exchange is possible.

Level 2 (Aligned) - it is possible to make changes and to adapt to common formats (imposed by partners), wherever possible significant standards are used.

Level 3 (Organised) - an organisation is well prepared for interoperability challenges, interoperability capabilities are extended to heterogeneous systems of partners.

Level 4 (Adopted) - organisations are prepared for the dynamic (on the fly) adaptation. Organisations are able to cooperate in a multilingual and multicultural, heterogeneous environment.

This article focuses on the role of the electronic form in the execution of the transaction as an element integrating different services (required for the execution of transactions) by the fact that the form is a carrier of readable and secure information. Therefore, one should consider how the features of an electronic document impact the ability to achieve higher levels of maturity. Following points of view should be taken into consideration:

1. **Data format.** To achieve the first level, it is necessary to ensure interoperability in terms of data formats being exchanged. In the case of the second level, this requirement is even stronger.
2. **Security (Legal effectiveness – signature).** To ensure the legal effectiveness (security) of the data, advanced use of the electronic signature is necessary. The use of "standard" (interoperable) solutions in this area will allow achieving the third level, that is, the execution of transactions in heterogeneous partners' environment.
3. **Exchange of messages.** The ability to exchange messages with an agreed/accepted (interoperable) format and syntax supports the achievement of the third level. The ability to dynamically define the content of a message and the way of providing data is necessary to reach the fourth level.
4. **Processing – implementation of services.** The use of universally recognised data formats and the resulting ability to quickly and easily implement the processing logic within the supporting services gives the opportunity to dynamically adapt to market requirements understood as the transaction execution environment. It is a necessary factor to reach the fourth level.
5. **Man – IT service interaction.** The possibility of human participation in transaction execution in any stage extends the interoperability. This extension comes from assembling the real human the world with virtual world of IT services and pushes the interoperability to a higher level.
6. **Multilingualism.** The form with the ability to express its semantics in many languages gives the opportunity to carry out transactions in an international, diverse (heterogeneous) linguistic environment, which is a condition for achieving the fourth level.

4   Electronic Form Interoperability

The electronic form as a mean of transfer information between the processing nodes is an important element affecting the maturity level of interoperability. There are several factors to consider before making certain implementation decisions.

4.1   Data Format

Regardless of the type of a processing node (man, machine), the processing must be able to read, interpret and process the document. The implementation of the document processing logic in the course of the transaction must anticipate this necessity, i.e. it must be able to read (parse) the document, recognise the physical and logical structure of the document and interpret its contents. Therefore, the structure (syntax of the document) must be known and accepted by the parties involved in the transaction (using, interpreting the document).

The electronic form is a file in XML format (W3C XML) [20] with syntax defined as XML Schema [21, 22]. Values (fields of the form) are held in XML nodes. The structure of XML is defined in a rigid way. This can be a source of uncomfortable constraint, because different IT services may need to interpret names of values in their own way. Thus it should be possible to use own specific attributes (XML nodes in their own, defined namespace [23]). This allows finding a value of the field by service specific attribute name using standard means like XPath [38]. Below in the Fig. 1 there is presented an example of the form consisting methods of XML element attribute identification Value (stringValue element) in the component textField is identified by the own attribute id="StringValueId" and/or/either external attribute other:id="OtherId", where otherId comes from xmlns:other = "http://other.org" namespace.

```
<textField X="0" Y="0" height="0" horizontalAlignment="Center"
width="1" xsi:type="stringField" xmlns:other="http://other.org">
    <value>
        <stringValue id="StringValueId" other:id="OtherId">Any
value</stringValue>
    </value>
</textField>
```

Fig. 1. Various identification of the value in the form field

All binary data stored in different parts of the form is encoded (converted) into Base64 form [36]. This gives a possibility to keep the binary data in the form in a secure way (it can be signed). Binary data can be transferred as a value of the form field, between transaction parties. It also can be processed automatically by IT services.

4.2   Legal Effectiveness (Electronic Signatures)

As stated before, the legal effectiveness is a critical feature of the form (information). Ensuring the legal effectiveness implies the usage of electronic signatures. Because the electronic form is transferred between multiple processing nodes and each node can make changes to parts of the form, it should be possible to submit multiple signatures (in one form) by multiple processing nodes signing different parts of the form. Figure 2 shows a document in which several signatures are defined for signing various parts of it.

```
<signature filterType="XPath" id="ClientSignature" name="Client
Signature" signatureQualification="shouldBeQualified">
      <dateTime
objectId="ClientSignature.SignatureDate"></dateTime>
      <countersignedBy>
            <signer>ServerSignature</signer>
      </countersignedBy>
</signature>
<signature filterType="XPath" id="ServerSignature" name="Server
Signature">
      <dateTime objectId="AuthenticationDate"></dateTime>
</signature>
```

Fig. 2. Many signatures on the form. Definition of two signatures in the form: ClientSignature and ServerSignature.

All transaction processing nodes must be able to generate and verify signatures themselves. Therefore, signatures and verification procedures used in forms must comply commonly available specifications. It is fulfilled by using W3C specifications: XMLdSig [24], XAdES [25, 26]. Public key applied in the signature is compatible with X.509 Certificate specification [27] with verification mechanisms based on certificate revocation lists (X.509 CRL) [28] or On Line Certificate Status Protocol (OCSP) [29].

Fragments of XML form shown in the Fig. 2 present the concept of signatures definition. There are two signatures ClientSignature and ServerSignature. The signature date time is defined in element dateTime and countersigning relation is defined in ClientSignature in counterSignBy element. (clientSignature is to be countersigned by the ServerSignature).

```
<group ... id="GroupHeader" width="280">
      <items>
            <textField> ... </textField>
            <text> ...</text>
            <textField> </textField>
      </items>
      <signedBy>
            <signer>ClientSignature</signer>
            <signer>ServerSignature</signer>
      </signedBy>
</group>
```

Fig. 3. Assignment of the signatures to the selected part of the document

Signatures assignment to the part of the form is presented in the Fig. 3. All elements contained in the group `GroupHeader` are to be signed by signatures `ClientSignature` and `ServerSignature`.

4.3   Exchange of Messages

Execution of transactions forces the use of services with specific protocols and message syntax. Thus it is necessary to build messages based on the current state of the form data (fields). On the other hand, there is a need to interpret and present data from messages received from service. Figures 4 and 5 present the mechanism of messages specification for sending and interpreting received messages in the Figs. 4 and 5.

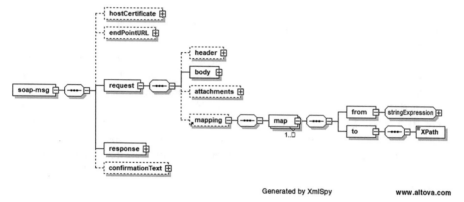

Fig. 4. SOAP message construction with mapping form fields values to SOAP request body (`map` element)

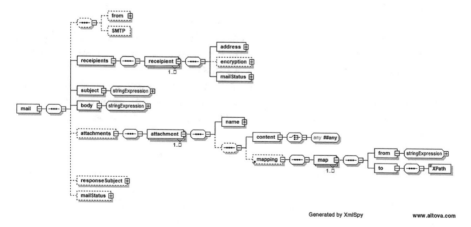

Fig. 5. Mail message (SMTP) constructions using concept presented in the Fig. 4 description.

Communication procedures being a part of the form reflect communication means:

- Simple Object Access Protocol (SOAP) [29–32] it is standard (de facto) protocol applied in many services.

- Representational State Transfer (REST) – the method of web services access, using JSON as a context syntax and HTTP protocol [37] for communication.
- Simple Mail Transfer Protocol (SMTP) – mail communication protocol [33].

The communication type and parameters are defined in the form as an XML Objects (Elements, DOM) [20], and application designated for using the form executes the communication.

XML messages which are to be sent as the content of SOAP-body request [29–32], REST request, SMTP [33] attachment can be constructed by an application when the request is being built, using the logic of the form. In this case, the logic holds the information of the method values stored in the form fields should be embedded in the XML structure which is to be sent to web service. And in the response case – values stored in a message, obtained from the service can be mapped and presented as elements of the form. All descriptions are elements of the form XML structure and can be done manually or by any software.

This approach enables asynchronous (no response expected) and synchronous (response expected) communication with services. A response may be a form or any XML structure specific for service taking a part in transaction execution.

Applying an electronic form as a mean of information transfer between services within the execution of one transaction increases the interoperability of the whole process of transaction execution, interoperability of services involved in the transaction, organisational interoperability as the benefit for all organisations executing the business.

In the Figs. 4 and 5 the syntax for a logic of SOAP and SMPT messages exchange is shown.

4.4 Processing on the Server

The use of a standard (de facto) document format and standard solutions (structures, syntax) in the scope of electronic signature, communication protocols, gives the possibility to build new automated services using "unified" software elements. It makes the development simple and allows to focus on the logic of the implemented part of the transaction and not on the technical details software.

The form is a standard XML structure and can be processed using standard XML parsers. Then the response is generated and returned to the originator (another service or human using application). Processing functions specific to the form processing can be limited to:

- parsing XML document for searching and setting, values,
- XML signature generation and verification (XMLdSig [24], XadES [25, 26], X.509 [27], OCSP) [28],
- Support for private key management (PKCS#12 [34], PKCS#11 [35]),
- Integrating the service of receiving/responding forms with other IT systems.

It is possible to define XML document syntax which describes the process of the form processing (process descriptor) by the service software. Such description in the declarative form consists of tasks which are to be executed after the service receives the request for processing. Such tasks are:

- Recognition of the form type by the content of the attribute set (element name, attribute name and their values),
- For recognised type of the form (it reflects the business case), following functions/procedures are sufficient for processing:
  - Set the value of the field of the form,
  - Selected signature for verification (CRL [27], OCSP [28]),
  - Generation of selected signature (the content of the signature is defined in the logic layer of the form) [24–26],
  - Generation of the XML message (for SOAP-body response) [29–32], or REST response (values are to be taken from the form),
  - Constructing (using values from the form) and sending a messages using communication protocols (SOAP [29–32], REST, SMTP [33], FTP [39] or local file system),
  - Integration with local systems (databases) with setting and getting values from and to the form).

Such simple descriptor can handle with presumably all cases and integrations with and between automatic services. The syntax of the descriptor is presented in Fig. 6.

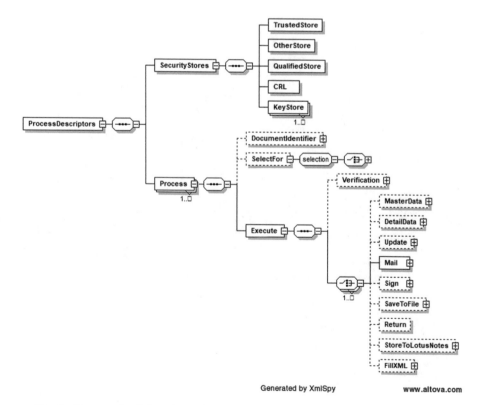

Generated by XmlSpy   www.altova.com

Fig. 6. The sample of the schema (syntax description) for the logic of forms processing

4.5 Man – IT Service

To use electronic form, the application to handle with defined electronic form is required. The application interprets the description of the form presentation and presents (displays) it to the user. In fact, the description of form visualization is a description of meaning (semantics) of all values that are held in the form. Thus the presentation (visualisation) layer can be named a semantic layer. In this way, all humans are conscious of the meaning data in the document.

The application can execute the task to deliver the form or message to the IT service and receive the response. Standard means are applied as described in previous chapters.

The document carrying legally binding information can also be transferred between humans (without electronic services) using SMTP (mail) or even transferring it manually as a common file. In this way, it is possible to implement the transaction without any web service basing only on mail and/or manual form transfer.

In this way, the human can be involved in the transaction execution as an active element. As a result, the transaction can be executed through the number of humans and electronic services applying a number of diverse communication protocols. All this is done with ensuring legal effectiveness (electronic signatures).

Figure 7 shows an example of human interaction with an automatic service. The document is comprehensible for both man and machine.

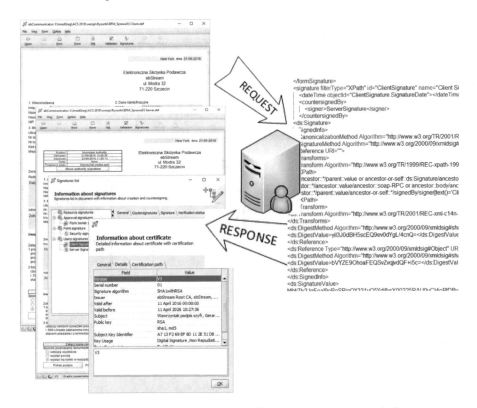

Fig. 7. Interaction between human and IT service using an electronic form

4.6   Multilingualism

In order to reach the level of interoperability above the level of human-machine cooperation, it should be possible to operate the form between people who speak different languages, i.e. the possibility of creating multilingual forms. In the category of logical structures of the form, it is necessary to create the possibility of expressing the semantics (visualisation layer) of the form in many languages. Without losing any of its features the same form may be used in one transaction by people speaking different languages.

The semantic is expressed it is the XML structure composed of a number of XML elements including texts being displayed in the form. It is possible to define a number of description texts with the same semantics (meaning) but expressed in different natural (human) languages (Fig. 8).

```
<text X="0" Y="20" height="16" horizontalAlignment="Left"
horizontalTextPosition="Left" verticalAlignment="Center"
verticalTextPosition="Center" width="80" xsi:type="stringText">
       <caption>
              <stringValue Language="PL">Ulica:</stringValue>
              <stringValue Language="EN">Street:</stringValue>
              <stringValue Language="GE">Strasse:</stringValue>
       </caption>
</text>
```

Fig. 8. The multilingualism concept expressed in XML form structure.

5   Discussion

The presented solutions of the electronic form reflect the requirements regarding the interoperability of the solutions supporting the execution of the transaction.

As shown in the article, the achievement of successive levels of interoperability, in the area of exchange of secure information, requires meeting specific levels of postulates. The electronic form, which is the carrier of secure information transported between IT services, including humans in the process of transaction execution, is an important element affecting the interoperability capabilities and their level. The presented features of the form in the context of the possibility of its processing by various IT systems show that the features of the form have a cardinal impact on the possibility of implementation of various services and integrating them into the transactions. The proposals for specific solutions presented above show that it is possible to construct a form that meets the postulates:

- document format (XML), its syntax (XSD) in terms of data, semantics and their structures,
- standard format of electronic signatures (XMLdSig, XAdES),
- message and data exchange protocols (SOAP, REST, SMTP, FTP),

- the ability to define the message building logic on the base of the status of data in the form fields,
- the use of the XML format and combining the data layer with the layer of semantics in one document that allows the form to be processed both in the environment of software information systems and man,
- ease of implementation of services processing and transferring the form,
- ability to express the semantic layer (presentation) in many languages, and thus implement and execute transactions in international and multilingual environments.

The use of the electronic form as an element facilitating the execution of transactions facilitates the achievement of a high level of maturity, that is:

- data exchange between computers,
- implementation of automatic integration,
- adapting to changes using commonly available standards,
- integration of heterogeneous participants environments, taking apart in the transaction,
- dynamic adaptation to changing requirements and cooperation in a multilingual environment of partners.

The abovementioned possibilities of the proposed electronic form prove that it meets the essential conditions for achieving the highest maturity level of interoperability.

6 Conclusion

The article presents the role of the form in the transaction. On the one hand, it is necessary to ensure the security/legal effectiveness of exchanged documents and their parts at particular stages of transaction execution, and on the other hand, the need to ensure a high level of interoperability of individual elements (processing nodes) of the transaction. The electronic form with the logic layer is an element enabling construction and mutual integration of transaction nodes.

The use of the form with the presented features allows for the construction of systems supporting the implementation of transactions at the highest (4 - Adapted) level of maturity. This level means that the organisation using this solution is able to dynamically adopt changes in the run and to interact in a heterogeneous, technical, multilingual and multicultural environment of partners.

Further works in the area of electronic form (as an instance of electronic document) should be conducted in the direction of stronger integration with other IT systems supporting various parts of transactions, like order systems, logistics systems, financial systems. Also integration with systems facilitating advanced communication (Voice Over IP, Session Initialisation Protocol – SIP), or new security trends, solutions like eIDAS and methods [40].

References

1. Online Etymology Dictionary. https://www.etymonline.com/word/transaction
2. Wiktionary. https://en.wiktionary.org/wiki/transact#English
3. BusinessDictionary. http://www.businessdictionary.com/definition/transaction.html
4. Gray, J.: The transaction concept: virtues and limitations. In: Proceedings of Seventh International Conference on Very Large Databases, September 1981. Published by Tandem Computers Incorporated (1981)
5. https://mfiles.pl/pl/index.php/Transakcja
6. Wawrzyniak, G., El Fray, I.: An electronic document for distributed electronic services. In: Saeed, K., Homenda, W. (eds.) CISIM 2016. LNCS, vol. 9842, pp. 617–630. Springer, Cham (2016). https://doi.org/10.1007/978-3-319-45378-1_54
7. Wawrzyniak, G., El Fray, I.: An electronic document for distributed electronic services. In: Saeed, K., Homenda, W. (eds.) CISIM 2017. LNCS, vol. 10244, pp. 697–708. Springer, Cham (2017). https://doi.org/10.1007/978-3-319-45378-1
8. https://www.merriam-webster.com/dictionary/transacted
9. Scholl, H.J., Kubicek, H., Cimander, R.: Interoperability, enterprise architectures, and IT governance in government. In: Janssen, M., Scholl, H.J., Wimmer, M.A., Tan, Y. (eds.) Electronic Government, EGOV 2011. LNCS, vol. 6846. Springer, Heidelberg (2011). https://doi.org/10.1007/978-3-642-22878-0_29
10. Council directive 91/250/EC, 14.5.1991 on the legal protection of computer programmes. Official Journal of the European Communities. No L 122, 17.05.91
11. Institute of electrical and electronics engineers, standard computer dictionary. IEEE Press, New York (1990)
12. European public administration network, e-government working group: key principles of an interoperability architecture, Brussels (2004)
13. European Telecommunications Standards Institute: achieving technical interoperability– the ETSI approach. ETSI white paper No. 3. By Hans van der Veer (Lucent Technologies) and Anthony Wiles (ETSI), October 2006. http://www.etsi.org/website/document/whitepapers/wp3_iop_final.pdf. Accessed 5 June 2018
14. ISO/IEC 2382–1:1993 Information Technology – Vocabulary – Part 1: Fundamental Terms, International Organization for Standardization (1993)
15. Commission of the European Communities: Communication from the Commission to the Council, the European Parliament, the European Economic and Social Committee and the Committee of the Regions, COM (2003) 567 final – The Role of eGovernment for Europe's Future, Brussels (2003)
16. Bugajski, J.M., Grossman, R.L., Vejcik, S.: A service oriented architecture supporting data interoperability for payments card processing systems. In: Dan, A., Lamersdorf, W. (eds.) Service-Oriented Computing – ICSOC 2006. LNCS, vol. 4294. Springer, Heidelberg (2006)
17. Guédria, W., Chen, D., Naudet, Y.: A maturity model for enterprise interoperability. In: Meersman, R., Herrero, P., Dillon, T. (eds.) On the Move to Meaningful Internet Systems: OTM 2009 Workshops, OTM 2009. LNCS, vol. 5872. Springer, Heidelberg (2009)
18. Method Integrated Team: Standard CMMI Appraisal Method for Process Improvement (SCAMPI), Version 1.1: Method Definition Document Members of the Assessment (2001)
19. Nešić, S.: Semantic document model to enhance data and knowledge interoperability. In: Devedžić, V., Gašević, D. (eds.) Web 2.0 & Semantic Web. Annals of Information Systems, vol. 6. Springer, Boston (2010)
20. Extensible Markup Language (XML) 1.0. https://www.w3.org/TR/xml/. Accessed 5 June 2018

21. W3C XML Schema Definition Language (XSD) 1.1 Part 1: Structures. https://www.w3.org/TR/xmlschema11-1/. Accessed 5 June 2018
22. W3C XML Schema Definition Language (XSD) 1.1 Part 2: Datatypes. https://www.w3.org/TR/xmlschema11-2/. Accessed 5 June 2018
23. Namespaces in XML 1.0 (Third Edition), W3C Recommendation 8 December 2009. https://www.w3.org/TR/xml-names/
24. XML Signature Syntax and Processing Version 2.0. https://www.w3.org/TR/xmldsig-core2/
25. XML Advanced Electronic Signatures (XAdES). https://www.w3.org/TR/XAdES/
26. ETSI TS 101 903 XAdES version 1.4.2 z 2010-12. https://portal.etsi.org/webapp/WorkProgram/Report_WorkItem.asp?WKI_ID=35243
27. RFC 5280, Internet X.509 Public Key Infrastructure Certificate and Certificate revocation List (CRL), IETF 2008, Profite (2008). https://tools.ietf.org/html/rfc5280
28. RFC 6960, X.509 Internet Public Key Infrastructure Online Certificate Status Protocol – OCSP,, IETF 2013. https://tools.ietf.org/html/rfc6960
29. SOAP Version 1.2 Part 0: Primer (Second Edition), W3C Recommendation 27 April 2007. https://www.w3.org/TR/2007/REC-soap12-part0-20070427/
30. SOAP Version 1.2 Part 1: Messaging Framework (Second Edition), W3C Recommendation 27 April 2007. https://www.w3.org/TR/2007/REC-soap12-part1-20070427/
31. SOAP Version 1.2 Part 2: Adjuncts (Second Edition), W3C Recommendation 27 April 2007. https://www.w3.org/TR/2007/REC-soap12-part2-20070427/
32. SOAP Version 1.2 Specification Assertions and Test Collection (Second Edition), W3C Recommendation 27 April 2007. https://www.w3.org/TR/2007/REC-soap12-testcollection-20070427/
33. RFC 5321, Simple Mail Transfer Protocol, IETF (2008). https://tools.ietf.org/html/rfc5321
34. PKCS #11: Cryptographic Token Interface Standard. RSA Laboratories
35. PKCS #12: Personal Information Exchange Syntax Standard. RSA Laboratories
36. RFC 4648, The Base16, Base32, and Base64 Data Encodings, IETF 2006
37. RFC 7230, Hypertext Transfer Protocol (HTTP/1.1): Message Syntax and Routing, IETF (2014)
38. XML Path Language (XPath) 3.1, W3C Recommendation 21 March 2017. https://www.w3.org/TR/2017/REC-xpath-31-20170321/
39. RFC 959, File Transfer Protocol (FTP), IETF (1985)
40. Hyla, T., Pejaś, J.: A practical certificate and identity based encryption scheme and related security architecture. In: Saeed, K., Chaki, R., Cortesi, A., Wierzchoń, S. (eds.) CISIM 2013. LNCS, vol. 8104, pp. 190–205. Springer, Heidelberg (2013)

Multimedia Systems

L-system Application to Procedural Generation of Room Shapes for 3D Dungeon Creation in Computer Games

Izabella Antoniuk$^{(\boxtimes)}$, Paweł Hoser, and Dariusz Strzęciwilk

Faculty of Applied Informatics and Mathematics, Department of Applied Informatics,
Warsaw Univesrity of Life Sciences, Warsaw, Poland
{izabella_antoniuk,pawel_hoser,dariusz_strzeciwilk}@sggw.pl

Abstract. In this paper we present a method for procedural generation of room shapes, using modified L-system algorithm and user-defined
properties. Existing solution dealing with dungeon creation usually focus
on generating entire systems (without giving considerable amount of control over layout of such constructions) and often don't consider three-
dimensional objects. Algorithms with such limitations are not suitable
for applications such as computer games, where structure of entire dungeon needs to be precisely defined and have a specific set of properties. We
propose a procedure, that can create interesting room shapes, with minimal user input, and then transfers those shapes to editable 3D objects.
Presented algorithm can be used both as part of bigger solution, as well
as separate procedure, able to create independent components. Output
objects can be used during design process, or as a base for dungeon creation, since all elements can be easily connected into bigger structures.

Keywords: Computer games · L-system
Procedural content generation · Procedural dungeon generation

1 Introduction

Designing maps and terrains for computer games represents a complex topic
with various challenges and requirements. Depending from computer game type,
properties of 3D terrain can greatly influence how such production is received
and to what degree player will be satisfied after finishing it. Since game world
is a place where all of the story happens, realistic and well considered locations
can enhance its reception, while unrealistic and defect ones can ruin it.

Among different areas, dungeons and underground structures hold a special place in computer games, with interesting challenges connected to structure
and layout of such spaces. We have areas of varying sizes, with sets of passages
between them and places where traps can be set or for enemies to hide. Finally
there is layout itself, that can provide a challenge, with overlapping structures
and complex connections. Especially in recent years, computer games grow more

© Springer Nature Switzerland AG 2019
J. Pejaś et al. (Eds.): ACS 2018, AISC 889, pp. 375–386, 2019.
https://doi.org/10.1007/978-3-030-03314-9_32

complex, with demanding graphics and elaborate objects. With such requirements it can take considerable amount of time to finish even simple underground system. At the same time, when created by human designer, such structures can become repeatable and boring for more advanced players.

Procedural content generation can be a solution to both of those problems. Different algorithms exist, adapted to generation of various objects and areas. That is also the case with dungeons and other underground structures, allowing creation of huge amounts of content, faster and with more diversity than any human designer can provide. At the same time, most of existing procedures either do not offer acceptable level of control over final object (which is an essential property, when it comes to incorporating obtained results in computer games), or produce complex shapes, without any supervision over generation process, and with no easy way to edit obtained elements after this process is completed [17–21]. Above problems often result in discarding procedural algorithms in favour of manual modelling.

While creating dungeons, the most challenging element is creating rooms, that have interesting layout, and are not repeatable. It is also important to remember, that any solution should consider both creation of 2D shapes (that can be later used i.e. for dungeon map, that player can refer to), as well as providing simple way to transfer those shapes to 3D objects, that preserve all required properties and transitions between regions (i.e. locations of doors and different obstacles such as columns).

In this work we present an improvement to room generation methods described in previous work (see [26]), used for room shape generation. We use similar method, based on modified L-system algorithm, to ensure that room shapes are interesting and not repeatable. At the same time we further expand it, obtaining more realistic shapes. Presented method is a part of bigger solution, allowing design and generation of complex underground systems. At the same time, it can be used separately, generating room shapes usable as components during design process, providing human designer with extensive base for creation of underground systems.

The rest of this work is organized as follows. In Sect. 2 we review some of existing solutions related to our approach. Section 3 outlines initial assumptions that led to our method in its current form, as well as describes properties of described algorithm. In Sect. 4 we present some of obtained results. Finally, in Sect. 5 we conclude our work, as well as outline some future research directions.

2   Existing Solutions

In recent years procedural content generation became a very popular topic, due to the possibilities that it brings [4, 11–13]. It is especially popular in applications, that require large amounts of high quality content. One of such areas are computer games, where the greatest challenge is providing acceptable level of control, without requiring that the designer will perform most of related work manually.

Existing solutions vary greatly in that aspect, from ensuring that object meets a series of strict properties [5] and using parameters to describe desired results [4], to using simplified elements as a base for generation [6] or employing story, to guide generation process [7]. Finally, we have some solutions, that use simplified maps, to assign different properties to final terrain and generate it accordingly [15,16,22,23].

When it comes to underground system generation, we can distinguish two main approaches. First one considers creating such systems in 2D, using such solutions as cellular automata [8], predefined shapes with fitness function used to connect them in various ways [9], checkpoints with fitness function applied to shape creation [10], or even simple maze generation [3]. Second group of solutions focuses on 3D shapes, and needs to consider additional problems and constraints (such as overlapping elements and multilevel layout of entire structure). Existing approaches focus on obtaining realistic features in caves [17–19], generating entire buildings [14], or in some cases, terrain playability [20,21]. Unfortunately, even for those approaches that consider computer games as their main application, designer usually has very limited influence over layout of generated system, while elements of final object are not easily separable (and therefore cannot be used as components in different system without additional actions). For detailed study of existing methods for procedural content generation see [1–4,11–13].

3 Procedural Room Generation

In previous approach to dungeon generation [26], system was divided into tiles and levels, where each level contained only structures that do not overlap, and each tile in single level could contain either large space, small space or corridor. In case of spaces, adopted approach produced results that were acceptable for smaller tiles, but for larger tiles they always created star-shaped elements, without enough variety (for example room shapes see Fig. 1). It was also noticed, that although room creation method was used as part of bigger solution, it could also be adapted to component generation, that designer could use as ready-made elements during modelling process.

Fig. 1. Example small (top) and large (bottom) rooms generated by previously used procedure. Red colour represents passages connecting room to neighbouring tiles

3.1   Initial Assumptions

Similarly to previous work on the subject, main focus of presented solution remains on creating objects intended for use in computer games and similar applications. Taking that into account, spaces generated by presented procedure need to meet series of properties, appropriate for such application:

– Generated rooms need to have interesting and not repeatable shapes.
– At least two types of spaces are required: small and large.
– Procedure needs to incorporate way to enforce either vertical or horizontal symmetry, as well as combination of both.
– Rooms should contain places where enemies or traps could be hidden.
– Room data should allow easy transition from 2D outlines to 3D objects.
– Generated components should be ready-made objects for dungeon creation.
– Solution should allow easy way to edit final objects.

Taking those properties into account, we decided to use, as in previous research [22–26], schematic maps as input, providing user with easy way to define type and number of generated rooms. We generate room shape in each tile using modified L-system algorithm and save obtained output as image files. Tile size is an user-defined parameter, that translates directly to image dimensions in pixels, for 2D maps representing shapes in each tile, and to number of vertices in 3D object. Using such property has this additional bonus, that mesh of the final object has fixed maximum complexity, that cannot be exceeded. Such characteristic is especially important in applications such as computer games and simulations, where computational complexity is an important factor.

3.2   L-system Settings

L-system can be defined as formal grammar, consisting of set of symbols (or alphabet), with rules for exchanging each symbol in a string of such characters. Alphabet elements can be defined either as terminal (when they have no further extensions) and nonterminal (when such extensions are defined).

   In approach presented in this work we use modified L-system algorithm with no terminal symbols. Exchange rules for each room set are randomly generated at the beginning of our procedure (although it is also possible to use predefined set of rules). Properties of L-system used for room shape generation are set according to two factors: type of space, and size of single tile. In that aspect we can distinguish following elements (all values are converted to integers):

– Number of alternative exchange rules for L-system keys (one set for each initial symbol). Value is set as maximum from set: [2; 10% of tile size].
– Length of starting string for L-system, also set at 10% of tile size.
– Number of L-system iterations. Since in our approach we use only nonterminal symbols, we define final set complexity only by limiting number of iterations. Value is set as floor from 5% of tile size for small space, and floor from 10% of tile size for large space.

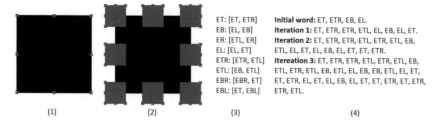

ET: [ET, ETR]   **Initial word:** ET, ETR, EB, EL.
EB: [EL, EB]   **Iteration 1:** ET, ETR, ETR, ETL, EL, EB, EL, ET.
ER: [ETL, ER]   **Iteration 2:** ET, ETR, ETR, ETL, ETR, ETL, EB,
EL: [EL, ET]   ETL, EL, ET, EL, EB, EL, ET, ET, ETR.
ETR: [ETR, ETL] **Itereation 3:** ET, ETR, ETR, ETL, ETR, ETL, EB,
ETL: [EB, ETL]   ETL, ETR, ETL, EB, ETL, EL, EB, EB, ETL, EL, ET,
EBR: [EBR, ET]   ET, ETR, EL, ET, EL, EB, EL, ET, ET, ETR, ET, ETR,
EBL: [ET, EBL]   ETR, ETL.

(1)   (2)   (3)   (4)

Fig. 2. Initial L-system settings: (1) initial extension points for shape drawing, (2) updated extension points, after inserting square shapes, (3) example rule set for basic keys and (4) extension sequence example for L-system word using rules from (3).

With given set of spaces to produce, we first proceed with room generation using L-system. With above properties, and initial set of keys, we generate shape placed in current tile. Initial key set is organized as follows:

- ET: extend top part of the room
- EB: extend bottom part of the room
- ER: extend right part of the room
- EL: extend left part of the room
- ETR: extend top right part of the room
- ETL: extend top left part of the room
- EBR: extend bottom right part of the room
- EBL: extend bottom left part of the room.

Using randomly generated key-sets, containing production rules with different initial keys, (i.e.: [ET → [ET, ET, ER], EB → [ER, EL, ETR], etc.), we first define final L-system word (that describes room shape when no symmetry is applied). Symmetry is then enforced, while drawing tile map representing current region (i.e. in case of horizontal symmetry, when ET symbol is present in final L-system world, both top and bottom parts of the room will be extended with the same shape). It is also the moment, when all extensions are done, and additional extension points are added.

In our approach we start by drawing square in the middle of the tile, with eight initial extension points, one for each basic key (see Fig. 2(1)). When basic key is chosen, we insert one of basic shapes at point in tile related to that key (currently we are using four shapes: square, circle, horizontal rectangle and vertical rectangle). Extension points are then updated, to include those contained by outer outline of new shape (see Fig. 2(2)). If next extension is done in that part of room, point at which additional shape will be added is chosen randomly from newly updated set of keys. Entire process is then repeated, until every character in final L-system string is addressed. Figure 2(3) and (4) present example rule sets and sequence for extending initial word.

While inserting shapes related to succeeding characters in generated string (representing room shape), we also check and enforce symmetry chosen for that particular space. As mentioned before, this parameter can have four values: no

symmetry, vertical symmetry, horizontal symmetry and both vertical and horizontal symmetry. For each key, if any type of symmetry is active, each shape is
first inserted in chosen place, and then reflected, to represent chosen symmetry
type. Finally, connections are drawn for defined neighbours (either to top, bottom right or left tile). For example of transferring L-system string to shape in
tile, along with influence symmetry has over final shape, see Fig. 3.

Such approach allows creation of some interesting room shapes, with niches
and obstacles that can represent columns. At the same time obtained designs are
not repeatable, with easy way to edit or regenerate them (i.e. all user needs to
do to regenerate tile/tiles is to change seed value for generation; another way for
modification is manual alteration of created tile maps in any 2D graphics editing
application). For full overview of shape generation procedure see Algorithm 1.

At this point user can choose which tiles will be forwarded to 3D modelling
application, discarding those elements that do not meet required properties.
Visualization of generated rooms at this point is greatly simplified, representing
only basic layout of produced shapes, without such improvements as placing
additional elements (such as enemies, treasure chests, traps, etc.). To translate
3D objects from generated room shapes, we use similar method as in previous
work (see [26]). Each pixel from room tile map is represented by single vertex
in initial Blender object (we use grid with number of vertices equal to squared
size of single tile). Room shape is obtained by removing vertices that are not

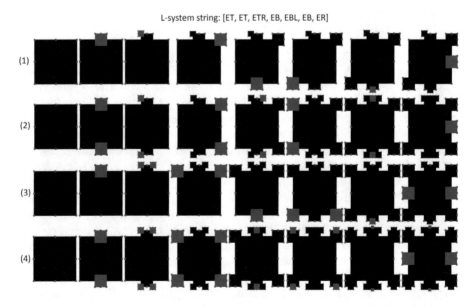

Fig. 3. Example of drawing the same L-system string defining room shape with: no
symmetry (1), horizontal symmetry (2), vertical symmetry (3) and both horizontal and
vertical symmetry (4). For the purpose of visualization only square shapes are included
during drawing process. Used L-system word is presented at the top.

classified as room interior (white on room tile map). After transferring room shape, walls are extruded, and volume of final object is increased, using in-build Blender functionality. At this point simple material and texture can also be added. For example results generated by described procedure (both 2D shapes, as well as 3D models with corresponding tile map), see Sect. 4.

4 Obtained Results

Algorithm presented in this paper was prepared as two separate procedures (both implemented using Python): first one for generating 2D shapes (using only basic language functionality) and second one for visualizing chosen shapes in 3D (created with Blender application using some of its inbuilt functionality; for documentation see [27]). Experiments were performed on a PC with Intel Core i7-4710HQ processor (2,5 GHz per core) and 8 GB of ram.

First checked element were actual shapes that algorithm could generate. For those rooms to serve their intended purpose, they needed to be interesting visually, as well as contain elements important to computer games, such as obstacles, and side spaces. As shown at Figs. 4 and 5, this goal was met, since algorithm can produce different rooms, that are not repeatable. Produced shapes also have spaces that can be hidden i.e. behind closed doors, or fake walls, as well as obstacles and partitions allowing such elements as hidden enemies or traps.

Algorithm 1. Procedural room shape generation with L-system.

```
algorithm generateRoom (numberOfRooms, roomTypes, connections, tileSize, keySet):
    definedRooms = getListOfRooms(numberOfRooms, roomType, connection)
    numberOfRules = calculateNumberOfRules(tileSize)
    startingFrazeLength = calculateStartingRuleLength(tileSize)
    rules = generateLSystemRules(numberOfRules, keySet)
    for room in definedRooms:
        symmetry = getRandomSymetryValue()
        initialFraze = generateStartingFraze(keySet, rules, startingFrazeLength)
        iterations = getNumberOfIterations(roomType, tileSize)
        finalFraze = extendLSystem(keySet, iterations, initialFraze)
        drawLSystem(keySet, symmetry, tileSize, finalFraze)
        connectTile(connection)

drawLSystem(keySet, symmetry, tileSize, finalFraze):
    extensionPoints = getBasicExtensionPoints(keySet)
    middle = integer(tileSize/2)
    for character in finalFraze:
        currentPoint = random(extensionPoints[character])
        drawLSystemCharacter(currentPoint, symmetry, middle)
        extensionPoints = updateExtensionPoints(character, symmetry, currentPoint)
```

Another important property concerned total generation times of room shapes in tiles. Since presented methods main use is either as a part of bigger solution, or as a component generator, those times should be short. At the same time, each iteration of presented procedure should return at least ten shapes, allowing designer to choose which shapes best meet his requirements. To confirm that

Table 1. Rendering times for tiles containing small rooms. Each run of the algorithm created 25 rooms of given type. The time is recorded in seconds [s].

Tile size	No symmetry	Vertical symmetry	Horizontal symmetry	Both symmetry types
11	1,843	1,891	1,875	1,942
21	2,574	2,628	2,580	2,701
31	3,724	3,812	3,876	3,915
41	5,082	5,247	5,153	5,199
51	7,307	7,502	7,590	8,209
71	14,714	14,918	14,987	15,098
91	25,869	26,189	26,299	26,925

Table 2. Rendering times for tiles containing large rooms. Each run of the algorithm created 25 rooms of given type. The time is recorded in seconds [s].

Tile size	No symmetry	Vertical symmetry	Horizontal symmetry	Both symmetry types
11	1,886	1,921	1,943	1,956
21	2,986	3,273	3,268	4,239
31	4,413	4,672	4,445	5,584
41	5,758	6,459	6,216	7,109
51	8,339	9,842	10,038	11,539
71	16,077	17,489	17,623	20,329
91	31,236	31,711	31,964	37,360

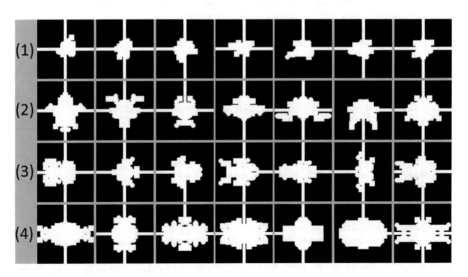

Fig. 4. Examples of small rooms generated by our procedure with different symmetry settings: no symmetry (1), vertical symmetry (2), horizontal symmetry (3) and both symmetry types (4). Tile size is set at 91.

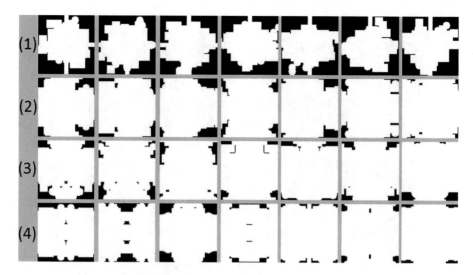

Fig. 5. Examples of large rooms generated by our procedure with different symmetry settings: no symmetry (1), vertical symmetry (2), horizontal symmetry (3) and both symmetry types (4). Tile size is set at 91.

obtained generation times are acceptable, series of experiments were performed, with different tile sizes, room types and symmetry setting. Obtained results are presented in Table 1 for small rooms and in Table 2 for large rooms. Each run of presented algorithm produced 25 rooms with given parameters. Although obtained times do not allow for interactive work (especially with tile size set at 41 and above), they are more than acceptable for component shape generation.

To ensure, that generated elements can be reused as many times as possible, all connection points are set at the same places in all tiles (middle of wall, connected to neighbouring tile). Because of that, all rooms with identical connections defined and same tile size, can be used interchangeably. The same property transfers to 3D shapes. Since designer can choose which elements to transfer, any final objects would meet defined requirements, forming an interchangeable component with interesting shape. For example 3D visualizations of room shapes generated by presented algorithm see Fig. 6.

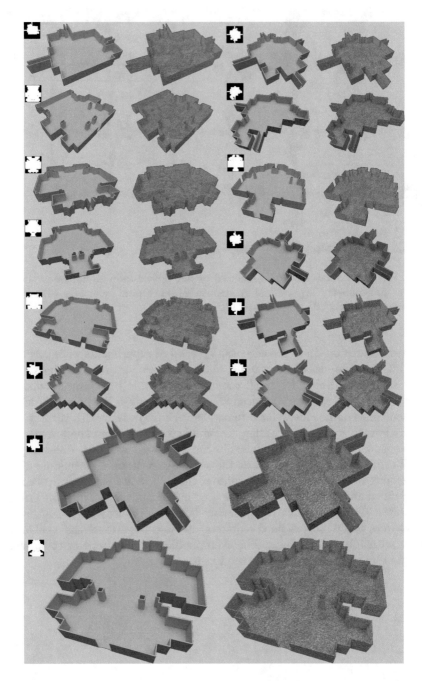

Fig. 6. Examples of rendered rooms, with corresponding tile shapes generated by our procedure. Each room is presented both as 3D object without modifications and model with assigned simple texture.

5   Conclusions and Future Work

In this paper we presented a method for procedural generation of rooms using modified L-system algorithm for shape creation. Our solution works in two main steps, first creating 2D maps, and then transferring shapes from tiles chosen by user to 3D objects. 2D shapes are created fast enough, to allow user large selection of potential space layouts in reasonable amount of time. Such approach maximizes the chance, that user will get elements meeting his requirements. In case that some changes are needed, obtained results (both 2D and 3D) can be easily edited, and since we ensure, that any entry/exit point is placed at the same place in each space (middle of tile edge with specified connection), they can also serve as components in bigger structures.

Our procedure still requires additional methods for placing different objects across generated rooms (such as doors, traps, torches, furniture and other elements commonly associated with dungeons). We plan to address that in future work.

Overall, presented approach can generate interesting elements, that can be instantly used, or further edited by graphic designers. Since complexity of each element can be defined by tile size parameter, it is easy to adjust it to requirements posed by different applications (i.e. different types of computer games). Elements generated by our procedure meet all specified requirements determined by computer games (i.e. creating spaces where enemies or traps can be hidden), and are not repeatable (creating rooms with different shapes, symmetries and overall layouts). Final objects can be used for visualization while designing dungeons, provide a basis for further shape editing, or be incorporated directly in simple computer game.

References

1. Shaker, N., Liapis, A., Togelius, J., Lopes, R., Bidarra, R.: Constructive generation methods for dungeons and levels. In: Procedural Content Generation in Games, pp. 31–55 (2015)
2. van der Linden, R., Lopes, R., Bidarra, R.: Procedural generation of dungeons. IEEE Trans. Comput. Intell. AI Games 6(1), 78–89 (2014)
3. Galin, E., Peytavie, A., Maréchal, N., Guérin, E.: Procedural generation of roads. Comput. Graph. Forum 29(2), 429–438 (2010)
4. Smelik, R., Galka, K., de Kraker, K.J., Kuijper, F., Bidarra, R.: Semantic constraints for procedural generation of virtual worlds. In: Proceedings of the 2nd International Workshop on Procedural Content Generation in Games, p. 9. ACM (2011)
5. Tutenel, T., Bidarra, R., Smelik, R.M., De Kraker, K.J.: Rule-based layout solving and its application to procedural interior generation. In: CASA Workshop on 3D Advanced Media in Gaming and Simulation (2009)
6. Merrell, P., Manocha, D.: Model synthesis: a general procedural modeling algorithm. IEEE Trans. Vis. Comput. Graph. 17(6), 715–728 (2011)
7. Matthews, E., Malloy, B.: Procedural generation of story-driven maps. In: CGAMES, pp. 107–112. IEEE (2011)

8. Johnson, L., Yannakakis, G.N., Togelius, J.: Cellular automata for real-time generation of infinite cave levels. In: Proceedings of the 2010 Workshop on Procedural Content Generation in Games, p. 10. ACM (2010)
9. Valtchanov, V., Brown, J.A.: Evolving dungeon crawler levels with relative placement. In: Proceedings of the 5th International C* Conference on Computer Science and Software Engineering, pp. 27–35. ACM (2012)
10. Ashlock, D., Lee, C., McGuinness, C.: Search-based procedural generation of maze-like levels. IEEE Trans. Comput. Intell. AI Games **3**(3), 260–273 (2011)
11. Hendrikx, M., Meijer, S., Van Der Velden, J., Iosup, A.: Procedural content generation for games: a survey. ACM TOMM **9**(1), 1 (2013)
12. Smelik, R.M., Tutenel, T., Bidarra, R., Benes, B.: A survey on procedural modelling for virtual worlds. Comput. Graph. Forum **33**(6), 31–50 (2014)
13. Ebert, D.S.: Texturing & Modeling: A Procedural Approach. Morgan Kaufmann, San Francisco (2003)
14. Pena, J.M., Viedma, J., Muelas, S., LaTorre, A., Pena, L.: emphDesigner-driven 3D buildings generated using variable neighborhood search. In: 2014 IEEE Conference on Computational Intelligence and Games, pp. 1–8. IEEE (2014)
15. Smelik, R.M., Tutenel, T., de Kraker, K.J., Bidarra, R.: A proposal for a procedural terrain modelling framework. In: EGVE, pp. 39–42 (20080
16. Smelik, R.M., Tutenel, T., de Kraker, K.J., Bidarra, R.: Declarative terrain modeling for military training games. Int. J. Comput. Games Technol. **2010** (2010). Article No. 2
17. Cui, J., Chow, Y.W., Zhang, M.: Procedural generation of 3D cave models with stalactites and stalagmites (2011)
18. Boggus, M., Crawfis, R.: Explicit generation of 3D models of solution caves for virtual environments. In: CGVR, pp. 85–90 (2009)
19. Boggus, M., Crawfis, R.: Procedural creation of 3D solution cave models. In: Proceedings of IASTED, pp. 180–186 (2009)
20. Santamaria-Ibirika, A., Cantero, X., Huerta, S., Santos, I., Bringas, P.G.: Procedural playable cave systems based on Voronoi diagram and delaunay triangulation. In: International Conference on Cyberworlds, pp. 15–22. IEEE (2014)
21. Mark, B., Berechet, T., Mahlmann, T., Togelius, J.: Procedural generation of 3D caves for games on the GPU. In: Foundations of Digital Games (2015)
22. Antoniuk, I., Rokita, P.: Procedural generation of adjustable terrain for application in computer games using 2D maps. In: Pattern Recognition and Machine Intelligence, pp. 75–84. Springer (2015)
23. Antoniuk, I., Rokita, P.: Generation of complex underground systems for application in computer games with schematic maps and L-systems. In: International Conference on Computer Vision and Graphics, pp. 3–16. Springer (2016)
24. Antoniuk, I., Rokita, P.: Procedural generation of adjustable terrain for application in computer games using 2D maps. In: Pattern Recognition and Machine Intelligence, pp. 75–84. Springer (2016)
25. Antoniuk, I., Rokita, P.: Procedural generation of underground systems with terrain features using schematic maps and L-systems. Challenges Modern Technol. **7**(3), 8–15 (2016)
26. Antoniuk, I., Rokita, P.: Procedural generation of multilevel dungeons for application in computer games using schematic maps and L-system. To be published in Studies in Big Data 40 Springer International Publishing
27. Blender application home page. https://www.blender.org/. Accessed 14 May 2018

Hardware-Efficient Algorithm
for 3D Spatial Rotation

Aleksandr Cariow$^{(\boxtimes)}$ and Galina Cariowa

Faculty of Computer Science and Information Technology, West Pomeranian
University of Technology, Żołnierska 52, 71-210 Szczecin, Poland
{acariow, gcariowa}@wi.zut.edu.pl

Abstract. In this paper, we have proposed a novel VLSI-oriented parallel
algorithm for quaternion-based rotation in 3D space. The advantage of our
algorithm is a reduction the number of multiplications through replacing part of
them by less costly squarings. The algorithm uses Logan's trick, which proposes
to replace the calculation of the product of two numbers on summing the squares
via the Binomial theorem. Replacing digital multipliers by squaring units
implies reducing power consumption as well as decreases hardware circuit
complexity.

Keywords: Quaternions · Rotation matrix · Fast algorithms

1 Introduction

The necessity of rotation from one coordinate system to another occurs in many areas
of science and technology including robotics, navigation, kinematics, machine vision,
computer graphics, animation, and image encoding [1–3]. Using quaternions is a useful
and elegant way to perceive rotation because every unit quaternion represents a rotation
in 3-dimensional vector spaces.

Suppose we have given a unit quaternion $q = (q_0, q_1, q_2, q_3)$ where q_0 is the real
part. A rotation from coordinate system x to coordinate system y in terms of the
quaternion can be accomplished as follows:

$$y = qxq^* \tag{1}$$

where $q^* = (q_0, -q_1, -q_2, -q_3)$ is a conjugation of q.

Performing of (1) requires 32 multiplications and 24 additions.

The alternative method introduces a rotation matrix, which enables the realization
of rotation via matrix-vector multiplication. Then we can represent a rotation in the
following form:

$$\mathbf{Y}_{3\times1} = \mathbf{R}_3\mathbf{X}_{3\times1} \tag{2}$$

where $\mathbf{X}_{3\times1} = [x_0, x_1, x_2]^\mathrm{T}$ and $\mathbf{Y}_{3\times1} = [y_0, y_1, y_2]^\mathrm{T}$- are vectors in coordinate system x
and y respectively, and is a rotation matrix corresponding to quaternion q. This matrix
is also called the direction cosine matrix (DCM) or attitude matrix.

© Springer Nature Switzerland AG 2019
J. Pejaś et al. (Eds.): ACS 2018, AISC 889, pp. 387–395, 2019.
https://doi.org/10.1007/978-3-030-03314-9_33

$$\mathbf{R}_3 = \begin{bmatrix} 2(q_0^2 + q_1^2) - 1 & 2(q_1q_2 - q_0q_3) & 2(q_1q_3 + q_0q_2) \\ \hline 2(q_0q_3 + q_1q_2) & 2(q_0^2 + q_2^2) - 1 & 2(q_2q_3 - q_0q_1) \\ \hline 2(q_1q_3 - q_0q_2) & 2(q_2q_3 + q_0q_1) & 2(q_0^2 + q_3^2) - 1 \end{bmatrix} \tag{3}$$

The direct realization of (2) requires only 15 conventional multiplications, 4 squarings, 18 additions and 9 trivial multiplications by two (which will not be taken into account). It is easily to calculate that this way to perform the rotation is preferable from the computation point of view. Below we show how to implement these calculations more efficiently from the point of view of hardware implementation.

2   The Algorithm

It easy to see, that relation (2) can be rewritten as follows:

$$\mathbf{Y}_{3\times 1} = \mathbf{P}_{3\times 6}^{(2)}[(\mathbf{R}_3^{(0)} \oplus (-\mathbf{I}_3)]\mathbf{P}_{6\times 3}^{(1)}\mathbf{X}_{3\times 1} \tag{4}$$

where

$$\mathbf{P}_{6\times 3}^{(1)} = \mathbf{1}_{2\times 1} \otimes \mathbf{I}_3 = \begin{bmatrix} 1 & & \\ & 1 & \\ & & 1 \\ \hline 1 & & \\ & 1 & \\ & & 1 \end{bmatrix}, \quad \mathbf{P}_{3\times 6}^{(2)} = [\mathbf{P}_{3\times 6}^{(1)}]^{\mathrm{T}}.$$

where $\mathbf{I}_3$ is the 3×3 identity matrix, $\mathbf{1}_{N\times M}$ is a unit matrix (an integer matrix consisting of all 1s), "$\otimes$", "$\oplus$" – denote the Kronecker product and direct sum of two matrices respectively [4], and

$$\mathbf{R}_3^{(0)} = \begin{bmatrix} 2(q_0^2 + q_1^2) & 2(q_0q_2 - q_0q_3) & 2(q_1q_3 + q_0q_2) \\ \hline 2(q_0q_3 + q_1q_2) & 2(q_0^2 + q_2^2) & 2(q_2q_3 - q_0q_1) \\ \hline 2(q_1q_3 - q_0q_2) & 2(q_2q_3 + q_0q_1) & 2(q_0^2 + q_3^2) \end{bmatrix} \tag{5}$$

Figure 1 shows a data flow diagram, which describes the computations in according to (4). In this paper, data flow diagrams are oriented from left to right. Straight lines in the figures denote the operations of data transfer. Points, where lines converge, denote summation. (The dotted lines indicate the subtractions). The rectangles indicate the matrix-vector multiplications by matrices inscribed inside rectangles. We use the usual lines without arrows on purpose, so as not to clutter the picture.

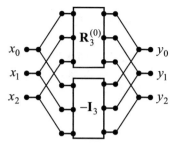

Fig. 1. Data flow diagram, which describes the decomposition of $\mathbf{R}_3$ matrix-vector multiplication in according to the procedure (4).

For a more compact representation, we introduce the following notation:

$$\mathbf{R}_3^{(0)} = \left[\begin{array}{c:c:c} c_{0,0} & c_{0,1} & c_{0,2} \\ \hdashline c_{1,0} & c_{1,1} & c_{1,2} \\ \hdashline c_{2,0} & c_{2,1} & c_{2,2} \end{array}\right],$$

where

$$c_{0,0} = 2(q_0^2 + q_1^2), c_{0,1} = 2(q_1 q_2 - q_0 q_3), c_{0,2} = 2(q_1 q_3 + q_0 q_2),$$

$$c_{1,0} = 2(q_1 q_2 + q_0 q_3), c_{1,1} = 2(q_0^2 + q_2^2), c_{1,2} = 2(q_2 q_3 - q_0 q_1),$$

$$c_{2,0} = 2(q_1 q_3 + q_0 q_2), c_{2,1} = 2(q_2 q_3 + q_0 q_1), c_{2,2} = 2(q_0^2 + q_3^2),$$

In 1971, Logan noted that the multiplication of two numbers can be performed using the following expression [5, 6]:

$$ab = \frac{1}{2}[(a+b)^2 - a^2 - b^2],$$

Using the Logan's identity we can write:

$$2(q_1 q_2 + q_0 q_3) = [(q_1 + q_2)^2 - (q_1^2 + q_2^2)] + [(q_0 + q_3)^2 - (q_0^2 + q_3^2)],$$
$$2(q_1 q_2 - q_0 q_3) = [(q_1 + q_2)^2 - (q_1^2 + q_2^2)] - [(q_0 + q_3)^2 - (q_0^2 + q_3^2)],$$
$$2(q_1 q_3 + q_0 q_2) = [(q_1 + q_3)^2 - (q_1^2 + q_3^2)] + [(q_0 + q_2)^2 - (q_2^2 + q_0^2)],$$
$$2(q_1 q_3 - q_0 q_2) = [(q_1 + q_3)^2 - (q_1^2 + q_3^2)] - [(q_0 + q_2)^2 - (q_2^2 + q_0^2)],$$
$$2(q_2 q_3 + q_0 q_1) = [(q_2 + q_3)^2 - (q_2^2 + q_3^2)] + [(q_0 + q_1)^2 - (q_1^2 + q_0^2)],$$
$$2(q_2 q_3 - q_0 q_1) = [(q_2 + q_3)^2 - (q_2^2 + q_3^2)] - [(q_0 + q_1)^2 - (q_1^2 + q_0^2)].$$

Then all entries of the matrix $\mathbf{R}_3^{(0)}$, that previously required performing the multiplications, can be calculated only with the help of squaring operations [7].

Therefore all entries of the matrix $\mathbf{R}_3^{(0)}$ can be calculated using the following vector–matrix procedure:

$$\mathbf{C}_{9\times1} = \mathbf{P}_9\mathbf{R}_9^{(4)}\mathbf{R}_{9\times12}^{(3)}\mathbf{R}_{12\times10}^{(2)}[\mathbf{R}_{10\times4}^{(1)}\mathbf{q}_{4\times1}]^2 \qquad (6)$$

where $\mathbf{C}_{9\times1} = [c_{0,0}, c_{1,0}, c_{2,0}, c_{0,1}, c_{1,1}, c_{2,1}, c_{0,2}, c_{1,2}, c_{2,2}]^{\mathrm{T}}$,

$\mathbf{q}_{4\times1} = [q_0, q_1, q_2, q_3]^{\mathrm{T}}$ is a vector containing components of the unit quaternion, and symbol $[\,\cdot\,]^2$ means squaring all the entries of the vector inscribed inside of the square brackets.

$$\mathbf{R}_{10\times4}^{(1)} = \begin{bmatrix} 1 & 1 & & \\ 1 & & 1 & \\ 1 & & & 1 \\ & 1 & 1 & \\ & 1 & & \\ & 1 & 1 & \\ \hline 1 & & & \\ & 1 & & \\ & & 1 & \\ & & & 1 \end{bmatrix}, \quad \mathbf{R}_{12\times10}^{(2)} = \begin{bmatrix} 1 & & & & & & & & & \\ & 1 & & & & & & & & \\ & & 1 & & & & & & & \\ & & & 1 & & & & & & \\ & & & & 1 & & & & & \\ \hline & & & & & 1 & 1 & & & \\ & & & & & 1 & & 1 & & \\ & & & & & 1 & & & 1 & \\ & & & & & & 1 & 1 & & \\ & & & & & & 1 & & 1 & \\ & & & & & & & 1 & 1 & \\ & & & & & & & & 1 & 1 \end{bmatrix},$$

$$\mathbf{R}_{9\times12}^{(3)} = \begin{bmatrix} & & & & 1 & & & & & & -1 & \\ & 1 & & & & & & & & -1 & & \\ & & 1 & & & & & & & & & -1 \\ & & & 1 & & & & & -1 & & & \\ & & & & & 1 & & & & & & -1 \\ 1 & & & & & & & -1 & & & & \\ & & & & & & 1 & & & & & \\ & & & & & & & 1 & & & & \\ & & & & & & & & 1 & & & \end{bmatrix},$$

$$\mathbf{R}_9^{(4)} = (\mathbf{I}_3 \otimes \mathbf{H}_2) \oplus \mathbf{I}_3 = \begin{bmatrix} 1 & 1 & \mathbf{0}_2 & \mathbf{0}_2 & \\ 1 & -1 & & & \\ \hline \mathbf{0}_2 & 1 & 1 & \mathbf{0}_2 & \mathbf{0}_{6\times3} \\ & 1 & -1 & & \\ \hline \mathbf{0}_2 & \mathbf{0}_2 & 1 & 1 & \\ & & 1 & -1 & \\ \hline & & & 1 & \\ \mathbf{0}_{3\times6} & & & 1 & \\ & & & & 1 \end{bmatrix},$$

$$\mathbf{P}_9 = \begin{bmatrix} & 1 & & & 1 & & & & \\ & & & 1 & & & & & \\ & & 1 & & & & & & \\ & & & & & 1 & & & \\ & & & & 1 & & & & \\ & & & 1 & & & & & \\ & & & & & 1 & & & \\ & & & & & & & & 1 \end{bmatrix}, \mathbf{H}_2 = \begin{bmatrix} 1 & 1 \\ 1 & -1 \end{bmatrix}$$

Figure 2 shows a data flow diagram of the process for calculating the $\mathbf{R}_3^{(0)}$ matrix entries, represented in its vectorized form (in form of the vector $\mathbf{C}_{9\times1}$). The small squares in this figure show the squaring operations, in turn, the big rectangles indicate the matrix–vector multiplications with the 2×2 Hadamard matrices.

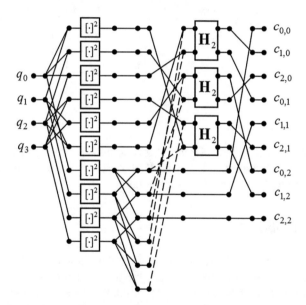

Fig. 2. Data flow diagram describing the process of calculating entries the vector $\mathbf{C}_{9\times1}$ in accordance with the procedure (6).

Taking into account the considerations and transformations given above, we can write the final computation procedure that describes the fully parallel algorithm for multiplying a vector by a rotation matrix:

$$\mathbf{Y}_{3\times1} = \boldsymbol{\Xi}_{3\times12}\mathbf{D}_{12}\mathbf{P}_{12\times3}\mathbf{X}_{3\times1} \tag{7}$$

where

$$\mathbf{D}_{12} = diag(c_{0,0}, c_{1,0}, c_{2,0}, c_{0,1}, c_{1,1}, c_{2,1}, c_{0,2}, c_{1,2}, c_{2,2},\ 1,\ 1,\ 1),$$

$$\Xi_{3\times12} = \mathbf{1}_{1\times3} \otimes \mathbf{I}_3 = \begin{bmatrix} 1 & & & 1 & & & 1 & & & 1 & & \\ & 1 & & & 1 & & & 1 & & & 1 & \\ & & 1 & & & 1 & & & 1 & & & 1 \end{bmatrix},$$

$$\mathbf{P}_{12\times3} = \begin{bmatrix} (\mathbf{1}_{1\times3} \otimes \mathbf{I}_3) \\ \hline -\mathbf{I}_3 \end{bmatrix} = \begin{bmatrix} 1 & & & & \\ 1 & & \mathbf{0}_{3\times1} & & \mathbf{0}_{3\times1} \\ 1 & & & & \\ \hline & & 1 & & \\ \mathbf{0}_{3\times1} & & 1 & & \mathbf{0}_{3\times1} \\ & & 1 & & \\ \hline & & & & 1 \\ \mathbf{0}_{3\times1} & & \mathbf{0}_{3\times1} & & 1 \\ & & & & 1 \\ \hline & & -\mathbf{I}_3 & & \end{bmatrix}.$$

Figure 3 shows a data flow diagram that describes the fully parallel algorithm for multiplying a vector by a rotation matrix. Each circle in this figure indicates a multiplication by the number inscribed inside the circle.

3   Implementation Complexity

Let us estimate the implementation complexity of our algorithm. We calculate how many dedicated blocks (multipliers, squarers and adders) are required for fully parallel implementation of the proposed algorithm, and compare it with the number of such blocks required for a fully parallel implementation of computation with correspondence to (2).

As already mentioned a fully parallel direct implementation of (2) requires 15 conventional two-input multipliers, 4 squarers, 18 adders. In contrast, the number of multipliers required using a fully parallel implementation of proposed algorithm is 9. In addition, a fully parallel implementation of our algorithm requires only 10 squarers and 35 adders.

In the other hand, the number of adders that required for a fully parallel implementation of our algorithm is 35. Thus, proposed algorithm saves 6 multipliers but increases 6 squarers and 17 adders compared with direct implementation of (2).

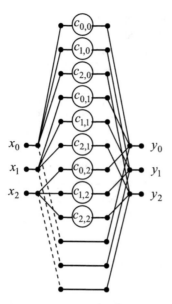

Fig. 3. Data flow diagram describing the fully parallel algorithm for multiplying a vector by a rotation matrix in accordance with the procedure (7).

So, using the proposed algorithm the number of multipliers is reduced. It should be noted that in low-power VLSI design, optimization must be primarily done at the level of logic gates amount. From this point of view, a multiplier requires much more intensive hardware resources than an adder. Moreover, a multiplier occupies much more area and consumes much more power than an adder. This is because the hardware complexity of a multiplier grows quadratically with operand size, while the implementation complexity of an adder increases linearly with operand size [8, 9]. Therefore, the algorithm containing as little as possible of real multiplications is preferable from point of view hardware implementation complexity.

On the other hand, it should be emphasized that squares are a special case of multiplication where both operands are identical. For this reason, designers often use general-purpose multipliers to implement the squaring units by connecting a multiplier's inputs together. Even though using general-purpose multipliers that are available as part of design packages reduces design time, it results in increased area and power requirements for the design [8]. Meanwhile, since the two operands are identical, some rationalizations can be made during the implementation of a dedicated squarer. In particular, unlike the general-purpose multiplier, a dedicated squaring unit will have only one input, which allows simplifying the circuit. The article [9] shows that the dedicated fully parallel squaring unit requires less than half whole amount of the logic gates as compared to the fully parallel general-purpose multiplier. Dedicated squarer is area efficient consumes less energy and dissipates less power as compared to the general-purpose multiplier. It should be noted that most modern FPGA's contain a number of embedded dedicated multipliers. If their number is sufficient, the constructing and using of additional squarers instead of multipliers is irrational. It makes, therefore, sense to try to exploit these multipliers. It would be unreasonable to refuse the possibility of using embedded multipliers. Nevertheless, the number of on-chip multipliers is always limited, and this number may sometimes not be enough. In this case, it is advisable to design the specialized squaring units using the existing field of logical gates.

Taking into account the reasoning given above, we introduce a number of factors that characterize the implementation complexity of the discussed algorithms. As a unit of measure, we take the number of logic gates required to realize of a certain arithmetic operation unit. Let $O_{\times} = n^2$, $O_{(\cdot)^2} = n^2/2$, $O_{\pm} = n$, are the implementation costs of the n-bit array multiplier, n-bit parallel squarer, and n-bit parallel adder/substractor, respectively. Taking into account the calculation of the entries of the matrix $\mathbf{R}_3$ the overall cost of implementing the algorithm corresponding to the expression (2) will be $O_1 = 17n^2 + 18n$. In turn, the cost of a hardware implementation of our algorithm will be $O_2 = 14n^2 + 33n$. Table 1 illustrates the overall hardware implementation complexity of compared algorithms for few examples. We can observe that with increasing n the complexity of our algorithm is reduced.

Table 1. Implementation complexity of compared algorithms

n	O_1	O_2	Hardware cost reduction, %
8	1232	1160	6.00%
16	4640	4112	11.00%
32	17984	15392	14.00%
64	70784	59458	16.00%

So, it is easy to estimate that our algorithm is more efficient in terms of the discussed parameters than the direct calculation of the rotation matrix entries in accordance with (2) and then multiplying this matrix by a vector $\mathbf{X}_{3\times1}$.

4   Conclusion

The article presents a new fully parallel hardware-oriented algorithm for 3D spatial rotation. To reduce the hardware complexity (number of two-operand multipliers), we exploit Logan's identity for number multiplication. This results in a reduction in hardware implementation cost and allows the effective use of parallelization of computations.

If the FPGA-chip already contains embedded hardwired multipliers, their maximum number is usually limited due to design constraints of the chip. This means that if the implemented algorithm contains a large number of multiplications, the developed processor may not always fit into the chip. So, the implementation of proposed in this paper algorithm on the base of FPGA chips, that have built-in binary multipliers, also allows saving the number of these blocks or realizing the whole processor with the use of a smaller number of simpler and cheaper FPGA chips. It will enable to design of data processing units using chips which contain a minimum required number of embedded multipliers and thereby consume and dissipate the least power.

How to implement a fully parallel dedicated processor for 3D spatial rotation on the base of concrete VLSI platform is beyond the scope of this paper, but it's a subject for follow-up articles.

References

1. Markley, F.L.: Unit quaternion from rotation matrix. J. Guid., Control. Dyn. **31**(2), 440–442 (2008). https://doi.org/10.2514/1.31730
2. Shuster, M.D., Natanson, G.A.: Quaternion computation from a geometric point of view. J. Astronaut. Sci. **41**(4), 545–556 (1993)
3. Doukhnitch, E., Chefranov, A., Mahmoud, A.: Encryption schemes with hyper-complex number systems and their hardware-oriented implementation. In: Elci, A. (ed.) Theory and Practice of Cryptography Solutions for Secure Information Systems, pp. 110–132. IGI Global, Hershey (2013)
4. Granata, J., Conner, M., Tolimieri, R.: The tensor product: a mathematical programming language for FFTs and other fast DSP operations. IEEE Signal Process. Mag. **9**(1), 40–48 (1992)
5. Logan, J.R.: A square-summing high-speed multiplier. Comput. Des., 67–70 (1971)
6. Johnson, E.L.: A digital quarter square multiplier. IEEE Trans. Comput. **C-29**(3), 258–261 (1980). https://doi.org/10.1109/tc.1980.1675558
7. Cariow, A., Cariowa, G.: A hardware-efficient approach to computing the rotation matrix from a quaternion, CoRR arXiv:1609.01585, pp. 1–5 (2016)
8. Deshpande, A., Draper, J.: Squaring units and a comparison with multipliers. In: 53rd IEEE International Midwest Symp. on Circuits and Systems (MWSCAS 2010), Seattle, Washington, 1st–4th August 2010, pp. 1266–1269 (2010). https://doi.org/10.1109/mwscas.2010.5548763
9. Liddicoat, A.A., Flynn, M.J.: Parallel computation of the square and cube function, Computer Systems Laboratory, Stanford University, Technical report No. CSL-TR-00-808, August (2000)

Driver Drowsiness Estimation by Means of Face Depth Map Analysis

Paweł Forczmański$^{(\boxtimes)}$ ⓘ and Kacper Kutelski

Faculty of Computer Science and Information Technology,
West Pomeranian University of Technology, Szczecin,
Żołnierska Str. 52, 71–210 Szczecin, Poland
pforczmanski@wi.zut.edu.pl, kkutelski@gmail.com,
http://pforczmanski.zut.edu.pl

Abstract. In the paper a problem of analysing facial images captured by depth sensor is addressed. We focus on evaluating mouth state in order to estimate the drowsiness of the observed person. In order to perform the experiments we collected visual data using standard RGB-D sensor. The imaging environment mimicked the conditions characteristic for driver's place of work. During the investigations we trained and applied several contemporary general-purpose object detectors known to be accurate when working in visible and thermal spectra, based on Haar-like features, Histogram of Oriented Gradients, and Local Binary Patterns. Having face detected, we apply a heuristic-based approach to evaluate the mouth state and then estimate the drowsiness level. Unlike traditional, visible light-based methods, by using depth map we are able to perform such analysis in the low level of even in the absence of cabin illumination. The experiments performed on video sequences taken in simulated conditions support the final conclusions.

Keywords: Depth map · Face detection · Haar–like features
Histogram of oriented gradients · Local binary patterns
Drowsiness evaluation

1 Introduction

There are many factors that affect the condition and behavior of motor vehicle operators and drivers. Detecting their undesirable psychophysical state is important in the context of the safety of road traffic. This problem has now become an important research issue. Such state can be estimated on the basis of subjective, physiological, behavioral, and vehicle-related factors. The analysis and evaluation the psychophysical condition of the driver can be based on observed external features and biomedical signals, i.e. face image and vital signs (pulse, body temperature, and blood pressure).

Existing driver fatigue assessment techniques rely largely on sensors and force the person to wear additional, often uncomfortable, elements. On the other hand,

© Springer Nature Switzerland AG 2019
J. Pejaś et al. (Eds.): ACS 2018, AISC 889, pp. 396–407, 2019.
https://doi.org/10.1007/978-3-030-03314-9_34

modern machine vision techniques allow continuous observation of the driver. Tired drivers show some observable behavior in head movement, movements of eyelids, or in general the way they look [18]. Vision systems for driver monitoring are the most convenient and non-invasive solution and some preliminary works of the authors confirms this fact [21].

Traditional imaging technique, namely capturing image in the visible lighting is the most straightforward and easy to implement method of visual data acquisition. Required hardware is not expensive and its operational parameters can be very high, in terms of spatial resolution, dynamic range and sensitivity. On the other hand, it should be remembered, that such devices can work only in good lighting conditions, namely during day. It would be impossible to light driver's face during driving with any sort of additional light source, since it could disturb his/her functioning.

Therefore, it is reasonable to equip the system with other capturing devices, working in different lighting spectra. Going beyond the visible spectrum offers a new perspective on this problem. Imaging technologies like X-ray, infrared, millimeter or submillimeter wave can be the examples here.

Since human face and its characteristics are one of the most obvious and adequate individual features, easy to capture, distinguish and identify [8], especially in the visible light spectrum. However, when environmental conditions are not fully controlled [9] or there is a need of increased security level beyond-visible-light imaging seems to be a good choice [4]. Images registered by infrared or thermal sensors can be used to perform face detection and recognition without the necessity to properly illuminate the subject. Moreover, it is resistant to spoofing attempts (e.g. using a photo or video stream [24]).

The authors assume that analysis of specific visual multispectral data (visible and infrared image, depth maps and thermal images of selected areas of the human body) may lead to an effective evaluation of psychophysical state of motor vehicle operator without the need of biomedical data analysis.

A depth map, in opposition to visible-light image, is an image which pixels represent the distance information from the scene objects to the camera. Depth information can be obtained applying the following techniques: stereo vision, structured-light, and time-of-flight. It is independent on the ambient temperature, general illumination and local shadows.

1.1 Existing Methods

Driver's fatigue estimation can be performed based on various techniques. In [20] the technique of the questionnaire was presented. In [22] the authors took up of the registration and evaluation of biometric parameters of the driver to determine the emotional state of the driver. For this purpose, a biomedical system concept based on three different mechanisms of measurement was proposed, namely recording vehicle speed, recording changes in the heartbeat of the driver and recording the driver's face. Similar, yet much simpler approach, was presented in [25].

Vision-based solutions provide an excellent mean for fatigue detection. The initial step in vision-based driver fatigue detection systems consist of detection of face and facial features [12]. Detected features are subsequently tracked to gather important temporal characteristics from which the appropriate conclusion of driver's fatigue can be drawn.

Detection of face and facial features are classical face recognition problems. By employing existing algorithms and image processing techniques it is possible to create an individual solution for driver fatigue/drowsiness detection based on eyes state. An example is presented in [19] where the OpenCV face and eye detectors are supported with the simple feature extractor based on the two dimensional Discrete Fourier Transform (DFT) to represent an eye region. Similarly, the fatigue of the driver determined through the duration of the eyes' blinks is presented in [6]. It operates in the visible and near infra-red (NIR) spectra allowing to analyse drivers state in the night conditions and poor visibility. A more complex, multimodal platform to identify driver fatigue and interference detection is presented in [5]. It captures audio and video data, depth maps, heart rate, steering wheel and pedals positions. The experimental results show that the authors are able to detect fatigue with 98.4% accuracy. There are solutions based on mobile devices, especially smartphones and tablets, or based on dedicated hardware [16,17,27]. In [1] the authors recognize the act of yawning using a simple webcam. In [14] the authors proposed a dynamic fatigue detection model based on Hidden Markov Model (HMM). This model can estimate driver fatigue in a probabilistic way using various physiological and contextual information. In a subsequent work [2] authors monitor information about the eyes and mouth of the driver. Then, this information is transmitted to the Fuzzy Expert System, which classifies the true state of the driver. The system has been tested using real data from various sequences recorded during the day and at night for users belonging to different races and genders. The authors claim that their system gives an average recognition accuracy of fatigue close to 100% for the tested video sequences.

The above analysis shows that many of current works is focused on the problem of recognizing driver's fatigue, yet there is no single methodology of acquisition of signals used to evaluate vehicle operator physical condition and fatigue level.

In this paper propose a simple system that works only with a single source of information, providing data about the state of the mouth, leading to the yawning detection. In contrast to one of the most complete and sophisticated research proposals [5] we capture and analyse video streams from single source only, namely depth sensor. The selection of such source makes it possible to increase the detection of face state in poor lighting conditions.

1.2   Problem Definition

The problem can be decomposed into two independent tasks: face detection (and tracking) and mouth state estimation (and drowsiness estimation).

Locating human faces in a static scene is a classical computer vision problem. Many methods employ so called sliding window approach where the detection is performed by the scanning of the image and matching the selected image parts with the templates collected in the training set. If there is no information about probable face position and size, the detection requires to perform search process in all possible locations, taking into consideration all probable window (or image) scales, which increases overall computational overhead.

The problems of human face detection and recognition in various spectra have been investigated many times, yet they still need some attention Since the visible-light imaging equipment is quite inexpensive and very wide spread, this is a source of the popularity of face detection and recognition in such spectrum. The other spectra (especially thermal) are not so popular. In this work we focus on a face detection in a depth maps (produced by RGB-D sensors) based on certain well-researched approaches, employing some general-purpose features extractors, namely Histogram of Oriented Gradients [7], Local Binary Patterns [23] and Haar-like features [26] combined with AdaBoost-based classifiers. Some preliminary investigations on these methods were presented in [10, 11].

The detection is performed iteratively over the whole scene and its effectiveness depends on the number of learning examples. During classification, an image is analysed using a sliding window approach. Features are calculated in all possible window locations. The window is slid with a varying step, which depends on the required accuracy and speed.

In the algorithm we also perform a simple face tracking in order to overcome the problem of face occlusion and changes of face orientation. It is based on predicted face position in subsequent frames providing certain low movement in short time interval.

The other task is mouth state estimation. It involves locating mouth part in the detected face and calculation of its features (geometrical, appearance-based, etc.) in order to detect yawning (as a determinant of drowsiness).

The mouth state analysis is performed using a heuristic-based rules, which are based on the proportion of pixel intensities in the binarized mouth image. It was observed, that closed and open mouth differ in terms of the number of black and white pixels. The additional rule counts these proportions over time to discriminate the act of speaking from the act of yawning (and further, continuous yawning).

2 Proposed Solution

2.1 General Overview

As it was mentioned previously, the algorithm consists of two main modules: face detection and tracking and mouth state analysis. It works in a loop iterated over the frames from the video stream. The algorithm is depicted in Fig. 1.

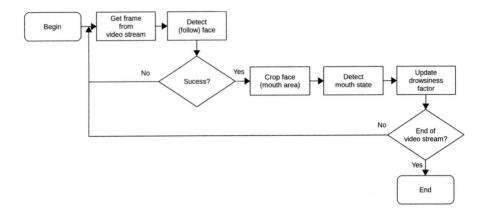

Fig. 1. Algorithm of drowsiness estimation

2.2   Data Collection

We have employed a simulation stand equipped with advanced vision sensors (video cameras, thermal imaging camera, depth sensors) described in our previous work [21]. The stand includes also some additional elements simulating the operating environment of the driver, realistically reflecting his working conditions and surrounding. The stand is used to gather video and complementary data from other sensors that can be processed in order to classify the psychophysical state. The RGB-D camera was Intel SR300 device (working in visible lighting and infrared NIR range) mounted near the steering wheel of simulated vehicle. It uses a short range, coded light and can provide up to 60 FPS at a resolution of 640×480. In order to capture a depth map, the Infra Red projector illuminates the scene with a set of predefined, increasing spatial frequency coded IR vertical bar patterns. These patterns are warped by the objects in the scene, reflected back and captured by the IR camera. Resulting pixels are then processed to generate a final depth map. According to the producer [15], the effective range of the camera is up to 1.5m, but it can be interpolated over an 8 m range (or 1/8 mm sub-pixel resolution). The scheme of data acquisition is presented in Fig. 2.

2.3   Face Detection and Tracking

In the algorithm presented in this paper we propose to use a standard sliding window object detector based on Viola-Jones algorithm employing AdaBoost [3,13]. In the beginning, we considered employing one of three low-level descriptors, namely Haar-like features, Histogram of Oriented Gradients and Local Binary Patterns. Each of them posses different properties. While Haar-like features effectively detect frontal faces, LBP and HOG allow for slight face angle variations. On the other hand, HOG and LBP can work on integer arithmetic and are much faster than Haar (at the learning stage, as well). The classifiers were implemented

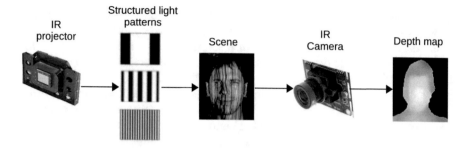

Fig. 2. Data acquisition flow (based on [15])

using Open Computer Vision library (OpenCV) on Intel i7 processor in Python environment.

In case of cascading classifier, during training, standard boosted cascade algorithm was applied, namely AdaBoost (Gentle AdaBoost variant). The detector was trained with the following set of parameters: window size equal to 59×51 pixels, positive samples number equal to 500, negative samples number equal to 1000. The detectors were trained on manually cropped faces that are presented in Fig. 3. The negative pool was collected from the Internet. It was extended with human torsos extracted from the sequences presenting human upper-body.

We tested the detectors on 984 images of size 640×480 pixels in grayscale, presenting 6 subjects (some of them wearing glasses). The manually marked faces have size of 99×96 pixels, with minimal and maximal value of 71×79 and 136×135, respectively.

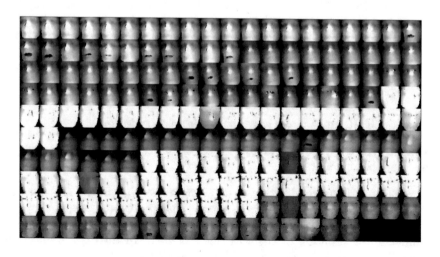

Fig. 3. Images used for learning

Since the mean accuracy and true positive rates of all the evaluated detectors are very high in case of frontal faces and semi-controlled conditions, we compared them based on other factor, namely Intersection over Union factor - IoU, as it is often used in object detection challenges. It is reported that IoU score higher than 0.5 is often considered an *acceptable* prediction.

From the practical point of view, in the approach presented here, the largest detected object in the scene is considered a face and it is a basis for IoU calculation. Analysing the results (see Table 1) one can see, the LBP gives the highest mean value of IoU, yet with the highest standard deviation. Hence, we decided to employ Haar-based detector, although having lower mean IoU, yet with the lowest standard deviation.

Table 1. The results of face detectors evaluation based on IoU

Detector	Haar	HOG	LBP
mean IoU	0.59	0.50	0.61
std. dev. IoU	0.25	0.27	0.31

It should be remembered, that the face sometimes may not be detected, because of occlusion or pose change. Therefore, the implemented face tracking is based on position approximation. It relies on the assumption that, under regular driving conditions, face position should not change significantly across a small number of frames. Therefore, the average coordinates of the face's bounding rectangle are calculated based on averaged 10 past detections. Although the implementation allows for a fair amount of leeway in the coordinates of the detected face, statistical methods are used to reject some visibly erroneous detections and to select the best candidate in the case of multiple faces being detected in a single frame. The averaged coordinates of the accepted, detected rectangles are used to allow the algorithm to run continuously without significant facial region change during the analysis.

2.4   Mouth State Estimation and Fatigue Prediction

The algorithm of mouth area analysis takes detected face as an input and performs the following steps:

1. input face submatrix detected at (x, y) of size $w \times h$, where x and y are the numbers of row and column in the image matrix, respectively;
2. crop mouth area located at $(x + h/2, y + w/4)$ of size $(h/2, 3 \cdot w/4)$;
3. binarize the resulting matrix with the threshold equal to 1/4 of maximum possible intensity (64 in case of 8-bit grayscale images);
4. invert the resulting image;
5. perform morphological closing with a kernel $k = [1, 1, 1; 1, 1, 1; 1, 1, 1]$;
6. invert the resulting image;

7. count black (0) pixels;
8. calculate normalized black pixels number (by dividing the result of previous step by the submatrix dimensions)
9. append the normalized black pixels number to the buffer representing last 30 frames;
10. if the two following conditions are satisfied, then the open mouth is detected and yawning is present:
    (a) normalized black pixels count is higher then 3.5% of the submatrix area (evaluation of current frame),
    (b) average normalized black pixels count in the buffer is higher then 5% of joint submatrix areas (favours intervals with larger mouth opening);
11. if yawning is detected, calculate the yawning duration:
    (a) if yawning duration is larger than 45 frames update the drowsiness status
        i. append the starting frame number and yawning duration to the double-ended queue (containing 20 elements)
        ii. if the number of frames with yawning in the above queue is larger than 200 in last 1000 frames, inform about continuous yawning (drowsiness alert)
    (b) otherwise go to step 1

The exemplary images, depicting the processing flow, are presented in Fig. 4.

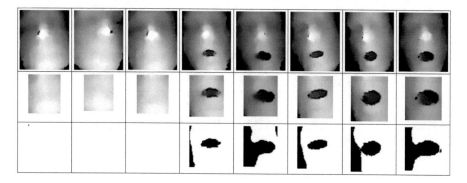

Fig. 4. Selected images showing the processing steps (in rows): detected face, cropped mouth area, binarized region

3 Experimental Results

The evaluation protocol is as follows. We manually marked the ground truth (the frames with mouth state change) in the validation video stream containing over 5100 frames extracted from original benchmark data [21]. They contained neutral poses as well as yawning occurrences. The original data contain the following actions performed by the observed humans: blinking eyes (opening and closing), squinting eyes, rubbing eyes, yawning, lowering the head, and shaking the head. We selected sequences with yawning only. In each case, four cameras observed driver's head and a fragment of his torso in three spectra: visible (VIS), near-infrared (NIR) and long-wavelength infrared (LWIR). It led to the five video streams: two normal (visible) sequences, thermal sequence, point cloud and depth map. In the experiments, only the NIR sequences with depth maps were taken into consideration. The spatial resolution of the video frame was 640×480 pixels stored in 8-bits grayscale.

The first experiment was aimed at the verification of mouth state change detection. It was designed to validate the basic capabilities of the algorithm, e.g. its ability to discern between images containing open and closed mouth regions. The second experiment was to check if the system is able to detect yawning as the indicator of drowsiness.

The numerical results of these experiments are presented in Tables 2 and 3. The "direct" column represents the results of mouth opening/closing detection (even if it is associated with speaking), while "corrected" represents actual yawning. Compared to the manually marked video, a simplified version of the process concerned only with grading the current frame's state managed to achieve an 85% sensitivity with a 99% specificity ratio when dealing with detecting an opening between the lips of the observed subjects.

As it can be seen, the second experiment gave slightly worse results (especially in terms of sensitivity). It is because of a very rigorous way of marking the testing video material. In such cases short mouth opening acts marked in the video are rejected by the algorithm.

Table 2. Mouth state estimation results

Mouth state	Detected			
	Direct		Corrected	
Actual	Opened	Closed	Opened	Closed
Opened	1179	41	1143	40
Closed	203	3758	239	3759

Table 3. Quality of mouth state estimation

	Direct	Corrected
Sensitivity	0.85	0.83
Specificity	0.99	0.99
Accuracy	0.95	0.95

The exemplary frames, containing detected acts of yawning, are presented in Fig. 5. The timeline of this video is presented in Fig. 6. As it can be seen, most of the yawning situations is detected, some of them, unfortunately, with a small

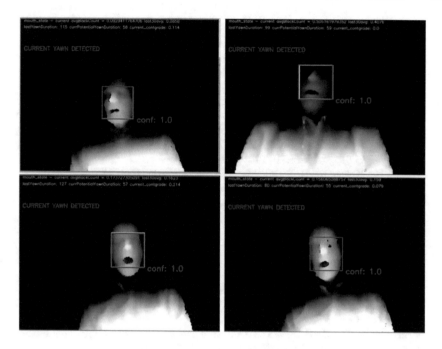

Fig. 5. Examples of yawning detection

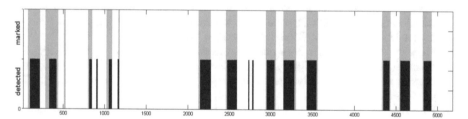

Fig. 6. The timeline of the validation sequence with yawning detection results

delay. It is caused by the applied buffer analysis. In three situations, the yawning was falsely detected, while in one case it was no detected at all.

4   Summary

In the paper we proposed an algorithm of driver's drowsiness detection based on depth map analysis. It consists of two modules: face detection and mouth state estimation. The detection uses Haar-like features and Viola-Jones detector, while mouth state is analysed using pixel intensity-based heuristic approach. The experiments showed that such a solution is capable of accurate drowsiness detection. It can work in complex lighting conditions, with an real-world application

References

1. Alioua, N., Amine, A., Rziza, M.: Driver's fatigue detection based on yawning extraction. Int. J. Veh. Technol., Article no. 678786 (2014). https://doi.org/10.1155/2014/678786
2. Azim, T., Jaffar, M.A., Mirza, A.M.: Fully automated real time fatigue detection of drivers through fuzzy expert systems. Appl. Soft Comput. **18**, 25–38 (2014)
3. Burduk, R.: The AdaBoost algorithm with the imprecision determine the weights of the observations. In: Intelligent Information and Database Systems, Part II, LNCS, vol. 8398, pp. 110–116 (2014)
4. Chang, H., Koschan, A., Abidi, M., Kong, S.G., Won, C.-H.: Multispectral visible and infrared imaging for face recognition. In: 2008 IEEE Computer Society Conference on Computer Vision and Pattern Recognition Workshops, pp. 1–6 (2008)
5. Craye, C., Rashwan, A., Kamel, M.S., Karray, F.: A multi-modal driver fatigue and distraction assessment system. Int. J. Intel. Transp. Syst. Res. **14**(3), 173–194 (2016)
6. Cyganek, B., Gruszczynski, S.: Hybrid computer vision system for drivers' eye recognition and fatigue monitoring. Neurocomputing **126**, 78–94 (2014)
7. Dalal, N., Triggs, B.: Histograms of oriented gradients for human detection. In: IEEE Computer Society Conference on Computer Vision and Pattern Recognition, vol. 1, pp. 886–893 (2005)
8. Forczmański, P., Kukharev, G.: Comparative analysis of simple facial features extractors. J. R. Time Image Process. **1**(4), 239–255 (2007)
9. Forczmański, P., Kukharev, G., Shchegoleva, N.: Simple and robust facial portraits recognition under variable lighting conditions based on two-dimensional orthogonal transformations. In: 7th International Conference on Image Analysis and Processing (ICIAP). LNCS, vol. 8156, pp. 602–611 (2013)
10. Forczmański, P.: Human face detection in thermal images using an ensemble of cascading classifiers. In: Hard and Soft Computing for Artificial Intelligence, Multimedia and Security, Advances in Intelligent Systems and Computing, vol. 534, pp. 205–215 (2016)
11. Forczmański, P.: Performance evaluation of selected thermal imaging-based human face detectors. In: Proceedings of the 10th International Conference on Computer Recognition Systems CORES 2017. Advances in Intelligent Systems and Computing, vol. 578, pp. 170–181 (2018)
12. Fornalczyk, K., Wojciechowski, A.: Robust face model based approach to head pose estimation. In: Proceedings of the 2017 Federated Conference on Computer Science and Information Systems, FedCSIS 2017, pp. 1291–1295 (2017)
13. Freund, Y., Schapire, R.E.: A decision-theoretic generalization of on-line learning and an application to boosting. In: Proceedings of the 2nd European Conference on Computational Learning Theory, pp. 23–37 (1995)
14. Fu, R., Wang, H., Zhao, W.: Dynamic driver fatigue detection using hidden Markov model in real driving condition. Exp. Syst. Appl. **63**, 397–411 (2016)
15. Intel RealSense Camera SR300 – Embedded Coded Light 3D Imaging System with Full High Definition Color Camera Product Datasheet, rev. 1 (2016). https://software.intel.com/sites/default/files/managed/0c/ec/realsense-sr300-product-datasheet-rev-1-0.pdf. Accessed 05 Oct 2018
16. Jo, J., Lee, S.J., Park, K.R., Kim, I.J., Kim, J.: Detecting driver drowsiness using feature-level fusion and user-specific classification. Exp. Syst. Appl. **41**(4), 1139–1152 (2014)

17. Kong, W., Zhou, L., Wang, Y., Zhang, J., Liu, J., Gao, S.: A system of driving fatigue detection based on machine vision and its application on smart device. J. Sens. **2015**, 11 pages (2015)
18. Krishnasree, V., Balaji, N., Rao, P.S.: A real time improved driver fatigue monitoring system. WSEAS Trans. Signal Process. **10**, 146–155 (2014)
19. Nowosielski, A.: Vision-based solutions for driver assistance. J. Theor. Appl. Comput. Sci. **8**(4), 35–44 (2014)
20. Makowiec-Dabrowska, T., Siedlecka, J., Gadzicka, E., Szyjkowska, A., Dania, M., Viebig, P., Kosobudzki, M., Bortkiewicz, A.: The work fatigue for drivers of city buses. Medycyna Pracy **66**(5), 661–677 (2015)
21. Małecki, K., Nowosielski, A., Forczmański, P.: Multispectral data acquisition in the assessment of driver's fatigue. In: Mikulski, J. (ed.) Smart Solutions in Today's Transport, TST 2017. Communications in Computer and Information Science, vol. 715. pp. 320–332 (2017)
22. Mitas, A., Czapla, Z., Bugdol, M., Ryguła, A.: Registration and evaluation of biometric parameters of the driver to improve road safety, pp. 71–79. Scientific Papers of Transport, Silesian University of Technology (2010)
23. Ojala, T., Pietikinen, M., Harwood, D.: Performance evaluation of texture measures with classification based on Kullback discrimination of distributions. In: Proceedings of the 12th International Conference on Pattern Recognition, vol. 1, pp. 582–585 (1994)
24. Smiatacz, M.: Liveness measurements using optical flow for biometric person authentication. Metrol. Meas. Syst. **19**(2), 257–268 (2012)
25. Staniucha, R., Wojciechowski, A.: Mouth features extraction for emotion classification. In: Proceedings of the 2016 Federated Conference on Computer Science and Information Systems, FedCSIS 2016, pp. 1685–1692 (2016)
26. Viola, P., Jones, M.J.: Robust real-time face detection. Int. J. Comput. Vis. **57**(2), 137–154 (2004)
27. Zhang, Y., Hua, C.: Driver fatigue recognition based on facial expression analysis using local binary patterns. Opt. Int. J. Light. Electron Opt. **126**(23), 4501–4505 (2015)

Vehicle Passengers Detection for Onboard eCall-Compliant Devices

Anna Lupinska-Dubicka[1], Marek Tabędzki[1(✉)], Marcin Adamski[1],
Mariusz Rybnik[2], Maciej Szymkowski[1], Miroslaw Omieljanowicz[1],
Marek Gruszewski[1], Adam Klimowicz[1], Grzegorz Rubin[3],
and Lukasz Zienkiewicz[1]

[1] Faculty of Computer Science, Bialystok University of Technology, Bialystok, Poland
{a.lupinska,m.tabedzki}@pb.edu.pl
[2] Faculty of Mathematics and Informatics, University of Bialystok, Bialystok, Poland
[3] Faculty of Computer and Food Science,
Lomza State University of Applied Sciences, Lomza, Poland

Abstract. The European eSafety initiative aims to improve the safety
and efficiency of road transport. The main element of eSafety is the pan
European eCall project – an in-vehicle system that informs about road
collisions or serious accidents. An on-board compact eCall device which
can be installed in used vehicle is being developed, partially with the
authors of the paper. The proposed system is independent of built-in
car systems, it is able to detect a road accident, indicate the number of
occupants inside the vehicle, report their vital functions and send those
information to dedicated emergency services via duplex communication
channel.

This paper focuses on an important functionality of such a device:
vehicle occupants detection and counting. The authors analyze a wide
variety of sensors and algorithms that can be used and present results of
their experiments based on video feed.

1 Introduction

According to the European Commission (EC) estimations approximately 25,500
people lost their lives on EU roads in 2016 and a further 135,000 people were
seriously injured [1]. Studies have shown that thanks to immediate information
about the location of a car accident, the response time of emergency services can
be reduced by 50% in rural areas and 60% in urban areas. Within the European
Union this can lead to saving 2,500 people a year [2,3].

The eCall system, the pan European emergency notification system, thanks
to such early alerting of the emergency services is expected to reduce the number
of fatalities as well as the severity of injuries caused by road accidents. In case
of an accident, an eCall-equipped car will automatically contact the nearest
emergency center. The operator will be able to decide which rescue services
should intervene at an accident scene. To make such a decision, the operator

© Springer Nature Switzerland AG 2019
J. Pejaś et al. (Eds.): ACS 2018, AISC 889, pp. 408–419, 2019.
https://doi.org/10.1007/978-3-030-03314-9_35

should obtain as much information as possible about the causes and effects of the accident and about the number of vehicle occupants and their health condition.

On 28 April 2015, the European Parliament adopted the legislation on eCall type approval requirements and made it mandatory for all new models of cars to be equipped with eCall technology from 1st April 2018 onward. Unfortunately only small part of cars driven in UE are brand new (about 3.7% brand new cars was sold in 2015).

An on-board compact eCall-compliant device can be installed as an additional unit in used vehicles at the owners' request. It is being developed as European Project 4.1 [4], partially with the authors of the paper. Proposed system will be able to detect a road accident, indicate the number of vehicle's occupants, report their vital functions and send those information to dedicated emergency services via duplex communication channel. This paper presents an important functionality of this device: vehicle occupants detection and counting.

There are many different approaches to human detection problem but only few of them are associated with the vehicles. Technology development enables usage of human detection methods in the intelligent transportation system in smart environment. However, it is still a challenge to implement an algorithm that will be robust, fast and could be used in automotive environment. In such systems, due to limited resources and space, low computational complexity is crucial.

The paper is organized as follows: in the second section the authors shortly describe the concept of compact device eCall system with its goals and requirements. In the third section, the authors propose a preliminary approach. The authors take into consideration many types of sensors, but focus on camera images and video capture. Fourth section describes array of sensors and algorithms, and evaluates their use for human presence and number detection, including various motion sensors, microphones, cameras, radars, algorithms for face detection, movement tracking and similar. Next, authors presents the results of their experiments based on video feed (Sect. 3.1) and sound separation (Sect. 3.2). Finally the conclusions and future work are given.

2   The Concept of Compact eCall Device

Not every road user wants or can afford a brand new car that would be equipped with eCall. Hence, the authors of the paper propose a compact system that could be installed in any vehicle. This would allow any car user to rely on the extra security that it provides, for a relatively small price. This chapter presents a general description of the device's operating concept and design.

The scheme of the system is depicted in Fig. 1. The main and the most important module is the accident detection module. Its task is to launch the entire notification procedure for the relevant service in case of road accident. The key problem here is determining how to identify the accident (using incorporated collision sensors). The system, as requested, should also allow manual triggering.

Another module is the communication module that upon request sends gathered data to the PSAP center, such as vehicle data, vehicle location and driving

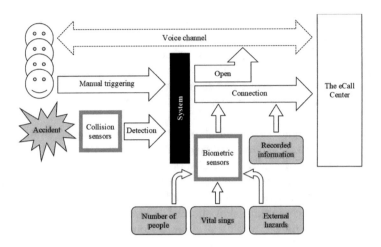

Fig. 1. System block diagram (Source: personal collection)

direction read from the GPS receiver. Additionally, the device should establish a voice call with the PSAP operators, allowing them to contact the victims.

The other modules of the system are designed to capture the situation inside the vehicle. The first of these modules detects presence and number of vehicle occupants. The task definitively should be performed before the accident occurs – periodically in order to accurately count as well as detect changes in passengers payload. Possibly post-event monitoring may also be provided to inform eCall operators if occupants have left the vehicle (or, for example, been thrown out if a collision has occurred and they have not fastened their seatbelts). Further work will examine the practicality of such a solution.

A very important requirement, that has to be taken into account, is the lack of interference in the construction of the vehicle, which makes it impossible to use the sensors installed in the vehicle or to gather data from the on-board computer. The authors do not exclude the use of seat pressure or weight sensors additionally installed in the seats. However, it should be taken into consideration that they cannot be the only source of passenger counting module due to the possibility of storing heavy objects in the seats and even fastening safety belts for those.

A separate, but crucial task is to identify the vital signs of occupants after an accident. Although this is not the subject of this article, it should be mentioned that the authors consider a number of sensors and methods for evaluating the vital signs, paying attention to the possibility of using them in the vehicle.

The proposed system should therefore include: a GPS vehicle positioning system, a set of sensors for accident detection (such as accelerometers, gyroscopes, pressure sensors, temperature sensors and sound detectors), a set of sensors to detect vehicle occupants presence (such as digital camera, digital infrared camera, radars or microphones) and a set of sensors for analyzing passengers' vital functions.

3   State of the Art in Human Presence Detection

In general, human presence detection may be based on intrinsic (static and dynamic) or extrinsic traits. Intrinsic traits are related to physical phenomena caused human presence that can be detected using various types of sensors. The information for algorithmic processing may be gathered using either distant sensors: camera (static photos or dynamic video), thermal imagery, radar based detection, sound; or contact sensors alike pressure sensors. Extrinsic traits make use of devices carried or worn by individuals such as portable communication devices (smartphones, smartbands) and wearable IDs. One may also use sensors that detect interaction with utilities present in the environment such as door and safety belts. Another option is to provide an interface for entering the number of persons, for example using console or voice recognition.

While extrinsic traits like wearable IDs or portable communication are becoming increasingly popular, they are however not universal or obligatory enough to rely on them. Universal intrinsic traits should rather be used in such an essential task. Fastening seat belts, although usually mandatory by law, is not always obeyed, therefore cannot be used as a reliable source of data. Requiring the driver to perform certain action to explicitly register number of persons is also not a good solution – the system should be able to work automatically.

The main advantages of intrinsic traits are universality and unattended manner of detection. One of the most commonly used techniques for detection of human presence is based on pressure sensors installed in car seats. Such technique is often used in cars together with safety belt engaging detector to inform the driver of unfasten belts. However, this approach cannot be used for reliable passenger counting due to fact that any object that occupies seat and inflicts certain pressure can result in a false positive detection.

The camera image is the natural source of data for determining the number of people in vehicle. There are many solutions available in the literature that detect people in camera images [5]. Different types of cameras can be used for the detection of people: visible light cameras, infrared cameras that register reflected light from external source, and thermal imaging cameras that record light emitted by objects with temperature above the absolute zero. The advantage of infrared cameras is the ability to work at night and, in case of thermal imaging, additional information in a form of temperature measurement helps to identify live objects. Another approach to person detection and counting is the use of radar sensors [6]. Their advantages are the ability to penetrate obstacles and reduction of privacy concerns.

In the literature there are many approaches using camera and radar sensors. However, only small number of them is associated with vehicles. In work [7] the human detection in car was performed using Viola-Jones face detection method applied to images from thermal camera, which registered electromagnetic radiation in infrared range. The main advantage of the proposed technique is the ability to use temperature measurement as additional factor to reduce false detections for objects that have face-like shape. Another concept can be found in [8], where a system for people counting in public transportation was presented. This concept was created due to the problem of monitoring the number of occupants getting in or out public transportations in order to improve vehicle's door control. This approach combines stereo video system with Disparity Image computation and 3D filtering.

3.1   Detection Using Camera/Video Feed

Analysis of video material opens up new possibilities, but it also brings other challenges. One can analyze and track the movement of objects using multiple subsequent frames. Additionally appearance or disappearance of the object of interest can be detected (a passenger entering or leaving the vehicle).

As a part of the proposed in-vehicle system, it is required to detect the number of occupants. In this chapter the literature review of the solutions related to human presence detection is presented. It should be noted that even algorithms that do not give satisfactory results alone can be applied in combination with others as so-called ensemble methods [9–11].

The basic tool that can be used for this purpose is the background subtraction. Assuming that the camera in the vehicle is stationary, and the only moving objects are people inside, one needs to find the difference between the image depicting the background and the image at a given time to record the movement.

However, this approach often faces multiple difficulties like shadows, variable lighting, reflections, etc. In that case, simple subtraction would not bring the expected results and more complex method is required. One of the considered approach is a Gaussian Mixture-based Background/Foreground Segmentation Algorithm [12]. This method is based on modeling each pixel as a mixture of Gaussians. Then, recursive equations are used to update the parameters and to select the required number of components per pixel. It provides improved segmentation, due to better adaptability to varying scenes. In [13] the modified algorithm is presented. It uses non-parametric adaptive density estimation method to provide more flexible model.

Different approach to object tracking in the video images are represented by Mean-Shift algorithm [14] and its extension CAMShift algorithm [15]. The Mean-Shift algorithm consists of four steps. At the beginning the window size and its initial location has to be chosen. Then computation of the mean location in the window is performed. As the third step the search window is centered at computed mean location. The second and the third step are repeated until calculated parameter moves less than assumed threshold. CAMShift, as an extension of Mean-Shift tries to solve its one critical issue – unchanging window size if the object is closer to the camera. In addition CAMShift calculates the orientation of the best fitting ellipse to the prepared window. Afterwards Mean-Shift is once again applied with the previously scaled window in the last known location. The whole process stops when accuracy is higher than the established threshold. Similarly to the case of Mean-Shift, there are also different modifications of CAMShift algorithm [16,17].

Another concept, based on an optical flow, allows for more precise movement tracking. Optical flow can be defined as the pattern of apparent motion of pixels in a visual scene caused by the movement of object in front of camera. It is described by 2D vector field, where each vector shows the movement of points from given frame to the next. Optical flow assumes that pixel intensities of an object do not change between consecutive frames and that neighboring pixels have similar motion. One of possible algorithms of optical flow estimation is Lucas-Kanade method [18]. This method assumes that the motion between two frames is small and constant within a 3x3 neighborhood around the point under consideration, and solves the optical flow equations by the least squares criterion. In contrast to point-wise methods it is less sensitive to image noise, however for large flow, it should be used on reduced-scale versions of images.

3.2   Detection Using Sound

Detecting people usually takes place using video-based techniques. However, video techniques require, among others, a direct line of sight and conditions with adequate lighting, while acoustics-based detection techniques do not require any of the above. On the other hand, they are susceptible to interference from background noise and interference from other signals that may occur simultaneously.

The human detection module of the proposed system could consist of two parts: source separation and signal detection. Each part would be meant to address a different kind of problem. Source separation part would be used to split mixed sounds into their constituent components, while detection part would be used to determine when a signal of interest (in this case human speech) is present in a recording.

Blind Source Separation (BSS) refers to a problem where both the sources and the mixing methodology are unknown, only mixture signals are available for further separation process. In case of proposed system the recording will be a combination of overlapping sounds coming from all vehicle's occupants and will include significant noise (such as traffic noise or engine sound). For its further usage it is strongly desirable to recover all individual sources, or at least to segregate a particular source.

Algorithms for blind source separation can be categorized taking into consideration the ratio of the number of receivers (microphones) to the number of signal sources. If multiple simultaneous recordings of the mixed signal are available then source separation can be performed using Principal Component Analysis (PCA) [19] or Independent Component Analysis (ICA) [20]. The main restriction is that a distinct recording is needed for every possible source signal. In case of eCall system that means that one need two, four or five (depending on the type of a passenger car) microphones inside a vehicle cabin – one for each potential vehicle's occupant. If the number of microphones used is less than the number of signal sources, then source separation techniques may be based on having a dictionary of signals of interest. The most common technique for single channel source separation is Non-negative Matrix Factorization (NMF) [21].

The next step after signal separation is to recognize whether the signal is a speech. This can be done using speech recognition algorithms to obtain words from an audio recording [22,23]. However, the main difficulty could be size or language of dictionary (training set) of words which these algorithms are able to recognize. Second approach can rely on methods referred to as Voice Activity Detection (VAD) [24]. These methods are able to recognize a human speech in the input signal on the basis of speech characteristics.

4   Results of Performed Experiments

In the previous work [25] the authors presented preliminary results of experiments on passengers' detection and counting based on face detection in static images taken from the camera installed inside a vehicle. Applied algorithm based on the Viola-Jones method [26] yielded results of 66.1%. The main obstacle that has been noted was the fact that detection algorithms usually give unsatisfactory results when the passenger's face is turned sideways in relation to the camera. In order to solve this problem, in the present work the authors focus on the analysis of continuous material – both video and audio recordings. In the case of a video recording, the applied algorithms analyze a series of frames from a given time interval. It should be noted that in order to correctly determine the number of people in the vehicle, correct detection in each frame is not required – the maximum reliable value from a given interval has been selected as the number of detected faces. This allows for partial elimination of false negatives. Audio analysis is to be carried out in a similar way.

4.1   Deep Neural Networks

Counting the number of people in the vehicle was carried out using deep neural networks. The method used is based on the detection of Single-Shot MultiBox Detector (SSD) [27]. In this approach, the process of locating the object and its classification is performed by means of one neural network, which significantly speeds up calculations and allows real-time video analysis. Deep Residual Network (ResNet) [28] was chosen as the architecture of the neural network.

The experiments were carried out for images recorded a camera inside a vehicle. Preliminary studies have shown that the results obtained from the SSD-ResNet detector are better than those obtained using the Viola-Jones method [7]. The algorithm correctly detects faces at a larger exposure range and suffers from less false detections. However, as previously the algorithm sometimes did not detect significantly obstructed faces. During the registration, the camera was installed close to rear-view mirror. As a result, some faces of the people in the rear row were obscured by the headrests and the front-seat passengers. The data was obtained in well illuminated garage, similarly to good weather conditions. The authors currently work on using IR cameras in low-illumination conditions.

Verification of the algorithm consisted in finding a difference (distance – where a discrete metric was adopted) between two functions. The first one described the number of people staying in the vehicle at a given moment (it was determined manually by the authors on the basis of photo analysis). The second one was the result of the algorithm. After initial research, it has been noticed that the results returned by the algorithm are sometimes subject to sudden changes – sometimes the face in individual frames is not detected correctly (or momentarily disappears from the frame), while in reality people do not appear and disappear so suddenly. In order to solve this problem, the following heuristics have been added to the evaluation function: the function that returns the number of vehicle's occupants does not return the value detected in a given frame of the recording, but the maximum value from a certain time window (it was initially assumed that its value will be ten seconds). Such assumption has been made because the second type of error (*false negative*) is more frequent than the error of the first type (*false positive*). If a person has been detected only in part of the frames of the analyzed window, it probably means that they are still there but have changed position or moved, which has been not correctly marked. This assumption has made it possible to get results closer to reality.

Algorithm based on deep neural networks correctly recognized number of people inside the vehicle in 72% of the cases.

While working with static images taken inside the stationary vehicle, the problem of person outside of the vehicle has been observed. However, it is virtually impossible to have such a situation while driving. The pedestrian's face caught while the vehicle is moving would most likely be blurred and hence undetectable by the algorithms. Even if the vehicle is moving so slowly that the face could be detected, it would only occur in a single frame of the shot. Such outliers would be eliminated by an algorithm that selects a number from passengers from a given time window. Detection of passengers from other vehicle is rather impossible, because the camera's position does not allow the observation of the passengers of the neighboring car, even when they approached each other very closely. The only possibility is to detect the pedestrian's face at a stop. It will therefore be necessary to identify whether the vehicle is moving and on this basis to determine the reliability of the collected data.

4.2   Sound

The purpose of this experiment was not to recognize human speech but only to say whether it exists in the recording. Therefore, the authors decided to use one of the VAD algorithms, namely the Sohn method [29]. As the separator of sources, one of the variations of the Independent Component Analysis algorithm – fastICA [30] was used. Due to the inaccessibility of sensors (microphones installed in the vehicle), the first experiments were carried out on the generated signals. Human speech recordings taken from the LibriVox library [31], sounds related to engine operation downloaded from the libraries of SoundJay [32] and SoundBible [33] as well as generated white and pink noise were taken into consideration. Twelve linearly mixed signals were created, each with a length of about 60 min. Two cases were considered:

- when the number of people (human speech signals) was equal to the number of microphones (output signals) and was equal to four;
- when the number of people (human speech signals) was smaller than the number of microphones (output signals) and was equal to two or three;

Each generated signal was divided into 10 more or less equal sections and subjected to separate analysis.

In the case of the situation for three or four persons in the vehicle, the selected methods correctly identified the source signals and recognized the speech signal in each case. The case of two people in a vehicle turned out to be problematic, for which the analysis of independent components was estimated in majority of cases by three source signals. The redundant signal was a weak and silent, but the speech detection algorithm also gave a positive result for it. However, the costs of error should be taken into account both ways: too few people per vehicle were detected. From the point of view of saving human life much more expensive is the case in which the passenger will not be identified by the detection algorithms.

It should be mentioned that detecting the number of people in the vehicle based on speech detection should be treated as an additional solution – one has to bear in mind that passengers might not always talk to each other or there might be disturbances such as radio broadcast that can falsely increase the number of people detected. However, the detection of the speech signal in the recording may be important if one tries to make the voice contact with a PSAP operator after the accident.

5   Conclusions and Future Works

eCall European directive seems to be an excellent initiative to save lives, however one should notice the important limitation: eCall system is mandatory only for the new cars sold in the European Union. New cars per year constitute only 3.7% of all cars driven in EU [34,35] (partially estimated). Thus the idea of compact and cheap device easily mounted in existing cars is surely a very interesting and practical solution.

In this article, the authors have shortly presented the idea of such a compact eCall-compliant device, and have concentrated on the single task of the proposed device: detecting and counting vehicle occupants. A variety of concerned algorithms has been presented with the focus on the most suitable solutions.

It is important to note that preferably the device should be independent of existing car systems, as in used older car they may be non-existent, hardly accessible or malfunctioning. Therefore the authors restrain from incorporating such (potentially efficient but unreliable) techniques as seat pressure sensors or safety-belt tension sensors. It is important to note that a number of people do not fasten their safety-belts, large object can as well falsely trigger a pressure sensor. Also for security reasons, interference with the existing car structure is not recommended.

In the previous work, the authors carried out research on a series of photos collected in-vehicle with camera close to rear-mirror. The algorithm based on the Viola-Jones method yielded correct passengers detection in 66.1% of cases. In this work, the authors have turned their attention to video materials and audio recordings. Algorithm based on deep neural networks and 10-s detection windows correctly recognized number of people inside the vehicle in 72% of cases. The new approach allowed to solve one of the biggest problems of the previous one – not recognizing the face turned sideways to the camera.

As a part of the further work the authors plan to use alternative image sources alike infrared and thermal vision. This will answer the problem of face detection in low light conditions (for under normal driving conditions at night, the interior of the vehicle is not illuminated). It is important to stress various nature of visual data for different cameras, for example watershed algorithm [36] could be efficiently used for thermal image segmentation. The authors are also investigating combining data from two cameras (one camera per sit row). This will answer the problem of detecting people in the back seats due to limited visibility resulted from head restraints. The authors plan also to establish ensemble of classifiers to efficiently combine various detection algorithms using video and sound.

It is important to bear in mind computational limitations of hardware that the portable device would be equipped with. The authors plan to use single-board computer with extensions like camera, microphone, GSM, etc. Such device should be capable of performing well in the versatility of tasks required for eCall constraints. It is important to note that video processing for detection of people is to be performed periodically (once for 15 s for example), similarly to vital signs detection (not in the scope of this paper). That manner of work is definitively less demanding for hardware than real-time operation.

Acknowledgments. The authors would like to sincerely thank Professor Khalid Saeed for content-related care, inspiration and motivation to work.

This work was supported by grant S/WI/1/2018 and S/WI/2/2018, and S/WI/3/2018 from Bialystok University of Technology and funded with resources for research by the Ministry of Science and Higher Education in Poland.

References

1. Road Safety: Encouraging results in 2016 call for continued efforts to save lives on EU roads. http://europa.eu/rapid/press-release_IP-17-674_en.htm. Accessed 24 Mar 2018
2. eCall: Time saved = lives saved. https://ec.europa.eu/digital-single-market/en/eCall-time-saved-lives-saved. Accessed 24 Mar 2018
3. European Parliament makes eCall mandatory from 2018. http://www.etsi.org/news-events/news/960-2015-05-european-parliament-makes-ecall-mandatory-from-2018. Accessed 24 Mar 2018

4. System sensorowy w pojazdach do rozpoznania stanu po wypadku z transmisja informacji do punktu przyjmowania zgloszen eCall. http://pb.edu.pl/projekty-pb/ ecall. Accessed 24 Mar 2018

5. Nguyen, D.T., Li, W., Ogunbona, P.O.: Human detection from images and videos: a survey. Pattern Recognit. **51**, 148–175 (2016)

6. Choi, J.W., Yim, D.H., Cho, S.H.: People counting based on an IR-UWB radar sensor. IEEE Sens. J. **17**, 5717–5727 (2017)

7. Zohn, Bc.L.: Detection of persons in a vehicle using IR cameras. Master's Thesis, Faculty of Transportation Sciences, Czech Technical University in Prague (2016)

8. Bernini, N., Bombini, L., Buzzoni, M., Cerri, P., Grisleri, P.: An embedded system for counting passengers in public transportation vehicles. In: 2014 IEEE/ASME 10th International Conference on Mechatronic and Embedded Systems and Applications Proceedings (2014)

9. Schapire, R.E.: The strength of weak learnability. Mach. Learn. **5**(2), 197–227 (1990)

10. Schapire, R.E., Freund, Y., Bartlett, P., Lee, W.S.: Boosting the margin: a new explanation for the effectiveness of voting methods. Ann. Stat. **26**(5), 1651–1686 (1998)

11. Zhihua, Z.: Ensemble Methods: Foundations and Algorithms. Chapman and Hall/CRC, Boca Raton (2012)

12. Zivkovic, Z.: Improved adaptive Gaussian mixture model for background subtraction. In: ICPR (2004)

13. Zivkovic, Z., van der Heijden, F.: Efficient adaptive density estimation per image pixel for the task of background subtraction. Pattern Recognit. Lett. **27**, 773 (2006)

14. Cheng, Y.: Mean shift, mode seeking, and clustering. IEEE Trans. Pattern Anal. Mach. Intel. **17**(8), 790–799 (1995)

15. Bradski, G.: Computer vision face tracking for use in a perceptual user interface. Intel Technol. J. **2**(2), 1–15 (1998)

16. Exner, D., Bruns, E., Kurz, D., Grundhöfer, A., Bimber, O.: Fast and robust CAMShift tracking. In: 2010 IEEE Computer Society Conference on Computer Vision and Pattern Recognition, San Francisco, USA, pp. 9–16 (2010)

17. Sooksatra, S., Kondo, T.: CAMShift-based algorithm for multiple object tracking. In: Proceedings of the 9th International Conference on Computing and Information Technology IC2IT 2013, Bangkok, Thailand, pp. 301–310 (2013)

18. Lucas, B.D., Kanade, T.: An iterative image registration technique with an application to stereo vision. In: Proceedings of Imaging Understanding Workshop (1981)

19. Abdi, H., William, L.J.: Principal component analysis. Wiley Interdiscip. Rev. Comput. Stat. **2**(4), 433–459 (2010)

20. Hyvarinen, A., Karhunen, J., Oja, E.: Independent component analysis: algorithms and applications. Neural Netw. **13**, 411–430 (2000)

21. Lee, D.D., Seung, H.S.: Algorithms for non-negative matrix factorization. In: Advances in Neural Information Processing Systems, vol. 13, pp. 556–562 (2001)

22. Rabiner, L.R.: A tutorial o hidden Markov models and selected applications in speech recognition. Proc. IEEE **77**, 257–286 (1989)

23. Deng, L., Yu, D.: Deep learning: methods and applications. Found. Trends Signal Process. **7**(3–4), 197–387 (2014)

24. Ramírez, J., Górriz, J.M., Segura, J.C.: Voice activity detection, fundamentals and speech recognition system robustness. In: Robust Speech Recognition and Understanding, pp. 1–22 (2007)

25. Lupinska-Dubicka, A., Tabedzki, M., Adamski, M., Rybnik, M., Omieljanowicz, M., Omieljanowicz, A., Szymkowski, M., Gruszewski, M., Klimowicz, A., Rubin, G., Saeed, K.: The concept of in-vehicle system for human presence and their vital signs detection. In: 5th International Doctoral Symposium on Applied Computation and Security Systems: ACSS2018 (2018)
26. Viola, P., Jones, M.: Rapid object detection using a boosted cascade of simple features. In: Proceedings of the 2001 IEEE Computer Society Conference on Computer Vision and Pattern Recognition, vol. 1, pp. 511–518 (2001)
27. Redmon, J., Divvala, S., Girshick, R., Farhadi, A.: You only look once: unified, real-time object detection. In: IEEE Conference on Computer Vision and Pattern Recognition (CVPR), pp. 779–788 (2016)
28. He, K., Zhang, X., Ren, S., Sun, J.: Deep residual learning for image recognition. In: IEEE Conference on Computer Vision and Pattern Recognition (CVPR), pp. 770–778 (2016)
29. Vadsohn. http://www.ee.ic.ac.uk/hp/staff/dmb/voicebox/doc/voicebox/vadsohn.html. Accessed 30 Apr 2018
30. Fastica. https://www.cs.helsinki.fi/u/ahyvarin/papers/fastica.shtml. Accessed 30 Apr 2018
31. Abercrombie, L.: Short poetry collection 091. http://librivox.org. Accessed 30 Apr 2018
32. Soundjay. https://www.soundjay.com/. Accessed 30 Apr 2018
33. Soundbible. http://soundbible.com/tags-driving.html. Accessed 30 Apr 2018
34. https://www.best-selling-cars.com/europe/2016-full-year-europe-best-selling-car-manufacturers-brands/ . Accessed 24 Mar 2018
35. Eurostat - Passenger cars in the EU. http://ec.europa.eu/eurostat/statistics-explained/index.php/Passenger_cars_in_the_EU. Accessed 24 Mar 2018
36. Bellucci, P., Cipriani, E.: Data accuracy on automatic traffic counting the smart project results. Eur. Transp. Res. Rev. **2**(4), 175–187 (2010)

An Algorithm for Computing the True Discrete Fractional Fourier Transform

Dorota Majorkowska-Mech$^{(\boxtimes)}$ and Aleksandr Cariow

Faculty of Computer Science and Information Technology, West Pomeranian University of Technology Szczecin, ul. Zolnierska 49, 71-210 Szczecin, Poland
{dmajorkowska,acariow}@wi.zut.edu.pl

Abstract. This paper proposes an algorithm for computing the discrete fractional Fourier transform. This algorithm takes advantages of a special structure of the discrete fractional Fourier transformation matrix. This structure allows to reduce the number of arithmetic operations required to calculate the discrete fractional Fourier transform.

Keywords: Discrete fractional transforms
Discrete fractional Fourier transform · Eigenvalue decomposition

1 Introduction

Fractional Fourier transform (FRFT) is a generalization of ordinary Fourier transform (FT) with one fractional parameter. This transform was first introduced in [1], but has become more popular after publication [2]. To compute the FRFT of any signal its discrete version was needed. It initiated the work for defining discrete FRFT (DFRFT) [3–5]. After DFRFT other discrete fractional transforms were defined [6–10]. These transforms have been found very useful for signal processing [11], digital watermarking [12], image encryption [13], image and video processing [14]. To date, a number of efficient algorithms for various discrete fractional transforms have been developed [3, 15, 16].

Among fractional transforms, the discrete fractional Fourier transform is the most commonly used. There exist a few types of DFRFT definition. In [17, 18] a comparative analysis of the best-known algorithms for all these types of DFRFTs was presented. Only DFRFT based on an eigenvalue decomposition [5, 19, 20] has all the properties which are required for DFRFT like unitarity, additivity, reduction to discrete Fourier transform when the power is equal to 1, and it is an approximation of the continuous FRFT [20]. We will call this type of DFRFT as "true". The major drawback of this DFRFT is that it cannot be written in a closed form. This DFRFT is an object of authors' interest. In work [9] the method to reduce the computational load of such DFRFT by about one half was described, but that method works only for signals of even length N.

In [21] a new approach to computation of DFRFT have been presented, but a full algorithm has not been given. Our goal is to complement this lack.

© Springer Nature Switzerland AG 2019
J. Pejaś et al. (Eds.): ACS 2018, AISC 889, pp. 420–432, 2019.
https://doi.org/10.1007/978-3-030-03314-9_36

2 Mathematical Foundations

The normalized discrete Fourier transform (DFT) matrix of size N is defined as follows:

$$\mathbf{F}_N = \frac{1}{\sqrt{N}} \begin{bmatrix} 1 & 1 & \cdots & 1 \\ 1 & w_N^1 & \cdots & w_N^{N-1} \\ \vdots & \vdots & \ddots & \vdots \\ 1 & w_N^{N-1} & \cdots & w_N^{(N-1)^2} \end{bmatrix}, \tag{1}$$

where $w_N = e^{-j\frac{2\pi}{N}}$ and j is the imaginary unit. The matrix $\mathbf{F}_N$ is symmetric and unitary. It follows that [22]: (1) all the eigenvalues of $\mathbf{F}_N$ are nonzero and have magnitude one, and (2) there exists a complete set of N orthonormal eigenvectors, so we can write

$$\mathbf{F}_N = \mathbf{Z}_N \mathbf{\Lambda}_N \mathbf{Z}_N^T, \tag{2}$$

where $\mathbf{\Lambda}_N$ is a diagonal matrix which diagonal entries are the eigenvalues of $\mathbf{F}_N$. The columns of $\mathbf{Z}_N$ are normalized mutually orthogonal eigenvectors of the matrix $\mathbf{F}_N$. For $N \geq 4$ the eigenvalues are degenerated and the eigenvectors can be chosen in many ways. However the eigenvectors of DFT matrix are either even or odd vectors [23].

The fractional power of matrix can be calculated from its eigenvalue decomposition and the power of eigenvalues. The definition of DFRFT was first introduced by Pei and Yeh [5, 19]

$$\mathbf{F}_N^a = \mathbf{Z}_N \mathbf{\Lambda}_N^a \mathbf{Z}_N^T, \tag{3}$$

where a is a real fractional parameter. For $a = 0$ the DFRFT matrix $\mathbf{F}_N^a$ is the identity matrix, and for $a = 1$ becomes the ordinary DFT matrix. Pei and Yeh defined the DFRFT using a particular set of eigenvectors [5]. This idea was developed in work [20].

3 Structure of DFRFT Matrix

In this paper we assume that the set of eigenvectors of the matrix $\mathbf{F}_N$ has already been calculated, as it was shown in [20], and the eigenvectors are ordered according to the increasing number of zero-crossings. After normalization, they form the matrix $\mathbf{Z}_N$ which occurs in Eqs. (2) and (3).

It is easy to check that the DFRFT matrix, calculated from (3), is symmetric. Moreover, the first row (and column) of the matrix $\mathbf{F}_N^a$ is an even vector and a matrix which we obtain after removing the first row and the first column from the matrix $\mathbf{F}_N^a$ is persymmetric [21]. These properties of the matrix $\mathbf{F}_N^a$ give it a special structure. Because of this structure, it is useful to write this matrix as a sum of three or two "special" matrices to reduce the number of arithmetical operations when we calculate its product by a vector [21]. The number of components of the sum is equal to three for even N or two for odd N. If N is even the matrix $\mathbf{F}_N^a$ can be written as a sum of three matrices

$$\mathbf{F}_N^a = \mathbf{A}_N^{(a)} + \mathbf{B}_N^{(a)} + \mathbf{C}_N^{(a)}, \tag{4}$$

where

$$\mathbf{A}_N^{(a)} = \begin{bmatrix} f_{0,0}^a & f_{0,1}^{(a)} & \cdots & f_{0,\frac{N}{2}-1}^{(a)} & f_{0,\frac{N}{2}}^{(a)} & f_{0,\frac{N}{2}-1}^{(a)} & \cdots & f_{0,1}^{(a)} \\ f_{0,1}^{(a)} & 0 & \cdots & 0 & 0 & 0 & \cdots & 0 \\ \vdots & & & & & & & \\ f_{0,\frac{N}{2}-1}^{(a)} & 0 & \cdots & 0 & 0 & 0 & \cdots & 0 \\ f_{0,\frac{N}{2}}^{(a)} & 0 & \cdots & 0 & 0 & 0 & \cdots & 0 \\ f_{0,\frac{N}{2}-1}^{(a)} & 0 & \cdots & 0 & 0 & 0 & \cdots & 0 \\ \vdots & & & & & & & \\ f_{0,1}^{(a)} & 0 & \cdots & 0 & 0 & 0 & \cdots & 0 \end{bmatrix}, \tag{5}$$

$$\mathbf{B}_N^{(a)} = \begin{bmatrix} 0 & 0 & \cdots & 0 & 0 & 0 & \cdots & 0 \\ 0 & f_{1,1}^{(a)} & \cdots & f_{1,\frac{N}{2}-1}^{(a)} & 0 & f_{1,\frac{N}{2}+1}^{(a)} & \cdots & f_{1,N-1}^{(a)} \\ \vdots & \vdots & \ddots & \vdots & \vdots & \vdots & \ddots & \vdots \\ 0 & f_{1,\frac{N}{2}-1}^{(a)} & \cdots & f_{\frac{N}{2}-1,\frac{N}{2}-1}^{(a)} & 0 & f_{\frac{N}{2}-1,\frac{N}{2}+1}^{(a)} & \cdots & f_{1,\frac{N}{2}+1}^{(a)} \\ 0 & 0 & \cdots & 0 & 0 & 0 & \cdots & 0 \\ 0 & f_{1,\frac{N}{2}+1}^{(a)} & \cdots & f_{\frac{N}{2}-1,\frac{N}{2}+1}^{(a)} & 0 & f_{\frac{N}{2}-1,\frac{N}{2}-1}^{(a)} & \cdots & f_{1,\frac{N}{2}-1}^{(a)} \\ \vdots & \vdots & \ddots & \vdots & \vdots & \vdots & \ddots & \vdots \\ 0 & f_{1,N-1}^{(a)} & \cdots & f_{1,\frac{N}{2}+1}^{(a)} & 0 & f_{1,\frac{N}{2}-1}^{(a)} & \cdots & f_{1,1}^{(a)} \end{bmatrix}, \tag{6}$$

$$\mathbf{C}_N^{(a)} = \begin{bmatrix} 0 & 0 & \cdots & 0 & 0 & 0 & \cdots & 0 \\ 0 & 0 & \cdots & 0 & f_{1,\frac{N}{2}}^{(a)} & 0 & \cdots & 0 \\ \vdots & \vdots & \ddots & \vdots & \vdots & \vdots & \ddots & \vdots \\ 0 & 0 & \cdots & 0 & f_{\frac{N}{2}-1,\frac{N}{2}}^{(a)} & 0 & \cdots & 0 \\ 0 & f_{1,\frac{N}{2}}^{(a)} & \cdots & f_{\frac{N}{2}-1,\frac{N}{2}}^{(a)} & f_{\frac{N}{2},\frac{N}{2}}^{(a)} & f_{\frac{N}{2}-1,\frac{N}{2}}^{(a)} & \cdots & f_{1,\frac{N}{2}}^{(a)} \\ 0 & 0 & \cdots & 0 & f_{\frac{N}{2}-1,\frac{N}{2}}^{(a)} & 0 & \cdots & 0 \\ \vdots & \vdots & \ddots & \vdots & \vdots & \vdots & \ddots & \vdots \\ 0 & 0 & \cdots & 0 & f_{1,\frac{N}{2}}^{(a)} & 0 & \cdots & 0 \end{bmatrix}. \tag{7}$$

If N is odd we can write the matrix $\mathbf{F}_N^a$ as a sum of only two matrices

$$\mathbf{F}_N^a = \mathbf{A}_N^{(a)} + \mathbf{B}_N^{(a)}, \tag{8}$$

where

$$
\mathbf{A}_N^{(a)} =
\begin{bmatrix}
f_{0,0}^a & f_{0,1}^{(a)} & \cdots & f_{0,\frac{N-1}{2}}^{(a)} & f_{0,\frac{N-1}{2}}^{(a)} & \cdots & f_{0,1}^{(a)} \\
f_{0,1}^{(a)} & 0 & \cdots & 0 & 0 & \cdots & 0 \\
\vdots & & & & & & \\
f_{0,\frac{N-1}{2}}^{(a)} & 0 & \cdots & 0 & 0 & \cdots & 0 \\
f_{0,\frac{N-1}{2}}^{(a)} & 0 & \cdots & 0 & 0 & \cdots & 0 \\
\vdots & & & & & & \\
f_{0,1}^{(a)} & 0 & \cdots & 0 & 0 & \cdots & 0
\end{bmatrix},
\tag{9}
$$

$$
\mathbf{B}_N^{(a)} =
\begin{bmatrix}
0 & 0 & \cdots & 0 & 0 & \cdots & 0 \\
0 & f_{1,1}^{(a)} & \cdots & f_{1,\frac{N-1}{2}}^{(a)} & f_{1,\frac{N+1}{2}}^{(a)} & \cdots & f_{1,N-1}^{(a)} \\
\vdots & \vdots & \ddots & \vdots & \vdots & \ddots & \vdots \\
0 & f_{1,\frac{N-1}{2}}^{(a)} & \cdots & f_{\frac{N-1}{2},\frac{N-1}{2}}^{(a)} & f_{\frac{N-1}{2},\frac{N+1}{2}}^{(a)} & \cdots & f_{1,\frac{N+1}{2}}^{(a)} \\
0 & f_{1,\frac{N+1}{2}}^{(a)} & \cdots & f_{\frac{N-1}{2},\frac{N+1}{2}}^{(a)} & f_{\frac{N-1}{2},\frac{N-1}{2}}^{(a)} & \cdots & f_{1,\frac{N-1}{2}}^{(a)} \\
\vdots & \vdots & \ddots & \vdots & \vdots & \ddots & \vdots \\
0 & f_{1,N-1}^{(a)} & \cdots & f_{1,\frac{N+1}{2}}^{(a)} & f_{1,\frac{N-1}{2}}^{(a)} & \cdots & f_{1,1}^{(a)}
\end{bmatrix}.
\tag{10}
$$

For example $\mathbf{F}_8^a$ and $\mathbf{F}_7^a$ have the following structures:

$$
\mathbf{F}_8^a =
\begin{bmatrix}
b & c & d & e & g & e & d & c \\
c & h & i & j & k & l & m & n \\
d & i & o & p & q & r & s & m \\
e & j & p & t & u & w & r & l \\
g & k & q & u & y & u & q & k \\
e & l & r & w & u & t & p & j \\
d & m & s & r & q & p & o & i \\
c & n & m & l & k & j & i & h
\end{bmatrix},
\quad
\mathbf{F}_7^a =
\begin{bmatrix}
b & c & d & e & e & d & c \\
c & g & h & i & j & k & l \\
d & h & m & n & o & p & k \\
e & i & n & q & r & o & j \\
e & j & o & r & q & n & i \\
d & k & p & o & n & m & h \\
c & l & k & j & i & h & g
\end{bmatrix},
\tag{11}
$$

where the entries: b, c, d, e, g, h, i, j, k, l, m, n, o, p, q, r, s, t, u, w, y are complex numbers, which are determined by N and the fractional parameter a. We can write $\mathbf{F}_8^a$ and $\mathbf{F}_7^a$ as the following sums:

$$
\mathbf{F}_8^a =
\begin{bmatrix}
b & c & d & e & g & e & d & c \\
c & 0 & 0 & 0 & 0 & 0 & 0 & 0 \\
d & 0 & 0 & 0 & 0 & 0 & 0 & 0 \\
e & 0 & 0 & 0 & 0 & 0 & 0 & 0 \\
g & 0 & 0 & 0 & 0 & 0 & 0 & 0 \\
e & 0 & 0 & 0 & 0 & 0 & 0 & 0 \\
d & 0 & 0 & 0 & 0 & 0 & 0 & 0 \\
c & 0 & 0 & 0 & 0 & 0 & 0 & 0
\end{bmatrix}
+
\begin{bmatrix}
0 & 0 & 0 & 0 & 0 & 0 & 0 & 0 \\
0 & h & i & j & 0 & l & m & n \\
0 & i & o & p & 0 & r & s & m \\
0 & j & p & t & 0 & w & r & l \\
0 & 0 & 0 & 0 & 0 & 0 & 0 & 0 \\
0 & l & r & w & 0 & t & p & j \\
0 & m & s & r & 0 & p & o & i \\
0 & n & m & l & 0 & j & i & h
\end{bmatrix}
+
\begin{bmatrix}
0 & 0 & 0 & 0 & 0 & 0 & 0 & 0 \\
0 & 0 & 0 & 0 & k & 0 & 0 & 0 \\
0 & 0 & 0 & 0 & q & 0 & 0 & 0 \\
0 & 0 & 0 & 0 & u & 0 & 0 & 0 \\
0 & k & q & u & y & u & q & k \\
0 & 0 & 0 & 0 & u & 0 & 0 & 0 \\
0 & 0 & 0 & 0 & q & 0 & 0 & 0 \\
0 & 0 & 0 & 0 & k & 0 & 0 & 0
\end{bmatrix},
\tag{12}
$$

$$
\mathbf{F}_7^a =
\begin{bmatrix}
b & c & d & e & e & d & c \\
c & 0 & 0 & 0 & 0 & 0 & 0 \\
d & 0 & 0 & 0 & 0 & 0 & 0 \\
e & 0 & 0 & 0 & 0 & 0 & 0 \\
e & 0 & 0 & 0 & 0 & 0 & 0 \\
d & 0 & 0 & 0 & 0 & 0 & 0 \\
c & 0 & 0 & 0 & 0 & 0 & 0
\end{bmatrix}
+
\begin{bmatrix}
0 & 0 & 0 & 0 & 0 & 0 & 0 \\
0 & g & h & i & j & k & l \\
0 & h & m & n & o & p & k \\
0 & i & n & q & r & o & j \\
0 & j & o & r & q & n & i \\
0 & k & p & o & n & m & h \\
0 & l & k & j & i & h & g
\end{bmatrix} .
\tag{13}
$$

4   Partial Products Calculation

We want to calculate the DFRFT for the input vector $\mathbf{x}_N$, generally complex. By $\mathbf{y}_N^{(a)}$ we denote the output vector calculated from the formula

$$
\mathbf{y}_N^{(a)} = \mathbf{F}_N^a \mathbf{x}_N .
\tag{14}
$$

We assume that the matrix $\mathbf{F}_N^a$ has been calculated in advance. To directly calculate the output vector it is necessary to perform N^2 multiplications and $N(N-1)$ additions of complex numbers. However, if we use the decompositions (4) or (8), the number of arithmetical operations can be significantly reduced. Let $\mathbf{y}_N^{(A,a)}$ denotes the product $\mathbf{A}_N^{(a)}\mathbf{x}_N$, $\mathbf{y}_N^{(B,a)}$ - the product $\mathbf{B}_N^{(a)}\mathbf{x}_N$ and, if N is even, $\mathbf{y}_N^{(C,a)}$ - the product $\mathbf{C}_N^{(a)}\mathbf{x}_N$. The partial products can be obtained using formulas presented below [21]. For even N the matrix $\mathbf{A}_N^{(a)}$ has the form (5) and

$$
\mathbf{y}_N^{(A,a)} = \mathbf{A}_N^{(a)}\mathbf{x}_N = \mathbf{T}_{N\times(N+1)}\mathbf{V}_{N+1}^{(a)}\mathbf{X}_{(N+1)\times N}\mathbf{x}_N ,
\tag{15}
$$

where

$$
\mathbf{X}_{(N+1)\times N} =
\left[
\begin{array}{c:c:c:c}
1 & \mathbf{0}_{1\times(\frac{N}{2}-1)} & 0 & \mathbf{0}_{1\times(\frac{N}{2}-1)} \\ \hdashline
\mathbf{0}_{(\frac{N}{2}-1)\times 1} & \mathbf{I}_{\frac{N}{2}-1} & \mathbf{0}_{(\frac{N}{2}-1)\times 1} & \mathbf{J}_{\frac{N}{2}-1} \\ \hdashline
0 & \mathbf{0}_{1\times(\frac{N}{2}-1)} & 1 & \mathbf{0}_{1\times(\frac{N}{2}-1)} \\ \hdashline
\mathbf{1}_{\frac{N}{2}\times 1} & \mathbf{0}_{\frac{N}{2}\times(\frac{N}{2}-1)} & \mathbf{0}_{\frac{N}{2}\times 1} & \mathbf{0}_{\frac{N}{2}\times(\frac{N}{2}-1)}
\end{array}
\right] .
\tag{16}
$$

In above equation, $\mathbf{1}_{m\times n}$ denotes a matrix of size $m \times n$ with all the entries equal to 1. $\mathbf{I}_k$ and $\mathbf{J}_k$ are the identity matrix and the exchange matrix of size k, respectively. The matrix $\mathbf{V}_{N+1}^{(a)}$, occurring in Eq. (15), is a diagonal matrix, which has the following form:

$$
\mathbf{V}_{N+1}^{(a)} = \mathrm{diag}(f_{0,0}^{(a)}, f_{0,1}^{(a)}, \ldots, f_{0,\frac{N}{2}}^{(a)}, f_{0,1}^{(a)}, \ldots, f_{0,\frac{N}{2}}^{(a)}) .
\tag{17}
$$

The last matrix $\mathbf{T}_{N\times(N+1)}$, which occurs in Eq. (15), has the form

$$
\mathbf{T}_{N\times(N+1)} =
\left[
\begin{array}{c:c:c}
\mathbf{1}_{1\times(\frac{N}{2}+1)} & \mathbf{0}_{1\times(\frac{N}{2}-1)} & 0 \\ \hdashline
\mathbf{0}_{(\frac{N}{2}-1)\times(\frac{N}{2}+1)} & \mathbf{I}_{\frac{N}{2}-1} & \mathbf{0}_{(\frac{N}{2}-1)\times 1} \\ \hdashline
\mathbf{0}_{1\times(\frac{N}{2}+1)} & \mathbf{0}_{1\times(\frac{N}{2}-1)} & 1 \\ \hdashline
\mathbf{0}_{(\frac{N}{2}-1)\times(\frac{N}{2}+1)} & \mathbf{J}_{\frac{N}{2}-1} & \mathbf{0}_{(\frac{N}{2}-1)\times 1}
\end{array}
\right] .
\tag{18}
$$

For odd N the matrix $\mathbf{A}_N^{(a)}$ has the form (9) and

$$\mathbf{y}_N^{(A,a)} = \mathbf{A}_N^{(a)}\mathbf{x}_N = \mathbf{T}_N \mathbf{V}_N^{(a)} \mathbf{X}_N \mathbf{x}_N, \tag{19}$$

where the matrices occurring in (19) are as follows:

$$\mathbf{X}_N = \left[\begin{array}{c|c|c} 1 & \mathbf{0}_{1\times\frac{N-1}{2}} & \mathbf{0}_{1\times\frac{N-1}{2}} \\ \hline \mathbf{0}_{\frac{N-1}{2}\times 1} & \mathbf{I}_{\frac{N-1}{2}} & \mathbf{J}_{\frac{N-1}{2}} \\ \hline \mathbf{1}_{\frac{N-1}{2}\times 1} & \mathbf{0}_{\frac{N-1}{2}} & \mathbf{0}_{\frac{N-1}{2}} \end{array} \right], \tag{20}$$

$$\mathbf{V}_N^{(a)} = \mathrm{diag}(f_{0,0}^{(a)}, f_{0,1}^{(a)}, \ldots, f_{0,\frac{N-1}{2}}^{(a)}, f_{0,1}^{(a)}, \ldots, f_{0,\frac{N-1}{2}}^{(a)}), \tag{21}$$

$$\mathbf{T}_N = \left[\begin{array}{c|c} \mathbf{1}_{1\times\frac{N+1}{2}} & \mathbf{0}_{1\times\frac{N-1}{2}} \\ \hline \mathbf{0}_{\frac{N-1}{2}\times\frac{N+1}{2}} & \mathbf{I}_{\frac{N-1}{2}} \\ \hline \mathbf{0}_{\frac{N-1}{2}\times\frac{N+1}{2}} & \mathbf{J}_{\frac{N-1}{2}} \end{array} \right]. \tag{22}$$

To calculate $\mathbf{y}_N^{(A,a)}$ it is necessary to perform $N-1$ additions of complex numbers. The number of multiplications is equal to $N+1$ for even N and N for odd N.

The next partial product is

$$\mathbf{y}_N^{(B,a)} = \mathbf{B}_N^{(a)}\mathbf{x}_N. \tag{23}$$

For even N the matrix $\mathbf{B}_N^{(a)}$ has the form (6). We can see that $y_0^{(B,a)} = y_{N/2}^{(B,a)} = 0$ and also entries x_0 and $x_{N/2}$ are not involved in this calculation, so we denote the vector $\mathbf{y}_N^{(B,a)}$ with removed the entries $y_0^{(B,a)}$ and $y_{N/2}^{(B,a)}$ by $\overline{\mathbf{y}}_{N-2}^{(B,a)}$, i.e.

$$\overline{\mathbf{y}}_{N-2}^{(B,a)} = [y_1^{(B,a)}, y_2^{(B,a)}, \ldots, y_{\frac{N}{2}-1}^{(B,a)}, y_{\frac{N}{2}+1}^{(B,a)}, \ldots, y_{N-1}^{(B,a)}]^T. \tag{24}$$

Similarly, we denote the vector $\mathbf{x}_N$ with removed the entries x_0 and $x_{N/2}$ by $\overline{\mathbf{x}}_{N-2}$. Then we can rewrite the Eq. (23) equivalently in the following form

$$\overline{\mathbf{y}}_{N-2}^{(B,a)} = \overline{\mathbf{B}}_{N-2}^{(a)} \overline{\mathbf{x}}_{N-2}, \tag{25}$$

where the matrix $\overline{\mathbf{B}}_{N-2}^{(a)}$ is the matrix $\mathbf{B}_N^{(a)}$ with removed all zero rows and columns. Calculation of the vector $\overline{\mathbf{y}}_{N-2}^{(B,a)}$ can be compactly described by the following matrix-vector procedure:

$$\overline{\mathbf{y}}_{N-2}^{(B,a)} = \mathbf{R}_{N-2} \mathbf{W}_{(N-2)\times\frac{(N-2)^2}{2}} \mathbf{Q}_{\frac{(N-2)^2}{2}}^{(a)} \mathbf{U}_{\frac{(N-2)^2}{2}\times(N-2)} \mathbf{M}_{N-2} \overline{\mathbf{x}}_{N-2}, \tag{26}$$

where $\mathbf{M}_{N-2}$ has the form

$$\mathbf{M}_{N-2} = \left[\mathbf{I}_{\frac{N-2}{2}} \otimes \begin{bmatrix} 1 \\ 1 \end{bmatrix} \middle| \mathbf{J}_{\frac{N-2}{2}} \otimes \begin{bmatrix} 1 \\ -1 \end{bmatrix} \right]. \tag{27}$$

The symbol $\otimes$ in above equation means the Kronecker product operation. The rest of matrices, which occur in Eq. (26), are as follows:

$$\mathbf{U}_{\frac{(N-2)^2}{2} \times (N-2)} = \mathbf{1}_{\frac{N-2}{2} \times 1} \otimes \mathbf{I}_{N-2}, \tag{28}$$

$$\mathbf{Q}^{(a)}_{\frac{(N-2)^2}{2}} = \text{diag}\left(\frac{f_{1,1}^{(a)}+f_{1,N-1}^{(a)}}{2}, \frac{f_{1,1}^{(a)}-f_{1,N-1}^{(a)}}{2}, \frac{f_{1,2}^{(a)}+f_{1,N-2}^{(a)}}{2}, \frac{f_{1,2}^{(a)}-f_{1,N-2}^{(a)}}{2}, \ldots, \right.$$

$$\left. \frac{f_{1,\frac{N}{2}-1}^{(a)}+f_{1,\frac{N}{2}+1}^{(a)}}{2}, \frac{f_{1,\frac{N}{2}-1}^{(a)}-f_{1,\frac{N}{2}+1}^{(a)}}{2}, \ldots, \frac{f_{\frac{N}{2}-1,\frac{N}{2}-1}^{(a)}+f_{\frac{N}{2}-1,\frac{N}{2}+1}^{(a)}}{2}, \frac{f_{\frac{N}{2}-1,\frac{N}{2}-1}^{(a)}-f_{\frac{N}{2}-1,\frac{N}{2}+1}^{(a)}}{2} \right), \tag{29}$$

$$\mathbf{W}_{(N-2) \times \frac{(N-2)^2}{2}} = \left[\begin{array}{c} \mathbf{I}_{\frac{N-2}{2}} \otimes \mathbf{1}_{1 \times \frac{N-2}{2}} \otimes [1 \quad 0] \\ \hline \mathbf{J}_{\frac{N-2}{2}} \otimes \mathbf{1}_{1 \times \frac{N-2}{2}} \otimes [0 \quad 1] \end{array} \right], \tag{30}$$

$$\mathbf{R}_{N-2} = \left[\begin{array}{c} \mathbf{I}_{\frac{N-2}{2}} \otimes [1 \quad 1] \\ \hline \mathbf{J}_{\frac{N-2}{2}} \otimes [1 \quad -1] \end{array} \right]. \tag{31}$$

If N is odd the matrix $\mathbf{B}_N^{(a)}$ has the form (10). We can see that $y_0^{(B,a)} = 0$ and also entry x_0 is not involved with this calculation, so we denote the vector $\mathbf{y}_N^{(B,a)}$ with removed the entry $y_0^{(B,a)}$ by $\tilde{\mathbf{y}}_{N-1}^{(B,a)}$, i.e.

$$\tilde{\mathbf{y}}_{N-1}^{(B,a)} = [y_1^{(B,a)}, y_2^{(B,a)}, \ldots, y_{N-1}^{(B,a)}]^T. \tag{32}$$

Similarly, we denote the vector $\mathbf{x}_N$ with removed the entry x_0 by $\tilde{\mathbf{x}}_{N-1}$. Then we can rewrite the Eq. (23) in the following form

$$\tilde{\mathbf{y}}_{N-1}^{(B,a)} = \tilde{\mathbf{B}}_{N-1}^{(a)} \tilde{\mathbf{x}}_{N-1} \tag{33}$$

where the matrix $\tilde{\mathbf{B}}_{N-1}^{(a)}$ denotes the matrix $\mathbf{B}_N^{(a)}$ with removed all zero row and column. For odd N the procedure for calculation of vector $\tilde{\mathbf{y}}_{N-1}^{(B,a)}$ is

$$\tilde{\mathbf{y}}_{N-1}^{(B,a)} = \mathbf{R}_{N-1} \mathbf{W}_{(N-1) \times \frac{(N-1)^2}{2}} \mathbf{Q}^{(a)}_{\frac{(N-1)^2}{2}} \mathbf{U}_{\frac{(N-1)^2}{2} \times (N-1)} \mathbf{M}_{N-1} \tilde{\mathbf{x}}_{N-1}, \tag{34}$$

where the matrices, occurring in the Eq. (34), have the forms

$$\mathbf{M}_{N-1} = \left[\mathbf{I}_{\frac{N-1}{2}} \otimes \left[\begin{array}{c} 1 \\ 1 \end{array} \right] \middle| \mathbf{J}_{\frac{N-1}{2}} \otimes \left[\begin{array}{c} 1 \\ -1 \end{array} \right] \right], \tag{35}$$

$$\mathbf{U}_{\frac{(N-1)^2}{2} \times (N-1)} = \mathbf{1}_{\frac{N-1}{2} \times 1} \otimes \mathbf{I}_{N-1}, \tag{36}$$

$$\mathbf{Q}^{(a)}_{\frac{(N-1)^2}{2}} = \text{diag}\left(\frac{f_{1,1}^{(a)}+f_{1,N-1}^{(a)}}{2}, \frac{f_{1,1}^{(a)}-f_{1,N-1}^{(a)}}{2}, \frac{f_{1,2}^{(a)}+f_{1,N-2}^{(a)}}{2}, \frac{f_{1,2}^{(a)}-f_{1,N-2}^{(a)}}{2}, \ldots, \right.$$

$$\left. \frac{f_{1,\frac{N-1}{2}}^{(a)}+f_{1,\frac{N+1}{2}}^{(a)}}{2}, \frac{f_{1,\frac{N-1}{2}}^{(a)}-f_{1,\frac{N+1}{2}}^{(a)}}{2}, \ldots, \frac{f_{\frac{N-1}{2},\frac{N-1}{2}}^{(a)}+f_{\frac{N-1}{2},\frac{N+1}{2}}^{(a)}}{2}, \frac{f_{\frac{N-1}{2},\frac{N-1}{2}}^{(a)}-f_{\frac{N-1}{2},\frac{N+1}{2}}^{(a)}}{2} \right), \tag{37}$$

$$\mathbf{W}_{(N-1)\times\frac{(N-1)^2}{2}} = \left[\begin{array}{c} \mathbf{I}_{\frac{N-1}{2}}\otimes\mathbf{1}_{1\times\frac{N-1}{2}}\otimes[1\ \ 0] \\ \hline \mathbf{J}_{\frac{N-1}{2}}\otimes\mathbf{1}_{1\times\frac{N-1}{2}}\otimes[0\ \ 1] \end{array}\right],\tag{38}$$

$$\mathbf{R}_{N-1} = \left[\begin{array}{c} \mathbf{I}_{\frac{N-1}{2}}\otimes[1\ \ 1] \\ \hline \mathbf{J}_{\frac{N-1}{2}}\otimes[1\ -1] \end{array}\right].\tag{39}$$

For even N to calculate the vector $\overline{\mathbf{y}}_{N-2}^{(B,a)}$ according to the procedure (26) it is necessary to perform $N(N-2)/2$ additions and $(N-2)^2/2$ multiplications of complex numbers. For odd N to calculate the vector $\tilde{\mathbf{y}}_{N-1}^{(B,a)}$ in accordance with the procedure (34) it is necessary to perform $(N+1)(N-1)/2$ additions and $(N-1)^2/2$ multiplications of complex numbers.

Now we will focus on the product

$$\mathbf{y}_N^{(C,a)} = \mathbf{C}_N^{(a)}\mathbf{x}_N,\tag{40}$$

which appears only for the even number N. The matrix $\mathbf{C}_N^{(a)}$ has the form (7). Since $y_0^{(C,a)} = 0$ and also entry x_0 is not involved in this calculation we denote the vector $\mathbf{y}_N^{(C,a)}$ with removed the entry $y_0^{(C,a)}$ by $\tilde{\mathbf{y}}_{N-1}^{(C,a)}$, i.e.

$$\tilde{\mathbf{y}}_{N-1}^{(C,a)} = [y_1^{(C,a)}, y_2^{(C,a)}, \ldots, y_{N-1}^{(C,a)}]^T.\tag{41}$$

Then we can rewrite the Eq. (40) equivalently in the following form:

$$\tilde{\mathbf{y}}_{N-1}^{(C,a)} = \tilde{\mathbf{C}}_{N-1}^{(a)}\tilde{\mathbf{x}}_{N-1},\tag{42}$$

where the matrix $\tilde{\mathbf{C}}_{N-1}^{(a)}$ is the matrix $\mathbf{C}_N^{(a)}$ with removed all zero row and column. Calculation of $\mathbf{y}_{N-1}^{(C,a)}$ can be compactly described by appropriate matrix-vector procedure. This procedure will be as follows:

$$\tilde{\mathbf{y}}_{N-1}^{(C,a)} = \mathbf{K}_{N-1}\mathbf{G}_{N-1}^{(a)}\mathbf{L}_{N-1}\tilde{\mathbf{x}}_{N-1},\tag{43}$$

where the matrices, occurring in Eq. (43), are as follows:

$$\mathbf{L}_{N-1} = \left[\begin{array}{ccc} \mathbf{I}_{\frac{N}{2}-1} & \mathbf{0}_{(\frac{N}{2}-1)\times 1} & \mathbf{J}_{\frac{N}{2}-1} \\ \hline \mathbf{0}_{\frac{N}{2}\times(\frac{N}{2}-1)} & \mathbf{1}_{\frac{N}{2}\times 1} & \mathbf{0}_{\frac{N}{2}\times(\frac{N}{2}-1)} \end{array}\right],\tag{44}$$

$$\mathbf{G}_{N-1}^{(a)} = \mathrm{diag}(f_{1,\frac{N}{2}}^{(a)}, f_{2,\frac{N}{2}}^{(a)}, \ldots, f_{\frac{N}{2},\frac{N}{2}}^{(a)}, f_{1,\frac{N}{2}}^{(a)}, f_{2,\frac{N}{2}}^{(a)}, \ldots, f_{\frac{N}{2}-1,\frac{N}{2}}^{(a)}),\tag{45}$$

$$\mathbf{K}_{N-1} = \left[\begin{array}{cc} \mathbf{0}_{(\frac{N}{2}-1)\times\frac{N}{2}} & \mathbf{I}_{\frac{N}{2}-1} \\ \hline \mathbf{1}_{1\times\frac{N}{2}} & \mathbf{0}_{1\times(\frac{N}{2}-1)} \\ \hline \mathbf{0}_{(\frac{N}{2}-1)\times\frac{N}{2}} & \mathbf{J}_{\frac{N}{2}-1} \end{array}\right].\tag{46}$$

To calculate $\tilde{\mathbf{y}}_{N-1}^{(C,a)}$ it is necessary to perform $N-2$ additions and $N-1$ multiplications of complex numbers.

5   The DFRFT Algorithm

If we want to obtain the final output vector $\mathbf{y}_N^{(a)}$, defined by (14), we have to add up vectors $\mathbf{y}_N^{(A,a)}$, $\mathbf{y}_N^{(B,a)}$ and also $\mathbf{y}_N^{(C,a)}$ if N is even.

For even N the matrix-vector procedure for calculating $\mathbf{y}_N^{(a)}$ will be as follows:

$$\mathbf{y}_N^{(a)} = \mathbf{\Omega}_{N\times(3N-3)}\mathrm{diag}(\mathbf{A}_N^{(a)}, \overline{\mathbf{B}}_{N-2}^{(a)}, \tilde{\mathbf{C}}_{N-1}^{(a)})\mathbf{\Psi}_{(3N-3)\times N}\mathbf{x}_N, \qquad (47)$$

where

$$\mathbf{\Psi}_{(3N-3)\times N} = \begin{bmatrix} \mathbf{I}_N \\ \mathbf{I}_{(N-2)\times N}^{(0,\frac{N}{2})} \\ \mathbf{I}_{(N-1)\times N}^{(0)} \end{bmatrix}, \quad \mathbf{\Omega}_{N\times(3N-3)} = \begin{bmatrix} \mathbf{I}_N \ \vdots \ \hat{\mathbf{I}}_{N\times(N-2)}^{(0,\frac{N}{2})} \ \vdots \ \hat{\mathbf{I}}_{N\times(N-1)}^{(0)} \end{bmatrix}. \quad (48)$$

The matrix $\mathbf{\Psi}_{(3N-3)\times N}$ is responsible for preparing the vector $[\mathbf{x}_N^T, \overline{\mathbf{x}}_{N-2}^T, \tilde{\mathbf{x}}_{N-1}^T]^T$, where the matrices $\mathbf{I}_{(N-2)\times N}^{(0,\frac{N}{2})}$ and $\mathbf{I}_{(N-1)\times N}^{(0)}$ are obtained from the identity matrix $\mathbf{I}_N$ by removing the rows with indexes 0 and $N/2$ or 0, respectively. The matrix $\mathbf{\Omega}_{N\times(3N-3)}$, occurring in (47), is responsible for summing up appropriate entries of vectors $\mathbf{y}_N^{(A,a)}$, $\overline{\mathbf{y}}_{N-2}^{(B,a)}$ and $\tilde{\mathbf{y}}_N^{(C,a)}$, where the matrices $\hat{\mathbf{I}}_{N\times(N-2)}^{(0,\frac{N}{2})}$ and $\hat{\mathbf{I}}_{N\times(N-1)}^{(0)}$ are obtained from the identity matrix $\mathbf{I}_N$ by removing the columns with indexes 0 and $N/2$ or 0, respectively. The matrix $\mathrm{diag}(\mathbf{A}_N^{(a)}, \overline{\mathbf{B}}_{N-2}^{(a)}, \tilde{\mathbf{C}}_{N-1}^{(a)})$ in the Eq. (47) is the block diagonal matrix and these blocks matrices are factorised as in (15), (26) and (43), respectively.

Figure 1 shows a graph-structural model and data flow diagram for calculation product $\mathbf{y}_N^{(a)}$ for the input vector of length 8. The graph-structural models and data flow diagrams are oriented from left to right. Points, where lines converge denote summation (or subtraction if the line is dotted). The rectangle show the operation of multiplication by a matrix inscribed inside and a circle show the operation of multiplication by a complex number inscribed inside a circle. In the Fig. 1 the numbers q_i are equal to: $q_0 = (h+n)/2$, $q_1 = (h-n)/2$, $q_2 = (i+m)/2$, $q_3 = (i-m)/2$, $q_4 = (j+l)/2$, $q_5 = (j-l)/2$, $q_6 = q_2$, $q_7 = q_3$, $q_8 = (o+s)/2$, $q_9 = (o-s)/2$, $q_{10} = (p+r)/2$, $q_{11} = (p-r)/2$, $q_{12} = q_4$, $q_{13} = q_5$, $q_{14} = q_{10}$, $q_{15} = q_{11}$, $q_{16} = (t+w)/2$, $q_{17} = (t-w)/2$, where $h, n, \ldots, w$ are the entries of the matrix $\mathbf{F}_8^{(a)}$ from (12).

For odd N the matrix-vector procedure for calculating $\mathbf{y}_N^{(a)}$ will be as follows:

$$\mathbf{y}_N^{(a)} = \mathbf{\Omega}_{N\times(2N-1)}\mathrm{diag}(\mathbf{A}_N^{(a)}, \tilde{\mathbf{B}}_{N-1}^{(a)})\mathbf{\Psi}_{(2N-1)\times N}\mathbf{x}_N, \qquad (49)$$

where the matrices on the right side of this equation have the form

$$\mathbf{\Psi}_{(2N-1)\times N} = \begin{bmatrix} \mathbf{I}_N \\ \mathbf{I}_{(N-1)\times N}^{(0)} \end{bmatrix}, \quad \mathbf{\Omega}_{N\times(2N-1)} = \begin{bmatrix} \mathbf{I}_N \ \vdots \ \hat{\mathbf{I}}_{N\times(N-1)}^{(0)} \end{bmatrix}. \quad (50)$$

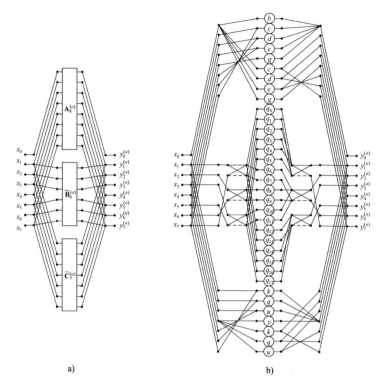

Fig. 1. Graph-structural model (a) and data flow diagram (b) for calculation $\mathbf{y}_8^{(a)}$.

Figure 2 shows a graph-structural model and data flow diagram for calculation product $\mathbf{y}_N^{(a)}$ for the input vector of length 7. In this figure the numbers q_i are equal to: $q_0 = (g + l)/2$, $q_1 = (g - l)/2$, $q_2 = (h + k)/2$, $q_3 = (h - k)/2$, $q_4 = (i + j)/2$, $q_5 = (i - j)/2$, $q_6 = q_2$, $q_7 = q_3$, $q_8 = (m + p)/2$, $q_9 = (m - p)/2$, $q_{10} = (n + o)/2$, $q_{11} = (n - o)/2$, $q_{12} = q_4$, $q_{13} = q_5$, $q_{14} = q_{10}$, $q_{15} = q_{11}$, $q_{16} = (q + r)/2$, $q_{17} = (q - r)/2$, where the numbers: $g, l, \ldots, r$ are the entries of the matrix $\mathbf{F}_7^{(a)}$ from (13).

6   Computational Complexity

Direct calculation of the discrete fractional Fourier transform for an input vector $\mathbf{x}_N$, assuming that the matrix $\mathbf{F}_N^a$ defined by (3) is given, requires N^2 multiplications and $N(N - 1)$ additions of complex numbers.

If we use the procedure (47) for even N or the procedure (49) for odd N the number of additions and multiplications will be smaller. For even N the total number of additions is equal to $N^2/2 + 3N - 6$ and the total number of multiplications is equal to $N^2/2 + 2$. For odd N these numbers are equal to $(N^2 - 1)/2 + 2N - 2$ and $(N^2 + 1)/2$, respectively. We can see that the number

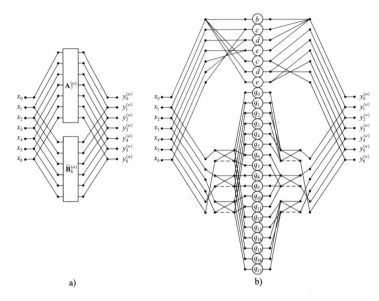

a)   b)

Fig. 2. Graph-structural model (a) and data flow diagram (b) for calculation $\mathbf{y}_7^{(a)}$.

of multiplications and additions in proposed algorithm is almost twice smaller than in the direct method of calculating DFRFT and it is truth for vectors of both even and odd lengths of the input vector.

7   Conclusion

In this paper, we propose an algorithm for "true" discrete fractional Fourier transform computation. The base of the proposed algorithm is the fact that the DFRFT matrix can be decomposed as a sum of a dense matrix and one or two sparse matrices. The dense matrix possesses a unique structure that allows us to perform its effective factorization and leads to accelerate computations by reducing the arithmetical complexity of a matrix-vector product. Based on the matrix factorization and Kronecker product, the effective algorithm for the DFRFT computation have been derived. The two examples of synthesis of such algorithms for $N = 8$ and $N = 7$ have been presented.

References

1. Wiener, N.: Hermitian polynomials and Fourier analysis. J. Math. Phys. **8**, 70–73 (1929)
2. Namias, V.: The fractional order Fourier transform and its application to quantum mechanics. J. Inst. Appl. Math. **25**, 241–265 (1980)
3. Ozaktas, H.M., Ankan, O., Kutay, M.A., Bozdagi, G.: Digital computation of the fractional Fourier transform. IEEE Trans. Signal Process. **44**(9), 2141–2150 (1996). https://doi.org/10.1109/78.536672

4. Santhanam, B., McClellan, J.H.: Discrete rotational Fourier transform. IEEE Trans. Signal Process. **44**(4), 994–998 (1996). https://doi.org/10.1109/78.492554
5. Pei, S.-C., Yeh, M.-H.: Discrete fractional Fourier transform. In: Proceedings of the IEEE International Symposium on Circuits and Systems, pp. 536–539 (1996)
6. Pei, S.-C., Tseng, C.-C., Yeh, M.-H., Shyu, J.-J.: Discrete fractional Hartley and Fourier transforms. IEEE Trans. Circuits Syst. II Analog. Digit. Signal Process. **45**(6), 665–675 (1998). https://doi.org/10.1109/82.686685
7. Pei, S.-C., Yeh, M.-H.: Discrete fractional Hadamard transform. In: Proceedings of the IEEE International Symposium on Circuits and Systems, vol. 3, pp. 179–182 (1999). https://doi.org/10.1109/ISCAS.1999.778814
8. Pei, S.-C., Yeh, M.-H.: Discrete fractional Hilbert transform. IEEE Trans. Circuits Syst. II Analog. Digit. Signal Process. **47**(11), 1307–1311 (2000). https://doi.org/10.1109/82.885138
9. Pei, S.-C., Yeh, M.H.: The discrete fractional cosine and sine transforms. IEEE Trans. Signal Process. **49**(6), 1198–1207 (2001). https://doi.org/10.1109/78.923302
10. Liu, Z., Zhao, H., Liu, S.: A discrete fractional random transform. Opt. Commun. **255**(4–6), 357–365 (2005). https://doi.org/10.1016/j.optcom.2005.06.031
11. Yetik, I.Ş., Kutay, M.A., Ozaktas, H.M.: Image representation and compression with the fractional Fourier transform. Opt. Commun. **197**, 275–278 (2001). https://doi.org/10.1016/S0030-4018(01)01462-6
12. Djurović, I., Stanković, S., Pitas, I.: Digital watermarking in the fractional Fourier transformation domain. J. Netw. Comput. Appl. **24**(2), 167–173 (2001). https://doi.org/10.1006/jnca.2000.0128
13. Hennelly, B., Sheridan, J.T.: Fractional Fourier transform-based image encryption: phase retrieval algorithm. Opt. Commun. **226**, 61–80 (2003). https://doi.org/10.1016/j.optcom.2003.08.030
14. Jindal, N., Singh, K.: Image and video processing using discrete fractional transforms. Signal Image Video Process. **8**(8), 1543–1553 (2014). https://doi.org/10.1007/s11760-012-0391-4
15. Tao, R., Liang, G., Zhao, X.: An efficient FPGA-based implementation of fractional Fourier transform algorithm. J. Signal Process. Syst. **60**(1), 47–58 (2010). https://doi.org/10.1007/s11265-009-0401-0
16. Cariow, A., Majorkowska-Mech, D.: Fast algorithm for discrete fractional Hadamard transform. Numer. Algorithms **68**(3), 585–600 (2015). https://doi.org/10.1007/s11075-014-9862-8
17. Bultheel, A., Martinez-Sulbaran, H.E.: Computation of the fractional Fourier transform. Appl. Comput. Harmon. Anal. **16**(3), 182–202 (2004)
18. Irfan, M., Zheng, L., Shahzad, H.: Review of computing algorithms for discrete fractional Fourier transform. Res. J. Appl. Sci. Eng. Technol. **6**(11), 1911–1919 (2013)
19. Pei, S.-C., Yeh, M.-H.: Improved discrete fractional Fourier transform. Opt. Lett. **22**(14), 1047–1049 (1997). https://doi.org/10.1364/OL.22.001047
20. Candan, Ç.C., Kutay, M.A., Ozaktas, H.M.: The discrete fractional Fourier transform. IEEE Trans. Signal Process. **48**(5), 1329–1337 (2000). https://doi.org/10.1109/78.839980

21. Majorkowska-Mech, D., Cariow, A.: A low-complexity approach to computation of the discrete fractional Fourier transform. Circuits Syst. Signal Process. **36**(10), 4118–4144 (2017). https://doi.org/10.1007/s00034-017-0503-z
22. Halmos, P.R.: Finite Dimensional Vector Spaces. Princeton University Press, Princeton (1947)
23. McClellan, J.H., Parks, T.W.: Eigenvalue and eigenvector decomposition of the discrete Fourier transform. IEEE Trans. Audio Electroacoust. **20**(1), 66–74 (1972). https://doi.org/10.1109/TAU.1972.1162342

Region Based Approach for Binarization of Degraded Document Images

Hubert Michalak and Krzysztof Okarma$^{(\boxtimes)}$ (iD)

Department of Signal Processing and Multimedia Engineering,
Faculty of Electrical Engineering,
West Pomeranian University of Technology, Szczecin,
26 Kwietnia 10, 71-126 Szczecin, Poland
{michalak.hubert,okarma}@zut.edu.pl

Abstract. Binarization of highly degraded document images is one of the key steps of image preprocessing, influencing the final results of further text recognition and document analysis. As the contaminations visible on such documents are usually local, the most popular fast global thresholding methods should not be directly applied for such images. On the other hand, the application of some typical adaptive methods based on the analysis of the neighbourhood of each pixel of the images is time consuming and not always leads to satisfactory results. To bridge the gap between those two approaches the application of region based modifications of some histogram based thresholding methods has been proposed in the paper. It has been verified for well known Otsu, Rosin and Kapur algorithms using the challenging images from Bickley Diary dataset. Experimental results obtained for region based Otsu and Kapur methods are superior in comparison to the use of global methods and may be the basis for further research towards combined region based binarization of degraded document images.

Keywords: Document images · Adaptive thresholding
Image binarization

1 Introduction

One of the most relevant operations, considered in many applications as an image preprocessing step, is image binarization. A significant decrease of the amount of data and simplicity of further analysis of shapes cause the popularity of binary image analysis in many applications related e.g. to Optical Character Recognition (OCR) [11] or some machine vision algorithms applied for robotic purposes, especially when the shape information is the most relevant. The choice of a proper image binarization methods influences strongly the results of further processing, being important also in many other applications e.g. recognition of vehicles' register plate numbers [27] or QR codes [16].

© Springer Nature Switzerland AG 2019
J. Pejaś et al. (Eds.): ACS 2018, AISC 889, pp. 433–444, 2019.
https://doi.org/10.1007/978-3-030-03314-9_37

Probably the most popular binarization methods has been proposed in 1979 by Otsu [17]. Is belongs to histogram based thresholding algorithms and utilizes the minimization of intra-class variance (being equivalent to maximizing the inter-class variance) between two classes of pixels representing foreground (resulting in logical "ones") and background ("zeros"). Such a global method allows achieving relatively good results for images having bi-modal histograms, however it usually fails in the case of degraded document images with many local distortions. Some modifications of this approach include multi-level thresholding as well as its adaptive version known as AdOtsu [14], which is computationally much more complex as it requires a separate analysis of the neighbourhood of each pixel with additional background estimation.

A similar global approach based on image entropy has been proposed by Kapur [6]. In this algorithm the two classes of pixels are described by two nonoverlapping probability distributions and the optimal threshold is set as the value minimizing the aggregated entropy (instead of variance used by Otsu). Another global histogram based method has been proposed by Rosin [19] which is dedicated for images with unimodal distributions and is based on the detection of a corner in the histogram plot.

An interesting method based on the application of Otsu's thresholding locally for blocks of 3×3 pixels has been proposed by Chou [2] with additional use of the Support Vector Machines (SVM) to improve the results obtained for regions containing only background pixels. Some other methods proposed recently include the use of Balanced Histogram Thresholding (BHT) for randomly chosen samples drawn according to the Monte Carlo method [9] and the use of the Monte Carlo approach for the iterative estimation of energy and entropy of the image for its fast binarization [10]. Another region based method proposed by Kulyuikin [8] is dedicated for barcodes recognition purposes whereas Wen has proposed [25] an approach based on Otsu's thresholding and Curvelet transform useful for unevenly lightened document images.

In contrast to fast global binarization algorithms, some more sophisticated and time-consuming adaptive methods have been introduced. The most popular of them have been proposed by Niblack [15] and Sauvola [21], further improved by Gatos [5]. The idea behind the Niblack's binarization is the analysis of the local average and variance of the image for local thresholding which has been further modified by Wolf [26] using the maximization of the local contrast, similarly as in another approach proposed by Feng [4] who has used median noise filtering with additional bilinear interpolation. An overview of some other modifications of adaptive methods based on Niblack's idea can be found in the papers written by Khurshid [7], Samorodova [20] and Saxena [22]. Some more detailed descriptions and comparisons of numerous recently proposed binarization methods can also be found in recent survey papers [12, 23].

Balancing the speed of the global methods with the flexibility of adaptive binarization, some possibilities of using the region based versions of three histogram based algorithms proposed initially by Otsu, Kapur and Rosin have been examined in the paper. The key issues in the conducted experimental research have been the correct choice of the block size and the additional threshold (vt) used for the local variance calculated for detection of the background blocks.

2   Proposed Region Based Approach and Its Verification

Considering the possible presence of some local distortions in the degraded historical document images, it has been verified that the application of the typical adaptive methods does not lead to satisfactory results, similarly as the use of popular global methods. To find a compromise between those two approaches the application of three histogram based thresholding methods introduced by Otsu, Kapur and Rosin is proposed.

Nevertheless, similarly as described in Chou's paper [2], one of the key issues is related with the presence of regions containing only background pixels which are incorrectly binarized. To simplify and speed-up the proposed algorithm, instead of SVM based approach proposed by Chou, a much more efficient calculation of the local variance has been proposed. Having determined a suitable size of the block (region) for each of three considered methods, the next step is the detection of "almost purely" background regions with the proper choice of the variance threshold (vt - equivalent to the maximum local variance considered as representing the background region further normalized to ensure its independence on the block size).

To compare the results obtained different binarization algorithms their comparison with the "ground-truth" images should be made. For this purpose the most commonly used F-Measure has been applied, known also as F1-score. Its value is defined as:

$$\text{FM} = 2 \cdot \frac{PR \cdot RC}{PR + RC} \cdot 100\% , \tag{1}$$

where Precision (PR) and Recall (RC) are calculated as the ratios of true positives to the sum of all positives (precision) and true positives to the sum of true positives and false negatives (recall).

The F-Measure values obtained for various block size (the square blocks have been assumed in the paper to simplify the experiments) and variance threshold using the region based Otsu method are illustrated in Fig. 1, whereas the results achieved for region based Kapur and Rosin methods are shown in Figs. 2 and 3 respectively. The best results have been achieved using 16×16 pixels block with $vt = 200$ for region based Otsu method (F-Measure equal to 0.7835), whereas region based Kapur algorithm requires larger blocks of 24×24 pixels with $vt = 225$ leading to F-Measure value of 0.7623 and for Rosin method blocks of 8×8 pixels with $vt = 200$ should be applied to achieve much lower F-Measure equal to 0.5171. All these results have been achieved for 7 images from the Bickley Diary dataset [3] shortly described below.

As in one of our earlier papers [13] the extension of Niblack's method for the region based approach has been presented and compared with popular adaptive methods, the results obtained using its further improved version has also been compared with the approach proposed in this paper. However, as shown

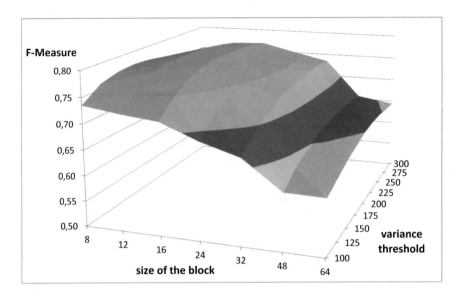

Fig. 1. Average F-Measure values for region based Otsu method for various block size and variance threshold.

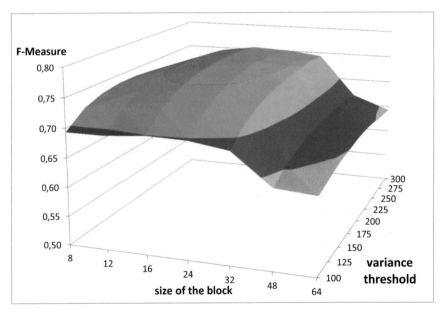

Fig. 2. Average F-Measure values for region based Kapur method for various block size and variance threshold.

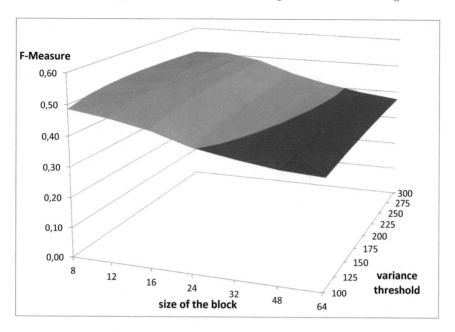

Fig. 3. Average F-Measure values for region based Rosin method for various block size and variance threshold.

in Table 1, better results for the considered demanding dataset can be obtained using the "classical" adaptive Niblack's method.

The main contribution of the proposed novel binarization method is related mainly to the optimization of the parameters, such as block size and variance threshold and verification of its usefulness for strongly distorted document images. The proposed solution extends the idea proposed by Chou [2] leading to much better results for them with comparable computational complexity - still much less than popular adaptive thresholding algorithms. In comparison with Chou's algorithm the proposed approach does not require the use of SVMs and the choice of its parameters can be made after the additional initial analysis of the image e.g. allowing to detect the size of the text lines.

Since the proposed approach has been developed for highly degraded historical document images, its verification using popular DIBCO datasets [18] has been replaced by much more challenging Bickley Diary dataset [3], similarly as e.g. in the paper written by Su et al. [24] where a robust image binarization based on adaptive image contrast is proposed. Although this method can be considered as state-of-the-art, its computational demands are quite high due to necessary image contrast construction, detection of stroke edge pixels using Otsu's global thresholding method followed by Canny's edge filtering, local threshold estimation and additional post-processing.

The Bickley Diary dataset contains 92 grayscale images being the photocopies of a diary written about 100 years ago by the wife of Bishop George H. Bickley

Table 1. F-Measure values obtained for 7 images from Bickley Diary dataset using various binarization methods

Binarization method	Image no.							Average F-Measure
	5	18	30	41	60	74	87	
Niblack	0.72	0.76	0.78	0.71	0.73	0.76	0.85	0.75
Sauvola	0.63	0.62	0.66	0.54	0.60	0.59	0.80	0.63
Wolf	0.60	0.58	0.61	0.46	0.53	0.55	0.77	0.59
Bradley (mean)	0.58	0.62	0.65	0.62	0.66	0.68	0.78	0.66
Bradley (Gaussian)	0.56	0.59	0.63	0.58	0.63	0.64	0.76	0.63
modified region Niblack	0.63	0.65	0.68	0.63	0.66	0.69	0.79	0.68
Chou [2]	0.52	0.51	0.57	0.46	0.50	0.57	0.71	0.55
global Otsu	0.47	0.48	0.54	0.43	0.45	0.48	0.67	0.50
global Kapur	0.47	0.50	0.54	0.39	0.44	0.50	0.65	0.50
global Rosin	0.32	0.28	0.30	0.25	0.28	0.28	0.31	0.29
region Otsu	**0.76**	**0.78**	**0.82**	**0.72**	0.77	**0.76**	**0.87**	**0.78**
region Kapur	0.71	0.74	0.79	0.71	**0.79**	**0.76**	0.83	0.76
region Rosin	0.50	0.51	0.53	0.52	0.53	0.51	0.52	0.52

- one of the first missionaries in Malaysia. The challenges in this dataset are related to discolorization and water stains, differences in ink contrast observed for different years as well as additional overall noise caused by photocopying. Nevertheless, only 7 of the images have been binarized manually and may be used as "ground-truth" images with additional annotations using the PixLabeler software. Therefore, all the results will be presented only for those 7 images (having their "ground truth" equivalents) to make them comparable with the other methods.

Analyzing the obtained results for the proposed region based Otsu method it can be noticed that achieved F-Measure value of 0.7835 is only slightly worse than the result reported by Su [24] (F-Measure equal to 0.7854) with much lower computational complexity of the proposed method.

The comparison of the F-Measure results obtained using the proposed methods with their global equivalents and some popular adaptive binarization methods introduced by Niblack [15], Sauvola [21], Wolf [26] and Bradley [1] together with its modification by using the Gaussian window is presented in Table 1.

Some results obtained for images from the Bickley Diary dataset are presented in Figs. 4, 5 and 6. Since the Bickley Diary dataset contains the additional 92 binary images prepared using the interactive Binarizationshop software [3], as the additional verification of similarity of the obtained results with them the F-Measure values have been calculated assuming the binary images provided in the dataset as the reference being "nearly ground truth" ones. Such obtained

Fig. 4. Binarization results obtained for image no. 5. - input image, "ground truth" (top), global Otsu, region Kapur (middle row), region Rosin and region Otsu (bottom).

Fig. 5. Binarization results obtained for image no. 18. - input image, "ground truth" (top), global Otsu, region Kapur (middle row), region Rosin and region Otsu (bottom).

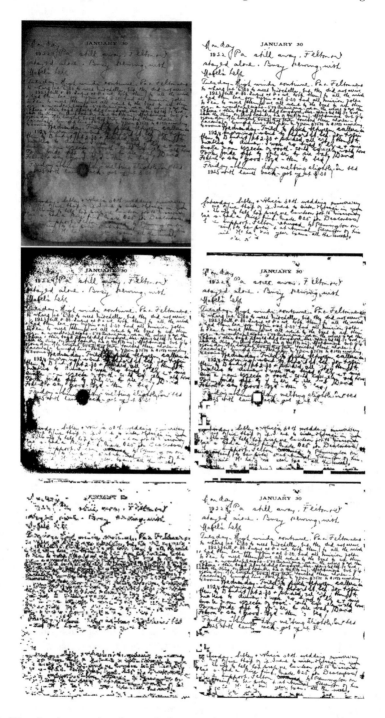

Fig. 6. Binarization results obtained the image no. 30. - input image, "ground truth" (top), global Otsu, region Kapur (middle row), region Rosin and region Otsu (bottom).

Table 2. Additional F-Measure values obtained for 92 images from Bickley Diary dataset assuming the provided binary images as the reference

Binarization method	Average F-Measure against "nearly ground truth" images
Niblack	**0.8441**
Sauvola	0.7305
Wolf	0.6973
Bradley (mean)	0.7097
Bradley (Gaussian)	0.6904
modified region Niblack	0.7467
Chou [2]	0.6456
global Otsu	0.5960
global Kapur	0.5891
global Rosin	0.3026
region Otsu	**0.8209**
region Kapur	**0.7688**
region Rosin	0.4980

average results for global and region based histogram thresholding are shown in Table 2.

Regardless of the non-optimality of the provided reference binary images the increase of the performance for the region based methods can be clearly visible as the obtained results are much closer to the reference ones. Analysing the output images provided by three considered region based methods, some disadvantages of the region based Rosin binarization can be noticed. Although the F-Measure values have increased in comparison to the application of the global Rosin thresholding, the shapes of individual characters on the images have been lost. The reason of such situation is the specificity of the algorithm dedicated for unimodal histogram images whereas the local distortion of image brightness is different. Therefore a reasonable choice is only the application of Otsu and Kapur methods with the proposed scheme. However, the closest results to the application of Binarizationshop have been achieved using Niblack's adaptive thresholding.

For further verification of the proposed algorithm for less demanding images, well known DIBCO datasets [18] have been used. The application of the proposed method for such images has led to results similar to those obtained using global Otsu, Niblack and Chou [2] methods. However, due to the optimization of parameters conducted using the Bickley Diary dataset as well as the presence of some images with much larger fonts and different types of usually slighter distortions, the results obtained for them are worse. The adaptation of the proposed method for various document images with recognition of text lines and their height would be much more computationally demanding and will be considered in our future research.

3 Concluding Remarks

The region based approach proposed in the paper allows to achieve good binarization performance in terms of F-Measure values, especially using Otsu's local thresholding with additional removal of low variance regions. The choice of the appropriate block size together with the variance threshold leads to the results close to state-of-the-art binarization algorithms preserving the low computational complexity of the proposed approach.

Since the results achieved applying region based approach for Kapur thresholding are only slightly worse and for some of the images can be even better, our further research will concentrate on the combination of both methods to develop a hybrid region based algorithm leading to even better binarization performance of highly degraded historical document images.

References

1. Bradley, D., Roth, G.: Adaptive thresholding using the integral image. J. Graph. Tools **12**(2), 13–21 (2007)
2. Chou, C.H., Lin, W.H., Chang, F.: A binarization method with learning-built rules for document images produced by cameras. Pattern Recognit. **43**(4), 1518–1530 (2010)
3. Deng, F., Wu, Z., Lu, Z., Brown, M.S.: Binarizationshop: a user assisted software suite for converting old documents to black-and-white. In: Proceedings of the Annual Joint Conference on Digital Libraries, pp. 255–258 (2010)
4. Feng, M.L., Tan, Y.P.: Adaptive binarization method for document image analysis. In: Proceedings of the 2004 IEEE International Conference on Multimedia and Expo (ICME), vol. 1, pp. 339–342, June 2004
5. Gatos, B., Pratikakis, I., Perantonis, S.: Adaptive degraded document image binarization. Pattern Recognit. **39**(3), 317–327 (2006)
6. Kapur, J., Sahoo, P., Wong, A.: A new method for gray-level picture thresholding using the entropy of the histogram. Comput. Vis. Graph. Image Process. **29**(3), 273–285 (1985)
7. Khurshid, K., Siddiqi, I., Faure, C., Vincent, N.: Comparison of Niblack inspired binarization methods for ancient documents. In: Document Recognition and Retrieval XVI, vol. 7247, pp. 7247–7247-9 (2009)
8. Kulyukin, V., Kutiyanawala, A., Zaman, T.: Eyes-free barcode detection on smartphones with Niblack's binarization and Support Vector Machines. In: Proceedings of the 16th International Conference on Image Processing, Computer Vision, and Pattern Recognition (IPCV 2012) at the World Congress in Computer Science, Computer Engineering, and Applied Computing WORLDCOMP, vol. 1, pp. 284–290. CSREA Press, July 2012
9. Lech, P., Okarma, K.: Fast histogram based image binarization using the Monte Carlo threshold estimation. In: Chmielewski, L.J., Kozera, R., Shin, B.S., Wojciechowski, K. (eds.) Computer Vision and Graphics. Lecture Notes in Computer Science, vol. 8671, pp. 382–390. Springer, Cham (2014)

10. Lech, P., Okarma, K.: Optimization of the fast image binarization method based on the monte carlo approach. Elektronika Ir Elektrotechnika **20**(4), 63–66 (2014)
11. Lech, P., Okarma, K.: Prediction of the optical character recognition accuracy based on the combined assessment of image binarization results. Elektronika Ir Elektrotechnika **21**(6), 62–65 (2015)
12. Leedham, G., Yan, C., Takru, K., Tan, J.H.N., Mian, L.: Comparison of some thresholding algorithms for text/background segmentation in difficult document images. In: Proceedings of the 7th International Conference on Document Analysis and Recognition, ICDAR 2003, pp. 859–864, August 2003
13. Michalak, H., Okarma, K.: Fast adaptive image binarization using the region based approach. In: Silhavy, R. (ed.) Artificial Intelligence and Algorithms in Intelligent Systems. Advances in Intelligent Systems and Computing, vol. 764, pp. 79–90. Springer, Cham (2019)
14. Moghaddam, R.F., Cheriet, M.: AdOtsu: an adaptive and parameterless generalization of Otsu's method for document image binarization. Pattern Recognit. **45**(6), 2419–2431 (2012)
15. Niblack, W.: An Introduction to Digital Image Processing. Prentice Hall, Englewood Cliffs (1986)
16. Okarma, K., Lech, P.: Fast statistical image binarization of colour images for the recognition of the QR codes. Elektronika Ir Elektrotechnika **21**(3), 58–61 (2015)
17. Otsu, N.: A threshold selection method from gray-level histograms. IEEE Trans. Syst. Man Cybern. **9**(1), 62–66 (1979)
18. Pratikakis, I., Zagoris, K., Barlas, G., Gatos, B.: ICDAR 2017 Document Image Binarization COmpetition (DIBCO 2017) (2017). https://vc.ee.duth.gr/dibco2017/
19. Rosin, P.L.: Unimodal thresholding. Pattern Recognit. **34**(11), 2083–2096 (2001)
20. Samorodova, O.A., Samorodov, A.V.: Fast implementation of the Niblack binarization algorithm for microscope image segmentation. Pattern Recognit. Image Anal. **26**(3), 548–551 (2016)
21. Sauvola, J., Pietikäinen, M.: Adaptive document image binarization. Pattern Recognit. **33**(2), 225–236 (2000)
22. Saxena, L.P.: Niblack's binarization method and its modifications to real-time applications: a review. Artif. Intell. Rev., 1–33 (2017)
23. Shrivastava, A., Srivastava, D.K.: A review on pixel-based binarization of gray images. Advances in Intelligent Systems and Computing, vol. 439, pp. 357–364. Springer, Singapore (2016)
24. Su, B., Lu, S., Tan, C.L.: Robust document image binarization technique for degraded document images. IEEE Trans. Image Process. **22**(4), 1408–1417 (2013)
25. Wen, J., Li, S., Sun, J.: A new binarization method for non-uniform illuminated document images. Pattern Recognit. **46**(6), 1670–1690 (2013)
26. Wolf, C., Jolion, J.M.: Extraction and recognition of artificial text in multimedia documents. Form. Pattern Anal. Appl. **6**(4), 309–326 (2004)
27. Yoon, Y., Ban, K.D., Yoon, H., Lee, J., Kim, J.: Best combination of binarization methods for license plate character segmentation. ETRI J. **35**(3), 491–500 (2013)

Partial Face Images Classification Using Geometrical Features

Piotr Milczarski[1]([⊠]) [iD], Zofia Stawska[1] [iD], and Shane Dowdall[2]

[1] Faculty of Physics and Applied Informatics, University of Lodz, Pomorska Str.
149/153, Lodz, Poland
{piotr.milczarski,zofia.stawska}@uni.lodz.pl
[2] Department of Visual and Human Centred Computing,
Dundalk Institute of Technology, Dundalk, Co. Louth, Ireland
shane.dowdall@dkit.ie

Abstract. In the paper, we have focused on the problem of choosing the best set of features in the task of gender classification/recognition. Choosing a minimum set of features, that can give satisfactory results, is important in the case where only a part of the face is visible. Then, the minimum set of features can simplify the classification process to make it useful in video analysis, surveillance video analysis as well as for IoT and mobile applications. We propose four partial view areas and show that the classification accuracy is lower by maximum 5% than in using full view ones and we compare the results using 5 different classifiers (SVM, 3NN, C4.5, NN, Random Forrest) and 2 test sets of images. That is why the proposed areas might be used while classifying or recognizing veiled or partially hidden faces.

Keywords: Geometric facial features · Image processing · Surveillance video analysis · Biometrics · Gender classification · Support vector machines K-Nearest neighbors · Neural networks · Decision tree · Random forrest

1 Introduction

In the facial images processing we have often problem of obscure or partially visible face. In the current paper, we search for points of the face that are the best for gender classification. We show the conditions for facial features to achieve higher accuracy in case of whole face and partial face visibility.

The problem of gender classification using only partial view was described by many authors. They was taking into account different parts of faces and acquisition conditions [5]. The authors used lower part of the face [8], top half of the face [2], veiled faces [9], periocular region [3, 14, 17] or they taking into account multiple facial parts such as lip, eyes, jaw, etc. [13]. The results reported by the authors are within 83.02–97.55% accuracy depending on the chosen method of classification and training dataset. The best results have been shown by Hassanat [9] for veiled faces. He obtained 97.55% accuracy using Random Forest and his own database. Some results with lower accuracy were shown by Lyle [14] and Hasnat [8]. The first one tested periocular region and obtained 93% accuracy, the second author used only lower part of the face

© Springer Nature Switzerland AG 2019
J. Pejaś et al. (Eds.): ACS 2018, AISC 889, pp. 445–457, 2019.
https://doi.org/10.1007/978-3-030-03314-9_38

and reported similar result about 93%. Both of them used SVM as a classification method. Using top half of the face Andreau [2] had got about 88% of accuracy using Near Neighbors method, and Kumari [13] reported 90% accuracy for the multiple facial parts (lip, eyes, jaw, etc.) using Neural Networks. They both used known FERET database as a training set. The worst results we can observed for the periocular region (Merkow [17] – 84.9% and Castrillon-Santana [3] – 83.02%). As a training set first author used web database and second one images of groups.

Gender can be recognized using many different human biometric features such as silhouette, gait, voice, etc. However, the most-used feature or part of the body is human face [11, 12, 20, 27]. We can distinguish two basic approaches for the gender recognition problem [11]. The first one takes into account the full facial image (set of pixels). Then, after pre-processing, that image is a training set for the classifier (appearance-based methods). In the second approach (the feature-based one), the set of face characteristic points is used as a training set.

In our research, we decided to use geometric face features to limit computational complexity. The tests confirmed that the acceptable (no different more than 5%) behavior will be observed in gender classification using the partial-view subsets of geometrical points/distances.

Many various classification methods can be used in a gender recognition task. The most popular classification methods include:

- neural networks [7],
- Gabor wavelets [26],
- Adaboost [23],
- Support Vector Machines [1],
- Bayesian classifiers [6, 24],
- Random Forest [9].

For our research we chose the most frequently used classification methods – Support Vector Machines (SVM), neural networks (NN), k-nearest neighbors (kNN), Random Forest (RF) and C 4.5.

The paper is organized as follows. In Sect. 2 datasets using in the research are described. In Sect. 3 the description of a facial model based on geometrical features scalable for the same person is presented. In Sect. 4 a general gender classification method description using different classifiers is described. In Sect. 5 the results of the research are shown. A deeper analysis of the obtained results can be found in Sect. 6 as well as the paper conclusions.

2 Datasets of Images Used in the Research

In the works presented above, it can be observed that the results may depend on the choice of the database. Some authors train their classifiers on the most popular databases as FERET database [22], others use their own databases sometimes built from e.g. web pictures. It can decide about obtained results.

In our research we decided to use:

- as a training set – a part of AR face database containing frontal facial images without expressions and a part of face dataset prepared by Angélica Dass in Humanæ project, jointly 120 cases,
- as a testing – 2 sets: 80 cases of Angélica Dass in Humanæ project and the Internet dataset consisting of 80 cases

The AR face database [16], prepared by Aleix Martinez and Robert Benavente, contains over 4,000 color images of 126 people's faces (70 men and 56 women). Images show frontal view faces with different illumination conditions and occlusions (sun glasses and scarfs). The pictures were taken at the laboratory under strictly controlled conditions, they are of 768×576 pixel resolution and of 24 bits of depth.

Humanæ project face dataset [10] contains over 1500 color photos of different faces (men, women and children). There are only frontal view faces, prepared in the same conditions, with resolution 756×756 pixels.

We have also created the Web repository for our research. The Web repository has been prepared from frontal facial images that are accessible in Internet. It contains 80 image files have different resolutions e.g. small ones: 300×358, 425×531, 546×720, 680×511, 620×420 etc., and big ones: 1920×1080, 2234×1676, etc. It is assumed that they have been taken in different conditions by different cameras.

In result, 92 frontal facial images from AR dataset, 108 images from Humanæ project dataset and 80 pictures taken from various random Internet pages were used in research.

The used classification data set consists of 280 images of females and males:

- 140 females – 49 from AR database, 51 from Humanæ project dataset and 40 pictures taken from various random Internet pages;
- 140 males – 43 from AR database, 57 from Humanæ project dataset and 40 pictures taken from various random Internet pages.

In the previous paper [28], we used 120 of the images as a training set. In that paper we used cross-validation as a testing method because of a small number of the cases.

In the current paper we will use additional eighty images from Humanæ project and eighty Internet images as two separate test sets to achieve more objective results.

3 Description of Facial Model

As we described in Sect. 2, we used a database of images made from two different available face databases: the AR database (92 cases), which we initially used, contains a small number of faces, therefore we extended this set by a number of cases from a Humanæ project dataset (28 cases). As Makinen pointed out in [15], training the classifier on photos from only one database, made in the same, controlled conditions, adjusts the classifier to a certain type of picture. As a result, we achieved more justified and objective classification results testing classifier with a set of photos from another source, e.g. from the Internet.

In our research we took into account 11 facial characteristic points (Fig. 1):

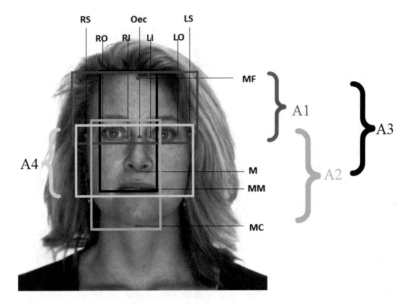

Fig. 1. Face characteristic points [18, 19] (image from AR database [16]).

- RI – right eye inner, LI - left eye inner,
- Oec, the anthropological facial point, that has coordinates derived as an arithmetical mean value of the points RI and LI [19],
- RO - right eye outer, LO - left eye outer,
- RS and LS – right and left extreme face point at eyes level,
- MF- forehead point – in the direction of facial vertical axis defined as in [18] or in [19].
- M – nose bottom point/philtrum point,
- MM – mouth central point,
- MC – chin point,

Points were marked manually on each image. These features were described in [18, 19] and are only a part of facial geometric features described in [6]. The coordinates are connected with the anthropological facial *Oec* point, as a middle point. The points and distance values are scaled in Muld [19] unit equal to the diameter of the eye. The diameter of the eye does not change in a person older than 4–5 years [21] and it measures:

$$1 \text{ Muld} = 10 \pm 0.5 \text{ mm} \tag{1}$$

That is why the facial model is always scalable, so the values for the person are always the same.

The chosen points allow us to define 11 distances which are used as the features in the classification process. The name and ordinal number are used interchangeably.

The names of the distances are identical with the name of the point not to complicate the issue, and they are:

1. MM – distance between anthropological point and mouth center.
2. MC – distance between anthropological point and chin point.
3. MC-MM – chin/jaw height.
4. MC-M – distance between nose-end point and chin point.
5. RSLS – face width at eye level.
6. ROLO – distance between outer eye corners.
7. MF-MC – face height.
8. M – distance between anthropological point and nose bottom point/philtrum point.
9. MF – distance between anthropological point and forehead point.
10. RILI – distance between inner eye corners.
11. MM-M – the distance between mouth center and philtrum point.

All the facial characteristic points were taken manually, in the same conditions, using the same feature point definitions and the same tool. The accuracy of the measurements is ± 1 px. The error of measurement is estimated as less than 5% because of the images resolution and the eyes' sizes.

The above 11 features have been chosen taking into account the average and variance values for males and females. That set of features have an anthropological invariances e.g. the outer eyes corners cannot be change, as well as the size of nose and MM distance, etc. To avoid diversions in chin distance (MC), we took closed-mouth faces only.

In the research:

- we have tested classification efficiency using subsets of the set of features described above,
- we have looked for a minimal set of features that give the best classification results,
- we want to check which of partial view areas A1, A2, A3 or A4 from Fig. 1 gives comparable accuracy with the full view area.

To achieve scalable face, the right or left eye, *Oec*, RI and LI have to be present in each partial image.

Below the numbers that are here corresponds to the ones from the list of the distances and are presented in the ascending order. The areas are defined as subsets of the facial distances (or the half of distances) as follows:

- A1 – {RILI, ROLO, RSLS, MF} or using numbers {5, 6, 9, 10};
- A2 – {RILI, ROLO, M, MM, MM-M, MC, MC-M, MC-MM} or {1, 2, 3, 4, 6, 8, 10, 11};
- A3 – {RILI, ROLO, MF, M, MM, MM-M} or {1, 6, 8, 9, 10, 11};
- A4 – {RILI, ROLO, RSLS, M, MM, MM-M} or {1, 5, 6, 8, 10, 11}.

That four areas have the same subset of the facial points and distances {RILI, ROLO}. The areas A2, A3 and A4 have the common subset of the facial points and distances {RILI, ROLO, M, MM, MM-M} or {1, 6, 8, 10, 11}. They consist of 1980 combinations and calculations altogether. That is why we used first of all SVM with RBF kernel function and k-fold cross-validation to search for the most efficient feature sets.

The classification accuracy is the ratio of correctly classified test examples to the total number of test examples, in our case it is similar to its general definition given by formula:

$$Accuracy = TP/(TP + FP) \qquad (2)$$

where TP and FP stands for true and false positive cases. We train and test the classifier on the different subsets defined in Sect. 3.

4 Classification Process Using Facial Model

In the initial research [28], we used Support Vector Machine (SVM) [4, 25] to find and choose the best feature subsets in gender classification. It appeared that the subsets consisting of 3–5 features gave 80% accuracy defined by (2) and 6 features give the best result 80.8% (see Table 1). At the beginning we have conducted several calculations to choose a proper kernel function. We tested our dataset using SVM with: radial basis function (RBF), linear, polynomial and sigmoid kernel functions. There were small differences between the results, but the RBF-kernel gave always the best results approx. at least 2% better than other kernels.

Table 1. The best sets of features for the whole image and the partial face view

No. of feat.	The best set of features for whole image	Accuracy (%)	The best set of partial view features	Accuracy (%)
1	(2) (4) (1)	68.3, 65.0, 64.1	(5) (6) (10)	54.2, 48.3, 55.8
2	(2,10) (1,9) (1,2) (1,10) (2,7)	74.2, 72.5, 71.7, 70.8, 70.8	(2,10) (1,2) (1,10)	74.17, 71.7, 70.8
3	(1,4,9) (1,2,7) (1,2,9) (1,4,7) (2,8,10)	80.0, 78.3, 77.5,77.5, 76.7	(1,8,10) A2 A3 A4 (5,6,10) A1 A4 (1,5,6) A4 (6,10,11) A2–4 (1,6,10) (5,6,8)	71.7, 65.0, 65.0, 65.0, 69.2, 68.3,
4	(1,2,4,9) (1,4,7,9) (1,2,8,9) (1,4,7,10)	80.0, 79.2, 78.3, 78.3	(1,6,8,10) A2 A3 A4 (6,8,10,11) A2–4 (5,6,9,10) A1 (5,6,10,11) A1A4 (1,5,6,10) (5,6,8,10) (6,8,9,10)	75, 74.2, 72.5, 70.8, 65.8, 64.1, 62.5

Table 1. (*continued*)

No. of feat.	The best set of features for whole image	Accuracy (%)	The best set of partial view features	Accuracy (%)
5	(1,4,7,8,9) (3,4,8,9,11) (1,2,4,7,9) (1,2,4,8,9) (1,3,4,8,9)	80.0, 80.0, 79.2, 78.3, 78.3	(1,4,6,8,10) A2 (1,6,8,10,11) A2, A3, A4 |(5,6,8,10,11) (1,5,6,8,10) (1,5,6,9,10) (5,6,8,9,10)	77.5, 73.3, 70.8, 68.2, 64.2, 62.5
6	(1,2,3,4,8,9) (1,4,5,9,10,11) (1,2,4,7,8,9) (1,3,4,7,8,9) (2,4,6,8,10,11)	80.0, 80.0, 79.2, 79.2, 79.2	(2,4,6,8,10,11)A2 (1,2,6,8,10,11) A2	79.2, 77.5

In the current paper, using Neural Networks (NN), decision tree C4.5, Random Forrest (RF) and *k*-Nearest Neighbour (*k*NN) methods, we will check the classification accuracy for the best SVM classifiers for the best previous feature subsets taking into account the whole and partial facial view. That would be the reference results for gender classification using previously defined partial view areas. We compare the classification results so as to choose the best method for the partial view images.

We build classifiers on j out of 11 features, where $1 \leq j \leq 11$ and systematically tried every combination of j features (the feature sets).

We also showed in [28] that the use of Leave-One-Out cross-validation or k-fold cross-validation gives the results that differ by 0.8%. Of course, Leave-One-Out cross-validation is much slower. That is why the following paper describes only the k-fold cross-validation method. It is defined as follows:

1. Take 5 female and 5 male cases from the entire data set and use these as the test set consisting of 120 cases (60 females, 60 males).
2. Use the remaining 110 cases (55 females and 55 males) as a training set.
3. A SVM classifier is then trained using the training set with the particular j features chosen and its Classification Rate, CR, is measured using the following:
   CR = (number of correctly classified cases in the test set)/10.
4. Steps 1, 2 and 3 are then repeated 12 times, each time with different elements in the test set. As a result, each element of the data-set is used in exactly one test-set.
5. The overall accuracy for a feature set is taken as the average of the 12 classification rates. The results are shown in Table 1.
6. During each round the Humanæ dataset consisting of 80 new cases and the Internet dataset consisting of 80 cases are used. The classification accuracy is counted in each round for both testing sets separately. The results are presented in Tables 1 and 2.
7. After training and testing the partial classifiers described in the steps 1–6, the general classifier from all 120 cases is built and tested on the new Humanæ and the

Table 2. Results of SVM, C4.5, RF, NN and 3NN classifications

No of feat.	The best feature sets	Class.	Acc. inner cv [%]	Acc. Web [%]	Acc. Hum. [%]	The best partial-view feature sets	Class.	Acc. inner cv [%]	Acc. Web [%]	Acc. Hum. [%]
2	2, 10	SVM	74.2	61.3	67.5	1, 10	SVM	70.8	61.3	75.0
		NN	85.0	72.5	75.0		NN	75.8	61.3	71.3
		RF	100	62.5	68.8		RF	100	60.0	73.8
		C4.5	85.8	65.0	76.3		C4.5	73.3	61.3	82.5
		3NN	83.3	65.0	70.0		3NN	86.7	56.3	70.0
3	1, 4, 9	SVM	80.0	67.5	76.3	1, 8, 10	SVM	71.7	62.5	72.5
		NN	94.2	72.5	71.3		NN	81.7	68.8	70.0
		RF	100	65.0	73.8		RF	100	66.3	63.8
		C4.5	82.5	62.5	81.3		C4.5	73.3	61.3	82.5
		3NN	90.8	67.5	73.8		3NN	86.7	68.8	67.5
4	1, 2, 4, 9	SVM	80.0	66.3	80.0	1, 6, 8, 10	SVM	75.0	63.8	78.8
		NN	90.0	75.0	72.5		NN	83.3	60.0	80.0
		RF	100	65.0	70.0		RF	100	61.3	76.3
		C4.5	81.7	67.5	78.8		C4.5	73.3	61.3	82.5
		3NN	90.0	67.5	75.0		3NN	83.3	60.0	72.5
5	1, 4, 7, 8, 9	SVM	80.0	72.5	77.5	1, 4, 6, 8, 10	SVM	77.5	66.3	80.0
		NN	95.8	73.8	62.5		NN	89.2	61.3	70.0
		RF	100	62.5	76.3		RF	100	65.0	77.5
		C4.5	82.5	62.5	81.3		C4.5	80.0	63.8	81.3
		3NN	90.0	67.5	76.3		3NN	82.5	63.8	73.8

(continued)

Table 2. (*continued*)

No of feat.	The best feature sets	Class.	Acc. inner cv [%]	Acc. Web [%]	Acc. Hum. [%]	The best partial-view feature sets	Class.	Acc. inner cv [%]	Acc. Web [%]	Acc. Hum. [%]
5	3, 4, 8, 9, 11	SVM	80.0	72.5	75.0	2, 6, 8, 10, 11	SVM	78.3	67.5	77.5
		NN	95.0	76.3	63.8		NN	84.2	62.5	70.0
		RF	100	67.5	78.8		RF	100	62.5	75.0
		C4.5	84.2	57.5	76.3		C4.5	90.8	65.0	76.3
		3NN	85.0	73.8	68.8		3NN	80.0	62.5	73.8
6	1, 2, 3, 4, 8, 9	SVM	80.8	66.3	77.5	2, 4, 6, 8, 10, 11	SVM	79.2	77.5	91.3
		NN	93.8	72.5	55.0		NN	90.8	66.3	77.5
		RF	100	65.0	73.8		RF	100	62.5	76.3
		C4.5	81.7	67.5	80.0		C4.5	90.8	65.0	77.5
		3NN	85.0	85.0	72.5		3NN	85.0	60.0	85.0
6	1, 4, 5, 9, 10, 11	SVM	80.0	72.5	75.0	1, 2, 6, 8, 10, 11	SVM	78.3	67.5	77.5
		NN	98.3	65.0	67.5		NN	94.2	60.0	77.5
		RF	100	61.3	77.5		RF	100	63.8	76.3
		C4.5	81.7	62.5	81.3		C4.5	90.8	65.0	76.3
		3NN	85.8	65.0	67.5		3NN	82.5	61.3	73.8

Internet datasets. Again, the classification accuracy is counted for both testing sets separately. The results are shown in Table 2.

8. Then, we choose the best feature subsets from all the combinations and for all combinations of the features defined for the partial areas A1, A2, A3 and A4. The results are shown in Tab. 2 in the right part.

9. In the final step, we use 3NN and NN classifiers to measure their accuracy on the subsets chosen in the Step. 8. The results are shown in Table 2.

5   Results of Classifications

5.1   Full Facial View Results of Classification

In Table 1 on the left, we show the best results of classification using k-fold cross-validation and SVM with radial basis kernel function (RBF).

We show in Table 1 that the best accuracy is achieved for six features 80.8%, but 3–5 element sets have only slightly lower accuracy 80.0%. It suggests that the classifier does not need a full set of features to achieve the best accuracy, so we can try to use some subsets of full facial features set.

SVM (RBF and k-fold cross-validation) results let us to pick the best classifier features. The results achieved using Random Forrest give always 100%. C4.5 NN are much better and sometimes they reach accuracy of 91–98%. 3NN results are usually better from 5 to 10.8% than SVM. They were usually worse than corresponding NN results.

While testing on external image sets (Web and Humanæ), SVM results are always worse than in the initial cross-validation by 0–6.7% in Humanæ case and 7.5–14.5% in Web case. The other classifiers, corresponding ones, usually gave bigger differences than in SVM case and they are worse even by almost 40% in RF case, 33% in NN case, 27% in C4.5 and 23% in 3NN case for Web and Humanæ cases.

The testing on Web subset shows usually the best accuracy NN up to 5 features than even in SVM case. The other classifier are rather worse than SVM.

We assumed before that classification on Humanæ subset will give better results than in Web case comparing with the works of Makinen.

5.2   Partial Facial View Results of Classification

In Table 1 in the right part, we show the best subsets of features based on partial facial views with their accuracy. Some of them might be used in the context of a partial view. Some of the best sets need the whole face, although it consists only few (e.g. 3) features.

From the Tables 1 and 2 it can be derived that:

- SVM (RBF and k-fold) cross-validation results were usually lower by 2–4% than SVM results for the full facial image. The classification rate usually is around 75–80%.

- The results achieved using Random Forrest show 100% accuracy on the training set. But the accuracy measured on Humanæ subsets varies from 69 to 79% for the whole view and 64–78% for the partial view. That is comparable with the results of SVM classification on the same Humanæ subset. The accuracy measured on the Web images is usually lower even up to 16% in a whole and partial view cases.
- C4.5 classifier gives the best or comparable classification results using Humanæ test sets. While using it on Web images the results are the worst in most of the full and partial cases.
- Neural Network (NN) classifies by 4–15.9% better than in SVM cases and sometimes they reach 94.2%. But SVM gives usually better results while testing on Humanæ and Web subsets.
- 3NN results of classification are behaving quite chaotically, e.g. in a case of up to 4 features they usually give higher accuracy than NN but sometimes they show the best accuracy. Otherwise, for 4 and bigger subsets they have lower accuracy.
- The pre-assumption that classification on Humanæ subsets might give better results than in Web case was true.

For the area A2 feature subset we have achieved the best results for partial view image subsets.

Table 2 shows only the results for 2–6 features subsets. One feature is too little to be taken into account.

6 Conclusions

In the paper, we have shown that it is possible to derive subsets of the features that show satisfactory results for classification of the partial-view images using geometrical points and testing on a good quality image subset like Humanæ one. The method described in the paper used support vector machine as a starting point in gender classification based on full facial view and on the partial one in four chosen areas. After that, we have analyzed the performance/accuracy of four additional classifiers (C4.5, Random Forrest, kNN, Neural Network) and datasets with features extracted in the same way (Humanæ, Web).

It can be concluded from the results shown in the paper that the choice of the classifier is very important. Some of them like Random Forrest and NN show almost 100% accuracy while training. The other like SVM show rather stable accuracy. But while testing on two independent datasets of images taken in very different conditions and having random resolution (like Web set) we achieved that usually the tests show smaller accuracy than in training case. But in the case where the test images have similar properties as the training ones the results in SVM case are close, in the case of the other classifiers they can behave randomly. In the case of testing on the Web repository the results are usually around 65–70%, similar to Makinen results [15].

References

1. Alexandre, L.A.: Gender recognition: a multiscale decision fusion approach. Pattern Recogn. Lett. **31**(11), 1422–1427 (2010)
2. Andreu, Y., Mollineda, R.A., Garcia-Sevilla, P.: Gender recognition from a partial view of the face using local feature vectors. In: Pattern Recognition and Image Analysis. Springer Verlag (2009)
3. Castrillon-Santana, M., Lorenzo-Navarro, J., Ramon-Balmaseda, E.: On using periocular biometric for gender classification in the wild. Pattern Recogn. Lett. **82**, 181–189 (2016)
4. Cortes, C., Vapnik, V.: Support-vector network. Mach. Learn. **20**(3), 273–297 (1995)
5. Demirkus, M., Toews, M., Clark, J.J., Arbel, T.: Gender classification from unconstrained video sequences. In: Computer Vision and Pattern Recognition Workshops (CVPRW), 2010 IEEE Computer Society Conference on Computer Vision and Pattern Recognition, pp. 55–62 (2010)
6. Fellous, J.M.: Gender discrimination and prediction on the basis of facial metric information. Vision. Res. **37**(14), 1961–1973 (1997)
7. Fok, T.H.C., Bouzerdoum, A.: A gender recognition system using shunting inhibitory convolutional neural networks. In: The 2006 IEEE International Joint Conference on Neural Network Proceedings, pp. 5336–5341 (2006)
8. Hasnat, A., Haider, S., Bhattacharjee, D., Nasipuri, M.: A proposed system for gender classification using lower part of face image. In: International Conference on Information Processing, pp. 581–585 (2015)
9. Hassanat, A.B., Prasath, V.B.S., Al-Mahadeen, B.M., Alhasanat, S.M.M.: Classification and gender recognition from veiled-faces. Int. J. Biometrics **9**(4), 347–364 (2017)
10. Humanæ project. http://humanae.tumblr.com. Accessed 15 Nove 2017
11. Jain, A., Huang, J., Fang, S.: Gender identification using frontal facial images in multimedia and expo. In: IEEE International Conference on ICME 2005, p. 4 (2005)
12. Kawano, T., Kato, K., Yamamoto, K.: An analysis of the gender and age differentiation using facial parts. In: IEEE International Conference on Systems Man and Cybernetics, vol. 4, 10–12 October, pp. 3432–3436 (2005)
13. Kumari, S., Bakshi, S., Majhi B.: Periocular gender classification using global ICA features for poor quality images. In: Proceedings of the International Conference on Modeling, Optimization and Computing (2012)
14. Lyle, J., Miller, P., Pundlik, S.: Soft biometric classification using periocular region features. In: Fourth IEEE International Conference Biometrics: Theory Applications and Systems (BTAS) (2010)
15. Mäkinen, E., Raisamo, R.: An experimental comparison of gender classification methods. Pattern Recogn. Lett. **29**, 1544–1556 (2008)
16. Martinez, A.M., Benavente, R.: The AR Face Database. CVC Technical report #24 (1998)
17. Merkow, J., Jou, B., Savvides, M.: An exploration of gender identification using only the periocular region. In: Proceedings 4th IEEE International Conference Biometrics Theory Application System BTAS, pp. 1–5 (2010)
18. Milczarski, P.: A new method for face identification and determining facial asymmetry. In: Katarzyniak, R. (ed.) Semantic Methods for Knowledge Management and Communication, Studies in Computational Intelligence, vol. 381, pp. 329–340 (2011)
19. Milczarski, P., Kompanets, L., Kurach, D.: An Approach to brain thinker type recognition based on facial asymmetry. In: Rutkowski, L., et al. (eds.) ICAISC 2010, Part I, LNCS 6113, pp. 643–650 (2010)

20. Moghaddam, B., Yang, M.H.: Learning gender with support faces. IEEE Trans. Pattern Anal. Mach. Intell. **24**(5), 707–711 (2002)
21. Muldashev, E.R.: Whom Did We Descend From?. OLMA Press, Moscow (2002). (In Russian)
22. Phillips, P.J., Moon, H., Rizvi, S.A., Rauss, P.J.: The FERET evaluation methodology for face-recognition algorithms. IEEE Trans. Pattern Anal. Mach. Intell. **22**(10), 1090–1104 (2000)
23. Shakhnarovich, G., Viola, P.A., Moghaddam, B.: A unified learning framework for real time face detection and classification. In: Proceedings International Conference on Automatic Face and Gesture Recognition (FGR 2002), pp. 14–21. IEEE (2002)
24. Sun, Z., Bebis, G., Yuan, X., Louis, S.J.: Genetic feature subset selection for gender classification: a comparison study. In: Proceedings IEEE Workshop on Applications of Computer Vision (WACV 2002), pp. 165–170 (2002)
25. Vapnik, V.N., Kotz, S.: Estimation of Dependences Based on Empirical Data. Springer, New York (2006)
26. Wiskott, L., Fellous, J.M., Krüger, N., von der Malsburg, C.: Face recognition by elastic bunch graph matching. In: Sommer, G., Daniilidis, K., Pauli, J. (eds.) 7th International Conference on Computer Analysis of Images and Patterns, CAIP 1997, Kiel, pp. 456–463. Springer, Heidelberg (1997)
27. Yamaguchi, M., Hirukawa, T., Kanazawa, S.: Judgment of gender through facial parts. Perception **42**, 1253–1265 (2013)
28. Milczarski, P., Stawska, Z., Dowdall, S.: Features selection for the most accurate SVM gender classifier based on geometrical features. In: Rutkowski, L., et al. (eds.) ICAISC 2018, LNCS 10842, pp. 191–206 (2018)

A Method of Feature Vector Modification in Keystroke Dynamics

Miroslaw Omieljanowicz[1]([⊠]), Mateusz Popławski[2], and Andrzej Omieljanowicz[3]

[1] Faculty of Computer Science,
Bialystok University of Technology, Bialystok, Poland
m.omieljanowicz@pb.edu.pl
[2] Walerianowskiego 25/68 Kleosin, Bialystok, Poland
mateusz.poplawski@gmail.com
[3] Faculty of Mechanical Engineering,
Bialystok University of Technology, Bialystok, Poland
andrzejom@gmail.com

Abstract. The aim of this paper is to conduct research which will investigate the impact of diverse features in vector on the identification and verification results. The selection of the features was based on the knowledge gained from scientific articles publish recently. One of the main goals of this paper is to probe the impact factor of weights in feature vector which will later serve in biometric authentication system based on keystroke dynamics. The unique application allows end-user to customize the vector parameters, such as: type of the feature and weight of the feature, additionally finding optimization for each custom feature vector.

Keywords: Biometrics · Keystroke dynamics · Feature extraction
Human recognition · Security

1 Introduction

Over the centuries people have recognized each other on the basis of many different features, for example, by seeing a familiar face, you can determine who this person is [1]. If there is not enough certainty, other features such as voice, height, or even style of walking are taken into account. Confirmation of identity can be done in traditional way, on the basis of more perceptive knowledge or a random object which is owned by a person. It can be keys, magnetic cards, or the acquaintance of a certain PIN code or a password. In this paper authors focus on the behavioral method. In common such methods are less expensive in implementation, commonly do not require specialized hardware and in addition operate in the background without disturbing the user. Drawback of such systems is low repeatability of features. It is hard for human to repeat action in explicitly same way. It creates information noise which decrease the effectiveness of the system.

The goal is to find the method which will work with very high accuracy despite the low repeatability of features. This work focuses on method which focalize on

© Springer Nature Switzerland AG 2019
J. Pejaś et al. (Eds.): ACS 2018, AISC 889, pp. 458–468, 2019.
https://doi.org/10.1007/978-3-030-03314-9_39

identification and verification of people based on the way how people tend to type on the keyboard. The dynamics of typing [2] is a process of analyzing not what the user writes, but, how he does it. The data is being quasi-rhythmically entered by the person (usually on a computer keyboard or a mobile device with a touch screen) and is monitored in order to create a unique user template.

User profile can be created by using many different properties, such as: pace of writing, time between keystrokes, finger placement on the key button, force which is applied on the key button. Recognition of people using this technique is non-invasive, cost-efficient and invisible to users (data can be collected without user cooperation or even without their awareness). In addition, it is very easy to obtain data, default tool is a computer keyboard and user do not need any additional hardware. As behavioral biometrics, the dynamics of typing are not stable due to transient factors such as stress or emotions, and other external factors such as using keyboard with a different key layout. The main disadvantage of this technique is low efficiency compared to other biometric systems. The authors made an attempt to increase efficiency by introducing weights in a feature vector and presented the results of their experiments.

2 Related Works

The analysis of the keyboard typing dynamics has been developed since the early 1980s, where the accuracy of assessment at level of 95% was obtained, seven people took part in the research at that time. Over the next years, further research was carried out depending on the selection of the database, features, data acquisition devices or the classification method. The results obtained by scientists are very diverse, where the work from 2014 reaches equal error rate (EER) from 5% [3] to 26% [4], while from previous years articles specifying EER below 1% [2]. Undoubtedly, the obtained results depend not only on the chosen solutions (base, features, classifier, etc.), but also on the objectives of the research being carried out. Although many years of work on biometric systems has shown that the use of the analysis of the way of writing on digital devices for text input does not provide enough accuracy to be able to identify or verification of people. The attractiveness of simple and cheap implementation conclude that work is still carried out in multimodal systems [5] or as an addition to the basic system most often based on physical features, most often fingerprint and the last appearance of the face near the device. Generally, all biometric systems usually consist of functional blocks as: data collection, feature extraction, classification, matching and decision making.

In systems based on the dynamics of typing on the keyboard, the extraction of features consists of registering time dependencies between operations of pressing and releasing keys in various combinations. Typical determinations of individual intervals [2] are shown in Fig. 1.

The data is usually collected in the form of the registration of the event and its occurrence in time line. The time axis is typically scaled in microseconds and the event is written in the form of one, two-letter or three-letter abbreviations, i.e. P - press, R - release with the relevant data about which key was used. In biometrics based on the analysis of the writing method, there are no standards indicating what thresholds are to

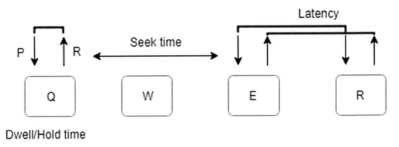

Fig. 1. Intervals naming in keystroke dynamics

be taken into account. Different authors use different feature vectors. The analogous situation concerns the selection of the classification method and decision making. The only common element is the assessment of the efficiency of such a system, without which it would be impossible to determine the practical suitability and the comparison to different solutions.

Various measures are used to assess the biometric system depending on its intended use. In the case of verification systems, it is common to use a number of efficiencies with the EER and two parameters, False Accept Rate (FAR) and False Rejection Rate (FRR). Identification systems are most often assessed using accuracy, understood as the ratio of correct identifications to the number of all attempts. The above-mentioned coefficients were used in this work to determine whether the introduction of weights into the feature vector allows improving the efficiency of verification and identification in systems based on the analysis of the typewriter style. To perform the experiments, the feature vectors proposed in the briefly further described selected literature items were used, starting with a feature vector based on two directly determined time types and ending with a vector of features based on four computed composite quantities.

2.1 Researches Carried Out by a Team from AGH University of Science and Technology

The main purpose of the described work [6] was to examine the impact of using different databases on the results achieved by biometric systems based on keystroke dynamics. In the article, the authors Piotr Panasiuk and Khalid Saeed present a general overview of the history of keyboard typos, describing selected methods and paying attention to modern solutions in the field of biometrics. Two databases (Ron Maxion and authors) and two biometric features were used during the tests:

- Time during which the button was hold (denoted as dwell).
- Time between releasing one and pressing the next key (labeled as flight).

The classification itself was carried out using the kNN classifier (k - nearest neighbors). Based on the selected number of neighbors k, a training and test set was created. In case where total number of samples of a given person is less than k + 1, the set it is not taken into account. The next step was the classification, during which the distance between the test sample and all samples of the training set was determined. To determine the distance, the authors used the Manhattan metric.

After determining the distance between the test sample and the training set samples, a decision-making process took place. Out of all results, the best ones were selected, and then the voting was performed. The shortest distance gets the highest weight equal to the number k, while the longest lowest one equals 1. Then the weights are added up within the same class. The sample is qualified to the person whose class received the most votes. The authors of the work [6] have concludes that results of the biometric systems based on keystroke are varied depending on the database and will rather not usable itself in practice unless will used in conjunction with some other biometric features.

2.2 Method Developed by Research Team from University of Buffalo

Researchers from the University of Buffalo: Hayreddin Çeker and Shambhu Upadhyaya presented in their work [7] a new adaptive classification mechanism, known as transferable learning. The main aim of the authors was to show that the use of adaptive techniques allows the identification of people at a later time using only a few samples from previous sessions. The work uses 31 values in a feature vector, where the following time values can be distinguished:

- Time during which the button was hold (H).
- Time between pressing one and pressing the next button (PP).
- Time between releasing one and pressing the next button (RP).

The main aim of the authors was to show that use of adaptive techniques allows the identification of people at a later time using only a few samples from previous sessions. The classification techniques proposed by the authors [7] are based on the Support Vectors machine (SVM). From conducted experiments it seems that adaptive techniques preponderate over classical methods, especially for small size samples, additionally it should be noted that the deviation values for adaptive algorithms are smaller than for the classical algorithm.

2.3 Method Proposed by Research Team from Kolhapur Technical University

The authors: Patil and Renke, in their work [8] point to the growing need to increase computer security in various types of Internet systems and show the simplicity of using the dynamics of writing in order to strengthen security at a low cost. When rewriting the given text, factors such as:

- Time interval between pressing and releasing the key
- Time interval from releasing one key to pressing the next
- Total time of pressing the key.

From such features, the authors [8] have built a vector consisting of four features. Two statistical values Mean (M) and Standard deviation (SD) were used to define these values. These features are:

- Average time interval between pressing and releasing the key - hold time H
- Average time interval from releasing one key to pressing the next
- Standard deviation from pressing to releasing the key
- Standard deviation from the release of one key to the next

In the classification process, the algorithm looks for differences between the current value in the database and the actual one obtained in the login process. The authors did not indicate a specific classifier used. In addition, the acceptance threshold is applied, which means that if the difference between the two samples falls within the threshold, it will be approved, otherwise the authentication will be rejected.

3 Experiments and Results

As part of the work, a dedicated software for data registration and the selection of features and their weights in the feature vector was created. The application registers a raw sample, what means - for each key struck, the time the key was pressed and the time the key was released, in msec. From the Windows-event clock. After that features time is calculated. The program allows user to examine the quality of classification. Research module offers a choice of more than 40 features with the ability to assign weights to generate a vector of features. The cross-validation method [9] was used to assess the quality of classification, while the classification itself was carried out using the weighted m-match, k-nearest-neighbors method as it was described in [10]. The distances between samples are determined by the Manhattan, Euclidean or Chebyshev metrics. The feature vector definition window is presented on Fig. 2.

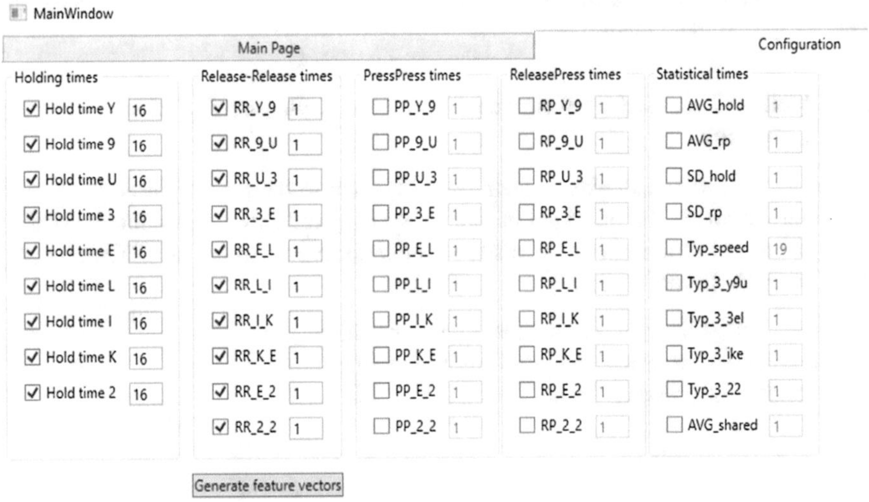

Fig. 2. Feature vector definition window

Based on the application prepared in this way, a series of experiments have been carried out to show whether the use of scales/weights in the feature vector will improve the quality assessment parameters of the identification process and the quality of the verification process. Using created application, a database of 770 raw samples of 16 persons was registered. The effectiveness of the classification was tested, the error of the FAR and FRR was checked using the cross-validation method of omitting individual elements. The classification was carried out using the weighted m-match k-NN algorithm [10] (where: used k was from 0 to 30 and m accordingly to k), while the Manhattan metric was used to determine the distance between the feature vectors. All tests were carried out using the same method, with a change only of the feature vector.

The main purpose of carried out tests is to investigate the impact of building the feature vector on the results. In addition, tests were carried out to analyze the application of the weighting system to the characteristics of the achieved results and also to find the optimal configuration. Three feature vectors used in the publications described in chapter 2 were selected for the study. They were used both in the tests of identification effectiveness and the verification process. The vectors that have been used are briefly characterized below.

3.1   Feature Vectors

Feature Vector Type 1 (Based on 2 Elements)
In the first approach, two properties were used to construct a feature vector, as presented in the work by Panasiuk-Saeed [6]. Both features are considered to be basic in the topic of the dynamics of typing, the first is the time in which the button is hold, in the test marked "H" (Hold). The second feature is the time between releasing one button and pressing the next one, marked with the "RP" (Release - Press) symbol. As a result of using these two quantities, the vector of traits consisted of 19 values.

$$FV = \{H_i, RP_{ij}, \ldots\ldots, H_2, RP_{22}\}, \tag{1}$$

where: H_i – time when button „i" is hold, RP_{ij} – time from releasing the "i" key till pressing "j" key.

Feature Vector Type 2 (Based on 3 Elements)
This feature vector is used by researchers from the University of Buffalo described in paper [7]. Compared to vector type 1 it contains one more feature. This is the time interval between pressing one and pressing the next button in tests marked as PP (Press - Press). Each sample is represented by 10 additional values. As a result of using these three properties, the feature vector consisted of 29 values.

$$FV = \{H_i, RP_{ij}, PP_{ij}, \ldots\ldots, H_2, RP_{22}, PP_{22}\};, \tag{2}$$

where: H_i – time when button "i" is hold, RP_{ij} – time from releasing the "i" key till pressing "j" key, PP_{ij} – time from pressing the "i" key till pressing "j" key.

Feature Vector Type 3 (Based on Statistical Values)
The third tested vector of features was used in the paper [8]. Its construction was based on statistical methods such as average value and standard deviation. These features can be described as:

- Average time intervals between pressing and releasing the key from all keys (labeled as avg_H)
- Standard deviation of time intervals between pressing and releasing the key among all keys (marked as sd_H)
- Average intervals between release and pressing the next key, all of the following keys (labeled as avg_RP)
- Standard deviation of the time intervals between release and pressing the next key, among all successive keys (marked as sd_RP).

As a result of using these four properties, the feature vector consisted of only four values.

$$FV = \{avg\_H, sd\_H, avg\_RP, sd\_RP\};\tag{3}$$

3.2   Identification Tests

The identification test consisted in determining the number of correctly classified samples in relation to the number of comparisons. Each input sample was compared to all samples in the database (i.e. more than 290 000 tests were made), the class obtaining the highest value is assigned to the classified sample, if the sample ID number agrees with the ID of the assigned class, the correct classification is considered. The experiments carried out for the abovementioned feature vectors are described below.

Identification Results for Feature Vector Type 1
The research began with determining the effectiveness of identification without applying the weights to the constituent vector components. The results achieved in this test are at level of 67,7% of proper identifications, which means that the selected vector of features is not suitable for effective identification. Subsequently, the weights of these features were manipulated. Increasing the weight of the H feature resulted in a significant improvement in the classification, weight increased to 5 resulted in a result of 79.38%. Raising the weight of the RP characteristic resulted in the reduction of correctly classified samples to just 61.74%. The best result for vector type 1 was achieved at 85.34%, with the weight of the H feature at level 16 and weight of the RP at level 1.

Identification Results for Feature Vector Type 2
Similarly, to type 1 vector, the tests began with determining the effectiveness of identification without introducing weights. The obtained result is 64.2% value. The best effect gave the weight increase of the time of pressing the key (H), the modification of the weight of this feature to 5 improved the efficiency of classification by 10.77% (to 74.97%). The highest classification efficiency was obtained by setting the weight of the H-time feature to 27. The best result was 85.34%, the same as with the type 1 vector, but this time it was necessary to significantly increase the weight of the feature.

Additional modifications of the weights of the remaining features did not give a better result. Increasing the weight of RP and PP features gave a negative effect. In the RP time interval, the best result was worse by 2.72% than that without changing the weights. By far the worst result was achieved by modifying the weight of the PP interval feature, as the best result with a weight 5 was only 59.27%, which is worse by 4.93% than that obtained in the basic configuration of features. Simultaneous modification of the weights of two features also gave mixed effects, in the case of increasing the weight to 5 of features RP and PP, the results deteriorated by 4.02%, the other modifications gave a slight improvement. Comparing the obtained results for vector type 2 with the type 1 vector, it can be concluded that the addition of the PP feature gave a negative effect.

Identification Results for Feature Vector Type 3

Tests made on the vector proposed by Patil and Renke [8] showed that using only these features does not give high results in the identification of people based on typing. The result obtained without using the weights of features is only 42.67%. Modification of the weights of individual traits gave mixed effects, we can observe a decrease in the classification efficiency by 9.86% by setting the weight of the sd_RP attribute (32.81%) to 5 or by improving the order of 3.63% by increasing the weight of the sd_H feature to 5 (46.3%). The best result was obtained by increasing the weight of the avg_H trait to 6 when 49.55% of correctly classified samples were obtained.

In general, it can be concluded that the introduction of weights into the feature vector allows for greater identification efficiency. However, improving the results requires choosing the size of the weight. At this stage of the work manual selection of the weight was made. The best effects occurred when introducing the weight into only one component of feature vector. The results are summarized in Table 1.

Table 1. Comparison of identification efficiency without and with the use of weights.

	The highest efficiency without using weights	The highest efficiency when using weights
Feature vector type 1	67,7%	85,34%
Feature vector type 2	64,2%	85,34%
Feature vector type 3	42,67%	49,55%

3.3   Verification Tests

During the verification tests, the number of incorrectly accepted (FAR) and incorrectly rejected (FRR) samples was examined. Each input sample was compared to all samples in the database (i.e. more that 290 000 tests were made), the class obtaining the highest value is compared with the currently set sensitivity threshold, if the threshold value is exceeded and the classes of both samples are the same the number of correct classifications is increased. If the class value is lower than the threshold and the sample classes are the same, the number of incorrectly rejected samples is increased. Trials treated as incorrectly accepted occur when the sensitivity threshold is reached, and the sample classes are different. During the tests, a sensitivity test was looked for at which the FAR and FRR error rates had a similar value and at the same time reached the minimum. In the first step, the values were determined without using weights. The values thus determined were then treated as a comparison/reference level for the situation using component weights in the feature vector. In experiments with the use of weights, the threshold value (sensitivity - s) was determined using the following formula: $s(s - 1)/2$, in the range $s = \langle 1, ..., 13 \rangle$ (over 13 worse results were obtained).

Verification Results for Feature Vector Type 1
To sum up the results achieved in the first verification test, significant differences in the achieved FAR and FRR values should be noted depending on the weight attributes assigned to them. Starting from the results without changing the weight of the features, where the FAR was obtained at 12.58% and the FRR of 12.32%, the modification of the weight of the RP feature caused an increase in both types of errors. Increasing the weight of the second characteristic (H) gave a very positive effect, the number of FAR and FRR errors decreased significantly, reaching FAR and FRR coefficients to 9.08% and 7.91%, respectively. The best result that could be obtained with feature vector type 1 was found when the weight of the H-feature was 17. The FAR was obtained with a value of 6.61% and the FRR of 6.49%. The tests clearly showed that the change in the weight of the trait can have both positive and negative effects on the results achieved but there is combination of weight where results are significantly better.

Verification Results for Feature Vector Type 2
Similarly, to the type 1 feature vector, the experiments were started with the determination of FAR and FRR values without using weights. The values were 12.58% and 12.32%, respectively. In experiments with the manipulation of weights of all characteristics, the best results were definitely obtained after increasing the weights of the H and PP features, leaving the RP feature in the weight 1. The lowest values were obtained at the threshold of 22, where FAR reached 11.8% and FRR 12.45%. In addition, we searched for balance settings, threshold sensitivity and the number of neighbors at which the FAR and FRR values were as small as possible. The best result obtained are both FRR and FAR values at 6.49%, with 7 neighbors and sensitivity threshold at level 16, which means a much better result than the situation of non-use of balances.

Verification Results for Feature Vector Type 3
Similarly, as in the case of the study of the feature vector type 1 and the feature vector type 2, the experiments were started with the determination of the FAR and FRR values without using weights. The values were respectively 14.79% and 15.56%. Further research was carried out with increasing the weights of each of the features. Increasing the weight for each of the features simultaneously gave a negative effect, in the case of each modification there was an increase in the number of errors of both types. The worst result was obtained after the weighting of the sd_H feature to 5, where the FAR increased by 2.33% to 17.12%, and the FRR increase by 1.95% to 17.51%. The best result that was obtained with this feature vector concerned the increase in the weight of the avg_H trait to 10, while the others with weights 1. A FAR value of 13.88% and an FRR of 15.05% were obtained. This

is an improvement of both parameters by 0.91% for FAR and 0.51% for FRR, respectively, compared to the situation without using weights.

Similarly, as in the identification process, it is possible to state that the introduction of weights into the feature vector allows for significant improvement (in some type of feature vectors) of verification systems based on keystroke dynamics. At this stage of the work manual selection of weighs was made. The results are summarized in Table 2.

Table 2. Comparison of FAR and FRR error values without and with the use of weights

	Lowest FAR/FRR without weights	Lowest FAR/FRR with weights
Feature vector type 1	12,58%/12,32%	6,61%/6,49%
Feature vector type 2	12,58%/12,58%	6,49%/6,49%
Feature vector type 3	14,79%/15,56%	13,88%/15,05%

4   Conclusion

Generally, it should be stated that the introduction of weights into the feature vector, both during the identification process and the verification process has a significant effect on the effectiveness of both processes. The results obtained during the identification tests are very divergent, depending on the applied vector of features or the attribution of appropriate weights to the traits one can notice a significant improvement or a worsening of the effectiveness of identification.

The analysis of the results obtained clearly shows the impact of the use of different vector features and the selection of appropriate weights on the effectiveness of the keystroke biometric system achieved. The most important conclusion is that the use of weights in the feature vector gives an improvement in the coefficients of the quality of identification and verification, as shown in Tables 1 and 2.

In the presented work, the selection of weights was performed manually until local maximum was noticed. An important conclusion is also that even with the manual manipulation of weights in the vector of features, it was possible to observe the occurrence of a local extreme. This indicates the direction of further work - introducing the algorithm (from machine learning or statistical methods) for selecting weights, which may allow to find a set of weights allowing further improvement of the quality of the identification and verification process.

Promising results of performed experiments also indicate the need to extend the research to a larger number of feature vectors as well as with a larger amount of processed data. The application made for the needs of the research will be used in further works to collect a larger amount of test data and supplemented by an automatic algorithm for selecting weights in the feature vector. The authors hope that it will also be possible to introduce learning mechanisms to the algorithm of selecting the weights of features and thus to further improve the quality of identification and verification systems based on keystroke dynamics.

Acknowledgements. The research has been done in the framework of the grant S/WI/3/2018 Bialystok University of Technology.

References

1. Ríha, Z., Matyáš, V.: Biometric Authentication Systems, Faculty of Informatics Masaryk University (2000)
2. Liakat, A. Md., Monaco, J.V., Tappert, C.C., Qiul, M.: Keystroke Biometric Systems for User Authentication. Springer Science Business Media, New York (2016)
3. Wankhede, S.B., Verma, S.: Keystroke dynamics authentication system using neural network. Int. J. Innovative Res. Dev. 3(1), 157–164 (2014)
4. Bours, P., Masoudian, E: Applying keystroke dynamics on one-time pincodes. In: International Workshop on Biometrics and Forensics (IWBF) (2014)
5. Szymkowski, M., Saeed, K.: A multimodal face and fingerprint recognition bio-metrics system. In: Lecture Notes in Computer Science, vol. 10244, pp. 131–140 (2017)
6. Panasiuk, P., Saeed, K.: Influence of database quality on the results of keystroke dynamics algorithms. In: Chaki, N., Cortesi, A. (eds.) Computer Information Systems – Analysis and Technologies. Communications in Computer and Information Science, vol. 245. Springer, Berlin, Heidelberg (2011)
7. Hayreddin, Ç., Upadhyaya, S.: Adaptive techniques for intra-user variability in keystroke dynamics. In: IEEE 8th International Conference Biometrics Theory. Applications and Systems (BTAS) (2016)
8. Patil, R.A., Renke, A.L.: Keystroke Dynamics for User Authentication and Identification by using Typing Rhythm. International Journal of Computer Applications (0975 – 8887), vol. 144 – No. 9, June 2016
9. Payam, R.Z., Lei, T., Huan, L.: Cross-Validation. In: Encyclopedia of Database Systems, pp. 532–538. Arizona State University, Springer, USA (2009)
10. Zack, R.S., Tappert, C.C., Cha, S.-H.: Performance of a long-text- input keystroke biometric authentication system using an improved k-nearest-neighbor classification method. In: Fourth IEEE International Conference on Biometrics: Theory, Applications and Systems, pp. 1–6 (2010)

Do-It-Yourself Multi-material 3D Printer for Rapid Manufacturing of Complex Luminaries

Dawid Paleń and Radosław Mantiuk[✉]

West Pomeranian University of Technology, Szczecin, Poland
rmantiuk@zut.edu.pl

Abstract. We present a do-it-yourself (DIY) 3D printer developed for rapid manufacturing of light fixtures (otherwise called luminaries) of complex and nonstandard shapes. This low-cost printer uses two individual extruders that can apply different filaments at the same time. The PLA (polylactic acid) filament is extruded for essential parts of the luminaire while the PVA (polyvinyl alcohol) filament is used to build support structures. PVA can be later effectively rinsed with water, leaving the luminaire with complex shape and diverse light channels. We provide a detailed description of the printer's construction including specification of the main modules: extruder, printer platform, positioning system, head with the nozzle, and controller based on the Arduino hardware. We explain how the printer should be calibrated. Finally, we present example luminaries printed using our DIY printer and evaluate the quality of these prints. Our printer provides low-cost manufacturing of single copies of the complex luminaries while maintaining sufficient print accuracy. The purpose of this work is to deliver the luminaries for the experimental augmented reality system, in which virtually rendered lighting should correspond to the characteristics of the physical light sources.

Keywords: Do-it-yourself 3D printer · Multi-material fabrication
Lighting luminaries manufacturing · Augmented reality

1 Introduction

The *light fixture* (called *lighting luminaire* in the lighting design literature) is a holder for the light source, which changes its lighting characteristic [11]. The more transparent the luminaire is, the higher is the efficacy of the lighting. Shading the luminaire will decrease efficiency but, at the same time, increase the directionality and the visual comfort probability. From a perceptual point of view, people prefer the luminaries of an interesting design and emanating a pleasant light. In the *augmented reality* (AR) systems people watch a physical environment augmented by the computer-generated objects [2]. In general, AR designers are limited by the regular shapes of the typical luminaries. They cannot use the luminaries of the unknown characteristic, because the physical lighting

© Springer Nature Switzerland AG 2019
J. Pejaś et al. (Eds.): ACS 2018, AISC 889, pp. 469–480, 2019.
https://doi.org/10.1007/978-3-030-03314-9_40

must interact with the rendered content [1]. Therefore, it is valuable to deliver a manufacturing technique, which fabricates complex luminaries but of shapes and transparency that strictly follow the computer-aided-design.

In this work, we describe the process of building a multi-material 3D printer, which was designed for a low cost and accurate manufacturing of the luminaries. The main feature of this printer is the use of two filaments: one for essential parts of the luminaries and the second one for the supporting structures that are further rinsed with water. This known technique allows printing of the complex luminaries with a diverse lighting characteristic.

The printer was built of cheap components available on the market. It works based on the fused filament fabrication (FFF) technology, in which melted filament is extruded on the platform in successive layers to form the object. The printer consists of two extruders for printing using both PLA and PVA filaments. Its head positioning system follows the CoreXY arrangement. The head is additionally equipped with the BLTouch sensor for levelling of the platform.

We present example luminaries printed by our DIY printer. The quality of these prints is evaluated and discussed to indicate the possibility of using the printer for producing luminaries for the augmented reality systems.

In Sect. 2 we introduce basic concepts related to the 3D printing, especially the fused filament fabrication technology. We also described the technological assumptions of the multi-material printing and the possibility of using this technique for the rapid manufacturing of the luminaries. In Sect. 3 a detailed description of the construction of our DIY printer is presented. In Sect. 4 we show example prints of the luminaries and discuss their quality.

2   Background and Previous Work

Fused filament fabrication (FFF) is an additive manufacturing technology commonly used for 3D printing. FFF printers lay down plastic *filament* to produce successive layers of the object. FFF begins with a software process, which mathematically slices and orientates the model for the build process. Additionally, *support structures* are generated to avoid unsupported stalactites. A filament is delivered as a thin wire unwound from a coil (see Fig. 1a). It is supplied to a *extruder* which can turn the flow on and off (Fig. 1b). An accurately controlled drive pushes the required amount of filament into the nozzle (Fig. 1e). The nozzle is heated to melt the filament well past their glass transition temperature (Fig. 1c). The material hardens immediately after extrusion from the nozzle when exposed to air. The *platform* is moved in vertical directions to built an object from the bottom up, one layer at a time (Fig. 1d). Same as the horizontal movement of the *head* (nozzle with the heating device) it is driven by stepper motors controlled by a computer-aided manufacturing (CAM) software package.

A number of *filaments* with different trade-offs between strength and temperature properties is available for FFF printing, such as Acrylonitrile Butadiene Styrene (ABS), Polylactic acid (PLA), Polycarbonate (PC), Polyamide (PA),

Polystyrene (PS), lignin, or rubber. There are *water-soluble filaments* that can be washed out from the object (e.g. polyvinyl alcohol (PVA)) to remove the support structures.

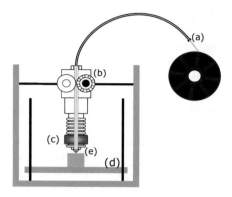

Fig. 1. Fused filament fabrication 3D printing technology.

Multi-material 3D Printers. Multi-material fabrication platforms simultaneously support more than one material (filament). They are used to create objects made of materials with different properties. Especially, these printers are used to manufacture the lighting luminaries of complex external and internal shape. Transparent materials are combined with the internal light tunnels. The support structures required to print the tunnels are fabricated of the material, which is later washed out using a solvent.

Some of the FFF printers available on the market support dual or triple extrusion (e.g. MakerBot Replicator 2X, Ultimaker 2 with Dual extruder upgrade, Zortrax Inventure, etc.). These printers can be used to print any multi-material objects including the luminaries. In our project, we develop a similar printer using inexpensive off-the-shelf components. However, the design and calibration of our printer have been focused on printing the luminaries. We use the PLA and PVA combination of filaments because these filaments are inexpensive and have good inter-adhesive properties.

There are the multi-material fabrication platforms built based on other technologies. Stereolithography has been adapted to support multiple materials using multiple vats with UV-curable polymers [8]. The printing process is slow because the material must be changed for each layer and the printed model must be cleaned from the previous resins. An additional disadvantage of this technique it is losing resin in cleaning time. Polyjet technology uses multiple inkjet printheads placed next to the lamp of UV lamp, which toughens polymer [5,9]. This technology ensures high-quality printing and large workspace. It is one of the most advanced multi-material printing technologies, but it is expensive. The multi-material inkjet printers are provided by 3D Systems and Stratasys. Selective laser sintering has been used with multiple powders [3,7]. This technology uses a laser as the power source to sinter powdered material in 3D space. On the commercial side, the multi-material printing is supported by the powder-based 3D printers developed by Z Corp.

Printing for Lighting Design. Lighting design [4] is concerned with the design of the environments in which people see clearly and without discomfort [10]. The objectives are not limited to meet the requirements of sufficient brightness of the lighting measured using the photometric techniques. The atmosphere resulting from interior design while keeping in mind issues of aesthetic, ergonomic, and energy efficiency is also important. The augmented reality systems support the lighting design projects in the evaluation of the perceptual notability of the designs.

Typically, the lighting designers use the luminaries of known photometric characteristic specified by the IES (Illuminating Engineering Society) data [11]. Lighting manufacturers publish IES files to describe their products. The program interprets the IES data to understand how to illuminate the environment. The IES file can also be used by the AR systems [6]. Variety of luminaries is limited to the products proposed by the manufacturers, which is not a large number because valuable IES data is delivered only by very few highly specialized producers. We argue that it is reasonable to manufacture own luminaries, especially if it is possible to adjust their structure to the designed IES data.

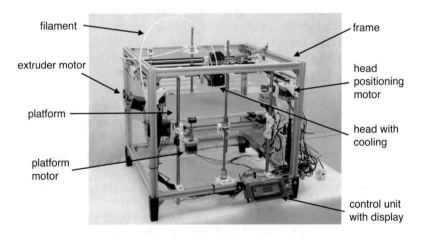

Fig. 2. DIY 3D printer.

3   Do-It-Yourself Printer

The general view of our 3D FFF printer is presented in Fig. 2. The *positioning system* (see Sect. 3.2) is mounted on the *printer frame* (see Sect. 3.1). It moves the *platform* (printing bed) and the *head* with nozzles and heating/cooling systems (see Sect. 3.4). The *material feeding system* (see Sect. 3.3) supplies filament to the head. The operation of the printer is controlled by the *Arduino module* (see Sect. 3.5). This module is also responsible for the *printer calibration* (see Sect. 3.6).

3.1 Frame and Platform

The printer external dimensions are $44 \times 58 \times 48$ cm (width, depth, and height respectively)(see Fig. 2). The frame is built of t-slot aluminium profiles (with a cross-section of 20×20 mm) that provide adequate structural strength and rigidity. Connections between rods are additionally stiffened with rectangular aluminium brackets. Other connections between printer elements (white plastic modules shown in Figs. 2 and 3) were printed based on custom models.

The platform frame is built of the same size aluminium profiles on which the 30×30 cm printing bed is mounted (see Fig. 3). The bed consists of three layers: a silicone hot pad responsible for heating the bed, a 4 mm thick aluminium sheet, which stiffens the structure and fixes it to the profiles, and a glass attached with clips, which allows to remove the printed object and gently separate it from the glass in the water. Vertical movement of the platform is stabilized by four stainless steel rods (10 mm in diameter) located in the corners of the platform frame.

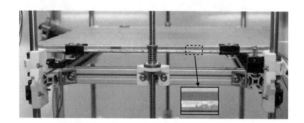

Fig. 3. The printer platform. Inset depicts layers of the printing bed.

3.2 Positioning System

The positioning system in our DIY printer is responsible for moving the head in horizontal XY directions and the platform in the vertical Z direction. The head movement is based on the CoreXY arrangement, which consists of two stepper motors (see Fig. 4) and two pulleys to equilibrate loads. In this arrangement, the head carriage stays always perpendicular without relying on the stiffness of the sliding mechanism.

The platform is moved by two motors attached to the bottom frame (see Fig. 5). They turn the trapezoidal 8 mm screw (tr8) through the clutch.

In both horizontal and vertical positioning systems, we use the same NEMA 17 stepper motors (model 17hs4401) with 1.8 deg step angle and 40 Ncm holding torque.

Fig. 4. Close-up of the stepper motor.

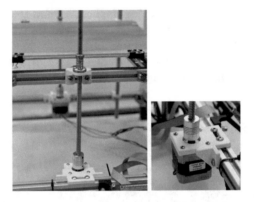

Fig. 5. Left: the screw and clutch of the platform positioning system. Right: Close-up of the bottom stepper motor.

3.3   Material Feeding System

We decided to use the Bowden filament feeding mechanism with the stepper motor attached to the printer frame (see Fig. 6). The motor pushes the filament through a teflon tube connected to the printer head. The advantage of this technique is a reduced weight of the element moving with the head. Actually, we use two heads to support multi-material printing. Two motors moving together with the head would significantly affect the quality and speed of the printing. For printing luminaries, we use the PLA and PVA filaments that are rigid enough and do not require a short connection between the stepper motor and the head. The feeding system is powered by stronger NEMA 17 stepper motors (model 17hs19-2004s1) with 1.8 deg step angle and 59 Ncm holding torque.

Fig. 6. Material feeding system with the stepper motor.

3.4 Head

The filament delivered to the printer head (see Fig. 7) is preheated to high temperatures of 150–250 °C controlled by the temperature sensor. An important part of the head is the *heat sink*, which prevents dissolution of plastic at the beginning of the head. Dissolved plastic is applied to the glass surface of the platform with the *nozzles* of an arbitrary diameter (using nozzles ranging from 0.2 mm to 0.8 mm is possible).

For multi-material printing, we decided to use two separate heads connected to each other (Chimera model). This solution allows printing simultaneously using two different filaments of different melting temperatures. The disadvantage is the nontrivial positioning of both nozzles in relation to the surface of the platform. Unwanted leakage of the filament from the second nozzle during printing is also possible.

Fig. 7. Printer head with the BLTouch sensor.

3.5 Control Module and Printing Pipeline

The entire hardware system of our printer (i.e. motors, temperature thermistors, extruder heaters, platform heater, BiTouch sensor) is controlled by the

Arduino Mega module with RAMPS 1.4. Figure 8 presents diagram of connections between modules.

The 3D model of the object to be printed is prepared in the Fusion 360 CAD/CAM software. Fusion automatically cuts the model into individual layers (slices) and generates the support structures. Finally, data, which controls the movement of the head and platform is delivered to the printer on the SD memory card.

Fusion 360 supports multi-material printing, i.e. it is possible to indicate that support structures should be printed by a different head than the main model.

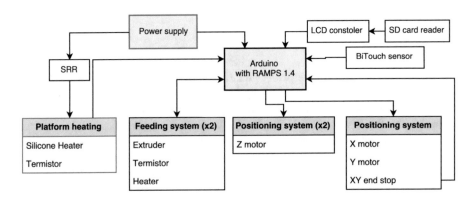

Fig. 8. Control module of the DIY printer.

3.6   Printer Calibration

Before connecting motors to the controller, an effective voltage V_{ref} for each motor must be calculated based on the following equation:

$$V_{ref} = A \cdot 8 \cdot RS. \tag{1}$$

A is the current required by the motor, and RS is the resistance located in the motor's stepstick. The actual voltage supplied to the motor should match V_{ref}. This voltage can be adjusted manually in the controller using the potentiometer.

The essential parameter is the number of motor steps per centimetre of the linear movement. This parameter must be calculated for all motors and deliver to the Arduino software.

For the XY positioning, the number of steps is calculated using the following formula:

$$XY_{steps} = \frac{MS \cdot MI}{PP \cdot PT}, \tag{2}$$

where MS is the number of motor steps per full rotation ($MS = 200$ for our printer), MI depicts number of microsteps per one motor step ($MI = 16$), PP is the stroke of the toothed belt ($PP = 2$), and PT is the number of teeth in the toothed belt ($PT = 20$). All listed values can be read from the motor and toothed belt parameters.

Positioning of the platform (Z-direction) requires the formula taking into account the thread parameters of the screw:

$$Z_{steps} = \frac{MS \cdot MI}{RP}, \tag{3}$$

where RP depicts pitch of the screw ($RP = 8$).

Calibration of the extruder motor is based on the following formula:

$$E_{steps} = \frac{MS \cdot MI \cdot WGR}{\pi \cdot HBC}, \tag{4}$$

where WGR is gearing on the gears of the extruder ($WGR = 1$), and HBC is diameter of the extruder screw at the point of contact with the filament ($HBC = 8$).

An important step of the printer calibration is the platform levelling. The distance between the head nozzles and the printing bed should be known for each location on the platform. Levelling can be performed manually by adjusting the height of each corner of the platform. However, the surface of the platform is not perfectly smooth and some irregularities can occur e.g. due to using liquids that improve the adhesion of the object to the surface or some mechanical defects. Therefore, in our printer we use Auto Bed Leveling (ABL) technique. In the ABL technique a number of measurements of the distance from nozzle to bed are performed using the BLTouch probe (see this sensor in Fig. 7) that emulates the servo through the retractable pin.

4   Test Prints

In this section, the accuracy of the multi-material prints with our DIY 3D printer is evaluated. We make test prints and check if their dimensions are consistent with the CAD model. Additionally, we discuss advantages and problems related to multi-material printing using the FFF technology.

The test prints are rather bi-material objects than luminaries (see Fig. 9). Both objects required many support structures that filled the whole empty interior of the objects (see Fig. 10). We used the PLA filament to print the white elements, while the supporting structures were printed with PVA. PVA was further rinsed with water. For the presented prints, it would be hardly feasible to remove the supporting structures printed with the same material as the main parts of the object. Most probably, this process would have to damage some part of the objects. On the other hand, rinsing with water is not a simple task. This process requires time and manual use of additional tools, especially inside the small objects like the interior of the tube.

We measured physical dimensions of the cube-shaped print. They are consistent with the CAD model of 49.5×49.5 mm dimensions with the accuracy of $+/-0.3$ mm. However, some parts of the object are deformed due to the rinsing

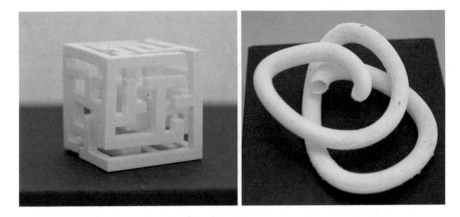

Fig. 9. From left: cube-shaped and tube-shaped objects.

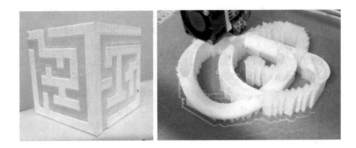

Fig. 10. From left: cube-shaped and tube-shaped objects with the support structures.

process (see the bottom left corner of this object in Fig. 11, left). These deformations are caused by a low adhesion between PLA and PVA filaments causing delamination of the printed object (see Fig. 11, centre). We managed to reduce this drawback by slowing down the printing process. In future work we plan to find filaments that would have better inter-adhesive properties.

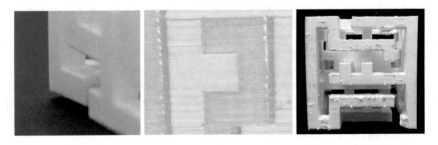

Fig. 11. Left: deformation of the object structure. Center: vertical delamination of the PLA and PVA filaments (darker lines between white PLA and light beige PVA).

We noticed that it is difficult to stop the leakage of the melted filament from the unused head completely. This leakage causes extruding of small amounts of PVA filament on the main parts of the object. After rinsing, there are micro-holes on the PLA surfaces. PLA filament is also extruded on the support structures forming unwanted structures (see Fig. 11, right). The solution to this problem would be a better head cooling system, however, these small structures should not substantially affect the characteristics of the luminaire.

In Fig. 12 the test prints have been illuminated to simulate the luminaries. In future work we plan to print the actual luminaries using the semi-transparent filaments.

Fig. 12. Test prints illuminated by the light source.

5   Conclusions and Future Work

Construction of a 3D printer is a challenging technical task, which requires specialized skills in the field of mechatronics. We have extended the typical FFF printer design by the dual-material module with separate extruders for each filament. Our low-cost DIY printer has been used to print luminaries of a complex shape. It was possible by rinsing in water the support structures printed using a PVA filament.

In future work we plan to print the luminaries of known photometric characteristic and evaluate if the printed objects follow these characteristics. We plan to use our DIY printer to prototype the complex luminaries that will be further used in the augmented reality system.

There are also possibilities to improve the printer itself through testing another printer heads that would reduce the unwanted leakage of the filament. Another type of filaments should also improve the quality of printed objects.

Acknowledgments. The project was partially funded by the Polish National Science Centre (decision number DEC-2013/09/B/ST6/02270).

References

1. Azuma, R., Baillot, Y., Behringer, R., Feiner, S., Julier, S., MacIntyre, B.: Recent advances in augmented reality. IEEE Comput. Graph. Appl. **21**(6), 34–47 (2001)
2. Bimber, O., Raskar, R.: Spatial Augmented Reality: Merging Real and Virtual Worlds. CRC press (2005)
3. Cho, W., Sachs, E.M., Patrikalakis, N.M., Troxel, D.E.: A dithering algorithm for local composition control with three-dimensional printing. Comput. Aided Des. **35**(9), 851–867 (2003)
4. Griffiths, A.: 21st Century Lighting Design. A&C Black (2014)
5. Khalil, S., Nam, J., Sun, W.: Multi-nozzle deposition for construction of 3D biopolymer tissue scaffolds. Rapid Prototyping J. **11**(1), 9–17 (2005)
6. Krochmal, R., Mantiuk, R.: Interactive prototyping of physical lighting. In: International Conference Image Analysis and Recognition, pp. 750–757. Springer (2013)
7. Kumar, P., Santosa, J.K., Beck, E., Das, S.: Direct-write deposition of fine powders through miniature hopper-nozzles for multi-material solid freeform fabrication. Rapid Prototyping J. **10**(1), 14–23 (2004)
8. Maruo, S., Ikuta, K., Ninagawa, T.: Multi-polymer microstereolithography for hybrid opto-mems. In: The 14th IEEE International Conference on Micro Electro Mechanical Systems MEMS 2001, pp. 151–154. IEEE (2001)
9. Sitthi-Amorn, P., Ramos, J.E., Wangy, Y., Kwan, J., Lan, J., Wang, W., Matusik, W.: Multifab: a machine vision assisted platform for multi-material 3D printing. ACM Trans. Graph. (TOG) **34**(4), 129 (2015)
10. Steffy, G.: Architectural Lighting Design. Wiley (2002)
11. Zumtobel: The Lighting Handbook. Zumtobel Lighting GmbH (2013)

Multichannel Spatial Filters for Enhancing SSVEP Detection

Izabela Rejer[(✉)]

Faculty of Computer Science and Information Technology,
West Pomeranian University of Technology Szczecin, Szczecin, Poland
irejer@wi.zut.edu.pl

Abstract. One of the procedures often used in an SSVEP-BCI (Steady State Evoked Potential Brain Computer Interface) processing pipeline is multichannel spatial filtering. This procedure not only improves SSVEP-BCI classification accuracy but also provides higher flexibility in choosing the localization of EEG electrodes on the user scalp. The problem is, however, how to choose the spatial filter that provides the highest classification accuracy for the given BCI settings. Although there are some papers comparing filtering procedures, the comparison is usually done in terms of one, strictly defined BCI setup [1, 2]. Such comparisons do not inform, however, whether some filtering procedures are superior to the others regardless of the experimental conditions. The research reported in this paper partially fills this gap. During the research four spatial filtering procedures (MEC, MCC, CCA, and FBCCA) were compared under 15 slightly different SSVEP-BCI setups. The main finding was that none of the procedures showed clear predominance in all 15 setups. By applying not-the-best procedure the classification accuracy dropped significantly, even of more than 30%.

Keywords: BCI · SSVEP · Brain Computer Interface · Spatial filter
CCA · MEC · MCC · FBCCA

1 Introduction

A BCI (Brain Computer Interface) is a communication system in which messages or commands that a user sends to the external world do not pass through the brain's normal output pathways of peripheral nerves and muscles [3]. There are three types of EEG-BCIs usually applied in practice: SSEP-BCI (Steady State Evoked Potentials BCI), P300-BCI (BCI based on P300 potentials), and MI-BCI (Motor Imagery BCI). They differ in the classes of mental states that are searched in the brain activity, and in the procedures used for evoking these states. In the two first BCI types the activity that is searched for is evoked by an external stimulation (periodic stimuli in the case of SSEP-BCI, and important vs no-important stimuli in the case of P300-BCI). On the contrary, in MI-BCI the desired activity is evoked directly by the user who is imagining movements of specific body parts.

A special type of SSEP-BCI is SSVEP-BCI (Steady State Visual Evoked Potentials BCI). With this type of BCI the periodic stimuli are delivered through a user visual system. Usually a flickering LEDs (Light Emitting Diodes) or flickering images

J. Pejaś et al. (Eds.): ACS 2018, AISC 889, pp. 481–492, 2019.
https://doi.org/10.1007/978-3-030-03314-9_41

displayed on a screen are used to evoke the brain response. The characteristic feature of a brain response evoked by a flickering image/LED is that its fundamental frequency is the same as the stimulus frequency [4, 5]. Hence, providing stimuli of different frequencies, different SSVEPs can be evoked. SSVEP is an automatic response of the visual cortex and hence SSVEP-BCI does not require the same amount of conscious attention as MI-BCI or even P300-BCI. Moreover, according to the neurobiological theory SSVEPs are stable brain responses [6], which means that the same stimulus frequency should induce similar response across time. That is why SSVEP-BCI are often applied in practice, even though they are rather tiring for their users.

A basic scheme of an SSVEP-BCI can be summarized as follows. A user is provided with a set of stimulation fields, each flickering with different frequency. The user task is to focus on one of the fields at a time. When the user is performing the task, his/her brain activity is recorded and then processed. The EEG signal processing pipeline is composed of four main stages: preprocessing, feature extraction, SSVEP detection, and classification. When this pipeline is completed and the class is known, the BCI sends the command associated to this class to the external environment and the whole process starts from the beginning, it is from the user focusing his attention on different (or the same) stimulus.

The most important stage of the BCI processing pipeline is the SSVEP detection stage. There are lots of papers that addresses the problem of SSVEP detection and lots of methods that can be used to deal with this task [7, 8]. However, when comes to build a BCI controlled with SSVEPs, it occurs that there are no clear guidelines discussing which method fits better to a designed setup. Of course, it is possible to point out the class of methods that usually perform better than others but the problem of choosing a specific method from this "winning" group still remains open. The problem is that at the moment there are no standards regarding the process of designing BCIs and hence each BCI can have entirely different setup. They can differ in: stimulation device (LEDs, LCD, CRT) [9], number of targets (from 2, via 48 [10] up to 84 [11] at the moment), number of electrodes used to acquire the EEG signal and their localization. There are also some more subtle differences such as: size and shape of the targets, distance between the targets, targets color and flickering pattern [12]. All these BCI design differences might have an impact on the performance of different detection procedures.

One issue regarding SSVEP detection methods on which most of scientists using SSVEPs agree is that BCI processing pipeline that involves multichannel spatial filtering (at the preprocessing or detection stage) is far better than the pipeline missing this procedure. Approaches using spatial filtering not only provide (usually) more true positives but also provide higher flexibility in choosing the localization of EEG electrodes on the user scalp. Spatial filters that linearly combines signals acquired from different EEG channels have the ability to extract and enhance SSVEP activity so usually it is enough to place the electrodes somewhere in the occipital and parieto-occipital areas instead of sticking strictly to 10–20 system locations such as: O1, O2, POz etc.

Among the approaches that use the multichannel spatial filtering procedure the most widely known are: MEC (Minimum Energy Combination), MCC (Maximum Contrast Combination), CSP (Common Spatial Patterns), ICA (Independent Component Analysis), PCA (Principal Component Analysis), and CCA (Canonical Correlation Analysis)

(CCA is not a pure spatial filter but plays similar role by linearly combining the information coming from different sources) [13]. In some of them, the spatial filtering process is performed simultaneously with the SSVEP detection process (CCA), in others both stages are separated and hence it is possible to use different detection methods after the filtering procedure (CSP, ICA, CCA). There are also approaches than theoretically support usage of different detection methods but in practice better detection rate is obtain when the method associated with the spatial filter is used (MEC, MCC).

To extract SSVEP activity, spatial filters linearly combines EEG signals acquired from all EEG channels. That means that in the detection process all the possible information is used simultaneously. The spatial filters are constructed either to directly extract the SSVEP activity from different channels and store it in one (or a few) components obtained after the filtering procedure (MEC, CSP, CCA) or to extract the non-SSVEP activity and remove most of it from the recorded EEG (MEC).

The aim of this paper is to compare four multichannel spatial filtering procedures (CCA, FBCCA, MCC, and MEC) in terms of SSVEP detection accuracy. The research question that gave an impulse to carry out such a comparison can be stated as follows: *is it possible to point out the multichannel spatial filtering procedure that will lead to more distinguishable SSVEPs regardless of a BCI setup?* To answer this question, 15 experiments were carried out, each experiment with a single subject. In all the experiments LEDs were used as stimulators. Each experiment differed in the BCI setup, specifically with: targets luminance, targets color, distance between targets, number and location of EEG electrodes, signal length, number of trials, and a set of targets' frequencies.

The rest of the paper is organized as follows: Sect. 2 shortly describes the multi-channel spatial filtering methods used in the paper, Sect. 3 provides setups of all the experiments, Sect. 4 presents the main results, and Sect. 5 concludes the paper.

2   Spatial Filtering Methods

2.1   Canonical Correlation Analysis

The Canonical Correlation Analysis (CCA) is a statistical method used for finding the correlation between two sets of variables. Since CCA uses the same correlation mea-sure as in the case of one-dimensional variables, it starts from transforming both multidimensional data sets into two one-dimensional linear combinations (called canonical variables), one combination for each set. The optimization criterion is to find canonical variables of the maximum correlation. Then, the next pair of canonical variables of the second highest correlation is searched for, then the third one, and so on. The whole process ends when the number of pairs is equal to the number of variables in the smaller set [14].

When CCA is used for SSVEP analysis, the EEG data set recorded for the given condition (it is recorded when a user focuses his/her attention on one of the flickering targets) is treated as the first of the two correlated variables ($Y \in R^{N \times M}$, N – number of time samples, M – number of EEG channels). The second variable is the matrix composed of reference signals created artificially for the target frequency ($X \in R^{N \times 2N_h}$,

N_h – number of harmonics). The reference matrix contains at least two columns, the sine and cosine wave of the stimulus fundamental frequency (f). Since SSVEP synchronization often occurs not only at the fundamental frequency, but also at succeeding harmonics, to enhance the recognition rate often also harmonic frequencies of the stimulus frequency are used for CCA coefficients calculation. If this is the case, two additional columns are added to the reference matrix per each harmonic:

$$Ref_f = \begin{pmatrix} \sin(2\pi ft) \\ \cos(2\pi ft) \\ * \\ * \\ \sin(2\pi(N_h)ft) \\ \cos(2\pi(N_h)ft) \end{pmatrix}, \tag{1}$$

where: f – stimulus fundamental frequency, t – sampling time ($t = 0, 1/Fs, ..., (N-1)/Fs$), Fs – sampling frequency, h – harmonic index ($h = 1... N_h$). In order to calculate CCA coefficients, it is enough to find eigenvalues for the following matrix:

$$R_{22}^{-1} R_{12} R_{11}^{-1} R_{12}^T, \tag{2}$$

where: R_{11}, R_{22}, R_{12} - matrix of correlation coefficients between variables in Y, X and (X, Y), respectively. The eigenvalues sorted in the descending order represent CCA coefficients found for pairs of canonical variables built as linear combinations of X and Y. Strictly speaking each CCA coefficient (r) is calculated as:

$$r = \sqrt{\lambda}, \tag{3}$$

where λ represents one of the matrix eigenvalues. Usually only the first CCA coefficient, it is the coefficient of the highest value, is used for SSVEP detection.

The classification based on CCA coefficients is straightforward. The coefficients calculated for all frequencies used as targets in the BCI are compared and the frequency of the highest coefficient is chosen as the winning one. Depending on the application, the class send by BCI to the external device or application is assigned at once when the winning frequency is chosen or the coefficient is compared with the given threshold and only when it exceeds this threshold, the class is assigned. When the winning frequency coefficient is under the threshold, the decision about lack of recognition is taken, and no class is send outside the interface.

2.2   Minimum Energy Combination

The starting point for the Minimum Energy Combination method is exactly the same as for CCA. There are two matrixes, the first (Y) contains EEG signal, the second (X) contains pure SSVEP components. Although the starting point is similar, the procedure is entirely opposite. While CCA directly looks for linear combinations enhancing the activity of interests, MEC starts from detecting and removing non-

SSVEP components from the recorded EEG. Only when the signal is cleaned of the most background EEG activity and external noise, the SSVEP detection starts.

The procedure can be summarized as follows. At first, the projection matrix (for projecting EEG signal onto the orthogonal complement of the space spanned by the vectors stored in X) is built. Since the vectors in X are linearly independent, the projection matrix can be written as:

$$Q = X(X^T X)^{-1} X^T \tag{4}$$

Next, EEG signal stored in matrix Y is projected with matrix Q onto the orthogonal complement of the space spanned by the SSVEP components from Y.

$$Y_n = Y - QY, \tag{5}$$

where Y_n - the matrix containing only nuisance components, it is EEG background activity and external noise.

In order to create a spatial filter, allowing for removing most of non-SSVEP activity from the recorded EEG, the matrix Y_n is decomposed into diagonal matrix of eigenvalues (Λ) and matrix of corresponding eigenvectors (V): $Y_n = V\Lambda V^{-1}$. The vectors forming columns of matrix V, sorted in an ascending order of their eigenvalues, show the directions of increasing amount of energy (variance). The first eigenvector (of the "noise" matrix Y_n) shows the direction of the smallest noise energy, and the last one - of the highest. The procedure assumes that 90% of the non-SSVEP activity should be filtered out from the signal to ensure the correct SSVEP detection. To this end, the eigenvalues are normalized to add to 1, and the spatial filter F is formed from the first s columns of V, where s is the smallest number fulfilling the condition [1]:

$$\frac{\sum_{i=1}^{s} \lambda_i}{\sum_{j=1}^{N} \lambda_j} > 0.1. \tag{6}$$

Finally, the last step of the procedure is to apply the spatial filter over the original matrix of EEG data (Y):

$$C = YF, \tag{7}$$

where: C – cleaned EEG data. To estimate the total SSVEP power contained in matrix C, usually the following formula is used:

$$P = \sum_{l=1}^{s} \sum_{h=1}^{2N_h + 2} \left\| X_h^T C_l \right\|^2 \tag{8}$$

The classification scheme is the same as in the case of CCA, it is the SSVEP power estimated for different frequencies used as targets in the BCI are compared and the decision on the class is taken using max rule (sometimes modified with the threshold).

2.3   Maximum Contrast Combination

The task set in the Maximum Contrast Combination method is to find the combination of input channels (stored in matrix Y) that simultaneously maximizes the energy in the SSVEP frequencies and minimizes the background EEG activity and other external noise [15]. Using matrix Y_n (containing nuisance components), defined in (5), we can calculate the noise energy (E_N) as:

$$E_N = (Y - QY)^T(Y - QY), \tag{9}$$

and the SSVEP energy as:

$$E_{SSVEP} = X^T X, \tag{10}$$

where: X - the reference SSVEP matrix defined in (1) and Q - the projection matrix defined in (4). To simultaneously minimize (9) and maximize (10) the generalized eigen decomposition of the matrixes E_N and E_{SSVEP} should be performed:

$$E_{SSVEP} = \Lambda E_N V \tag{11}$$

After solving (11), the eigenvector from V corresponding to the largest element in Λ contains the coefficients of the spatial filter (it is the coefficients of the maximum contrast combination). The rest of procedure is the same as in the MEC method, it is the spatial filter is applied over the original matrix of EEG data (7) and the SSVEP power of the filtered EEG data is calculated (8). The procedure is repeated for each target and the decision of the class is taken under max rule.

2.4   Filter Bank Canonical Correlation Analysis (FBCCA)

One of the well-known facts about the SSVEP phenomenon is that the synchronization usually takes place not only at the stimulus fundamental frequency but also at its harmonics. This fact is utilized in all spatial filters described so far by introducing harmonic terms to the SSVEP matrix defined as (1). In the FBCCA approach, the harmonic frequencies are used in a more explicit way by applying a spectral filter bank on the EEG signal before applying the spatial filtering procedure (here: CCA) [2]. A short summary of the algorithm is as follows. First, a filter bank is designed and applied on each EEG channel. Assuming that the bank is composed of K bandpass filters, during the filtering process K matrixes ($K \in R^{NxM}$) are created. Each matrix contains data from all original EEG channels filtered with one of K filters. Next, the CCA coefficients are calculated for succeeding targets. For a single target, the standard CCA algorithm is applied K times, correlating the SSVEP matrix (created for this target), with each of the K matrixes. The K CCA coefficients obtained for the given target are then aggregated together. The process is repeated for all targets. The decision on the target attended by the user is taken under the max rule, it is the target of the maximum value of the aggregated CCA coefficient is chosen.

Although the algorithm looks quite straightforward, there are two issues that have to be carefully designed. First is the choice of the filters forming the filter bank, and second is the method used to aggregate individual CCA coefficients. There are many different approaches to deal with both tasks. One of the simplest is to use constant step during designing filter bank and aggregate the individual CCA coefficients with a sum operation.

3   Experimental Setup

Fifteen subjects (12 men, 3 women; mean age: 21.8 years; range: 20–24 years) participated in the experiments (each subject took part only in one experiment). All subjects had normal or corrected-to-normal vision and were right-handed. None of the subjects had previous experiences with SSVEP-BCI and none reported any mental disorders. Written consent was obtained from all subjects. The study was approved by the Bioethics Committee of Regional Medical Chamber (consent no. OIL-Sz/MF/KB/452/20/05/2017).

The BCI system used in the experiments was composed of three modules: control module, EEG recording module, and signal processing module. The main part of the control module was a square frame with two sets of LEDs: stimulation LEDs and control LEDs (each set was composed of 4 LEDs). The stimulation LEDs were flickering all the time with the frequencies set at the beginning of the experiment; each LED was flickering with another frequency. The control LEDs were used to draw the user attention to the stimulation LEDs to which he/she should attend to at the succeeding moments.

During the experiment, EEG data were recorded from four monopolar channels at a sampling frequency of 256 Hz. From 4 to 8 passive electrodes were used in the experiments. The reference and ground electrodes were located at left and right mastoid, respectively, and the remaining electrodes were attached over the occipital and parieto-occipital areas in positions established according to the Extended International 10–20 System [16]. The impedance of the electrodes was kept below 5 kΩ. The EEG signal was acquired with a Discovery 20 amplifier (BrainMaster) and recorded with OpenViBE Software [17]. EEG data were filtered with a Butterworth band-pass filter of the fourth order in the 4–50 Hz band. Apart from this preliminarily broad-band filtering, the EEG signals gathered during the experiments were not submitted to any artifact control procedure.

The detailed scheme of the experiment with one subject was as follows. The subject was placed in a comfortable chair and EEG electrodes were applied on his or her head. The control module with LED frame was placed approximately 70 cm in front of his/her eyes. To make the experimental conditions more realistic, the subjects were not instructed to sit still without blinking and moving – the only requirements for them was to stay at the chair and observe the targets pointed by the control LEDs.

The start of the experiment was announced by a short sound signal, and 5 s later, EEG recording started. During the experiment only one control LED was active at a time, pointing to one stimulation LED. The control LEDs changed each t seconds (depending on the experiments t was equal from 1.25 to 4).

To compare the four methods described in Sect. 2 against different experimental setup, 15 experiments were performed. Each experiment was carried out with another subject and with slightly changed setup. The detailed description of all 15 experimental setups is gathered in Table 1. For each method the same four SSVEP reference matrixes were used, one per target. Only fundamental frequency was used to build each reference matrix. Three methods, CCA, MEC, and MCC did not require any additional settings. Only for the FBCCA method, the approach to create filter bank and the procedure for aggregating the individual CCA coefficients determined for the given target after applying individual filters had to be established. According to [9] the highest detection accuracy can be obtained when the filters in the filter bank have similar high cutoff frequency and increasing low cutoff frequency. Following this remark, the high cutoff frequency for all the filters was set to 50 Hz and low cutoff

Table 1. The description of experimental setups.

Exp. no.	Color	Luminance [lx]	Frequencies	Distance [cm]	No. of trials	Signal length [s]	Channels
Exp.1	White	4000	30, 30.5, 31, 31.5	13	20	4	O1, O2, Oz, Pz, POz
Exp.2	White	4000	26, 27, 28, 29	13	20	1.5	O1, O2, Oz
Exp.3	Green	1000	5.9, 6.7, 7.7, 10.4	10	35	2	O1, O2, Pz, Cz
Exp.4	White	4000	17, 18, 19, 20	13	20	1.25	O1, O2, Oz
Exp.5	Green	1000	6.1, 7.1, 7.9, 9.6	10	35	4	O1, O2, Pz, Cz
Exp.6	Green	2000	15, 17, 18, 19	13	20	3	O1, O2, Pz, Cz
Exp.7	White	4000	15, 16, 17, 18	13	20	2	O1, O2, Oz
Exp.8	White	1000	28, 29, 29.5, 30	16.5	60	1.5	O1, O2, Oz, Pz, POz
Exp.9	Green	1000	6.9, 8.7, 12.2, 13.2	10	40	4	O1, O2, Pz, Cz, C3, C4
Exp.10	White	2000	5.5, 8.5, 9, 9.5	16.5	80	1.5	O1, O2, Oz, Pz, POz
Exp.11	White	2000	6, 8.5, 9, 9.5	16.5	80	1.25	O1, O2, Oz
Exp.12	Green	1000	6.6, 8.2, 9, 14.3	10	40	3	O1, O2, Oz
Exp.13	Blue	350	5.8, 6.8, 7.9, 9	10	100	5	O1, O2, Pz, Cz
Exp.14	White	1000	5, 6, 7, 8	16	40	2	O1, O2, Oz, Pz, POz
Exp.15	White	2000	5, 6, 7, 8	13	48	1.5	O1, O2, Oz, Pz, POz

frequencies were set to: $L_1 = 5$ Hz, $L_2 = 10$ Hz, ... $L_K = 45$ Hz. Regarding the aggregation operation, the sum of individual CCA coefficients was applied.

4 Results

Table 2 presents the SSVEP detection accuracy obtained in each experiment after applying one of the analyzed multichannel spatial filtering procedures. The accuracy was calculated as the number of trials with correctly recognized target divided by the total number of trials. Two last columns of the table present the results aggregated for each experiment over the four applied methods. As it can be noticed in the table, the

Table 2. The SSVEP detection accuracy obtained in each experiment after applying the analyzed filtering procedures.

Exp. no.	CCA	FBCCA	MCC	MEC	Mean	Max
Exp.1	**0.85**	0.80	**0.85**	**0.85**	0.84	0.85
Exp.2	0.75	**0.95**	0.75	0.90	0.84	0.95
Exp.3	**0.86**	0.80	**0.86**	0.70	0.81	0.86
Exp.4	0.95	0.95	0.95	**1.00**	0.96	1.00
Exp.5	0.74	**0.80**	0.74	0.77	0.76	0.80
Exp.6	0.85	**0.90**	0.85	0.80	0.85	0.90
Exp.7	0.90	**0.95**	0.90	0.85	0.90	0.95
Exp.8	0.90	**0.93**	0.90	0.73	0.87	0.93
Exp.9	**0.90**	0.78	**0.90**	0.75	0.83	0.90
Exp.10	**0.90**	0.80	**0.90**	**0.90**	0.88	0.90
Exp.11	**0.95**	0.90	**0.95**	**0.95**	0.94	0.95
Exp.12	**0.83**	0.73	**0.83**	**0.83**	0.81	0.83
Exp.13	0.80	0.79	0.80	**0.81**	0.80	0.81
Exp.14	0.93	0.75	0.93	**0.95**	0.89	0.95
Exp.15	**1.00**	0.75	**1.00**	0.98	0.93	1.00
Mean	**0.87**	**0.84**	**0.87**	**0.85**	**0.86**	**0.91**

detection accuracy was quite high - regardless of the experimental setup and the method used for spatial filtering and SSVEP detection, it was always equal or higher than 73%.

Analyzing the results gathered in Table 2 it is quite easy to answer the question posed in the first section of the paper: *is it possible to point out the multichannel spatial filtering procedure that will lead to more distinguishable SSVEPs regardless of a BCI setup?* The answer cannot be affirmative because none of the methods showed clear predominance in all 15 experiments (CCA – 7, FBCCA – 5, MCC – 7, MEC – 7). The analysis of the mean accuracy shown in the last row of Table 2 also does not allow to rank the methods - the mean values calculated over all 15 experiments are almost the same.

This does not mean, however, that the choice of the spatial filtering procedure does not matter. Just the opposite, the specific filter can highly deteriorate or boost the detection accuracy. The problem is that although the mean detection accuracy is quite similar, the individual results differ significantly (Fig. 1). For example, if CCA or MCC was applied (instead of FBCCA) for Exp. 2, the loss of accuracy would be more than 25% (0.75 for CCA or MCC vs. 0.95 for FBCCA). Similarly, if in Exp. 15, FBCCA was applied instead of any other method, the loss in the detection accuracy would exceed 30% (0.75 for FBCCA vs. 1 for CCA or MCC, or 0.98 for MEC). Hence, although on average all four methods provided the same accuracy, in individual BCI

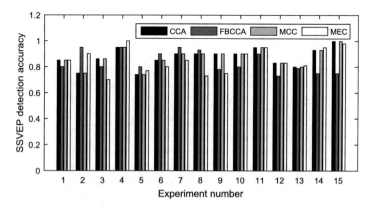

Fig. 1. The SSVEP detection accuracy.

setups some of them worked significantly better than the others. The question now is whether it is possible to find out what are the reasons of these differences? It other words, is it possible to define which spatial filter should provide the highest accuracy in the given BCI setup.

Of course, to fully answer this question a lot of additional experiments should be performed. However, it seems that the differences in the detection accuracy do not stem from the features of the subjects or the stimulation power (regulated by targets' color and luminance or distances between targets). The most probable reasons for such differences are signal parameters such as: signal length, number of sensors, or frequency resolution.

5   Conclusion

The study whose results were reported in this paper shows that it is not enough to apply any of the spatial filtering procedures in the SSVEP-BCI processing pipeline to enhance the SSVEP detection. What is really important is the correct choice of the procedure. Only when the procedure fits the BCI setup, it will provide the true benefits, it is a significant increase of the classification accuracy.

The question how to choose the filtering procedure best fitted to the given BCI setup still remains open. Of course, always the calibration session can be run before the online experiments and the filtering procedure can be chosen via the offline analysis of the calibration data. However, much better solution would be to find out which features of the BCI setup influence the performance of different spatial filters. If such features were defined than the calibration session would not be necessary.

References

1. Friman, O., Friman, O., Volosyak, I., Volosyak, I., Gräser, A., Gräser, A.: Multiple channel detection of steady-state visual evoked potentials for brain-computer interfaces. IEEE Trans. Biomed. Eng. **54**, 742–750 (2007)
2. Chen, X., Wang, Y., Gao, S., Jung, T.P., Gao, X.: Filter bank canonical correlation analysis for implementing a high-speed SSVEP-based brain-computer interface. J. Neural Eng. **12**, 46008 (2015)
3. Wolpaw, J.R., Birbaumer, N., McFarland, D.J., Pfurtscheller, G., Vaughan, T.M.: Brain Computer Interfaces for communication and control. Front. Neurosci. **4**, 767–791 (2002)
4. Regan, D.: Human Brain Electrophysiology: Evoked Potentials and Evoked Magnetic Fields in Science and Medicine. Elsevier, New York (1989)
5. Herrmann, C.S.: Human EEG responses to 1-100 Hz flicker: resonance phenomena in visual cortex and their potential correlation to cognitive phenomena. Exp. Brain Res. **137**, 346–353 (2001)
6. Vialatte, F.B., Maurice, M., Dauwels, J., Cichocki, A.: Steady-state visually evoked potentials: focus on essential paradigms and future perspectives. Prog. Neurobiol. **90**, 418–438 (2010)
7. Oikonomou, V.P., Liaros, G., Georgiadis, K., Chatzilari, E., Adam, K., Nikolopoulos, S., Kompatsiaris, I.: Comparative Evaluation of State-of-the-Art Algorithms for SSVEP-Based BCIs, pp. 1–33 (2016)
8. Kołodziej, M., Majkowski, A., Oskwarek, Ł., Rak, R.J.: Comparison of EEG signal preprocessing methods for SSVEP recognition. In: 2016 39th International Conference on Telecommunication and Signal Processing TSP 2016, pp. 340–345 (2016)
9. Zhu, D., Bieger, J., Garcia Molina, G., Aarts, R.M.: A survey of stimulation methods used in SSVEP-based BCIs. In: Computational Intelligence and Neuroscience 2010 (2010)
10. Gao, X., Xu, D., Cheng, M., Gao, S.: A BCI-based environmental controller for the motion-disabled. IEEE Trans. Neural Syst. Rehabil. Eng. **11**, 137–140 (2003)
11. Gembler, F., Stawicki, P., Volosyak, I.: Exploring the possibilities and limitations of multitarget SSVEP-based BCI applications. In: Proceedings of Annual International Conference of IEEE Engineering in Medicine and Biology Society EMBS 2016–October, pp. 1488–1491 (2016)
12. Duszyk, A., Bierzyńska, M., Radzikowska, Z., Milanowski, P., Kuś, R., Suffczyński, P., Michalska, M., Labęcki, M., Zwoliński, P., Durka, P.: Towards an optimization of stimulus parameters for brain-computer interfaces based on steady state visual evoked potentials. PLoS One **9** (2014)
13. Liu, Q., Chen, K., Ai, Q., Xie, S.Q.: Review: Recent development of signal processing algorithms for SSVEP-based brain computer interfaces. J. Med. Biol. Eng. **34**, 299–309 (2014)
14. Lin, Z., Zhang, C., Wu, W., Gao, X.: Frequency recognition based on canonical correlation analysis for SSVEP-Based BCIs. IEEE Trans. Biomed. Eng. **54**, 1172–1176 (2007)

15. Zhu, D., Molina, G.G., Mihajlović, V., Aartsl, R.M.: Phase synchrony analysis for SSVEP-based BCIs. In: Proceedings of ICCET 2010 - 2010 International Conference on Computer Engineering and Technology, vol. 2, pp. 329–333 (2010)
16. Jasper, H.H.: The ten-twenty electrode system of the international federation in electroencephalography and clinical neurophysiology. EEG J. **10**, 371–375 (1958)
17. Renard, Y., Lotte, F., Gibert, G., Congedo, M., Maby, E., Delannoy, V., Bertrand, O., Lecuyer, A.: OpenViBE: an open-source software platform to design, test, and use brain-computer interfaces in real and virtual environments. Presence-Teleoperators Virtual Environ. **19**, 35–53 (2010)

Author Index

© Springer Nature Switzerland AG 2019
J. Pejaś et al. (Eds.): ACS 2018, AISC 889, pp. 493–494, 2019.
https://doi.org/10.1007/978-3-030-03314-9

Printed in the United States
By Bookmasters